NON-EQUILIBRIUM THERMODYNAMICS

NON-EQUILIBRIUM THERMODYNAMICS

S. R. DE GROOT
Professor of Theoretical Physics
University of Amsterdam, The Netherlands

P. MAZUR
Professor of Theoretical Physics
University of Leiden, The Netherlands

DOVER PUBLICATIONS, INC., NEW YORK

Published in Canada by General Publishing Company, Ltd., 30 Lesmill Road, Don Mills, Toronto, Ontario.
Published in the United Kingdom by Constable and Company, Ltd., 10 Orange Street, London WC2H 7EG.

This Dover edition, first published in 1984, is an unabridged, corrected republication of the work originally published by the North-Holland Publishing Company, Amsterdam, 1962. A new Preface has been specially prepared for this edition by the authors.

Manufactured in the United States of America
Dover Publications, Inc., 31 East 2nd Street, Mineola, N.Y. 11501

Library of Congress Cataloging in Publication Data

Groot, S. R. de (Sybren Ruurds de), 1916–
Non-equilibrium thermodynamics.

Originally published: Amsterdam : North-Holland Pub. Co., 1962.
Includes index.
1. Irreversible processes. 2. Statistical thermodynamics.
I. Mazur, P. (Peter) II. Title.
QC318.I7G76 1984 536'.7 84-7956
ISBN 0-486-64741-2 (pbk.)

PREFACE TO THE DOVER EDITION

The original edition of this monograph appeared more than twenty years ago. But the material presented in this book has, it seems to us, retained its usefulness and has played its role in the development of statistical physics.

On the occasion of this republication it is perhaps worthwhile to dwell once more on the domain of validity of the theory presented. This domain of validity is essentially the one for which the hypothesis of local equilibrium is valid, and includes, as far as transport processes are concerned, those phenomena for which the *dissipative thermodynamic fluxes are linear functions of the gradients of the thermodynamic-state variables*. It is for this reason that the theory in question is sometimes referred to as "linear thermodynamics of irreversible processes." Nevertheless, this theory gives rise to partial differential equations for the state variables which are *non-linear* for a variety of reasons, such as (i) the presence of convection terms and of (ii) quadratic source terms in, e.g., the energy equation, (iii) the non-linear character of the equations of state and (iv) the dependence of the phenomenological transport coefficients on the state variables. The name "linear thermodynamics of irreversible processes" is therefore slightly misleading. Of course, the formalism may be fully linearized and is then applicable to situations in which one is not only not too far from local equilibrium, but even near to overall, global, equilibrium. Some parts of this monograph deal with the fully linearized version of non-equilibrium thermodynamics; mainly, however, the theory is of wider applicability, as outlined. We hope that these remarks, which the reader will also encounter disseminated throughout the book in one form or another, will help to bring its subject matter into proper perspective.

SYBREN R. DE GROOT and PETER MAZUR

Amsterdam, Leiden
July 1983

PREFACE TO THE FIRST EDITION

The aim of the theory of non-equilibrium thermodynamics is to englobe within a single scheme a large class of phenomenological treatments for irreversible processes. In this book we have attempted to acquaint the reader with the general theoretical framework of this theory and to give a survey of its applications. In the first part we have also included a discussion of the statistical foundations of non-equilibrium thermodynamics primarily with the purpose of justifying the Onsager reciprocal relations and the use of thermodynamic functions outside equilibrium.

We wish to express our gratitude to Professor E. P. Wigner and Professor L. Van Hove for criticism and advice, and to Dr. J. Vlieger and Mr. J. van der Linden for their help in preparing some of the sections of this book. We are also very much indebted to Dr. C. D. Hartogh, Messrs. J. van der Linden, R. H. Terwiel, H. W. Capel, F. A. Berends, P. J. M. Bongaarts, F. Niemeijer and especially to Dr. J. Vlieger for reading the manuscript with great care.

SYBREN R. DE GROOT AND PETER MAZUR

Leyden, May 1961

TABLE OF CONTENTS

CHAPTER V

THE STATIONARY STATES

CHAPTER VI

PROPERTIES OF THE PHENOMENOLOGICAL EQUATIONS AND THE ONSAGER RELATIONS

CHAPTER VII

DISCUSSION OF THE STATISTICAL FOUNDATIONS

CHAPTER VIII

THE FLUCTUATION DISSIPATION THEOREM

CHAPTER IX

DISCUSSION OF FOUNDATIONS BY MEANS OF KINETIC THEORY

PART B. APPLICATIONS

Chapter X

CHEMICAL REACTIONS AND RELAXATION PHENOMENA

Chapter XI

HEAT CONDUCTION, DIFFUSION AND CROSS-EFFECTS

Chapter XII

VISCOUS FLOW AND RELAXATION PHENOMENA

Chapter XIII

ELECTRICAL CONDUCTION

CHAPTER XIV

IRREVERSIBLE PROCESSES IN POLARIZED SYSTEMS

CHAPTER XV

DISCONTINUOUS SYSTEMS

CHAPTER I

INTRODUCTION

§ 1. *Historical Background of Non-Equilibrium Thermodynamics*

Thermodynamic considerations were first applied to the treatment of irreversible processes by W. Thomson* in 1854. He analysed the various thermo-electric phenomena and established the famous two relations which bear his name. The first of these relations follows from conservation of energy. The second relation, which relates the thermo-electric potential of a thermocouple to its Peltier heat, was obtained from the two laws of thermodynamics, and an additional assumption about the so-called "reversible" contributions to the process. Later Boltzmann** attempted without success to justify the Thomson hypothesis. We now know that no basis exists for this hypothesis. Thomson's second relation was finally derived correctly by Onsager who showed that this relation was a consequence of the invariance of the microscopic equations of motion under time reversal. Thomson's method was applied with varying success to a number of other irreversible phenomena, but a coherent scheme for the macroscopic description of irreversible processes could not be developed in this way.

Independently of the theoretical development described above, a number of physicists undertook, at the turn of the century, to give more refined formulations of the second law of thermodynamics for non-equilibrium situations. As early as 1850 Clausius introduced the concept of "non-compensated heat" as a measure of irreversibility (in systems which need not be thermally insulated from their surroundings). Duhem, Natanson, Jaumann and Lohr and later Eckart† attempted to

* W. Thomson (Lord Kelvin) Proc. roy. Soc. Edinburgh **3** (1854) 225; *Ibid.* Trans. **21** I (1857) 123; Math. phys. Papers **1** (1882) 232.

** L. Boltzmann, Sitz. ber. Akad. Wiss. Wien, Math.-Naturw. Kl. Abt. II **96** (1887) 1258; Wiss. Abh. **3** (1909) 321.

† P. Duhem, Energétique (2 vol.) (Gauthier-Villars, Paris, 1911).

L. Natanson, Z. phys. Chem. **21** (1896) 193.

G. Jaumann, Sitz. ber. Akad. Wiss. Wien, Math.-Naturw. Kl. Abt. II A **120** (1911) 385; Denkschr. Akad. Wiss. Wien, Math.-Naturw. Kl. **95** (1918) 461.

obtain expressions for the rate of change of the local entropy in non-uniform systems by combining the second law of thermodynamics with the macroscopic laws of conservation of mass, momentum and energy. In this way they derived formulae which related irreversibility to the non-uniformity of the system. Similarly De Donder* was able to relate the "non-compensated heat" in a chemical reaction to the affinity, a thermodynamic variable characterizing the state of the system. The systematic discussion of irreversible processes along these lines however was not completed until much later.

In the meantime, in 1931, Onsager** established his celebrated "reciprocal relations" connecting the coefficients, which occur in the linear phenomenological laws that describe the irreversible processes. These reciprocal relations, of which Thomson's second relation is an example, reflect on the macroscopic level the time reversal invariance of the microscopic equations of motion. In 1945 Casimir*** reformulated the reciprocal relations, so that they would be valid for a larger class of irreversible phenomena than had been previously considered by Onsager.

Finally Meixner† in 1941 and the following years, and somewhat later Prigogine††, set up a consistent phenomenological theory of irreversible processes, incorporating both Onsager's reciprocity theorem and the explicit calculation for a certain number of physical situations of the so-called entropy source strength (which is in fact the non-compensated heat of Clausius). In this way a new field of "thermodynamics of irreversible processes" was born. It developed rapidly in various directions.

Coupled with the recent growing interest in the statistical mechanical theory of irreversible processes are some significant studies on the statistical basis of the thermodynamics of irreversible processes. Thus special attention is paid to the validity on one hand of thermodynamic

E. Lohr, Denkschr. Akad. Wiss. Wien, Math.-Naturw. Kl. **93** (1916) 339; **99** (1924) 11, 59; Festschr. Techn. Hochsch. Brünn (1924) 176.

C. Eckart, Phys. Rev. **58** (1940) 267, 269, 919, 924.

* Th. De Donder, L'affinité (Gauthier-Villars, Paris, 1927).

** L. Onsager, Phys. Rev. **37** (1931) 405; **38** (1931) 2265.

*** H. B. G. Casimir, Rev. mod. Phys. **17** (1945) 343; or Philips Res. Rep. **1** (1945) 185.

† J. Meixner, Ann. Physik [5] **39** (1941) 333; **41** (1942) 409; **43** (1943) 244; Z. phys. Chem. B **53** (1943) 235.

†† I. Prigogine, Etude thermodynamique des phénomènes irréversibles (Dunod, Paris and Desoer, Liège, 1947).

relations outside equilibrium and, on the other hand, to the Onsager reciprocal relations. Many of these studies have employed methods and concepts borrowed from the theory of stochastic processes.

§ 2. *Systematic Development of the Theory*

The field of non-equilibrium thermodynamics provides us with a general framework for the macroscopic description of irreversible processes. As such it is a branch of macroscopic physics, which has connexions with other macroscopic disciplines such as fluid dynamics and electromagnetic theory, insofar as the latter fields are also concerned with non-equilibrium situations. Thus the thermodynamics of irreversible processes should be set up from the start as a continuum theory, treating the state parameters of the theory as field variables, *i.e.*, as continuous functions of space coordinates and time. Moreover one would like to formulate the basic equations of the theory in such a way that they only contain quantities referring to a single point in space at one time, *i.e.* in the form of local equations. This is also the way in which fluid dynamics and the Maxwell theory are formulated. In equilibrium thermodynamics such a local formulation is generally not needed, since the state variables are usually independent of the space coordinates.

In non-equilibrium thermodynamics the so-called balance equation for the entropy plays a central role. This equation expresses the fact that the entropy of a volume element changes with time for two reasons. First it changes because entropy flows into the volume element, second because there is an entropy source due to irreversible phenomena inside the volume element. The entropy source is always a non-negative quantity, since entropy can only be created, never destroyed. For reversible transformations the entropy source vanishes. This is the local formulation of the second law of thermodynamics. The main aim is to relate the entropy source explicitly to the various irreversible processes that occur in a system. To this end one needs the macroscopic conservation laws of mass, momentum and energy, in local, *i.e.* differential form. These conservation laws contain a number of quantities such as the diffusion flows, the heat flow and the pressure tensor, which are related to the transport of mass, energy and momentum. The entropy source may then be calculated if one makes use of the thermodynamic Gibbs relation which connects, in an isotropic multi-component fluid for instance, the rate of change of entropy in each mass element, to the rate of change of energy and the rates of change in composition. It turns out that the entropy source has a very simple appearance: it is a sum of terms each being a product of a flux characterizing an irreversible process, and a quantity, called thermodynamic force, which is related to the non-uniformity of the system (the gradient of the temperature for instance) or to the deviations of some internal state variables from their equilibrium values (the chemical affinity for instance). The entropy source strength can thus serve as a basis for the systematic description of the irreversible processes occurring in a system.

As yet the set of conservation laws, together with the entropy balance equation and the equations of state is to a certain extent empty, since this set of equations contains the irreversible fluxes as unknown parameters and can therefore not be solved with given initial and boundary conditions for the state of the system. At this point we must therefore supplement our equations by an additional set of phenomenological equations which relate the irreversible fluxes and the thermodynamic forces appearing in the entropy source strength. In first approximation the fluxes are linear functions of the thermodynamic forces. Fick's law of diffusion, Fourier's law of heat conduction, and Ohm's law of electric conduction, for instance, belong to this class of linear phenomenological laws. It also contains in addition possible cross-effects between various phenomena, since each flux may in principle be a linear function of all the thermodynamic forces which are needed to characterize the entropy source strength. The Soret effect, which results from diffusion in a temperature gradient is an example of such a cross-effect. Many others exist such as the thermoelectric effects, the group of thermomagnetic and galvanomagnetic effects, and also the electrokinetic effects. Non-equilibrium thermodynamics, in its present form, is mainly restricted to the study of such linear phenomena. Very little of a sufficiently general nature is known outside this linear domain. This is not a very serious restriction however, since even in rather extreme physical situations, transport processes, for example, are still described by linear laws. Together with the phenomenological equations the original set of conservation laws may be said to be complete in the sense that one now has at one's disposal a consistent set of partial differential equations for the state parameters of a material system, which may be solved with the proper initial and boundary conditions.

Some rather important statements of a macroscopic nature may be made concerning the matrix of phenomenological coefficients which relate the fluxes and the thermodynamic forces, appearing in the entropy source strength. In the first place the Onsager–Casimir reciprocity theorem gives rise to a number of relations amongst these coefficients, thus reducing the number of independent quantities and relating distinct physical effects to each other. It is one of the aims of non-equilibrium thermodynamics to study the physical consequences of the reciprocal relations in applications of the theory to various physical situations. Apart from the reciprocity theorem, which is based on the time reversal invariance of the microscopic equations of motion, possible spatial symmetries of a material system may further simplify the scheme of phenomenological coefficients. Thus in an isotropic fluid a scalar phenomenon like a chemical reaction cannot be coupled to a vectorial phenomenon like heat conduction. This reduction of the scheme of phenomenological coefficients, which results from invariance of the phenomenological equations under special orthogonal transformations, goes under the name of the Curie principle, but should more appropriately be called Curie's theorem.

The reciprocal relations have transformation properties which have been studied extensively. Thus Meixner showed that the Onsager relations are invariant under certain transformations of the fluxes and the thermodynamic forces. There exist a number of other general theorems, which are of use in non-equilibrium thermodynamics: one can show that at mechanical equilibrium the entropy production has some additional invariance properties. It can also be shown that stationary non-equilibrium states are characterized by a minimum property: under certain restrictive conditions the entropy production has, in the stationary state, a minimum value compatible with given boundary conditions. Both of these theorems have first been obtained by Prigogine.

Just as the principles of equilibrium thermodynamics must be justified by means of statistical mechanical methods, so the principles of thermodynamics of irreversible processes require a discussion of their microscopic basis. In the present state of theoretical development a microscopic discussion of irreversibility itself from first principles would lie outside the scope of this treatise. However, even if the irreversible behaviour of macroscopic systems is taken for granted, one still has the problem of discussing the remaining foundations of the theory. On these premises Onsager's reciprocity theorem can indeed be derived from microscopic properties of a mechanical many-particle system. Concepts of fluctuation theory and the theory of stochastic processes play an essential role in such a discussion of the foundations, to which Onsager and Machlup have contributed by using a Brownian motion type model for the regression of fluctuations. Such a model can also serve for a justification of the use of thermodynamic relations outside equilibrium. Furthermore the methods of the theory of stochastic processes are used in relating the spontaneous fluctuations in equilibrium to the macroscopic response of a system under external driving forces. The relation thus obtained is known as the fluctuation-dissipation theorem and is due to Callen, Greene and Welton. It represents in fact a generalization of the famous Nyquist formula in the theory of electric noise.

The problem of justifying the principles of non-equilibrium thermodynamics can alternatively be approached from the viewpoint of the kinetic theory of gases. Such a method is more limited since it only applies to gaseous systems at low density, however it permits one to express those macroscopic quantities which pertain to irreversible processes in terms of molecular parameters. The irreversibility itself is already contained in the fundamental equation of the kinetic theory of gases, the Boltzmann integro-differential equation. One may then justify the use of thermodynamic relations for gases outside equilibrium (as was first done by Prigogine), and derive the Onsager reciprocal relations.

The theory of non-equilibrium thermodynamics has found a great variety of applications in physics and chemistry. For a systematic classification of these applications one may group the various irreversible phenomena according to their "tensorial character".

First one has "scalar phenomena". These include chemical reactions and structural relaxation phenomena. Onsager relations are of help in this case in solving the set of ordinary differential equations which describe the simultaneous relaxation of a great number of internal variables.

A second group of phenomena is formed by "vectorial processes" such as diffusion, heat conduction, and their cross-effects.

Viscous phenomena can be considered as a third group to which methods of non-equilibrium thermodynamics have been applied. In particular the theory of acoustical relaxation has been consistently developed within this framework by Meixner.

Altogether new aspects arise when an electromagnetic field acts on a material system. Then the continuity laws for electromagnetic energy and momentum which follow from the Maxwell equations must also be taken into account. One must therefore reformulate the theory to suit the need of this case with its numerous applications to both polarized and unpolarized media.

PART A

GENERAL THEORY

CHAPTER II

CONSERVATION LAWS

§ 1. *Introduction*

Thermodynamics is based on two fundamental laws: the first law of thermodynamics or law of conservation of energy, and the second law of thermodynamics or entropy law. A systematic macroscopic scheme for the description of non-equilibrium processes (*i.e.* the scheme of thermodynamics of irreversible processes) must also be built upon these two laws. However, it is necessary to formulate these laws in a way suitable for the purpose at hand.

In this chapter we shall be concerned with the first law of thermodynamics. Since we wish to develop a theory applicable to systems of which the properties are continuous functions of space coordinates and time, we shall give a local formulation of the law of conservation of energy. As the local momentum and mass densities may change in time, we shall also need local formulations of the laws of conservation of momentum and mass.

In the following sections these conservation laws will be written down for a multi-component system in which chemical reactions may occur and on which conservative external forces are exerted.

It may be remarked that the macroscopic conservation laws of matter, momentum and energy are, from a microscopic point of view, consequences of the mechanical laws governing the motions of the constituent particles of the system.

§ 2. *Conservation of Mass*

Let us consider a system consisting of n components amongst which r chemical reactions are possible.

The rate of change of the mass of component k within a given volume V is

$$\frac{\mathrm{d}}{\mathrm{d}t}\int^{V} \rho_k \, \mathrm{d}V = \int^{V} \frac{\partial \rho_k}{\partial t} \, \mathrm{d}V \, , \tag{1}$$

where ρ_k is the density (mass per unit volume) of k. This quantity is equal to the sum of the material flow of component k into the volume V through its surface Ω and the total production of k in chemical reactions which occur inside V

$$\int^V \frac{\partial \rho_k}{\partial t} \mathrm{d}V = - \int^\Omega \rho_k \boldsymbol{v}_k \cdot \mathrm{d}\boldsymbol{\Omega} + \sum_{j=1}^{r} \int^V \nu_{kj} J_j \, \mathrm{d}V \,, \tag{2}$$

where $\mathrm{d}\boldsymbol{\Omega}$ is a vector with magnitude $\mathrm{d}\Omega$ normal to the surface and counted positive from the inside to the outside. Furthermore $\boldsymbol{v}_k$ is the velocity of k, and $\nu_{kj}J_j$ the production of k per unit volume in the jth chemical reaction. The quantity ν_{kj} divided by the molecular mass M_k of component k is proportional to the stoichiometric coefficient with which k appears in the chemical reaction j. The coefficients ν_{kj} are counted positive when components k appear in the second, negative when they appear in the first member of the reaction equation. The quantity J_j is called the chemical reaction rate of reaction j. It represents a mass per unit volume and unit time. The quantities ρ_k, $\boldsymbol{v}_k$ and J_j occurring in (2) are all functions of time and of space coordinates.

Applying Gauss' theorem to the surface integral occurring in (2), we obtain

$$\frac{\partial \rho_k}{\partial t} = - \operatorname{div} \rho_k \boldsymbol{v}_k + \sum_{j=1}^{r} \nu_{kj} J_j \,, \quad (k = 1, 2, \ldots, n) \,, \tag{3}$$

since (2) is valid for an arbitrary volume V. This equation has the form of a so-called balance equation: the local change of the left-hand side is equal to the negative divergence of the flow of k and a source term giving the production (or destruction) of substance k.

Since mass is conserved in each separate chemical reaction we have

$$\sum_{k=1}^{n} \nu_{kj} = 0 \,, \quad (j = 1, 2, \ldots, r) \,. \tag{4}$$

Summing equation (3) over all substances k one obtains then the *law of conservation of mass*:

$$\frac{\partial \rho}{\partial t} = - \operatorname{div} \rho \boldsymbol{v} \,, \tag{5}$$

where ρ is the total density

$$\rho = \sum_{k=1}^{n} \rho_k \tag{6}$$

and $\boldsymbol{v}$ the centre of mass ("barycentric") velocity

$$\boldsymbol{v} = \sum_{k=1}^{n} \rho_k \boldsymbol{v}_k / \rho \,. \tag{7}$$

Equation (5) expresses the fact that the total mass is conserved, *i.e.* that the total mass in any volume element of the system can only change if matter flows into (or out of) the volume element.

The mass equations can be written in an alternative form by introducing the (barycentric) substantial time derivative

$$\frac{\mathrm{d}}{\mathrm{d}t} = \frac{\partial}{\partial t} + \boldsymbol{v} \cdot \mathrm{grad} \tag{8}$$

and the "diffusion flow" of substance k defined with respect to the barycentric motion

$$\boldsymbol{J}_k = \rho_k(\boldsymbol{v}_k - \boldsymbol{v}) \,. \tag{9}$$

With the help of (8) and (9), equations (3) become

$$\frac{\mathrm{d}\rho_k}{\mathrm{d}t} = -\rho_k \,\mathrm{div}\, \boldsymbol{v} - \mathrm{div}\, \boldsymbol{J}_k + \sum_{j=1}^{r} \nu_{kj} J_j \,, \quad (k = 1, 2, \ldots, n) \tag{10}$$

and equation (5)

$$\frac{\mathrm{d}\rho}{\mathrm{d}t} = -\rho \,\mathrm{div}\, \boldsymbol{v} \,. \tag{11}$$

If mass fractions c_k:

$$c_k = \rho_k / \rho \,, \quad \left(\sum_{k=1}^{n} c_k = 1 \right) \tag{12}$$

are employed, equations (10) take the simple form

$$\rho \frac{\mathrm{d}c_k}{\mathrm{d}t} = -\mathrm{div}\, \boldsymbol{J}_k + \sum_{j=1}^{r} \nu_{kj} J_j \,, \quad (k = 1, 2, \ldots, n) \,, \tag{13}$$

where (11) has been used also.

With the specific volume $v = \rho^{-1}$ formula (11) may also be written as

$$\rho \frac{\mathrm{d}v}{\mathrm{d}t} = \operatorname{div} \boldsymbol{v} . \tag{14}$$

We note that it follows from (7) and (9) that

$$\sum_{k=1}^{n} \boldsymbol{J}_k = 0 , \tag{15}$$

which means that only n–1 of the n diffusion flows are independent. Similarly only n–1 of the n equations (13) are independent. In fact by summing (13) over all k, both members vanish identically as a result of (4), (12) and (15). The nth independent equation describing the change of mass density within the system is now equation (14).

We note finally that the following relation, valid for an arbitrary local property a, (which may be a scalar or a component of a vector or tensor, etc.):

$$\rho \frac{\mathrm{d}a}{\mathrm{d}t} = \frac{\partial a\rho}{\partial t} + \operatorname{div} a\rho\boldsymbol{v} \tag{16}$$

is a consequence of the mass equation (5) and of (8).

§ 3. *The Equation of Motion*

The equation of motion of the system is

$$\rho \frac{\mathrm{d}v_\alpha}{\mathrm{d}t} = - \sum_{\beta=1}^{3} \frac{\partial}{\partial x_\beta} P_{\beta\alpha} + \sum_{k=1}^{n} \rho_k F_{k\alpha} , \quad (\alpha = 1, 2, 3) , \tag{17}$$

where v_α $(\alpha = 1,2,3)$ is a Cartesian component of $\boldsymbol{v}$, and where x_α $(\alpha = 1,2,3)$ are the Cartesian coordinates. The derivative $\mathrm{d}v_\alpha/\mathrm{d}t$ is a component of the acceleration of the centre of gravity motion.

The quantities $P_{\alpha\beta}$ $(\alpha,\beta = 1,2,3)$ and $F_{k\alpha}$ $(\alpha = 1,2,3)$ are the Cartesian components of the pressure (or stress) tensor P of the medium and of the force per unit mass $\boldsymbol{F}_k$ exerted on the chemical component k respectively. We shall assume here* that the pressure tensor P is symmetric,

* This assumption is usually made in hydrodynamics, but is rigorously only justified for systems consisting of spherical molecules or at very low densities. For other systems however the pressure tensor may contain an antisymmetric part.

$$P_{\alpha\beta} = P_{\beta\alpha}, \quad (\alpha, \beta = 1, 2, 3). \tag{18}$$

In tensor notation equations (17) are written as

$$\rho \frac{\mathrm{d}\boldsymbol{v}}{\mathrm{d}t} = -\operatorname{Div} \mathsf{P} + \sum_k \rho_k \boldsymbol{F}_k . \tag{19}$$

From a microscopic point of view one can say that the pressure tensor P results from the short-range interactions between the particles of the system, whereas $\boldsymbol{F}_k$ contains the external forces as well as a possible contribution from long-range interactions in the system.

For the moment we shall restrict the discussion to the consideration of conservative forces which can be derived from a potential ψ_k independent of time

$$\boldsymbol{F}_k = -\operatorname{grad} \psi_k, \quad \frac{\partial \psi_k}{\partial t} = 0. \tag{20}$$

Using relation (16), the equation of motion (19) can also be written as

$$\frac{\partial \rho \boldsymbol{v}}{\partial t} = -\operatorname{Div}(\rho \boldsymbol{v}\boldsymbol{v} + \mathsf{P}) + \sum_k \rho_k \boldsymbol{F}_k , \tag{21}$$

where $\boldsymbol{v}\boldsymbol{v}$ is an ordered (dyadic) product, (*cf.* Appendix I on matrix and tensor notation). This equation has the form of a balance equation for the momentum density $\rho\boldsymbol{v}$. In fact it is seen that one can interpret the quantity $\rho\boldsymbol{v}\boldsymbol{v} + \mathsf{P}$ as a momentum flow, with a convective part $\rho\boldsymbol{v}\boldsymbol{v}$, and the quantity $\sum_k \rho_k \boldsymbol{F}_k$ as a source of momentum.

It is possible to derive from (17) a balance equation for the kinetic energy of the centre of gravity motion by multiplying both members with the component v_α of the barycentric velocity and summing over α

$$\rho \frac{\mathrm{d}\frac{1}{2}v^2}{\mathrm{d}t} = -\sum_{\alpha,\beta} \frac{\partial}{\partial x_\beta}(P_{\beta\alpha} v_\alpha) + \sum_{\alpha,\beta} P_{\beta\alpha} \frac{\partial}{\partial x_\beta} v_\alpha + \sum_{k,\alpha} \rho_k F_{k\alpha} v_\alpha , \tag{22}$$

Inclusion of such terms into the general formalism gives rise to a slight modification of the hydrodynamics of the systems: *v.* J. Frenkel, Kinetic Theory of Liquids (Oxford University Press, 1946) Ch. V, § 7; H. Grad, Comm. pure and appl. Math. (New-York) **5** (1952) 455; C. F. Curtiss, J. chem. Phys. **24** (1956) 225; *cf.* also Chapter XII, § 1 of this book.

or, in tensor notation,

$$\rho \frac{\mathrm{d}\frac{1}{2}\boldsymbol{v}^2}{\mathrm{d}t} = -\operatorname{div}(\boldsymbol{P}\cdot\boldsymbol{v}) + \boldsymbol{P} : \operatorname{Grad}\boldsymbol{v} + \sum_k \rho_k \boldsymbol{F}_k\cdot\boldsymbol{v}\,, \tag{23}$$

where

$$\boldsymbol{P} : \operatorname{Grad}\boldsymbol{v} = \sum_{\alpha,\beta} P_{\alpha\beta}\frac{\partial}{\partial x_\beta}v_\alpha = \sum_{\alpha,\beta} P_{\beta\alpha}\frac{\partial}{\partial x_\beta}v_\alpha\,. \tag{24}$$

With the help of (16), equation (23) becomes

$$\frac{\partial\frac{1}{2}\rho\boldsymbol{v}^2}{\partial t} = -\operatorname{div}(\tfrac{1}{2}\rho\boldsymbol{v}^2\boldsymbol{v} + \boldsymbol{P}\cdot\boldsymbol{v}) + \boldsymbol{P} : \operatorname{Grad}\boldsymbol{v} + \sum_k \rho_k\boldsymbol{F}_k\cdot\boldsymbol{v}\,. \tag{25}$$

We wish to establish now an equation for the rate of change of the potential energy density $\rho\psi \equiv \sum_k \rho_k\psi_k$. In fact it follows from (3), (9) and (20) that

$$\frac{\partial\rho\psi}{\partial t} = -\operatorname{div}(\rho\psi\boldsymbol{v} + \sum_{k=1}^{n}\psi_k\boldsymbol{J}_k) - \sum_{k=1}^{n}\rho_k\boldsymbol{F}_k\cdot\boldsymbol{v} - \sum_{k=1}^{n}\boldsymbol{J}_k\cdot\boldsymbol{F}_k + \sum_{k=1}^{n}\sum_{j=1}^{r}\psi_k\nu_{kj}J_j\,. \tag{26}$$

The last term vanishes if the potential energy is conserved in a chemical reaction

$$\sum_k \psi_k\nu_{kj} = 0\,,\quad (j = 1, 2, \ldots, r)\,. \tag{27}$$

This is the case if the property of the particles, which is responsible for the interaction with a field of force, is itself conserved. Examples for this case are the mass in a gravitational field and the charge in an electric field. Equation (26) then reduces to

$$\frac{\partial\rho\psi}{\partial t} = -\operatorname{div}(\rho\psi\boldsymbol{v} + \sum_k\psi_k\boldsymbol{J}_k) - \sum_k\rho_k\boldsymbol{F}_k\cdot\boldsymbol{v} - \sum_k\boldsymbol{J}_k\cdot\boldsymbol{F}_k\,. \tag{28}$$

Let us now add the two equations (25) and (28) for the rate of change of the kinetic energy $\frac{1}{2}\rho\boldsymbol{v}^2$ and the potential energy $\rho\psi$:

$$\frac{\partial \rho(\frac{1}{2}\boldsymbol{v}^2 + \psi)}{\partial t} = - \operatorname{div} \{ \rho(\tfrac{1}{2}\boldsymbol{v}^2 + \psi)\boldsymbol{v} + \boldsymbol{P}\cdot\boldsymbol{v} + \sum_k \psi_k \boldsymbol{J}_k \} + \boldsymbol{P} : \operatorname{Grad} \boldsymbol{v} - \sum_k \boldsymbol{J}_k \cdot \boldsymbol{F}_k . \quad (29)$$

This equation shows that the sum of kinetic and potential energy is not conserved, since a source term appears at the right-hand side.

§ 4. *Conservation of Energy*

According to the principle of conservation of energy the total energy content within an arbitrary volume V in the system can only change if energy flows into (or out of) the volume considered through its boundary Ω

$$\frac{\mathrm{d}}{\mathrm{d}t} \int^{V} \rho e \, \mathrm{d}V = \int^{V} \frac{\partial \rho e}{\partial t} \mathrm{d}V = - \int^{\Omega} \boldsymbol{J}_e \cdot \mathrm{d}\boldsymbol{\Omega} , \quad (30)$$

Here e is the energy per unit mass, and $\boldsymbol{J}_e$ the energy flux per unit surface and unit time. We shall refer to e as the total specific energy, because it includes all forms of energy in the system. Similarly we shall call $\boldsymbol{J}_e$ the total energy flux. With the help of Gauss' theorem we obtain the differential or local form of the law of conservation of energy

$$\frac{\partial \rho e}{\partial t} = - \operatorname{div} \boldsymbol{J}_e . \quad (31)$$

In order to relate this equation to the previously obtained result (29) for the kinetic and potential energy, we must specify which are the various contributions to the energy e and the flux $\boldsymbol{J}_e$.

The total specific energy e includes the specific kinetic energy $\frac{1}{2}\boldsymbol{v}^2$, the specific potential energy ψ and the specific internal energy u:

$$e = \tfrac{1}{2}\boldsymbol{v}^2 + \psi + u . \quad (32)$$

From a macroscopic point of view this relation can be considered as the definition of internal energy u. From a microscopic point of view u represents the energy of thermal agitation as well as the energy due to the short-range molecular interactions. Similarly the total energy flux includes a convective term $\rho e \boldsymbol{v}$, an energy flux $\boldsymbol{P}\cdot\boldsymbol{v}$ due to the mechanical work performed on the system, a potential energy flux $\sum_k \psi_k \boldsymbol{J}_k$ due to the diffusion of the various components in the field of force, and finally a "heat flow" $\boldsymbol{J}_q$:

$$\boldsymbol{J}_e = \rho e \boldsymbol{v} + \mathsf{P} \cdot \boldsymbol{v} + \sum_k \psi_k \boldsymbol{J}_k + \boldsymbol{J}_q \, . \tag{33}$$

This relation may be considered as defining the heat flow $\boldsymbol{J}_q$. If we subtract equation (29) from equation (31), we obtain, using also (32) and (33), the balance equation for the internal energy u:

$$\frac{\partial \rho u}{\partial t} = - \operatorname{div} (\rho u \boldsymbol{v} + \boldsymbol{J}_q) - \mathsf{P} : \operatorname{grad} \boldsymbol{v} + \sum_k \boldsymbol{J}_k \cdot \boldsymbol{F}_k \, . \tag{34}$$

It is apparent from this equation that the internal energy u is not conserved. In fact a source terms appears, which is equal but of opposite sign to the source term of the balance equation (29) for kinetic and potential energy.

The equation (34) may be written in an alternative form. We can split the total pressure tensor into a scalar* hydrostatic part p and a tensor $\mathsf{\Pi}$:

$$\mathsf{P} = p \mathsf{U} + \mathsf{\Pi} \, , \tag{35}$$

where U is the unit matrix with elements $\delta_{\alpha\beta}$ ($\delta_{\alpha\beta} = 1$ if $\alpha = \beta$, $\delta_{\alpha\beta} = 0$ if $\alpha \neq \beta$). With this relation and (16), equation (34) becomes

$$\begin{aligned} \rho \frac{\mathrm{d}u}{\mathrm{d}t} &= - \operatorname{div} \boldsymbol{J}_q - p \operatorname{div} \boldsymbol{v} - \mathsf{\Pi} : \operatorname{Grad} \boldsymbol{v} + \sum_k \boldsymbol{J}_k \cdot \boldsymbol{F}_k \\ &= \rho \frac{\mathrm{d}q}{\mathrm{d}t} - p \operatorname{div} \boldsymbol{v} - \mathsf{\Pi} : \operatorname{Grad} \boldsymbol{v} + \sum_k \boldsymbol{J}_k \cdot \boldsymbol{F}_k \, , \end{aligned} \tag{36}$$

where use has been made of the equality

$$\mathsf{U} : \operatorname{Grad} \boldsymbol{v} = \sum_{\alpha, \beta = 1}^{3} \delta_{\alpha\beta} \frac{\partial}{\partial x_\beta} v_\alpha = \sum_{\alpha = 1}^{3} \frac{\partial}{\partial x_\alpha} v_\alpha = \operatorname{div} \boldsymbol{v} \tag{37}$$

* In assuming that the equilibrium part of the total tensor is a scalar we restrict the discussion to non-elastic fluids. For an elastic medium the equilibrium "pressure" tensor is the elastic stress *tensor*.

and where

$$\rho \frac{\mathrm{d}q}{\mathrm{d}t} = - \operatorname{div} \boldsymbol{J}_q \tag{38}$$

defines dq, the "heat" added per unit of mass.

With (14) equation (36), the "first law of thermodynamics", can finally be written in the form

$$\frac{\mathrm{d}u}{\mathrm{d}t} = \frac{\mathrm{d}q}{\mathrm{d}t} - p \frac{\mathrm{d}v}{\mathrm{d}t} - v\boldsymbol{\Pi} : \operatorname{Grad} \boldsymbol{v} + v \sum_k \boldsymbol{J}_k \cdot \boldsymbol{F}_k , \tag{39}$$

where $v \equiv \rho^{-1}$ is the specific volume.

As stated in the preceding section we have restricted ourselves in this chapter to the consideration of conservative forces $\boldsymbol{F}_k$ of the type (20). The more general case, which arises for instance when electromagnetic forces are considered, will be treated in Chapter XIII.

CHAPTER III

ENTROPY LAW AND ENTROPY BALANCE

§ 1. *The Second Law of Thermodynamics*

According to the principles of thermodynamics one can introduce for any macroscopic system a state function S, the entropy of the system, which has the following properties.

The variation of the entropy $\mathrm{d}S$ may be written as the sum of two terms

$$\mathrm{d}S = \mathrm{d_e}S + \mathrm{d_i}S\,, \tag{1}$$

where $\mathrm{d_e}S$ is the entropy supplied to the system by its surroundings, and $\mathrm{d_i}S$ the entropy produced inside the system. The second law of thermodynamics states that $\mathrm{d_i}S$ must be zero for reversible (or equilibrium) transformations and positive for irreversible transformations of the system:

$$\mathrm{d_i}S \geqslant 0\,. \tag{2}$$

The entropy supplied, $\mathrm{d_e}S$, on the other hand may be positive, zero or negative, depending on the interaction of the system with its surroundings. Thus for an adiabatically insulated system (*i.e.* a system which can exchange neither heat nor matter with its surroundings) $\mathrm{d_e}S$ is equal to zero, and it follows from (1) and (2) that

$$\mathrm{d}S \geqslant 0 \text{ for an adiabatically insulated system.} \tag{3}$$

This is a well-known form of the second law of thermodynamics.

For a so-called closed system, which may only exchange heat with its surroundings, we have according to the theorem of Carnot–Clausius:

$$\mathrm{d_e}S = \frac{\mathrm{d}Q}{T}\,, \tag{4}$$

where $\mathrm{d}Q$ is the heat supplied to the system by its surroundings and T

the absolute temperature at which heat is received by the system. From (1) and (2) it follows for this case that

$$\mathrm{d}S \geqslant \frac{\mathrm{d}Q}{T} \quad \text{for a closed system}\,, \tag{5}$$

which is also a well-known form of the second law of thermodynamics.

For open systems, *i.e.* systems which may exchange heat as well as matter with their surroundings $\mathrm{d_e}S$ contains also a term connected with the transfer of matter (*cf.* also § 2 of this chapter). The theorem of Carnot–Clausius, which is contained in formulae (1), (2) and (4), does not apply to such systems. However the very general statements contained in (1) and (2) alone remain valid.

We may remark at this point that thermodynamics in the customary sense is concerned with the study of the reversible transformations for which the equality in (2) holds. In thermodynamics of irreversible processes, however, one of the important objectives is to relate the quantity $\mathrm{d_i}S$, the entropy production, to the various irreversible phenomena which may occur inside the system. Before calculating the entropy production in terms of the quantities which characterize the irreversible phenomena, we shall rewrite (1) and (2) in a form which is more suitable for the description of systems in which the densities of the extensive properties (such as mass and energy, considered in the previous chapter) are continuous functions of space coordinates. Let us write

$$S = \int^{V} \rho s \mathrm{d}V\,, \tag{6}$$

$$\frac{\mathrm{d_e}S}{\mathrm{d}t} = -\int^{\Omega} \boldsymbol{J}_{s,\mathrm{tot}} \cdot \mathrm{d}\boldsymbol{\Omega}\,, \tag{7}$$

$$\frac{\mathrm{d_i}S}{\mathrm{d}t} = \int^{V} \sigma\, \mathrm{d}V\,, \tag{8}$$

where s is the entropy per unit mass, $\boldsymbol{J}_{s,\,\mathrm{tot}}$ the total entropy flow per unit area and unit time, and σ the entropy source strength or entropy production per unit volume and unit time.

With (6), (7) and (8), formula (1) may be rewritten, using also Gauss' theorem, in the form

$$\int^{V} \left(\frac{\partial \rho s}{\partial t} + \operatorname{div} \boldsymbol{J}_{s,\,\mathrm{tot}} - \sigma \right) \mathrm{d}V = 0 \,. \tag{9}$$

From this relation it follows, since (1) and (2) must hold for an arbitrary volume V, that

$$\frac{\partial \rho s}{\partial t} = - \operatorname{div} \boldsymbol{J}_{s,\,\mathrm{tot}} + \sigma \,, \tag{10}$$

$$\sigma \geqslant 0 \,. \tag{11}$$

These two formulae are the local forms of (1) and (2), *i.e.* the local mathematical expression for the second law of thermodynamics. Equation (10) is formally a balance equation for the entropy density ρs, with a source term σ which satisfies the important inequality (11). With the help of relation (II.16), equation (10) can be rewritten in a slightly different form,

$$\rho \frac{\mathrm{d}s}{\mathrm{d}t} = - \operatorname{div} \boldsymbol{J}_s + \sigma \,, \tag{12}$$

where the entropy flux $\boldsymbol{J}_s$ is the difference between the total entropy flux $\boldsymbol{J}_{s,\,\mathrm{tot}}$ and a convective term $\rho s \boldsymbol{v}$

$$\boldsymbol{J}_s = \boldsymbol{J}_{s,\,\mathrm{tot}} - \rho s \boldsymbol{v} \,. \tag{13}$$

In obtaining (10) and (11) we have assumed that the statements (1) and (2) also hold for infinitesimally small parts of the system, or in other words, that the laws which are valid for macroscopic systems remain valid for infinitesimally small parts of it. This is in agreement with the point of view currently adopted in a macroscopic description of a continuous system. It implies, on a microscopic model, that the local macroscopic measurements performed on the system, are really measurements of the properties of small parts of the system, which still contain a large number of the constituting particles. Such small parts of the system one might call physically infinitesimal. With this in mind it still makes sense to speak about the local values of such fundamentally macroscopic concepts as entropy and entropy production.

§ 2. *The Entropy Balance Equation*

We must now relate the variations in the properties of systems studied in Chapter II to the rate of change of the entropy. This will enable us to obtain more explicit expressions for the entropy flux $\boldsymbol{J}_s$ and the entropy source strength σ which appear in (12).

From thermodynamics we know that the entropy per unit mass s is, for a system in equilibrium, a well-defined function of the various parameters which are necessary to define the macroscopic state of the system completely. For the systems considered in Chapter II these are the internal energy u, the specific volume v, and the mass fractions c_k:

$$s = s\,(u, v, c_k)\,. \tag{14}$$

This is also expressed by the fact that, in equilibrium, the total differential of s is given by the Gibbs relation (*cf.* Appendix II):

$$T\mathrm{d}s = \mathrm{d}u + p\mathrm{d}v - \sum_{k=1}^{n} \mu_k \mathrm{d}c_k\,, \tag{15}$$

where p is the equilibrium pressure, and μ_k the thermodynamic or chemical potential of component k (partial specific Gibbs function).

It will now be assumed that, although the total system is not in equilibrium, there exists within small mass elements a state of "local" equilibrium, for which the local entropy s is the same function (14) of u, v and c_k as in real equilibrium. In particular we assume that formula (15) remains valid for a mass element followed along its centre of gravity motion:

$$T\frac{\mathrm{d}s}{\mathrm{d}t} = \frac{\mathrm{d}u}{\mathrm{d}t} + p\frac{\mathrm{d}v}{\mathrm{d}t} - \sum_{k=1}^{n} \mu_k \frac{\mathrm{d}c_k}{\mathrm{d}t}\,, \tag{16}$$

where the time derivatives are given by (II.8). This hypothesis of "local" equilibrium can, from a macroscopic point of view, only be justified by virtue of the validity of the conclusions derived from it. For special microscopic models it can indeed be shown that the relation (16) is valid for deviations from equilibrium which are not "too large". Criteria specifying how far from equilibrium (16) can be used may also be derived from these microscopic considerations. We shall come back to this point in Chapters VII and IX. It may already be stated here that for most familiar transport phenomena the use of (16) is justified.

In order to find the explicit form of the entropy balance equation (12) we have to insert the expressions (II.39), with (II.38), for $\mathrm{d}u/\mathrm{d}t$ and (II.13) for $\mathrm{d}c_k/\mathrm{d}t$ into formula (16). This gives

$$\rho \frac{\mathrm{d}s}{\mathrm{d}t} = -\frac{\operatorname{div} \boldsymbol{J}_q}{T} - \frac{1}{T} \Pi : \operatorname{Grad} \boldsymbol{v} + \frac{1}{T} \sum_{k=1}^{n} \boldsymbol{J}_k \cdot \boldsymbol{F}_k + \frac{1}{T} \sum_{k=1}^{n} \mu_k \operatorname{div} \boldsymbol{J}_k - \frac{1}{T} \sum_{j=1}^{r} J_j A_j , \quad (17)$$

where we have introduced the so-called chemical affinities of the reactions j $(= 1, 2, \ldots, r)$ defined by

$$A_j = \sum_{k=1}^{n} \nu_{kj} \mu_k , \quad (j = 1, 2, \ldots, r) . \quad (18)$$

It is easy to cast equation (17) into the form (12) of a balance equation:

$$\rho \frac{\mathrm{d}s}{\mathrm{d}t} = -\operatorname{div} \left(\frac{\boldsymbol{J}_q - \sum_k \mu_k \boldsymbol{J}_k}{T} \right) - \frac{1}{T^2} \boldsymbol{J}_q \cdot \operatorname{grad} T - \frac{1}{T} \sum_{k=1}^{n} \boldsymbol{J}_k \cdot \left(T \operatorname{grad} \frac{\mu_k}{T} - \boldsymbol{F}_k \right) - \frac{1}{T} \Pi : \operatorname{Grad} \boldsymbol{v} - \frac{1}{T} \sum_{j=1}^{r} J_j A_j . \quad (19)$$

From comparison with (12) it follows that the expressions for the entropy flux and the entropy production are given by

$$\boldsymbol{J}_s = \frac{1}{T} \left(\boldsymbol{J}_q - \sum_{k=1}^{n} \mu_k \boldsymbol{J}_k \right) , \quad (20)$$

$$\sigma = -\frac{1}{T^2} \boldsymbol{J}_q \cdot \operatorname{grad} T - \frac{1}{T} \sum_{k=1}^{n} \boldsymbol{J}_k \cdot \left(T \operatorname{grad} \frac{\mu_k}{T} - \boldsymbol{F}_k \right) - \frac{1}{T} \Pi : \operatorname{Grad} \boldsymbol{v} - \frac{1}{T} \sum_{j=1}^{r} J_j A_j \geqslant 0 . \quad (21)$$

The way in which the separation of the right-hand side of (17) into the divergence of a flux and a source term has been achieved may at first sight seem to be to some extent arbitrary. The two parts of (19) must, however, satisfy a number of requirements which determine this separation uniquely. Thus we know that the entropy source strength σ must be zero if the thermodynamic equilibrium conditions are satisfied within the system. Another requirement which (21) must satisfy is that it be invariant under a Galilei transformation, since the notions of reversible and irreversible behaviour must be invariant under such a transformation. It is seen that (21) satisfies automatically this requirement. Finally it may be noted that by integrating (19) over the volume V of a closed system one obtains, with the inequality of (21),

$$\frac{\mathrm{d}S}{\mathrm{d}t} \geqslant -\int^{\Omega} \frac{\boldsymbol{J}_q}{T} \cdot \mathrm{d}\boldsymbol{\Omega}\,, \tag{22}$$

which is equivalent with the Carnot–Clausius theorem (5) as it should be.

Let us consider in more detail the expressions (20) and (21) for the entropy flow $\boldsymbol{J}_s$ and the entropy production σ. The first formula shows that for open systems the entropy flow consists of two parts: one is the "reduced" heat flow $\boldsymbol{J}_q/T$, the other is connected with the diffusion flows of matter $\boldsymbol{J}_k$. The second formula demonstrates that the entropy production contains four different contributions. The first term at the right-hand side of (21) arises from heat conduction, the second from diffusion, the third is connected to the gradients of the velocity field, giving rise to viscous flow, and the fourth is due to chemical reactions. The structure of the expression for σ is that of a bilinear form: it consists of a sum of products of two factors. One of these factors in each term is a flow quantity (heat flow $\boldsymbol{J}_q$, diffusion flow $\boldsymbol{J}_k$, momentum flows or viscous pressure tensor Π, and chemical reaction rate J_j) already introduced in the conservation laws of Chapter II. The other factor in each term is related to a gradient of an intensive state variable (gradients of temperature, chemical potential and velocity) and may contain the external force $\boldsymbol{F}_k$; it can also be a difference of thermodynamic state variables, *viz.* the chemical affinity A_j. These quantities which multiply the fluxes in the expression for the entropy production are called "thermodynamic forces" or "affinities".

§ 3. *Alternative Expressions for the Entropy Production; on different Definitions of the Heat Flow*

It is convenient for a number of applications to write the entropy production (21) in a different form. The thermodynamic force which multiplies the diffusion flow $\boldsymbol{J}_k$ includes a part which is proportional to the gradient of the temperature. By using the thermodynamic relation

$$T\,\mathrm{d}\left(\frac{\mu_k}{T}\right) = (\mathrm{d}\mu_k)_T - \frac{h_k}{T}\,\mathrm{d}T\,, \tag{23}$$

where the index T indicates that the differential has to be taken at constant temperature, and where h_k is the partial specific enthalpy of component k, and by introducing a new flux, defined as

$$\boldsymbol{J}_q' = \boldsymbol{J}_q - \sum_{k=1}^{n} h_k \boldsymbol{J}_k\,, \tag{24}$$

the entropy production (21) can be written as

$$\sigma = -\frac{1}{T^2}\boldsymbol{J}_q'\cdot\mathrm{grad}\,T - \frac{1}{T}\sum_{k=1}^{n}\boldsymbol{J}_k\cdot\{(\mathrm{grad}\,\mu_k)_T - \boldsymbol{F}_k\}$$

$$-\frac{1}{T}\boldsymbol{\Pi}:\mathrm{Grad}\,\boldsymbol{v} - \frac{1}{T}\sum_{j=1}^{r} J_j A_j\,. \tag{25}$$

In this way the thermodynamic force conjugate to the diffusion flow $\boldsymbol{J}_k$ does not contain a term in grad T. However, the flow which is conjugate to the temperature gradient is now $\boldsymbol{J}_q'$ of formula (24) instead of $\boldsymbol{J}_q$. From (24) it is clear that the difference between $\boldsymbol{J}_q$ and $\boldsymbol{J}_q'$ represents a transfer of heat due to diffusion. Therefore the quantity $\boldsymbol{J}_q'$ also represents an irreversible heat flow. In fact in diffusing mixtures the concept of heat flow can be defined in different ways. Obviously a different definition of the notion of heat flux leaves all physical results unchanged. But to any particular choice corresponds a special form of the entropy production σ. It is a matter of expediency which choice is the most suitable in a particular application of the theory. The freedom of defining the heat flow in various ways, of which the possibility was indicated here in the framework of a macroscopic treatment, exists also in the microscopic theories of transport phenomena in mixtures.

With the definition (24) the entropy flow gets the form

$$\boldsymbol{J}_s = \frac{1}{T}\boldsymbol{J}_q' + \sum_{k=1}^{n} s_k \boldsymbol{J}_k\,, \tag{26}$$

where $s_k = -(\mu_k - h_k)/T$ is the partial specific entropy of component k. Written in this way the entropy flux contains the heat flow $\boldsymbol{J}_q'$ and a transport of partial entropies with respect to the barycentric velocity $\boldsymbol{v}$.

Still a different form of the entropy production can be obtained by using the equality

$$T \operatorname{grad} \left(\frac{\mu_k}{T}\right) = \operatorname{grad} \mu_k - \left(\frac{\mu_k}{T}\right) \operatorname{grad} T \qquad (27)$$

and the definition (20) of the entropy flow:

$$T\sigma = -\boldsymbol{J}_s \cdot \operatorname{grad} T - \sum_{k=1}^{n} \boldsymbol{J}_k \cdot (\operatorname{grad} \mu_k - \boldsymbol{F}_k) - \boldsymbol{\Pi} : \operatorname{Grad} \boldsymbol{v} - \sum_{j=1}^{r} J_j A_j . \qquad (28)$$

It is seen that in this way the force conjugate to the diffusion flow $\boldsymbol{J}_k$ contains simply a gradient of the chemical potential μ_k. Since [*cf.* (II.20)]

$$\boldsymbol{F}_k = -\operatorname{grad} \psi_k , \qquad (29)$$

we may write, by introducing the quantity

$$\tilde{\mu}_k = \mu_k + \psi_k , \qquad (30)$$

instead of (28)

$$T\sigma = -\boldsymbol{J}_s \cdot \operatorname{grad} T - \sum_{k=1}^{n} \boldsymbol{J}_k \cdot \operatorname{grad} \tilde{\mu}_k - \boldsymbol{\Pi} : \operatorname{Grad} \boldsymbol{v} - \sum_{j=1}^{r} J_j A_j . \qquad (31)$$

In the case of an electrostatic potential energy, ψ_k is equal to $z_k \varphi$ with z_k the charge per unit mass of component k, and φ the electrostatic potential, and $\tilde{\mu}_k$ is then the electrochemical potential. Quite in general it can be said that in the form (28) of the entropy production σ, where the entropy flow $\boldsymbol{J}_s$ is employed, the thermodynamic force conjugate to the diffusion flow can be written as the gradient of a single quantity, if the force $\boldsymbol{F}_k$ is conservative (*e.g.* an electrostatic or a gravitational force). This is the reason why the form (28) is of special advantage in applications, dealing with electric processes.

§ 4. *Kinetic Energy of Diffusion*

In Chapter II, § 4, we have defined the internal energy u by equation (II.32), *i.e.* by subtracting from the total energy e the potential energies of all components $\psi = \sum_k c_k \psi_k$ and the barycentric kinetic energy $\frac{1}{2}\boldsymbol{v}^2$. This means that the internal energy u still contains the macroscopic kinetic energy of the components with respect to the centre of gravity motion. It is possible to define a different internal energy per unit mass u^*, by subtracting from the total energy e, the potential energies and the kinetic energies of all components

$$u^* = e - \sum_k c_k \psi_k - \sum_k \tfrac{1}{2} c_k \boldsymbol{v}_k^2$$

$$= e - \psi - \tfrac{1}{2}\boldsymbol{v}^2 - \sum_k \tfrac{1}{2} c_k (\boldsymbol{v}_k - \boldsymbol{v})^2$$

$$= u - \sum_k \tfrac{1}{2} c_k (\boldsymbol{v}_k - \boldsymbol{v})^2 , \tag{32}$$

where (II.7) and (II.32) have been used. Since the internal energy should only contain contributions from the thermal agitation and the short-range molecular interactions, the quantity u^* has perhaps more right to this name than the quantity u. In equilibrium the Gibbs relation (15) is in fact a relation between the entropy s and the quantities u^*, v and c_k, since in equilibrium diffusion fluxes must vanish. Therefore (15) should read

$$T \, \mathrm{d}s = \mathrm{d}u^* + p \mathrm{d}v - \sum_{k=1}^{n} \mu_k^* \mathrm{d}c_k , \tag{33}$$

where we have introduced the chemical potential μ_k^* related to u^* by

$$\sum_k c_k \mu_k^* = u^* - Ts + pv . \tag{34}$$

In agreement with the hypothesis of local equilibrium one should therefore assume (33) to hold outside equilibrium in the form

$$T \frac{\mathrm{d}s}{\mathrm{d}t} = \frac{\mathrm{d}u^*}{\mathrm{d}t} + p \frac{\mathrm{d}v}{\mathrm{d}t} - \sum_{k=1}^{n} \mu_k^* \frac{\mathrm{d}c_k}{\mathrm{d}t} , \tag{35}$$

instead of (16).

Introducing into this equation the relation (32), one obtains, with (II.9)

$$T\frac{\mathrm{d}s}{\mathrm{d}t} = \frac{\mathrm{d}u}{\mathrm{d}t} + p\frac{\mathrm{d}v}{\mathrm{d}t} - \sum_{k=1}^{n} \mu_k \frac{\mathrm{d}c_k}{\mathrm{d}t} - \rho^{-1}\sum_{k=1}^{n} \boldsymbol{J}_k \cdot \frac{\mathrm{d}(\boldsymbol{v}_k - \boldsymbol{v})}{\mathrm{d}t}, \tag{36}$$

where μ_k is related to μ_k^* by

$$\mu_k = \mu_k^* + \tfrac{1}{2}(\boldsymbol{v}_k - \boldsymbol{v})^2 \tag{37}$$

and to u by the relation analogous to (34)

$$\sum_k c_k \mu_k = u - Ts + pv\,. \tag{38}$$

It is seen that (36) and (16) are identical if $\mathrm{d}(\boldsymbol{v}_k - \boldsymbol{v})/\mathrm{d}t$ vanishes, *i.e.* if the substantial time derivative of the velocities of the various components with respect to the barycentric motion may be neglected. We shall see later that frequently this may indeed be done. The use of (16) is then justified.

From (36) one obtains, using (II.39), (II.38) and (II.13), the entropy balance equation, which reads now

$$\begin{aligned}\rho\frac{\mathrm{d}s}{\mathrm{d}t} = &- \operatorname{div}\left(\frac{\boldsymbol{J}_q - \sum_k \mu_k \boldsymbol{J}_k}{T}\right) - \frac{1}{T^2}\boldsymbol{J}_q\cdot\operatorname{grad} T \\ &- \frac{1}{T}\sum_k \boldsymbol{J}_k \cdot \left\{T\operatorname{grad}\frac{\mu_k}{T} - \boldsymbol{F}_k + \frac{\mathrm{d}(\boldsymbol{v}_k - \boldsymbol{v})}{\mathrm{d}t}\right\} - \frac{1}{T}\Pi : \operatorname{Grad}\boldsymbol{v} \\ &- \frac{1}{T}\sum_j J_j A_j\,.\end{aligned} \tag{39}$$

This equation is identical with (19) except for the inclusion of an "inertia term" in the thermodynamic force of diffusion. Examples in which such "inertia terms" must be retained will be considered later.

CHAPTER IV

THE PHENOMENOLOGICAL EQUATIONS

§ 1. *The Linear Laws*

In the preceding chapter it has already been mentioned that the expression for the entropy production σ vanishes, when the thermodynamic equilibrium conditions are satisfied, *i.e.* when the (independent) thermodynamic forces are zero. In conformity with the concept of equilibrium we also require that all fluxes in σ vanish simultaneously with the thermodynamic forces.

It is known empirically that for a large class of irreversible phenomena and under a wide range of experimental conditions, the irreversible flows are linear functions of the thermodynamic forces, as expressed by the phenomenological laws which are introduced *ad hoc* in the purely phenomenological theories of irreversible processes. Thus, *e.g.* Fourier's law for heat conduction expresses that the components of the heat flow are linear functions of the components of the temperature gradient, and Fick's law establishes a linear relation between the diffusion flow of matter and the concentration gradient. Also included in this kind of description are the laws for such cross-phenomena as thermal diffusion, in which the diffusion flow depends linearly on both the concentration and temperature gradients. If we restrict ourselves to this linear region we may write quite generally

$$J_i = \sum_k L_{ik} X_k \,, \tag{1}$$

where J_i and X_i are any of the Cartesian components of the independent fluxes and thermodynamic forces appearing in the expression for the entropy production [*cf. e.g.* (III.21)], which is of the form $\sigma = \sum_i J_i X_i$. The quantities L_{ik} are called the phenomenological coefficients and the relations (1) will be referred to as the phenomenological equations. It is clear that this scheme includes the examples mentioned above.

If one introduces the phenomenological equations into the expression for the entropy production σ, one gets a quadratic expression in the thermodynamic forces of the form $\sum_{i,k} L_{ik} X_i X_k$ which, since one has $\sigma \geqslant 0$, must be positive definite or at least non-negative definite. A sufficient condition for this is that all principal co-factors of the symmetric matrix with elements $L_{ik} + L_{ki}$ are positive (or at least non-negative). This implies that all diagonal elements are positive whereas the off-diagonal elements must satisfy, for instance, conditions of the form $L_{ii} L_{kk} \geqslant \frac{1}{4} (L_{ik} + L_{ki})^2$.

With the help of the relations (1) it is now possible, using the conservation laws and balance equations of Chapters II and III, to determine in principle the evolution in time of *all local* thermodynamic state variables of the system. This is one of the advantages of the systematic formulation of thermodynamics of irreversible processes. On the other hand this formulation will also enable us to derive some important relationships which exist between the phenomenological coefficients (*cf.* § 3).

It is very well possible that some irreversible processes must be described by non-linear phenomenological laws. Such processes lie outside the scope of the present theory. However, even for such processes one may assume the linear relations to be valid within a very limited range close to equilibrium. Thus ordinary transport phenomena like heat conduction and electric conduction are linear even under rather extreme experimental conditions, whereas chemical reactions must nearly always be described by non-linear laws.

In the following sections the linear laws (1) will be given in explicit form for the systems studied in the preceding chapters, and the general properties of the matrix L_{ik} of phenomenological coefficients will be studied.

§ 2. *Influence of Symmetry Properties of Matter on the Linear Laws; Curie Principle*

Before stating in this section the influence of the symmetry properties of matter on the phenomenological equations (1), we wish to write the entropy production (III.21), (III.25) or (III.31) in a slightly different form.

Let us split up the symmetric viscous pressure tensor Π and the tensor Grad $\boldsymbol{v}$ in the following way

$$\Pi = \Pi U + \overset{\circ}{\Pi}, \tag{2}$$

$$\text{Grad}\, \boldsymbol{v} = \tfrac{1}{3} (\text{div}\, \boldsymbol{v}) U + \overset{\circ}{\text{Grad}}\, \boldsymbol{v}, \tag{3}$$

where the quantity Π is given by

$$\Pi = \tfrac{1}{3}\mathsf{\Pi}:\mathsf{U} = \tfrac{1}{3}\sum_{\alpha=1}^{3} \Pi_{\alpha\alpha}, \tag{4}$$

that is, as one third of the trace of the viscous pressure tensor. Similarly div $\boldsymbol{v}$ is the trace of Grad $\boldsymbol{v}$ [*cf.* (II.37)]:

$$\text{div}\, \boldsymbol{v} = (\text{Grad}\, \boldsymbol{v}) : \mathsf{U} = \sum_{\alpha=1}^{3} \frac{\partial v_\alpha}{\partial x_\alpha}. \tag{5}$$

The tensors $\mathring{\mathsf{\Pi}}$ and $\mathring{\text{Grad}}\, \boldsymbol{v}$ defined by (2) and (3) have zero trace according to (4) and (5):

$$\mathring{\mathsf{\Pi}}:\mathsf{U} = \sum_{\alpha=1}^{3} \mathring{\Pi}_{\alpha\alpha} = 0, \tag{6}$$

$$(\mathring{\text{Grad}}\, \boldsymbol{v}):\mathsf{U} = \sum_{\alpha=1}^{3} \left(\frac{\partial v_\alpha}{\partial x_\alpha} - \frac{1}{3}\sum_{\beta} \frac{\partial v_\beta}{\partial x_\beta}\right) = 0. \tag{7}$$

For the scalar product of (2) and (3), one finds with the help of (6) and (7)

$$\mathsf{\Pi} : \text{Grad}\, \boldsymbol{v} = \mathring{\mathsf{\Pi}} : \mathring{\text{Grad}}\, \boldsymbol{v} + \Pi\, \text{div}\, \boldsymbol{v}. \tag{8}$$

The tensor $\mathring{\text{Grad}}\, \boldsymbol{v}$ can be split into a symmetric and an anti-symmetric part

$$\mathring{\text{Grad}}\, \boldsymbol{v} = (\mathring{\text{Grad}}\, \boldsymbol{v})^{s} + (\mathring{\text{Grad}}\, \boldsymbol{v})^{a}, \tag{9}$$

with

$$(\mathring{\text{Grad}}\, \boldsymbol{v})^{s}_{\alpha\beta} = \frac{1}{2}\left(\frac{\partial v_\beta}{\partial x_\alpha} + \frac{\partial v_\alpha}{\partial x_\beta}\right) - \tfrac{1}{3}\,\delta_{\alpha\beta} \sum_{\gamma=1}^{3} \frac{\partial v_\gamma}{\partial x_\gamma}, \quad (\alpha, \beta = 1, 2, 3), \tag{10}$$

$$(\mathring{\text{Grad}}\, \boldsymbol{v})^{a}_{\alpha\beta} = \frac{1}{2}\left(\frac{\partial v_\beta}{\partial x_\alpha} - \frac{\partial v_\alpha}{\partial x_\beta}\right). \tag{11}$$

Using (9) the result (8) becomes

$$\mathsf{\Pi} : \text{Grad}\, \boldsymbol{v} = \mathring{\mathsf{\Pi}} : (\mathring{\text{Grad}}\, \boldsymbol{v})^{s} + \Pi\, \text{div}\, \boldsymbol{v}, \tag{12}$$

since the doubly contracted product of a symmetric and an anti-symmetric tensor vanishes.

If one introduces (12) into the form (III.25) of the entropy production and eliminates $\boldsymbol{J}_n$ with the help of (II.15), one obtains

$$\sigma = -\frac{1}{T^2}\boldsymbol{J}'_q \cdot \text{grad}\, T - \frac{1}{T}\sum_{k=1}^{n-1} \boldsymbol{J}_k \cdot [\{\text{grad}\,(\mu_k - \mu_n)\}_T - \boldsymbol{F}_k + \boldsymbol{F}_n]$$

$$-\frac{1}{T}\,\mathring{\Pi} : (\text{Gr}\mathring{\text{a}}\text{d}\,\boldsymbol{v})^s - \frac{1}{T}\,\Pi\,\text{div}\,\boldsymbol{v} - \frac{1}{T}\sum_{j=1}^{r} J_j A_j \geqslant 0\,. \quad (13)$$

The total contribution of viscous phenomena to the entropy production has thus been split up into two parts. The second part, — $(1/T)\,\Pi\,\text{div}\,\boldsymbol{v}$, is related to the rate of change of specific volume. This is the part which is due to bulk viscosity.

We shall now establish the phenomenological equations (1) between the independent fluxes and thermodynamic forces of this expression. In principle any Cartesian component of a flux can be a linear function of the Cartesian components of all thermodynamic forces. We note, however, that the fluxes and the thermodynamic forces of (13) do not all have the same tensorial character: some are scalars, some are vectors and one is a tensor (of second rank). This means that under rotations and reflections the Cartesian components of these quantities transform in different ways. As a consequence symmetry properties of the material system considered may have the effect that the components of the fluxes do not depend on all components of the thermodynamic forces. This fact is often referred to as the Curie symmetry principle. Thus, in particular for an isotropic system (*i.e.* a system of which the properties at equilibrium are the same in all directions) it can be shown that fluxes and thermodynamic forces of different tensorial character do not couple. The proof of this statement will be given in Chapter VI, where we shall study in a more formal way the influence of symmetry elements on the coupling of fluxes and thermodynamic forces. For an isotropic system the phenomenological equations read

$$\boldsymbol{J}'_q = -L_{qq}\,(\text{grad}\,T)/T^2 - \sum_{k=1}^{n-1} L_{qk}[\{\text{grad}\,(\mu_k - \mu_n)\}_T - \boldsymbol{F}_k + \boldsymbol{F}_n]/T\,, \quad (14)$$

$$\boldsymbol{J}_i = -L_{iq}\,(\text{grad}\,T)/T^2 - \sum_{k=1}^{n-1} L_{ik}[\{\text{grad}\,(\mu_k - \mu_n)\}_T - \boldsymbol{F}_k + \boldsymbol{F}_n]/T\,, \quad (15)$$

$$\mathring{\Pi}_{\alpha\beta} = -\frac{L}{T}(\mathrm{Gr\mathring{a}d}\,\boldsymbol{v})^s_{\alpha\beta} = -\frac{L}{2T}\left(\frac{\partial v_\beta}{\partial x_\alpha}+\frac{\partial v_\alpha}{\partial x_\beta}-\frac{2}{3}\delta_{\alpha\beta}\sum_\gamma \frac{\partial v_\gamma}{\partial x_\gamma}\right), (\alpha, \beta = 1, 2, 3), \quad (16)$$

$$\Pi = -l_{vv}(\mathrm{div}\,\boldsymbol{v})/T - \sum_{m=1}^{r} l_{vm}A_m/T\,, \quad (17)$$

$$J_j = -l_{jv}(\mathrm{div}\,\boldsymbol{v})/T - \sum_{m=1}^{r} l_{jm}A_m/T\,, \quad (j = 1, 2, \ldots, r)\,. \quad (18)$$

Equations (14) and (15) describe the vectorial phenomena of heat conduction, diffusion and their cross-effects. The coefficients L_{qq}, L_{iq}, L_{qk} and L_{ik} $(i, k = 1,2, \ldots, n)$ are scalar quantities. This is also a consequence of the isotropy of the system. Equations (16) relate the Cartesian components of the pressure tensor $\mathring{\Pi}$ to the components of the symmetric tensor $(\mathrm{Gr\mathring{a}d}\,\boldsymbol{v})^s$. Due to the isotropy of the system only corresponding tensor components α, β are linearly related with each other by means of the same coefficient L. Finally equations (17) and (18) describe the scalar processes of bulk viscosity and chemistry and their possible cross-phenomena.

Another consequence of the fact that in isotropic media fluxes and thermodynamic forces of different tensorial character do not interfere, is that the entropy production (13) falls apart into three contributions, which are separately positive definite

$$\sigma_0 = -\frac{1}{T}\Pi\,\mathrm{div}\,\boldsymbol{v} - \frac{1}{T}\sum_{j=1}^{r} J_j A_j \geqslant 0\,, \quad (19)$$

$$\sigma_1 = -\frac{1}{T^2}\boldsymbol{J}'_q\cdot\mathrm{grad}\,T - \frac{1}{T}\sum_{k=1}^{n-1}\boldsymbol{J}_k\cdot[\{\mathrm{grad}(\mu_k-\mu_n)\}_T - \boldsymbol{F}_k + \boldsymbol{F}_n] \geqslant 0\,, \quad (20)$$

$$\sigma_2 = -\frac{1}{T}\mathring{\Pi} : (\mathrm{Gr\mathring{a}d}\,\boldsymbol{v})^s \geqslant 0\,. \quad (21)$$

This can be concluded when the phenomenological equations (14)–(18) are substituted into (13).

We shall also write down the general form of the phenomenological equations in anisotropic crystals in which no chemical reactions occur. Since in such systems no viscous flows exist, we are left with the phenomena of heat conduction, diffusion and their cross-effects. The phenomenological equations corresponding to this case are

$$\boldsymbol{J}_q' = -\mathsf{L}_{qq}\cdot(\operatorname{grad} T)/T^2 - \sum_{k=1}^{n-1}\mathsf{L}_{qk}\cdot[\{\operatorname{grad}(\mu_k-\mu_n)\}_T - \boldsymbol{F}_k + \boldsymbol{F}_n]/T\,, \quad (22)$$

$$\boldsymbol{J}_i = -\mathsf{L}_{iq}\cdot(\operatorname{grad} T)/T^2 - \sum_{k=1}^{n-1}\mathsf{L}_{ik}\cdot[\{\operatorname{grad}(\mu_k-\mu_n)\}_T - \boldsymbol{F}_k + \boldsymbol{F}_n]/T\,,$$

$$(i = 1, 2, \ldots, n-1)\,. \quad (23)$$

The quantities L_{qq}, L_{qk}, L_{iq} and L_{ik} are tensors. For instance L_{qq} is related to the heat conduction tensor. The form of these tensors depends on the symmetry elements of the system. We have seen above that in isotropic media all tensors in (22) and (23) reduce to scalar multiples of the unit tensor. This is also the case in crystals with cubic symmetry.

Since the isotropic fluid and the anisotropic crystal are in actual physical applications the two most frequently encountered types of systems, we have confined the discussion of the influence of symmetry properties of matter on the phenomenological laws to these two cases.

§ 3. *The Onsager Reciprocal Relations*

In the preceding section we have considered the influence of spatial symmetry on the phenomenological equations. In this section we shall discuss the influence of the property of "time reversal invariance" of the equations of motion of the individual particles, of which the system consists, on the phenomenological equations. This property of "time reversal invariance" expresses the fact that the mechanical equations of motion (classical as well as quantum mechanical) of the particles are symmetric with respect to the time. It implies that the particles retrace their former paths if all velocities are reversed.

From this microscopic property one may conclude to a macroscopic theorem, due to Onsager. In this section we shall state the content of this theorem. In Chapter VII the derivation of this theorem is discussed.

Let us consider an adiabatically insulated system. We shall first take the case that no external magnetic field acts on the system. The state of the system can be described by a number of independent parameters. These parameters may be of two types. Some of these are even functions of the particle velocities (one may think, for instance, of local energies, concentrations, etc.). These are denoted by $A_1, A_2, \ldots, A_n$.

The other parameters are odd functions of the particle velocities (*e.g.* momentum densities), and are denoted by $B_1, B_2, \ldots, B_m$. The equilibrium values of these variables are $A_1^0, A_2^0, \ldots, A_n^0$ and B_1^0, $B_2^0, \ldots, B_m^0$. The deviations of all these parameters from their equilibrium values are given by

$$\alpha_i = A_i - A_i^0, \quad (i = 1, 2, \ldots, n), \tag{24}$$

$$\beta_i = B_i - B_i^0, \quad (i = 1, 2, \ldots, m). \tag{25}$$

At equilibrium the entropy has a maximum, and the state variables $\alpha_1, \alpha_2, \ldots, \alpha_n$ and $\beta_1, \beta_2, \ldots, \beta_m$ are zero by definition; this means that for a non-equilibrium state one can write for the deviation ΔS of the entropy from its equilibrium value, as a first approximation, a quadratic expression in the state variables $\alpha_1, \alpha_2, \ldots, \alpha_n$ and β_1, $\beta_2, \ldots, \beta_m$:

$$\Delta S = -\tfrac{1}{2} \sum_{i,k=1}^{n} g_{ik}\alpha_k\alpha_i - \tfrac{1}{2} \sum_{i,k=1}^{m} h_{ik}\beta_k\beta_i, \tag{26}$$

where g_{ik} $(i, k = 1, 2, \ldots, n)$ and h_{ik} $(i, k = 1, 2, \ldots, m)$, the second derivatives of ΔS with respect to the α- and β-variables, are positive definite matrices. In the absence of an external magnetic field no cross-terms between α- and β-type variables occur in (26) since ΔS must be an even function of the particle velocities.

It is assumed that the time behaviour of the state parameters can be described by linear phenomenological equations of the type

$$\frac{d\alpha_i}{dt} = -\sum_{k=1}^{n} M_{ik}^{(\alpha\alpha)} \alpha_k - \sum_{k=1}^{m} M_{ik}^{(\alpha\beta)} \beta_k, \quad (i = 1, 2, \ldots, n), \tag{27}$$

$$\frac{d\beta_i}{dt} = -\sum_{k=1}^{n} M_{ik}^{(\beta\alpha)} \alpha_k - \sum_{k=1}^{m} M_{ik}^{(\beta\beta)} \beta_k, \quad (i = 1, 2, \ldots, m), \tag{28}$$

where the $M_{ik}^{(\alpha\alpha)}$, $M_{ik}^{(\alpha\beta)}$, $M_{ik}^{(\beta\alpha)}$ and $M_{ik}^{(\beta\beta)}$ are the phenomenological coefficients. Onsager's theorem establishes a number of relations between these coefficients, *viz.*,

$$\sum_{k=1}^{n} M_{ik}^{(\alpha\alpha)} g_{kj}^{-1} = \sum_{k=1}^{n} M_{jk}^{(\alpha\alpha)} g_{ki}^{-1}, \quad (i, j = 1, 2, \ldots, n), \tag{29}$$

$$\sum_{k=1}^{m} M_{ik}^{(\alpha\beta)} h_{kj}^{-1} = - \sum_{k=1}^{n} M_{jk}^{(\beta\alpha)} g_{ki}^{-1}, \quad (i = 1, 2, \ldots, n; j = 1, 2, \ldots, m), \tag{30}$$

$$\sum_{k=1}^{m} M_{ik}^{(\beta\beta)} h_{kj}^{-1} = \sum_{k=1}^{m} M_{jk}^{(\beta\beta)} h_{ki}^{-1}, \quad (i, j = 1, 2, \ldots, m), \tag{31}$$

where the g_{ik}^{-1} and h_{ik}^{-1} are the reciprocal matrices of the g_{ik} and h_{ik} These relations, which express the content of Onsager's theorem, can be written in a somewhat more transparant form, by writing the phenomenological equations (27) and (28) in a different fashion. To this purpose let us introduce the following linear combinations of the state parameters

$$X_i = \frac{\partial \Delta S}{\partial \alpha_i} = - \sum_{k=1}^{n} g_{ik} \alpha_k, \quad (i = 1, 2, \ldots, n). \tag{32}$$

$$Y_i = \frac{\partial \Delta S}{\partial \beta_i} = - \sum_{k=1}^{m} h_{ik} \beta_k, \quad (i = 1, 2, \ldots, m). \tag{33}$$

Solving for the α_i and β_i we obtain

$$\alpha_i = - \sum_{k=1}^{n} g_{ik}^{-1} X_k, \quad (i = 1, 2, \ldots, n), \tag{34}$$

$$\beta_i = - \sum_{k=1}^{m} h_{ik}^{-1} Y_k, \quad (i = 1, 2, \ldots, m). \tag{35}$$

Introducing (34) and (35) into (27) and (28), these relations become

$$\frac{d\alpha_i}{dt} = \sum_{k=1}^{n} L_{ik}^{(\alpha\alpha)} X_k + \sum_{k=1}^{m} L_{ik}^{(\alpha\beta)} Y_k, \quad (i = 1, 2, \ldots, n), \tag{36}$$

$$\frac{d\beta_i}{dt} = \sum_{k=1}^{n} L_{ik}^{(\beta\alpha)} X_k + \sum_{k=1}^{m} L_{ik}^{(\beta\beta)} Y_k, \quad (i = 1, 2, \ldots, m), \tag{37}$$

where the coefficients are given by

$$L_{ik}^{(\alpha\alpha)} = \sum_{j=1}^{n} M_{ij}^{(\alpha\alpha)} g_{jk}^{-1}, \quad (i, k = 1, 2, \ldots, n), \tag{38}$$

$$L_{ik}^{(\alpha\beta)} = \sum_{j=1}^{m} M_{ij}^{(\alpha\beta)} h_{jk}^{-1}, \quad (i = 1, 2, \ldots, n; \quad k = 1, 2, \ldots, m), \tag{39}$$

$$L_{ik}^{(\beta\alpha)} = \sum_{j=1}^{n} M_{ij}^{(\beta\alpha)} g_{jk}^{-1}, \quad (i = 1, 2, \ldots, m; k = 1, 2, \ldots, n), \tag{40}$$

$$L_{ik}^{(\beta\beta)} = \sum_{j=1}^{m} M_{ij}^{(\beta\beta)} h_{jk}^{-1}, \quad (i, k = 1, 2, \ldots, m). \tag{41}$$

With the help of these quantities, the Onsager relations (29)–(31) *become*

$$L_{ik}^{(\alpha\alpha)} = L_{ki}^{(\alpha\alpha)}, \qquad (i, k = 1, 2, \ldots, n), \tag{42}$$

$$L_{ik}^{(\alpha\beta)} = -L_{ki}^{(\beta\alpha)}, \quad (i = 1, 2, \ldots, n; k = 1, 2, \ldots, m), \tag{43}$$

$$L_{ik}^{(\beta\beta)} = L_{ki}^{(\beta\beta)}, \qquad (i, k = 1, 2, \ldots, m). \tag{44}$$

In this simple form they are usually referred to as Onsager's reciprocal relations.

To summarize the results it can be said that the Onsager relations (42)–(44) *are valid for the coefficients of the phenomenological equations, if the independent "fluxes"* J_i *and* I_i

$$J_i \equiv \frac{\mathrm{d}\alpha_i}{\mathrm{d}t}, \quad (i = 1, 2, \ldots, n), \tag{45}$$

$$I_i \equiv \frac{\mathrm{d}\beta_i}{\mathrm{d}t}, \quad (i = 1, 2, \ldots, m), \tag{46}$$

are written as linear functions of the independent "thermodynamic forces" X_i *and* Y_i *which are the derivatives of the entropy with respect to* α_i *and* β_i *respectively*

$$X_i = \frac{\partial \Delta S}{\partial \alpha_i}, \quad (i = 1, 2, \ldots, n), \tag{47}$$

$$Y_i = \frac{\partial \Delta S}{\partial \beta_i}, \quad (i = 1, 2, \ldots, m). \tag{48}$$

The Onsager relations hold in the form (42)–(44) if no external magnetic field $\boldsymbol{B}$ is present. In the presence of an external magnetic field the property of "time reversal invariance" implies that the particles retrace their former paths only if both the particle velocities and the magnetic field are reversed. This follows from the form of the expression for the Lorentz force, which is proportional to the vector product of the particle velocity and the magnetic field. A similar situation arises in rotating systems. Then the particles retrace their former paths if both their velocities and the angular velocity $\boldsymbol{\omega}$ are reversed, since the particles are then subjected to the so-called coriolis force which is proportional to the vector product of the particle velocity and the angular velocity. As a consequence the Onsager relations (42)–(44) must be modified to read*

$$L_{ik}^{(\alpha\alpha)}(\boldsymbol{B}, \boldsymbol{\omega}) = L_{ki}^{(\alpha\alpha)}(-\boldsymbol{B}, -\boldsymbol{\omega}), \qquad (i, k = 1, 2, \ldots, n), \tag{49}$$

$$L_{ik}^{(\alpha\beta)}(\boldsymbol{B}, \boldsymbol{\omega}) = -L_{ki}^{(\beta\alpha)}(-\boldsymbol{B}, -\boldsymbol{\omega}), \quad (i = 1, 2, \ldots, n; k = 1, 2, \ldots, m), \tag{50}$$

$$L_{ik}^{(\beta\beta)}(\boldsymbol{B}, \boldsymbol{\omega}) = L_{ki}^{(\beta\beta)}(-\boldsymbol{B}, -\boldsymbol{\omega}), \qquad (i, k = 1, 2, \ldots, m). \tag{51}$$

It is interesting to write down the time derivative of the entropy (26), *i.e.* the entropy production, due to the irreversible processes occurring in the system:

$$\frac{\mathrm{d}\Delta S}{\mathrm{d}t} = -\sum_{i,k=1}^{n} g_{ik}\alpha_k \frac{\mathrm{d}\alpha_i}{\mathrm{d}t} - \sum_{i,k=1}^{m} h_{ik}\beta_k \frac{\mathrm{d}\beta_i}{\mathrm{d}t}, \tag{52}$$

and therefore with (32), (33) and (45), (46):

$$\frac{\mathrm{d}\Delta S}{\mathrm{d}t} = \sum_{i=1}^{n} J_i X_i + \sum_{i=1}^{m} I_i Y_i. \tag{53}$$

The entropy production is therefore a bilinear expression in the fluxes and thermodynamic forces appearing in the phenomenological equations for which the Onsager relations hold. The calculations of the entropy production therefore affords a means of finding the proper "conjugate" irreversible fluxes and thermodynamic forces necessary for the establishment of phenomenological equations of which the coefficients obey the Onsager relations (42)–(44) or (49)–(51).

* It should be noted that in the presence of a magnetic field the thermodynamic forces (47) and (48) are not given by the last members of (32) and (33) since the entropy ΔS may then contain cross-terms between α- and β-variables (the entropy must be invariant for a reversal of both the particle velocities and the magnetic field, *cf.* Chapter VII).

Although the fluxes in the local entropy production σ, calculated in the preceding chapter and used in § 2 of this chapter, are not necessarily time derivatives of state variables as the fluxes (45) and (46) in (53), or in other words, although the local entropy production σ is not a total time derivative such as (52) is, it can be shown that the phenomenological coefficients appearing in the linear laws established between fluxes and thermodynamic forces of the *local entropy production* also obey the reciprocal relations (42)–(44) or (49)–(51). The formal proof of this statement will be given in Chapter VI.

Thus the following relations exist amongst the coefficients of the phenomenological laws (14)–(18) of the isotropic fluid (in the absence of a magnetic field)

$$L_{qi} = L_{iq}\,, \qquad (i = 1, 2, \ldots, n-1)\,, \tag{54}$$

$$L_{ik} = L_{ki}\,, \qquad (i, k = 1, 2, \ldots, n-1)\,, \tag{55}$$

$$l_{vj} = -\,l_{jv}\,, \quad (j = 1, 2, \ldots, r)\,, \tag{56}$$

$$l_{jm} = l_{mj}\,, \qquad (j, m = 1, 2, \ldots, r)\,. \tag{57}$$

Relation (56) is an example of (43) since it describes a cross-effect between an α- and a β-type variable; the chemical affinity A and the divergence of the velocity $\boldsymbol{v}$ respectively. The symmetry relations (54)–(57) establish a number of connections between otherwise independent irreversible processes. One of the objectives of non-equilibrium thermodynamics is to study the physical consequences of such relations (see part B).

For the coefficients of the phenomenological laws (22) and (23) of the anisotropic crystal the reciprocal relations, in the presence of a magnetic field, are

$$\mathsf{L}_{qq}(\boldsymbol{B}) = \tilde{\mathsf{L}}_{qq}(-\boldsymbol{B})\,, \tag{58}$$

$$\mathsf{L}_{qi}(\boldsymbol{B}) = \tilde{\mathsf{L}}_{iq}(-\boldsymbol{B})\,, \quad (i = 1, 2, \ldots, n-1)\,, \tag{59}$$

$$\mathsf{L}_{ik}(\boldsymbol{B}) = \tilde{\mathsf{L}}_{ki}(-\boldsymbol{B})\,, \quad (i, k = 1, 2, \ldots, n-1)\,, \tag{60}$$

where the tildas mean transposing of Cartesian components μ and ν of a tensor, for instance

$$\tilde{L}_{iq,\,\mu\nu}(\boldsymbol{B}) = L_{iq,\,\nu\mu}(\boldsymbol{B})\,, \quad (\mu, \nu = 1, 2, 3)\,. \tag{61}$$

We note that for the anisotropic case, the Onsager relations, in the absence of a magnetic field ($\boldsymbol{B} = 0$), have as a consequence that the tensors L_{qq} and L_{ii} ($i = 1, 2, \ldots, n-1$) are symmetric. In the presence of a magnetic field the relations (58)–(60) yield also some information about the parity of certain coefficients with respect to reversal of the magnetic field.

The Onsager relations have been written down here for the phenomenological equations involving the fluxes and thermodynamic forces occurring in the form (13) of the entropy production. Any of the alternative forms of the entropy production derived in Chapter III involving other fluxes and thermodynamic forces would have led to phenomenological laws with other coefficients for which, however, reciprocal relations still hold. In fact it can easily be seen that the transformations of Chapter III, from the description with one set of fluxes and thermodynamic forces to another, preserve the validity of the Onsager relations (*cf.* also Chapter VI, § 5).

§ 4. *The Differential Equations*

If one substitutes the phenomenological equations (14)–(18) into the conservation laws for matter (II.13), the momentum equation (II.19) and the balance equation of internal energy (II.36), one obtains with (II.5) a set of $n+4$ partial differential equations for the $n+4$ independent variables: the density ρ, the $n-1$ concentrations c_1, $c_2, \ldots, c_{n-1}$, the three Cartesian components v_x, v_y and v_z of the velocity $\boldsymbol{v}$, and the temperature T. The equations of state of the system allow to express the energy u, the equilibrium pressure p and the chemical potentials μ_k, occurring in the partial differential equations, in terms of those independent variables.

For a one-component isotropic fluid these partial differential equations are (in the absence of external forces):

$$\frac{\partial \rho}{\partial t} = -\operatorname{div} \rho \boldsymbol{v}, \tag{62}$$

$$\rho \frac{\mathrm{d}\boldsymbol{v}}{\mathrm{d}t} = -\operatorname{grad} p + \eta \triangle \boldsymbol{v} + (\tfrac{1}{3}\eta + \eta_v) \operatorname{grad} \operatorname{div} \boldsymbol{v}, \tag{63}$$

$$\rho \frac{\mathrm{d}u}{\mathrm{d}t} = \lambda \triangle T - p \operatorname{div} \boldsymbol{v} + 2\eta\, (\mathring{\mathrm{Grad}}\, \boldsymbol{v})^s : (\mathring{\mathrm{Grad}}\, \boldsymbol{v})^s + \eta_v\, (\operatorname{div} \boldsymbol{v})^2 . \tag{64}$$

The first of these equations is simply the equation of conservation of mass (II.5). The second is found by substituting (2), (16) and (17) (without chemical terms) into (II.9) with (II.35). The coefficients η and η_v, defined as $\eta = L/2T$ and $\eta_v = l_{vv}/T$, are called the shear viscosity and the bulk viscosity respectively. It has been assumed here that the viscosity coefficients are constants. The third equation follows from (II.36) with (14) (without diffusion terms), (16) and (17). The coefficient λ, defined as L_{qq}/T^2, is called the heat conductivity, and has also been assumed to be a constant. The symbol $\triangle$ stands for the Laplace operator. These equations must be supplemented by the equations of state

$$p = p\,(\rho, T)\,, \tag{65}$$

$$u = u\,(\rho, T)\,. \tag{66}$$

Equations (62)–(66) describe completely the time behaviour of the one-component isotropic fluid for specified initial and boundary conditions. It is customary to limit the field of hydrodynamics to equations (62), (63) and (65) alone, by assuming that either isothermal or isentropic conditions are fulfilled. In both cases pressure is a function of density only, so that the hydrodynamic behaviour is completely described by (62) and (63). In the more general case the complete set of equations (62)–(66) is necessary to describe the behaviour of the system. One might call the theory based on this complete set of equations "thermo-hydrodynamics" which is thus found to be part of the more general theory of non-equilibrium thermodynamics. On the other hand the theory of heat conduction is also contained in these equations.

We note that (63) is the well-known Navier-Stokes equation. The last two terms of (64) represent the Rayleigh dissipation function. Equation (64) becomes Fourier's differential equation for heat conduction

$$\rho c_v \frac{\partial T}{\partial t} = \lambda \triangle T \tag{67}$$

for a medium in which the velocity $\boldsymbol{v}$ is zero; ($c_v = (\partial u/\partial T)_v$ is the specific heat at constant volume per unit mass).

For more general cases, for instance in a multi-component system where diffusion occurs, the set of simultaneous differential equations becomes more complicated. It may be said that non-equilibrium thermodynamics has the purpose to study various irreversible processes as heat conduction, diffusion and viscous flow from a single point of view. It englobes a number of phenomenological theories such as the hydrodynamics of viscous fluids, the theory of diffusion and the theory of heat conduction.

CHAPTER V

THE STATIONARY STATES

§ 1. *Introduction*

Stationary states, *i.e.* states in which the state parameters are independent of time, play an important role in the applications of non-equilibrium thermodynamics. These stationary states can be either equilibrium or non-equilibrium states depending on the boundary conditions, which are imposed on the system.

Before the stationary states are studied some properties of the state of mechanical equilibrium will be discussed in this chapter. We shall then prove that under certain conditions, of which the most important are the supposition of constancy of the phenomenological coefficients and the validity of the Onsager reciprocal relations, the stationary states are also states of minimum entropy production compatible with the external constraints. If the conditions referred to are not fulfilled a theorem of a more general character can still be derived. This theorem is discussed in the last section of this chapter.

It will also be shown that stationary non-equilibrium states are stable with respect to perturbations. This fact constitutes a generalization of the Le Chatelier–Braun principle for equilibrium states.

§ 2. *Mechanical Equilibrium*

For the state of mechanical equilibrium a theorem can be derived which may simplify the description of some irreversible processes, in particular of diffusion phenomena.

The mechanical equilibrium state is the state in which the acceleration $\mathrm{d}\boldsymbol{v}/\mathrm{d}t$ vanishes. We shall be interested more specifically in such mechanical equilibrium states in which not only the acceleration vanishes, but in which velocity gradients and therefore also the viscous pressure tensor $\boldsymbol{\Pi}$ may be neglected. For such states the equation of motion (II.19) gets the form

$$0 = -\operatorname{grad} p + \sum_k \rho_k \boldsymbol{F}_k \, . \tag{1}$$

In a number of important cases the mechanical equilibrium state described by (1) is indeed established very quickly in comparison with the thermodynamical processes. Then virtually at the beginning of the irreversible process studied such a state is reached. In the most general case this needs certainly not always be true; it depends entirely on the particular physical situation. One can imagine oscillating systems in which acceleration terms subsist in the course of time. But, for instance, in the cases of diffusion or thermal diffusion phenomena in closed vessels one can safely assume that a state of mechanical equilibrium obeying (1) is quickly realized to a sufficient approximation. It is true that in diffusion experiments the acceleration $\mathrm{d}\boldsymbol{v}/\mathrm{d}t$ may be somewhat different from zero, for instance if the molecular masses of the components differ. This acceleration is however very small and the resulting pressure gradients (assuming no external forces to be present) are completely negligible. An initially imposed pressure difference would also cause acceleration phenomena, which would be damped out by viscosity, long before the diffusion phenomena reach stationary states. Thus, again supposing vanishing external forces $\boldsymbol{F}_k$, the pressure gradients would become negligible almost at the beginning of the diffusion process.

For mechanical equilibrium (1) Prigogine* has proved the theorem that in the entropy production, written in the form (IV.13), the barycentric velocity $\boldsymbol{v}$, occurring in the definition of the diffusion flow $\boldsymbol{J}_k$ (II.9), can be replaced by an arbitrary other velocity $\boldsymbol{v}^a$.

The proof of this theorem is based on the validity of the following equality:

$$\sum_{k=1}^{n} \rho_k \{ (\operatorname{grad} \mu_k)_T - \boldsymbol{F}_k \} = 0 . \tag{2}$$

This equality is most easily derived by noting first that for the specific Gibbs function g,

$$g = \sum_{k=1}^{n} c_k \mu_k , \tag{3}$$

one has

$$\delta g = - s\delta T + v\delta p + \sum_{k=1}^{n} \mu_k \delta c_k . \tag{4}$$

* I. Prigogine, Etude thermodynamique des phénomènes irréversibles (Desoer, Liège, 1947).

From (3) and (4) follows the Gibbs–Duhem relation

$$\sum_{k=1}^{n} \rho_k \delta\mu_k = - \rho s \delta T + \delta p \tag{5}$$

or

$$\sum_{k=1}^{n} \rho_k (\mathrm{grad}\, \mu_k)_T = \mathrm{grad}\, p \,. \tag{6}$$

Introduction of grad p from the equation of motion for the case of mechanical equilibrium (1) into the last relation yields (2).

Prigogine's theorem, as stated above, follows now quite easily. In fact the diffusion term σ_D of (III.25) with the explicit form (II.9) of $\boldsymbol{J}_k$ is

$$\sigma_D = - \frac{1}{T} \sum_{k=1}^{n} \rho_k(\boldsymbol{v}_k - \boldsymbol{v}) \cdot \{ (\mathrm{grad}\, \mu_k)_T - \boldsymbol{F}_k \} \,. \tag{7}$$

This expression is equal to

$$\sigma_D = - \frac{1}{T} \sum_{k=1}^{n} \rho_k(\boldsymbol{v}_k - \boldsymbol{v}^a) \cdot \{ (\mathrm{grad}\, \mu_k)_T - \boldsymbol{F}_k \} \,, \tag{8}$$

where $\boldsymbol{v}^a$ is an arbitrary reference velocity, since the difference of (7) and (8) vanishes according to (2). The equality of (7) and (8) proves Prigogine's theorem. We shall need this theorem for the discussion of a great variety of phenomena connected with diffusion processes.

Finally we remark that the external force $\boldsymbol{F}_k$ was supposed to be a conservative force of the form (II.20). The case of velocity dependent external forces will be treated in Chapters XI and XIII.

§ 3. *Stationary States with Minimum Entropy Production*

Stationary non-equilibrium states have the important property that, under certain conditions, they are characterized by a *minimum of the entropy production*, compatible with the external constraints imposed on the system. This property is valid only if the phenomenological coefficients are supposed to be *constants*. Since in real systems this is not true in general, it means that overall gradients of the thermodynamic parameters over the complete system have to be small enough so that the assumption of constancy of the phenomenological coefficients holds approximately. (In the next section the case in which this condition is not fulfilled will be studied). For the derivation of the property stated above one needs furthermore the linear phenomenological equations and the Onsager reciprocal relations. It will be supposed that the system is subject to external constraints at its surface, which fix some physical parameters in such a way that they are time independent. The property of minimum entropy production was first derived* for the so-called "discontinuous systems" (*cf.* Chapter XV), but will be obtained** here for continuous systems, such as considered in the preceding chapters. Two examples will be worked out: (i) thermal conduction. In this case of one irreversible process Onsager relations play no role. (ii) thermal conduction, diffusion and chemical reactions.

In both cases not only the property of minimum entropy production in stationary states will be proved, but also the fact that these states have a stable character with respect to perturbations of local state variables. This last property constitutes a generalization of the Le Chatelier–Braun principle for the stability of equilibrium states to stationary states.

(i) *Thermal conduction*

Let us consider a one-component isotropic system of which the temperatures have fixed time independent non-uniform values at the walls of the vessel in which the system is enclosed. It will be assumed that no viscous phenomena take place inside the system. The local entropy production is then according to (III.21)

$$\sigma = \boldsymbol{J}_q \cdot \operatorname{grad} \frac{1}{T}, \tag{9}$$

where $\boldsymbol{J}_q$ is the heat flow and T the temperature. The phenomenological equation is [*cf.* (IV.14)]

$$\boldsymbol{J}_q = L_{qq} \operatorname{grad} \frac{1}{T}, \tag{10}$$

where L_{qq}, which is connected with the heat conductivity coefficient λ by $L_{qq} = \lambda T^2$, will be supposed to be approximately constant over the system, and a function only of the overall equilibrium temperature.

* I. Prigogine, Etude thermodynamique des phénomènes irréversibles (Desoer, Liège, 1947).

** P. Mazur, Bull. Acad. roy. Belgique Cl. Sc. **38** (1952) 182.

If we suppose for simplicity that our system is a solid of which the thermal expansion may be neglected, the energy equation (II.36) can be written in the form

$$\rho \frac{\partial u}{\partial t} = \rho c_v \frac{\partial T}{\partial t} = -\operatorname{div} \boldsymbol{J}_q , \tag{11}$$

where c_v is the specific heat at constant volume.

The total entropy production P in the system is the volume integral of (9):

$$P = \int^{V} \sigma \mathrm{d}V = \int^{V} \boldsymbol{J}_q \cdot \operatorname{grad} \frac{1}{T} \mathrm{d}V = \int^{V} L_{qq} \left(\operatorname{grad} \frac{1}{T} \right)^2 \mathrm{d}V , \tag{12}$$

where (10) has been introduced. We now wish to find the temperature distribution for which the total entropy production has a minimum value, *i.e.* for which

$$\delta P = \delta \left\{ L_{qq} \int^{V} \left(\operatorname{grad} \frac{1}{T} \right)^2 \mathrm{d}V \right\} = 0 . \tag{13}$$

The solution of this variational problem, for which the variations δT are zero at the boundary, is given by the Euler equation

$$\operatorname{div} \operatorname{grad} \frac{1}{T} = 0 . \tag{14}$$

With the phenomenological equation (10) this can be written as

$$\operatorname{div} \boldsymbol{J}_q = 0 . \tag{15}$$

From (11) it then follows that (15) corresponds to a stationary state

$$\frac{\partial T}{\partial t} = 0 . \tag{16}$$

This completes the proof of the theorem that the stationary state is characterized by a minimum of the entropy production, compatible with the imposed temperature distribution at the walls of the system.

We now wish to show that this state is stable with respect to local perturbations of the temperature. If we differentiate (12) with respect to time we obtain

$$\frac{\partial P}{\partial t} = 2 \int^{V} L_{qq} \, \mathrm{grad} \frac{1}{T} \cdot \mathrm{grad} \left(\frac{\partial}{\partial t} \frac{1}{T} \right) \mathrm{d}V \,, \tag{17}$$

or, with (10), and after partial integration

$$\frac{\partial P}{\partial t} = 2 \int^{V} \boldsymbol{J}_q \cdot \mathrm{grad} \left(\frac{\partial}{\partial t} \frac{1}{T} \right) \mathrm{d}V = 2 \int^{\Omega} \left(\frac{\partial}{\partial t} \frac{1}{T} \right) \boldsymbol{J}_q \cdot \mathrm{d}\boldsymbol{\Omega} - 2 \int^{V} \left(\frac{\partial}{\partial t} \frac{1}{T} \right) \mathrm{div} \, \boldsymbol{J}_q \, \mathrm{d}V \,. \tag{18}$$

The surface integral in this expression vanishes since the temperature is constant at the walls. With (11) formula (18) becomes

$$\frac{\partial P}{\partial t} = -2 \int^{V} \rho \frac{c_v}{T^2} \left(\frac{\partial T}{\partial t} \right)^2 \mathrm{d}V \leqslant 0 \,. \tag{19}$$

Since c_v is positive, expression (19) is negative, *i.e.* the entropy production P diminishes in the course of time, or, in other words, the stationary state (of minimum entropy production) is indeed a stable state.

(ii) *Thermal conduction, diffusion, chemical reactions and cross-effects*

In this second example, where cross-effects between irreversible phenomena occur, the Onsager reciprocal relations are required for the derivation of the theorem of minimum entropy production in stationary states. We consider an isotropic mixture of n chemical components k ($k = 1, 2, \ldots, n$) amongst which r chemical reactions j ($j = 1,2,\ldots,r$) are possible. We shall assume that the system is in mechanical equilibrium ($\mathrm{d}\boldsymbol{v}/\mathrm{d}t \simeq 0$) and also that the barycentric velocity $\boldsymbol{v}$ is approximately zero. This implies that changes of the total density ρ are small and negligible in the course of time. (Viscous phenomena and the effect of external forces could easily be accounted for in the derivations, but they are not considered here).

The mass equations (II.13), the equation of motion (II.19) and the energy equation (II.36) now have the forms

$$\rho \frac{\partial c_k}{\partial t} = - \operatorname{div} \boldsymbol{J}_k + \sum_{j=1}^{r} \nu_{kj} J_j , \quad (k = 1, 2, \ldots, n) , \tag{20}$$

$$0 = \operatorname{grad} p , \tag{21}$$

$$\rho \frac{\partial u}{\partial t} = - \operatorname{div} \boldsymbol{J}_q . \tag{22}$$

The entropy balance equation (III.12) reads here

$$\rho \frac{\partial s}{\partial t} = - \operatorname{div} \boldsymbol{J}_s + \sigma , \tag{23}$$

with the entropy flow (III.20)

$$\boldsymbol{J}_s = \frac{1}{T} (\boldsymbol{J}_q - \sum_{k=1}^{n} \mu_k \boldsymbol{J}_k) \tag{24}$$

and the entropy source strength [*cf.* (III.21)]

$$\sigma = \boldsymbol{J}_q \cdot \operatorname{grad} \frac{1}{T} - \sum_{k=1}^{n-1} \boldsymbol{J}_k \cdot \operatorname{grad} \frac{\mu_k - \mu_n}{T} - \frac{1}{T} \sum_{j=1}^{r} J_j A_j , \tag{25}$$

where the identity $\sum_{k=1}^{n} \boldsymbol{J}_k = 0$ has been used.

The phenomenological equations, written as linear relations between the independent fluxes and forces occurring in (25), are

$$\boldsymbol{J}_q = L_{qq} \operatorname{grad} \frac{1}{T} - \sum_{k=1}^{n-1} L_{qk} \operatorname{grad} \frac{\mu_k - \mu_n}{T} , \tag{26}$$

$$\boldsymbol{J}_i = L_{iq} \operatorname{grad} \frac{1}{T} - \sum_{k=1}^{n-1} L_{ik} \operatorname{grad} \frac{\mu_k - \mu_n}{T} , \quad (i = 1, 2, \ldots, n-1) , \tag{27}$$

$$J_j = - \sum_{m=1}^{r} l_{jm} \frac{A_m}{T} = - \sum_{m=1}^{r} l_{jm} \sum_{k=1}^{n-1} \nu_{km} \frac{\mu_k - \mu_n}{T} , \quad (j = 1, 2, \ldots, r) , \tag{28}$$

where (III.18) and (II.4) have been introduced.

The Onsager reciprocal relations are

$$L_{qk} = L_{kq}\,, \quad (k = 1, 2, \ldots, n-1)\,, \tag{29}$$

$$L_{ik} = L_{ki}\,, \quad (i, k = 1, 2, \ldots, n-1)\,, \tag{30}$$

$$l_{jm} = l_{mj}\,, \quad (j, m = 1, 2, \ldots, r)\,. \tag{31}$$

The total entropy production P is

$$P = \int^{V} \sigma \mathrm{d}V = \int^{V} \Big\{ L_{qq} \left(\operatorname{grad}\frac{1}{T}\right)^2 - \sum_{k=1}^{n-1} (L_{qk} + L_{kq}) \operatorname{grad}\frac{\mu_k - \mu_n}{T} \cdot \operatorname{grad}\frac{1}{T}$$

$$+ \sum_{i,k=1}^{n-1} L_{ik} \operatorname{grad}\frac{\mu_i - \mu_n}{T} \cdot \operatorname{grad}\frac{\mu_k - \mu_n}{T} + \sum_{j,m=1}^{r} l_{jm} \frac{A_j}{T}\frac{A_m}{T} \Big\} \mathrm{d}V\,, \tag{32}$$

according to (25) with (26), (27) and (28). The integrand of this expression is a function of T and $\mu_k - \mu_n$, or, more conveniently, of $1/T$ and $(\mu_k - \mu_n)/T$. The state of minimum entropy production follows from the condition

$$\delta P = 0 \tag{33}$$

for arbitrary variations $\delta(1/T)$ and $\delta\{(\mu_k - \mu_n)/T\}$. (The pressure cannot be varied independently and is determined by the condition (21) of mechanical equilibrium.) The Euler equations of this variational problem are

$$2L_{qq} \operatorname{div}\operatorname{grad}\frac{1}{T} - \sum_{k=1}^{n-1} (L_{qk} + L_{kq}) \operatorname{div}\operatorname{grad}\frac{\mu_k - \mu_n}{T} = 0\,, \tag{34}$$

$$(L_{qi} + L_{iq}) \operatorname{div}\operatorname{grad}\frac{1}{T} - \sum_{k=1}^{n-1} (L_{ik} + L_{ki}) \operatorname{div}\operatorname{grad}\frac{\mu_k - \mu_n}{T}$$

$$+ \sum_{j\ m=1}^{r} l_{jm} \frac{A_j}{T} \frac{\partial(A_m/T)}{\partial\{(\mu_i - \mu_n)/T\}} + \sum_{j,m=1}^{r} l_{jm} \frac{A_m}{T} \frac{\partial(A_j/T)}{\partial\{(\mu_i - \mu_n)/T\}} = 0\,,$$

$$(i = 1, 2, \ldots, n-1)\,. \tag{35}$$

With the help of the Onsager relations (29), (30) and (31), and with

$$\sum_{j,m=1}^{r} l_{jm} \frac{A_j}{T} \frac{\partial(A_m/T)}{\partial\{(\mu_i-\mu_n)/T\}} = \sum_{j,m=1}^{r} l_{jm} \frac{A_m}{T} \frac{\partial(A_j/T)}{\partial\{(\mu_i-\mu_n)/T\}}$$

$$= \sum_{j,m=1}^{r} l_{jm} \frac{A_m}{T} \nu_{ij}, \quad (36)$$

equations (34) and (35) may be written as

$$L_{qq} \operatorname{div} \operatorname{grad} \frac{1}{T} - \sum_{k=1}^{n-1} L_{qk} \operatorname{div} \operatorname{grad} \frac{\mu_k - \mu_n}{T} = 0, \quad (37)$$

$$L_{iq} \operatorname{div} \operatorname{grad} \frac{1}{T} - \sum_{k=1}^{n-1} L_{ik} \operatorname{div} \operatorname{grad} \frac{\mu_k - \mu_n}{T} + \sum_{j,m=1}^{r} l_{jm} \frac{A_m}{T} \nu_{ij} = 0. \quad (38)$$

These equations become with the phenomenological equations (26), (27) and (28)

$$\operatorname{div} \boldsymbol{J}_q = 0, \quad (39)$$

$$\operatorname{div} \boldsymbol{J}_i - \sum_{j=1}^{r} \nu_{ij} J_j = 0. \quad (40)$$

With these results we can conclude from the conservation laws (20) and (22),

$$\partial c_i/\partial t = 0, \quad (i = 1, 2, \ldots, n), \quad (41)$$

$$\partial u/\partial t = 0. \quad (42)$$

Thus again the state of minimum entropy production turns out to be a *stationary state.*

We may remark that in general the specific volume v (or the density $\rho = v^{-1}$) should also be considered as an independent variable. But in the example treated above the barycentric velocity $\boldsymbol{v}$ was supposed to vanish, which implied that the specific volume was kept constant. If the phenomenon of viscous flow had been included we would also have obtained a condition for this additional variable.

Another remark is that the Onsager relations had to be used in their form (29)–(31), *i.e.* in the absence of a magnetic field. Otherwise the Onsager relations would have been of the type (IV.49) and the theorem proved above would not have been valid.

Let us prove, as in the first example, that this stationary state is stable. Differentiation of expression (32) for the total entropy production P gives, using the phenomenological equations (26), (27) and (28), and the Onsager relations (29), (30) and (31),

$$\frac{\partial P}{\partial t} = 2 \int^{V} \left(\boldsymbol{J}_q \cdot \operatorname{grad} \frac{\partial}{\partial t} \frac{1}{T} - \sum_{k=1}^{n-1} \boldsymbol{J}_k \cdot \operatorname{grad} \frac{\partial}{\partial t} \frac{\mu_k - \mu_n}{T} \right.$$

$$\left. - \sum_{j=1}^{r} J_j \frac{\partial (A_j/T)}{\partial t} \right) \mathrm{d}V \,. \tag{43}$$

Partial integration of the first two terms on the right-hand side and introduction of (III.18) for the affinity and of (II.4) leads to

$$\frac{\partial P}{\partial t} = 2 \int^{\Omega} \left(\boldsymbol{J}_q \frac{\partial}{\partial t} \frac{1}{T} - \sum_{k=1}^{n-1} \boldsymbol{J}_k \frac{\partial}{\partial t} \frac{\mu_k - \mu_n}{T} \right) \cdot \mathrm{d}\boldsymbol{\Omega}$$

$$- 2 \int^{V} \left\{ (\operatorname{div} \boldsymbol{J}_q) \frac{\partial}{\partial t} \frac{1}{T} - \sum_{k=1}^{n-1} \left(\operatorname{div} \boldsymbol{J}_k - \sum_{j=1}^{r} \nu_{kj} J_j \right) \frac{\partial}{\partial t} \frac{\mu_k - \mu_n}{T} \right\} \mathrm{d}V \,. \tag{44}$$

Owing to the boundary conditions (*e.g.*, T independent of time and $\boldsymbol{J}_k = 0$) the surface integrals vanish, so that

$$\frac{\partial P}{\partial t} = -2 \int^{V} \left\{ (\operatorname{div} \boldsymbol{J}_q) \frac{\partial}{\partial t} \frac{1}{T} - \sum_{k=1}^{n-1} \left(\operatorname{div} \boldsymbol{J}_k - \sum_{j=1}^{r} \nu_{kj} J_j \right) \frac{\partial}{\partial t} \frac{\mu_k - \mu_n}{T} \right\} \mathrm{d}V \,. \tag{45}$$

From the energy law (22) and the fact that the pressure is independent of time in the present problem and also because the velocity $\boldsymbol{v}$ has been assumed to vanish, we obtain the following balance equation for the specific enthalpy $h = u + pv$:

$$\rho \frac{\partial h}{\partial t} = - \operatorname{div} \boldsymbol{J}_q \,. \tag{46}$$

With this equation and the mass law (20) the expression (45) can be written as

$$\frac{\partial P}{\partial t} = 2 \int^{V} \left(\rho \frac{\partial h}{\partial t} \frac{\partial}{\partial t} \frac{1}{T} - \sum_{k=1}^{n-1} \rho \frac{\partial c_k}{\partial t} \frac{\partial}{\partial t} \frac{\mu_k - \mu_n}{T} \right) \mathrm{d}V \,. \tag{47}$$

Since the pressure is constant, the specific enthalpy depends only on the temperature T and $n-1$ of the mass fractions c_i, say $c_1, c_2, \ldots, c_{n-1}$. Therefore we have

$$\frac{\partial h}{\partial t} = \left(\frac{\partial h}{\partial T}\right)_{p, c_1, \ldots, c_n} \frac{\partial T}{\partial t} + \sum_{i=1}^{n-1} \left(\frac{\partial h}{\partial c_i}\right)_{p, T, c_1, c_2, \ldots, c_{i-1}, c_{i+1}, \ldots, c_{n-1}} \frac{\partial c_i}{\partial t}$$

$$= c_p \frac{\partial T}{\partial t} + \sum_{i=1}^{n-1} (h_i - h_n) \frac{\partial c_i}{\partial t}, \tag{48}$$

where c_p is the specific heat at constant pressure and where h_i is the partial specific enthalpy of chemical component i (*cf.* Appendix II). Furthermore we have the thermodynamic relation [*cf.* (III.23)]:

$$\frac{\partial \{ (\mu_k - \mu_n)/T \}}{\partial t} = -\frac{h_k - h_n}{T^2} \frac{\partial T}{\partial t} + \frac{1}{T} \sum_{i=1}^{n-1} \frac{\partial (\mu_k - \mu_n)}{\partial c_i} \frac{\partial c_i}{\partial t}. \tag{49}$$

Introducing (48) and (49) into (47), we find that

$$\frac{\partial P}{\partial t} = -2 \int^{V} \frac{\rho}{T} \left\{ \frac{c_p}{T} \left(\frac{\partial T}{\partial t}\right)^2 + \sum_{i,k=1}^{n-1} \frac{\partial (\mu_k - \mu_n)}{\partial c_i} \frac{\partial c_i}{\partial t} \frac{\partial c_k}{\partial t} \right\} \mathrm{d}V. \tag{50}$$

This expression is negative because c_p is positive and because

$$\sum_{i,k=1}^{n-1} \frac{\partial (\mu_k - \mu_n)}{\partial c_i} \frac{\partial c_i}{\partial t} \frac{\partial c_k}{\partial t} \geqslant 0. \tag{51}$$

This thermodynamic stability condition is proved in Appendix II.

The fact that (50) is negative proves again the stability of the stationary state.

Finally it may be remarked that thermodynamic equilibrium is a special case of a stationary state which is reached if the boundary conditions are compatible with the equilibrium conditions.

§ 4. *Stationary States without Minimum Entropy Production*

If the conditions, necessary to derive the theorem of the preceding section, are not fulfilled one can still derive a theorem of a more general character for the stationary state*. The phenomenological equations will not be needed in order to derive this more general theorem. Therefore no use will be made of the Onsager reciprocal relations, nor will it be necessary to assume that the phenomenological coefficients (of linear phenomenological laws) are constants.

The theorem can be formulated in the following way. Let us write again the total entropy production

$$P = \int \sigma \, \mathrm{d}V = \int \sum_i J_i X_i \, \mathrm{d}V \,, \tag{52}$$

where J_i and X_i are thermodynamic flows and forces respectively. Then one has for the time derivative of P

$$\frac{\partial P}{\partial t} = \int^V \sum_i J_i \frac{\partial X_i}{\partial t} \mathrm{d}V + \int^V \sum_i \frac{\partial J_i}{\partial t} X_i \, \mathrm{d}V \,. \tag{53}$$

which will be denoted as

$$\frac{\partial P}{\partial t} = \int^V \frac{\partial_X \sigma}{\partial t} \mathrm{d}V + \int^V \frac{\partial_J \sigma}{\partial t} \mathrm{d}V \equiv \frac{\partial_X P}{\partial t} + \frac{\partial_J P}{\partial t} \,. \tag{54}$$

The quantity $\partial_X P / \partial t$ now obeys the inequality

$$\frac{\partial_X P}{\partial t} \equiv \int^V \frac{\partial_X \sigma}{\partial t} \mathrm{d}V \equiv \int^V \sum_i J_i \frac{\partial X_i}{\partial t} \mathrm{d}V \leqslant 0 \,, \tag{55}$$

whereas nothing can be said about the sign of (54). This means that the stationary state does not necessarily correspond to a state of minimum entropy production. However, we may still infer that in so far as the change of the entropy production is due to the rate of change of the thermodynamic forces, this change will be negative. The proof of (55) will be given for the example of heat conduction in a non-expanding solid. For this case the left-hand side of (55) reads according to formula (9)

$$\frac{\partial_X P}{\partial t} = \int^V \boldsymbol{J}_q \cdot \frac{\partial}{\partial t} \operatorname{grad} \frac{1}{T} \, \mathrm{d}V \,. \tag{56}$$

* P. Glansdorff and I. Prigogine, Physica **20** (1954) 773.

This expression leads, with partial integration, to

$$\frac{\partial_X P}{\partial t} = \int^{\Omega} \left(\frac{\partial}{\partial t} \frac{1}{T} \right) \boldsymbol{J}_q \cdot \mathrm{d}\boldsymbol{\Omega} - \int^{V} \left(\frac{\partial}{\partial t} \frac{1}{T} \right) \operatorname{div} \boldsymbol{J}_q \, \mathrm{d}V . \tag{57}$$

The surface integral vanishes because the temperature is independent of time at the boundary. The last term may be transformed with the help of the energy law (11), so that (57) becomes

$$\frac{\partial_X P}{\partial t} = - \int^{V} \frac{\rho c_v}{T^2} \left(\frac{\partial T}{\partial t} \right)^2 \mathrm{d}V \leqslant 0 . \tag{58}$$

This is evidently negative, because c_v is positive, and thus the inequality (55) is established for this case.

In the preceding section (§ 3) the use of the constancy of phenomenological constants, the Onsager relations and the phenomenological equations, led to inequality (19). Comparison of this formula with the result (58) shows that under the restrictive conditions of § 3 we have

$$\frac{\partial_X P}{\partial t} = \frac{\partial_J P}{\partial t} = \frac{1}{2} \frac{\partial P}{\partial t} \leqslant 0 . \tag{59}$$

This result holds quite generally under the restrictive assumptions of § 3. Indeed if

(i) between the fluxes and forces of (52) linear phenomenological relations

$$J_i = \sum_k L_{ik} X_k \tag{60}$$

are assumed to hold,

(ii) the Onsager reciprocal relations have the form

$$L_{ik} = L_{ki} , \tag{61}$$

(iii) the phenomenological coefficients L_{ik} are constants, we have

$$\frac{\partial_J P}{\partial t} = \int^{V} \sum_i \frac{\partial J_i}{\partial t} X_i \, \mathrm{d}V = \int^{V} \sum_{i,k} L_{ik} X_i \frac{\partial X_k}{\partial t} \mathrm{d}V$$

$$= \int^{V} \sum_k J_k \frac{\partial X_k}{\partial t} \mathrm{d}V = \frac{\partial_X P}{\partial t} = \frac{1}{2} \frac{\partial P}{\partial t} \leqslant 0 , \tag{62}$$

where (53), (54) and the theorem (55) have been used. Thus with conditions (i), (ii) and (iii) the results of this section agree with those of § 3 and the stationary state corresponds to a minimum of the entropy production.

However, even under the more general conditions of this section (*e.g.* if the phenomenological equations are non-linear as in chemical reactions), the entropy production will decrease in the course of time until the stationary state is reached, if independently of (55),

$$\frac{\partial_J P}{\partial t} < 0 . \tag{63}$$

This may be the case for special kinetic laws*.

* H. C. Mel, Bull. Acad. roy. Belgique Cl. Sc. **40** (1954) 834.

CHAPTER VI

PROPERTIES OF THE PHENOMENOLOGICAL EQUATIONS AND THE ONSAGER RELATIONS

§ 1. *Introduction*

In Chapter IV we have stated some properties of the scheme of phenomenological coefficients occurring in the linear laws without actual proof.

Thus we have mentioned in Chapter IV, § 2 the influence of symmetry properties of matter on the linear laws (Curie principle).

In § 3 of Chapter IV a statement is given of Onsager's theorem of reciprocity. The theorem in this form strictly applies only to the coefficients of linear laws of which the fluxes are time derivatives of thermodynamic state variables. We mentioned, however, that the reciprocal relations also hold for the coefficients occurring in laws of vectorial or tensorial character where the fluxes are not time derivatives of state variables. In Chapter IX, § 7 this fact is directly established for a special system on the basis of the kinetic theory of gases. In general, however, one must start from the original formulation of Onsager's theorem (*cf.* also Chapter VII) and deduce from it the more general validity of the reciprocal relations for vectorial and tensorial phenomena.

In the present chapter this proof will be given as well as a formal discussion of the Curie principle. In addition we shall be concerned with general transformation properties of the Onsager relations.

§ 2. *The Curie Principle*

As already stated in Chapter IV, § 2, the existence of spatial symmetry properties in a system may simplify the form of the phenomenological equations in such a way, that the Cartesian components of the fluxes do not depend on all the Cartesian components of the thermodynamic forces. This statement is called the Curie principle*.

We shall study in this section the influence of the spatial symmetry properties of the system on the scheme of phenomenological coefficients.

* P. Curie, Oeuvres (Paris, 1908) p. 118.

Amongst other things we shall show that in an isotropic system fluxes and thermodynamic forces of different tensorial character do not couple.

In the macroscopic expressions which were derived in Chapter III for the entropy source strength, no fluxes and thermodynamic forces occurred of tensorial order higher than the second. We shall therefore restrict the following considerations to an entropy production of the form

$$\sigma = \sum_{i=1}^{m_0} J_i X_i + \sum_{i=1}^{m_1} \boldsymbol{J}_i \cdot \boldsymbol{X}_i + \sum_{i=1}^{m_2} \mathsf{J}_i : \mathsf{X}_i , \tag{1}$$

which contains fluxes and thermodynamic forces of the zeroth, first and second tensorial order (scalars J_i and X_i, vectors $\boldsymbol{J}_i$ and $\boldsymbol{X}_i$ and tensors J_i and X_i). If we split a tensor (of the second order) in the following way

$$\mathsf{T} = \tfrac{1}{3}\, \mathsf{U} \operatorname{Tr} \mathsf{T} + \mathsf{T}^{(a)} + \overset{\circ}{\mathsf{T}}{}^{(s)} , \tag{2}$$

where U is the unit tensor, $\operatorname{Tr} \mathsf{T}$ the trace of T, $\mathsf{T}^{(a)}$ its antisymmetric part and $\overset{\circ}{\mathsf{T}}{}^{(s)}$ the symmetric part of $\mathsf{T} - \tfrac{1}{3}\mathsf{U}\operatorname{Tr}\mathsf{T}$, then scalar products of tensors T and V, such as occur in the last sum of (1) can be written as

$$\mathsf{T} : \mathsf{V} = \tfrac{1}{3} (\operatorname{Tr} \mathsf{T}) (\operatorname{Tr} \mathsf{V}) + \mathsf{T}^{(a)} \cdot \mathsf{V}^{(a)} + \overset{\circ}{\mathsf{T}}{}^{(s)} : \overset{\circ}{\mathsf{V}}{}^{(s)} \tag{3}$$

The first term on the right-hand side is a product of scalars. The second term can alternatively be written as the scalar product of two axial vectors. So finally from (3) it is clear that four types of terms occur in (1): products of scalars J^s and X^s, products of (polar) vectors $\boldsymbol{J}^v$ and $\boldsymbol{X}^v$, products of axial vectors $\boldsymbol{J}^a$ and $\boldsymbol{X}^a$ and products of symmetric tensors with zero trace which we shall denote by J^t and X^t. If for conveniency we take one single term of each category only, we have for the entropy production

$$\sigma = J^s X^s + \boldsymbol{J}^v \cdot \boldsymbol{X}^v + \boldsymbol{J}^a \cdot \boldsymbol{X}^a + \mathsf{J}^t : \mathsf{X}^t . \tag{4}$$

The phenomenological equations will in general have the form

$$\begin{aligned}
J^s &= L^{ss} X^s + \boldsymbol{L}^{sv} \cdot \boldsymbol{X}^v + \boldsymbol{L}^{sa} \cdot \boldsymbol{X}^a + \mathsf{L}^{st} : \mathsf{X}^t , \\
\boldsymbol{J}^v &= \boldsymbol{L}^{vs} X^s + \mathsf{L}^{vv} \cdot \boldsymbol{X}^v + \mathsf{L}^{va} \cdot \boldsymbol{X}^a + \mathsf{L}^{vt} : \mathsf{X}^t , \\
\boldsymbol{J}^a &= \boldsymbol{L}^{as} X^s + \mathsf{L}^{av} \cdot \boldsymbol{X}^v + \mathsf{L}^{aa} \cdot \boldsymbol{X}^a + \mathsf{L}^{at} : \mathsf{X}^t , \\
\mathsf{J}^t &= \mathsf{L}^{ts} X^s + \mathsf{L}^{tv} \cdot \boldsymbol{X}^v + \mathsf{L}^{ta} \cdot \boldsymbol{X}^a + \mathsf{L}^{tt} : \mathsf{X}^t .
\end{aligned} \tag{5}$$

The phenomenological coefficients occurring in these equations have apparently tensorial order and polar or axial character as marked in the following table:

tensorial order	0	1	2	3	4
polar tensors	L^{ss}	L^{sv}, L^{vs}	L^{vv}, L^{aa}, L^{st}, L^{ts}	L^{vt}, L^{tv}	L^{tt}
axial tensors	—	L^{sa}, L^{as}	L^{va}, L^{av}	L^{at}, L^{ta}	—

Since X^t is a symmetric tensor with zero trace the tensors L^{st}, L^{vt}, L^{at} and L^{tt} must be symmetric and have zero trace with respect to their last pair of Cartesian indices. Thus for L^{vt} we have

$$L^{vt}_{\alpha\beta\gamma} = L^{vt}_{\alpha\gamma\beta}, \quad (\alpha, \beta, \gamma = 1, 2, 3) \tag{6}$$

and

$$\sum_{\beta=1}^{3} L^{vt}_{\alpha\beta\beta} = 0, \quad (\alpha = 1, 2, 3). \tag{7}$$

The tensors L^{ts}, L^{tv}, L^{ta} and L^{tt} must have the same properties with respect to their first pair of Cartesian indices because J^t is a symmetric tensor with zero trace.

Each quantity T of tensorial order n transforms in the following way under an orthogonal transformation A (with determinant value $|A| = \pm 1$)

$$T'_{i_1 i_2 \ldots i_n} = |A|^{\varepsilon} \sum_{j_1, j_2, \ldots, j_n} A_{i_1 j_1} A_{i_2 j_2} \ldots A_{i_n j_n} T_{j_1 j_2 \ldots j_n}, \tag{8}$$

where $i_1, i_2, \ldots, i_n$ and $j_1, j_2, \ldots, j_n$ are indices for the Cartesian components of T and A, and where

$$\varepsilon = 0 \quad \text{for polar tensors},$$

$$\varepsilon = 1 \quad \text{for axial tensors}.$$

Formula (8) will be written in a symbolic form as

$$T' = |A|^{\varepsilon} A^n(\cdot)\, T. \tag{9}$$

If the system has no symmetry properties whatsoever, then under orthogonal transformations, the transformed phenomenological tensors of order 1 and higher will in general be different from the untransformed ones:

$$L' \neq L. \tag{10}$$

If, however, the system has a symmetry property, this means that a transformation A corresponding to this symmetry property leaves the phenomenological tensors unaltered, *i.e.*

$$L' = L, \tag{11}$$

or with (9)

$$|A|^{\varepsilon} A^{n}(\cdot)\, L = L. \tag{12}$$

For any symmetry property conclusions on the form of the tensor L can be drawn from (12).

The isotropic system. Let us consider an isotropic system with an entropy production of the form (4). One of the symmetry properties which has such a system is invariance under an inversion I:

$$I = \begin{pmatrix} -1 & 0 & 0 \\ 0 & -1 & 0 \\ 0 & 0 & -1 \end{pmatrix}, \quad |I| = -1. \tag{13}$$

From (12) with $A = I$, we obtain

$$|I|^{\varepsilon} I^{n}(\cdot)\, L = L, \tag{14}$$

which gives with (13)

$$(-1)^{\varepsilon+n}\, L = L. \tag{15}$$

We can immediately conclude that all coefficients with odd $\varepsilon + n$ must vanish since then (15) yields

$$L = 0. \tag{16}$$

This eliminates the polar tensors ($\varepsilon = 0$) $\boldsymbol{L}^{sv}$, $\boldsymbol{L}^{vs}$ ($n = 1$), L^{vt}, L^{tv} ($n = 3$), and the axial tensors ($\varepsilon = 1$) L^{av} and L^{va} ($n = 2$) from the coefficient scheme.

Furthermore the properties of the isotropic system are also invariant under an arbitrary rotation R, $(\mid \mathsf{R} \mid = 1)$. For the scalar coefficient L^{ss} equation (12) is satisfied in a trivial way. For the vectors $\boldsymbol{L}^{as}$ and $\boldsymbol{L}^{sa}$ we have from (12) with $\mathsf{A} = \mathsf{R}$

$$\mathsf{R}\cdot\boldsymbol{L}^{as} = \boldsymbol{L}^{as}, \quad \mathsf{R}\cdot\boldsymbol{L}^{sa} = \boldsymbol{L}^{sa} \tag{17}$$

This relation can clearly only be satisfied if

$$\boldsymbol{L}^{as} = 0\,, \quad \boldsymbol{L}^{sa} = 0\,. \tag{18}$$

In order to draw conclusions on the remaining phenomenological tensors, we use the fact that a scalar is invariant under an arbitrary rotation. Thus, for instance, with the tensor L^{vv} we can construct the scalar quantity

$$\mathsf{L}^{vv} : \boldsymbol{ab} = \mathsf{L}'^{vv} : \boldsymbol{a}'\boldsymbol{b}'\,, \tag{19}$$

where $\boldsymbol{a}$ and $\boldsymbol{b}$ are arbitrary vectors. Since in the isotropic system L^{vv} is invariant under an arbitrary rotation

$$\mathsf{L}^{vv} : \boldsymbol{ab} = \mathsf{L}^{vv} : \boldsymbol{a}'\boldsymbol{b}'\,. \tag{20}$$

This relation shows that a bilinear expression in $\boldsymbol{a}$ and $\boldsymbol{b}$ becomes after transformation a bilinear form in $\boldsymbol{a}'$ and $\boldsymbol{b}'$ with the same coefficients L^{vv}. The expression (20) is therefore linear in the bilinear invariants of $\boldsymbol{a}$ and $\boldsymbol{b}$. Since the only invariant of this kind is $(\boldsymbol{a}\cdot\boldsymbol{b})$, we can conclude that (20) is equal to $L^{vv}\,(\boldsymbol{a}\cdot\boldsymbol{b})$, where L^{vv} is a scalar. Then, however, L^{vv} must have the form

$$\mathsf{L}^{vv} = L^{vv}\,\mathsf{U}\,. \tag{ 1}$$

Similarly we have

$$\mathsf{L}^{aa} = L^{aa}\,\mathsf{U}\,, \tag{22}$$

and

$$\mathsf{L}^{st} = L^{st}\,\mathsf{U}\,, \quad \mathsf{L}^{ts} = L^{ts}\,\mathsf{U}\,. \tag{23}$$

Since the last two tensors L^{st} and L^{ts} have zero trace, it follows that

$$\mathsf{L}^{st} = 0\,, \quad \mathsf{L}^{ts} = 0\,. \tag{24}$$

For the third order tensors L^{ta} (and L^{at}) we can write analogous to (20)

$$\sum_{\alpha,\beta,\gamma=1}^{3} L^{ta}_{\alpha\beta\gamma}\, a_\gamma T_{\beta\alpha} = \sum_{\alpha,\beta,\gamma=1}^{3} L^{ta}_{\alpha\beta\gamma}\, a'_\gamma T'_{\beta\alpha}\,, \tag{25}$$

where $\boldsymbol{a}$ is an arbitrary axial vector and T an arbitrary tensor. The expression (25) must now be proportional to the only invariant (of third degree) of $\boldsymbol{a}$ and T, *viz.* $\sum_{\text{cycl.}} a_\alpha T^{(a)}_{\beta\gamma}$, in which case (25) has the form $c \sum_{\text{cycl.}} a_\alpha T^{(a)}_{\beta\gamma}$, where c is a scalar constant and $T^{(a)}$ is the anti-symmetric part of the tensor T. Then L^{ta} must have components

$$L^{ta}_{321} = L^{ta}_{132} = L^{ta}_{213} = \tfrac{1}{2}c\,, \quad L^{ta}_{312} = L^{ta}_{231} = L^{ta}_{123} = -\tfrac{1}{2}c, \tag{26}$$

and the others zero, since only then (25) has the form $c \sum_{\text{cycl.}} a_\alpha T^{(a)}_{\beta\gamma}$ as can easily be checked. Since furthermore $L^{ta}_{\alpha\beta\gamma}$ is symmetric in α and β, it must vanish:

$$L^{ta} = 0\,, \quad L^{at} = 0\,. \tag{27}$$

In the scheme of tensors all those connecting fluxes and forces of different character (L^{sv}, L^{sa}, L^{st}, L^{vs}, L^{va}, L^{vt}, L^{as}, L^{av}, L^{at}, L^{ts}, L^{tv} and L^{ta}) have disappeared now. This proves our previous statement for the isotropic systems.

We still wish to discuss the form of L^{tt}. To that purpose we write

$$\sum_{\alpha,\beta,\gamma,\delta=1}^{3} L^{tt}_{\alpha\beta\gamma\delta} T_{\delta\gamma} V_{\beta\alpha} = \sum_{\alpha,\beta,\gamma,\delta=1}^{3} L^{tt}_{\alpha\beta\gamma\delta} T'_{\delta\gamma} V'_{\beta\alpha}\,, \tag{28}$$

where T and V are arbitrary tensors. The form (28) must be a linear combination of the invariants (of the fourth degree) of the ordered product TV and thus equal to

$$L^{tt}\, \mathring{T}^{(s)} : \mathring{V}^{(s)} + c_1\, T^{(a)} : V^{(a)} + c_2\, (\mathrm{Tr}\, T)\, (\mathrm{Tr}\, V)\,, \tag{29}$$

where L^{tt}, c_1 and c_2 are scalar constants, and the index (s) denotes a symmetric tensor, the index (a) an antisymmetric tensor, the symbol $\circ$ a tensor with zero trace, and Tr the trace. Then the fourth order tensor L^{tt} must have the components

$$L^{tt}_{\alpha\beta\gamma\delta} = L^{tt}\, \{ \tfrac{1}{2}(\delta_{\alpha\delta}\delta_{\beta\gamma} + \delta_{\alpha\gamma}\delta_{\beta\delta}) - \tfrac{1}{3}\delta_{\alpha\beta}\delta_{\gamma\delta} \}$$
$$+ c_1\, (\tfrac{1}{2}\delta_{\alpha\delta}\delta_{\beta\gamma} - \tfrac{1}{2}\delta_{\alpha\gamma}\delta_{\beta\delta}) + c_2\, \delta_{\alpha\beta}\delta_{\gamma\delta}\,, \quad (\alpha,\beta,\gamma,\delta = 1,2,3)\,, \tag{30}$$

since only with this form inserted into (28) one obtains (29). Since L^{tt} must be symmetric and have zero trace in its first two and in its last two components

$$L^{tt}_{\alpha\beta\gamma\delta} = L^{tt}_{\beta\alpha\gamma\delta} = L^{tt}_{\alpha\beta\delta\gamma}\,, \tag{31}$$

$$\sum_{\alpha} L_{\alpha\alpha\gamma\delta} = 0\,, \quad \sum_{\gamma} L_{\alpha\beta\gamma\gamma} = 0\,, \tag{32}$$

the constants c_1 and c_2 in (30) must vanish. So the form of L^{tt} is finally

$$L^{tt}_{\alpha\beta\gamma\delta} = L^{tt}\left\{\tfrac{1}{2}(\delta_{\alpha\delta}\delta_{\beta\gamma} + \delta_{\alpha\gamma}\delta_{\beta\delta}) - \tfrac{1}{3}\delta_{\alpha\beta}\delta_{\gamma\delta}\right\}, \quad (\alpha,\beta,\gamma,\delta = 1,2,3)\,, \tag{33}$$

containing one single scalar constant L^{tt}.

Of the 81 components 21 are different from zero.

$$\begin{aligned}
&L^{tt}_{1111} = L^{tt}_{2222} = L^{tt}_{3333} = \tfrac{2}{3}L^{tt}\,,\\
&L^{tt}_{1122} = L^{tt}_{2233} = L^{tt}_{3311} = L^{tt}_{2211} = L^{tt}_{3322} = L^{tt}_{1133} = -\tfrac{1}{3}L^{tt}\,,\\
&L^{tt}_{1212} = L^{tt}_{2323} = L^{tt}_{3131} = L^{tt}_{2121} = L^{tt}_{3232} = L^{tt}_{1313} = \tfrac{1}{2}L^{tt}\,,\\
&L^{tt}_{1221} = L^{tt}_{2332} = L^{tt}_{3113} = L^{tt}_{2112} = L^{tt}_{3223} = L^{tt}_{1331} = \tfrac{1}{2}L^{tt}\,.
\end{aligned} \tag{34}$$

The phenomenological equation (5) for J^t reads with the results (16), (24) and (27)

$$J^t_{\alpha\beta} = \sum_{\gamma,\delta=1}^{3} L^{tt}_{\alpha\beta\gamma\delta}\; X^t_{\gamma\delta}\,, \quad (\alpha,\beta = 1,2,3)\,. \tag{35}$$

With (33) inserted, this becomes simply

$$J^t_{\alpha\beta} = L^{tt}\, X^t_{\alpha\beta}\,, \quad (\alpha,\beta = 1,2,3)\,. \tag{36}$$

We can, with all the results obtained above, write, instead of (5) as phenomenological equations

$$\begin{aligned}
J^s &= L^{ss}X^s\,,\\
\mathbf{J}^v &= L^{vv}\mathbf{X}^v\,,\\
\mathbf{J}^a &= L^{aa}\mathbf{X}^a\,,\\
J^t &= L^{tt}X^t\,,
\end{aligned} \tag{37}$$

where L^{ss}, L^{vv}, L^{aa} and L^{tt} are all scalars. If more than one vectorial phenomenon exists, as in the first example of Chapter IV, § 2, we have cross-coefficients which couple the various vectorial processes, but which have, according to the theory outlined here, also scalar character. The form of the phenomenological equations written down in the first example of Chapter IV, § 2, is thus completely justified.

We may finally remark that in cases with lower symmetry than the isotropic system, formula (12) will always lead to the properties of the phenomenological coefficient scheme, if as transformations A the symmetries of the system are inserted. Thus, for instance, in a cubic system we have the following symmetries: inversion, rotation of 90° around, say, the x_1-axis and rotation of 120° around a body diagonal. One can then check in this case that *e.g.* L^{vv} and L^{aa} again reduce to scalar multiples of the unit tensor, but that L^{tt} contains now two essentially different coefficients instead of a single one in the isotropic case. (The result is that in the first two lines of (34) a new constant appears which is different from the L^{tt} of the last line.)

§ 3. *Dependent Fluxes and Thermodynamic Forces*

Let us now consider another property of the scheme of phenomenological coefficients.

In Chapter IV, § 3 it was stated that the Onsager relations (IV.49), (IV.50) and (IV.51) were valid if in the phenomenological equations (IV.36) and (IV.37) independent fluxes (IV.45) and (IV.46) and independent thermodynamic forces (IV.47) and (IV.48) were employed.

In this section we wish to investigate the influence of linear dependencies between the fluxes or between the thermodynamic forces on the phenomenological coefficients and the Onsager reciprocal relations*. Such cases are frequently dealt with in the applications of thermodynamics of irreversible processes. We shall prove two theorems, the first (i) referring to the case where a dependency exists amongst the fluxes only, and the second (ii) referring to the case where both fluxes and thermodynamic forces are dependent. These theorems will be formulated for the case that the system is described by α-type variables only, and that no external magnetic field is present. They may, however, be generalized to include the case where also β-variables are needed and an external magnetic field is applied. The formalism is written

* G. J. Hooyman and S. R. de Groot, Physica **21** (1955) 73.

down for scalar irreversible processes, but it is equally valid for vectorial and tensorial processes.

(i) THEOREM I. *A linear homogeneous dependency amongst the fluxes leaves the validity of the Onsager reciprocal relations unimpaired.*

Proof: Let us write the entropy production as

$$\sigma = \sum_{i=1}^{n} J_i X_i \,. \tag{38}$$

As stated in the introductory paragraph of this section the phenomenological equations, which connect the fluxes J_i $(i = 1, 2, \ldots, n)$ and the thermodynamic forces X_i $(i = 1, 2, \ldots, n)$,

$$J_i = \sum_{k=1}^{n} L_{ik} X_k \,, \quad (i = 1, \; , \ldots, n) \,, \tag{39}$$

contain a set of n^2 phenomenological coefficients L_{ik} which obey the Onsager reciprocal relations

$$L_{ik} = L_{ki} \,, \quad (i, k = 1, 2, \ldots, n) \,, \tag{40}$$

if both the fluxes and the thermodynamic forces constitute sets of independent variables. We wish to show here that the relations (40) remain valid, even if a dependency between the fluxes exists. Let us suppose that the fluxes are connected in the following homogeneous linear way

$$\sum_{i=1}^{n} a_i J_i = 0 \,, \tag{41}$$

where the coefficients a_i may still be functions of the (local) equilibrium state variables. Using this relation the flow J_n can be eliminated from (38), when $a_n \neq 0$. This gives

$$\sigma = \sum_{i=1}^{n-1} J_i \{ X_i - (a_i/a_n) X_n \} \,, \tag{42}$$

so that we are now left with $n - 1$ *independent* fluxes and thermodynamic forces. With the use of these quantities the phenomenological equations will have the form

$$J_i = \sum_{k=1}^{n-1} l_{ik} \{ X_k - (a_k/a_n) X_n \}, \quad (i = 1, 2, \ldots, n-1), \tag{43}$$

where now the reciprocal relations

$$l_{ik} = l_{ki}, \quad (i, k = 1, 2, \ldots, n-1) \tag{44}$$

are valid according to the Onsager theorem. If we solve J_n from (41) and insert (43) for the fluxes J_i $(i = 1, 2, \ldots, n-1)$ we obtain (changing the notation of some dummy indices)

$$J_n = - \sum_{i,k=1}^{n-1} \{ (a_k/a_n) l_{ki} X_i - (a_i a_k/a_n^2) l_{ik} X_n \}. \tag{45}$$

From comparison of the coefficients in (39) and those in (43) and (45) it follows that

$$L_{ik} = l_{ik}, \qquad (i, k = 1, 2, \ldots, n-1), \tag{46}$$

$$L_{in} = - \sum_{k=1}^{n-1} (a_k/a_n) l_{ik}, \quad (i = 1, 2, \ldots, n-1), \tag{47}$$

$$L_{ni} = - \sum_{k=1}^{n-1} (a_k/a_n) l_{ki}, \quad (i = 1, 2, \ldots, n-1), \tag{48}$$

$$L_{nn} = \sum_{i,k=1}^{n-1} (a_i a_k/a_n^2) l_{ik}. \tag{49}$$

From these relations by which the phenomenological coefficients L_{ik} $(i, k = 1, 2, \ldots, n)$ are expressed in terms of the coefficients l_{ik} $(i, k = 1, 2, \ldots, n-1)$ it follows that

$$\sum_{k=1}^{n} a_k L_{ik} = 0, \quad (i = 1, 2, \ldots, n), \tag{50}$$

$$\sum_{i=1}^{n} a_i L_{ik} = 0, \quad (k = 1, 2, \ldots, n). \tag{51}$$

These relations between the coefficients L_{ik} $(i, k = 1, 2, \ldots, n)$ are a consequence of the connection (41). We note that only $2n - 1$ of the relations are independent, because from (50) as well as from (51) it can be concluded that

$$\sum_{i,k=1}^{n} a_i a_k L_{ik} = 0 . \tag{52}$$

Finally it is seen that the Onsager relations (44) imply, according to formulae (46)–(49), the validity of the reciprocal relations (40), as stated in the theorem.

(ii) THEOREM II. *If linear homogeneous relationships exist between the fluxes and also between the thermodynamic forces, then the phenomenological coefficients are not uniquely defined, and the Onsager relations are not necessarily fulfilled. It can, however, be shown that the coefficients can always be chosen in such a way that the Onsager relations hold.*

Proof: In addition to (41) we shall assume the linear relation

$$\sum_{i=1}^{n} b_i X_i = 0 \tag{53}$$

to hold between the thermodynamic forces, with $b_n \neq 0$. Eliminating both J_n and X_n from (38) we obtain

$$\sigma = \sum_{i=1}^{n-1} J_i \{ X_i + (a_i/a_n) \sum_{j=1}^{n-1} (b_j/b_n) X_j \} . \tag{54}$$

The phenomenological equations can therefore be written as

$$J_i = \sum_{k=1}^{n-1} l_{ik} \{ X_k + (a_k/a_n) \sum_{j=1}^{n-1} (b_j/b_n) X_j \} , \quad (i = 1, 2, \ldots, n-1) , \tag{55}$$

or, interchanging dummy indices,

$$J_i = \sum_{k=1}^{n-1} \{ l_{ik} + (b_k/b_n) \sum_{j=1}^{n-1} (a_j/a_n) l_{ij} \} X_k , \quad (i = 1, 2, \ldots, n-1) . \tag{56}$$

From this expression and (41) it follows that

$$J_n = - \sum_{i,k=1}^{n-1} (a_i/a_n) \{ l_{ik} + (b_k/b_n) \sum_{j=1}^{n-1} (a_j/a_n) l_{ik} \} X_k . \tag{57}$$

On the other hand, we can eliminate X_n from (39) with the help of (53). The flows are then expressed in terms of independent thermodynamic forces:

$$J_i = \sum_{k=1}^{n-1} \{ L_{ik} - (b_k/b_n)L_{in} \} X_k , \quad (i = 1, 2, \ldots, n) . \tag{58}$$

Comparison of (56) and (57) with (58) gives

$$L_{ik} - (b_k/b_n)L_{in} = l_{ik} + (b_k/b_n) \sum_{j=1}^{n-1} (a_j/a_n)l_{ij} , \quad (i, k = 1, 2, \ldots, n-1) , \tag{59}$$

$$L_{nk} - (b_k/b_n)L_{nn} = - \sum_{i=1}^{n-1} (a_i/a_n) \{ l_{ik} + (b_k/b_n) \sum_{j=1}^{n-1} (a_j/a_n)l_{ij} \} ,$$

$$(k = 1, 2, \ldots, n-1) , \tag{60}$$

a set of $n(n-1)$ relations for the n^2 coefficients L_{ik} $(i, k = 1, 2, \ldots, n)$. It is clear that one is therefore left with an n-fold arbitrariness, and that consequently the scheme of coefficients L_{ik} need not be symmetric. On the other hand it is also seen that a possible solution of (59) and (60) is given by the formulae (46)–(49), which, as we saw above, ensured the symmetry of the L_{ik}-scheme, since the coefficients l_{ik} are subject to the Onsager relations (44). In other words it is always possible to dispose in such a way of the liberty which is left in the definition of the phenomenological coefficients, that the latter fulfill the Onsager relations in the form (40). This completes the proof of the statements of theorem II.

Finally a few additional remarks can be made. From the relations (59) and (60), or from (58) with the conditions (41), a number of $n-1$ relations

$$\sum_{i=1}^{n} a_i \{ L_{ik} - (b_k/b_n)L_{in} \} = 0 , \quad (k = 1, 2, \ldots, n-1) \tag{61}$$

is seen to exist between the phenomenological coefficients. These $n-1$ relations and the n-fold arbitrariness show again that one is left with a sensible set of $(n-1)^2$ coefficients L_{ik} only. This is also the number of independent coefficients l_{ik} $(i, k = 1, 2, \ldots, n-1)$.

One can dispose in an elegant way of the n-fold arbitrariness by choosing a certain arbitrary value of L_{nn} and by imposing the $n-1$ conditions

$$L_{in} = L_{ni}\,, \quad (i = 1, 2, \ldots, n-1)\,. \tag{62}$$

Then a general symmetric solution of (59) and (60) can be obtained:

$$L_{ik} = L_{ki} = l_{ik} + (b_i b_k / b_n^2) \{ L_{nn} - \sum_{p,q=1}^{n-1} (a_p a_q / a_n^2) l_{pq} \}\,,$$
$$(i,k = 1, 2, \ldots, n-1)\,, \tag{63}$$

where use has been made of the symmetry (44) of the l_{ik}-scheme.

The two theorems derived show that in applications of the phenomenological theory one may use the Onsager relations even when linear relationships exist between the fluxes and between the thermodynamic forces.

§ 4. *Onsager Relations for Vectorial (and Tensorial) Phenomena*

In Chapter IV, § 3, Onsager reciprocal relations have been written down for the phenomenological coefficients, occurring in the phenomenological equations which connect fluxes and thermodynamic forces in a linear way. These fluxes and forces were taken from the local entropy production. In doing so a difficulty may arise, which requires special attention in a number of cases*. The point is that in the derivation of Chapter VII of Onsager's relations the fluxes are required to be time derivatives of state variables. This is, however, not the case for the vectorial fluxes (heat flow, diffusion flow) and neither for the tensorial fluxes (viscous pressure tensor) which were employed in our phenomenological laws, valid in "continuous systems" (*i.e.* with state variables which are continuous functions of space coordinates and time). It will be shown here that the Onsager relations must remain valid in the usual form for measurable phenomenological coefficients of vectorial or tensorial laws. (On the basis of the kinetic theory of gases and therefore for a special class of systems this fact will be established independently in Chapter IX, § 7.)

We shall prove this for two examples: (i) heat conduction in an anisotropic solid (of which the thermal expansion may be neglected), under the influence of an external magnetic field $\boldsymbol{B}$, (ii) heat conduction, diffusion and cross-effects in an isotropic fluid in the absence of a magnetic field.

* H. B. G. Casimir, Rev. mod. Phys. **17** (1945) 343; or Philips Res. Rep. **1** (1946) 185.

(i) *Heat conduction in an anisotropic solid*

In order to find the fluxes and thermodynamic forces to which the Onsager theorem, as stated in Chapter IV, § 3, applies we write down the total entropy production of an adiabatically insulated solid system in which only heat conduction occurs

$$\frac{\mathrm{d}S}{\mathrm{d}t} = \int^{V} \sigma \,\mathrm{d}V = \int^{V} \boldsymbol{J}_q \cdot \operatorname{grad} \frac{1}{T} \,\mathrm{d}V \,, \tag{64}$$

where (V.9) has been inserted. After partial integration, this gives

$$\frac{\mathrm{d}S}{\mathrm{d}t} = - \int^{V} \frac{1}{T} \operatorname{div} \boldsymbol{J}_q \,\mathrm{d}V \,. \tag{65}$$

Neglecting thermal expansion this becomes with the help of (V.11)

$$\frac{\mathrm{d}S}{\mathrm{d}t} = \rho \int^{V} \frac{1}{T} \frac{\partial u}{\partial t} \,\mathrm{d}V \,, \tag{66}$$

where ρ is the (constant, uniform) density and u the specific energy. Owing to the fact that the energy of an adiabatically insulated system of constant volume and shape is constant, we also have

$$\int^{V} \frac{\partial u}{\partial t} \,\mathrm{d}V = 0 \,. \tag{67}$$

As a consequence of this relation we may rewrite (66) in the form

$$\frac{\mathrm{d}S}{\mathrm{d}t} = \rho \int^{V} \Delta \frac{1}{T} \frac{\partial u}{\partial t} \,\mathrm{d}V \,, \tag{68}$$

where $\Delta T^{-1} \equiv T^{-1}(\boldsymbol{r}) - T_0^{-1}$, where T_0 is the temperature at an arbitrary point $\boldsymbol{r}_0$. Up to second order in $\Delta T \equiv T(\boldsymbol{r}) - T_0$ and u (68) becomes

$$\frac{\mathrm{d}S}{\mathrm{d}t} = - \frac{\rho}{T_0^2} \int^{V} \Delta T \frac{\partial u}{\partial t} \,\mathrm{d}V \,. \tag{69}$$

We have here a bilinear expression in the fluxes $\partial u/\partial t$ and the thermodynamic forces $-\rho T_0^{-2}\,\Delta T$ of the form (IV.53). The fluxes are here indeed proper time derivatives of thermodynamic state variables, while the thermodynamic forces are related to the entropy in the manner of (IV.47).

The phenomenological equations (IV.36) between the fluxes and thermodynamic forces of (69) may be written here in the form

$$\frac{\partial u(\boldsymbol{r})}{\partial t} = -\frac{\rho}{T_0^2}\int^{V} K(\boldsymbol{r},\boldsymbol{r}'\,;\{\boldsymbol{B}\})\,\Delta T(\boldsymbol{r}')\,\mathrm{d}\boldsymbol{r}'\,, \tag{70}$$

where for the differential of the volume element the notation $\mathrm{d}\boldsymbol{r}'$ has been employed. Equation (70) is valid at all points in the system except the arbitrary point $\boldsymbol{r}_0$. The phenomenological coefficients $K(\boldsymbol{r},\boldsymbol{r}'\,;\{\boldsymbol{B}\})$ are functions of the two positions $\boldsymbol{r}$ and $\boldsymbol{r}'$ and may be functionals of the external magnetic field $\boldsymbol{B}$, which may be non-uniform in the system. The Onsager relations

$$K(\boldsymbol{r},\boldsymbol{r}'\,;\{\boldsymbol{B}\}) = K(\boldsymbol{r}',\boldsymbol{r}\,;\{-\boldsymbol{B}\}) \tag{71}$$

are valid for these coefficients at all points $\boldsymbol{r}$ and $\boldsymbol{r}'$ excepting $\boldsymbol{r} \neq \boldsymbol{r}_0$ and $\boldsymbol{r}' = \boldsymbol{r}_0$.

We must investigate which consequences these reciprocal relations have for the heat conductivity tensor $\mathsf{L}(\boldsymbol{r};\boldsymbol{B})$ which appears in the ordinary Fourier law for anisotropic crystals:

$$\boldsymbol{J}_q(\boldsymbol{r}) = -\,\mathsf{L}(\boldsymbol{r}\,;\boldsymbol{B})\cdot\operatorname{grad}\Delta T\,. \tag{72}$$

Again $\Delta T = T(\boldsymbol{r}) - T_0$, while the equation is valid at all points except at the arbitrary chosen point $\boldsymbol{r}_0$. If this expression for the heat flow is inserted into the energy equation (V.11), one obtains, neglecting thermal expansion

$$\rho\frac{\partial u(\boldsymbol{r})}{\partial t} = \operatorname{div}\,\mathsf{L}\cdot\operatorname{grad}\Delta T(\boldsymbol{r})\,, \tag{73}$$

or, using the nabla notation,

$$\rho\frac{\partial u(\boldsymbol{r})}{\partial t} = \nabla\cdot\mathsf{L}\cdot\nabla\,\Delta T(\boldsymbol{r})\,. \tag{74}$$

It is easy to cast this equation into the form (70) by writing first

$$\rho \frac{\partial u(\boldsymbol{r})}{\partial t} = \int (\boldsymbol{\nabla}' \cdot \mathsf{L}' \cdot \boldsymbol{\nabla}' \, \Delta T') \, \delta(\boldsymbol{r} - \boldsymbol{r}') \, \mathrm{d}\boldsymbol{r}' \,, \tag{75}$$

where dashes indicate dependency on $\boldsymbol{r}'$, and where $\delta(\boldsymbol{r} - \boldsymbol{r}')$ is a three-dimensional delta-function, and by performing two successive partial integrations

$$\rho \frac{\partial u(\boldsymbol{r})}{\partial t} = - \int (\mathsf{L}' \cdot \boldsymbol{\nabla}' \, \Delta T') \cdot \boldsymbol{\nabla}' \, \delta(\boldsymbol{r} - \boldsymbol{r}') \, \mathrm{d}\boldsymbol{r}'$$

$$= \int [\boldsymbol{\nabla}' \cdot \{ \tilde{\mathsf{L}}' \cdot \boldsymbol{\nabla}' \, \delta(\boldsymbol{r} - \boldsymbol{r}') \}] \, \Delta T' \, \mathrm{d}\boldsymbol{r}' \,, \tag{76}$$

where $\tilde{\mathsf{L}}$ is the transposed matrix of L. Since this has indeed the form (70), the coefficients obey the Onsager relations (71)

$$\boldsymbol{\nabla}' \cdot \{ \tilde{\mathsf{L}}'(\boldsymbol{B}') \cdot \boldsymbol{\nabla}' \, \delta(\boldsymbol{r} - \boldsymbol{r}') \} = \boldsymbol{\nabla} \cdot \{ \tilde{\mathsf{L}}(- \boldsymbol{B}) \cdot \boldsymbol{\nabla} \, \delta(\boldsymbol{r} - \boldsymbol{r}') \} \tag{77}$$

for all $\boldsymbol{r}$ and $\boldsymbol{r}'$, excepting $\boldsymbol{r} = \boldsymbol{r}_0$ and $\boldsymbol{r}' = \boldsymbol{r}_0$. However, since $\boldsymbol{r}_0$ is an arbitrary point, this result holds everywhere. Applying the rule

$$g' \boldsymbol{\nabla}' \, \delta(\boldsymbol{r} - \boldsymbol{r}') \equiv g \boldsymbol{\nabla}' \, \delta(\boldsymbol{r} - \boldsymbol{r}') - (\boldsymbol{\nabla} g) \, \delta(\boldsymbol{r} - \boldsymbol{r}') \,, \tag{78}$$

(where g is an arbitrary function of $\boldsymbol{r}$), to the left-hand side, and performing the differentiation at the right-hand side, relation (77) becomes

$$\mathsf{L}(\boldsymbol{B}) : \boldsymbol{\nabla\nabla} \delta(\boldsymbol{r} - \boldsymbol{r}') + \{ \boldsymbol{\nabla} \cdot \mathsf{L}(\boldsymbol{B}) \} \cdot \boldsymbol{\nabla} \delta(\boldsymbol{r} - \boldsymbol{r}')$$

$$= \mathsf{L}(- \boldsymbol{B}) : \boldsymbol{\nabla\nabla} \delta(\boldsymbol{r} - \boldsymbol{r}') + \{ \boldsymbol{\nabla} \cdot \tilde{\mathsf{L}}(- \boldsymbol{B}) \} \cdot \boldsymbol{\nabla} \delta(\boldsymbol{r} - \boldsymbol{r}') \,. \tag{79}$$

Eliminating the δ-functions (by multiplying both members with an arbitrary function $f(\boldsymbol{r}')$ and integrating over $\boldsymbol{r}'$) this relation gets the form

$$\mathsf{L}(\boldsymbol{B}) : \boldsymbol{\nabla\nabla} f + \{ \boldsymbol{\nabla} \cdot \mathsf{L}(\boldsymbol{B}) \} \cdot \boldsymbol{\nabla} f = \mathsf{L}(- \boldsymbol{B}) : \boldsymbol{\nabla\nabla} f + \{ \boldsymbol{\nabla} \cdot \tilde{\mathsf{L}}(- \boldsymbol{B}) \} \cdot \boldsymbol{\nabla} f \,. \tag{80}$$

In the first terms of both sides of (79) and (80) only the symmetric part $\mathsf{L}^s = \frac{1}{2}(\mathsf{L} + \tilde{\mathsf{L}})$ of the tensor L remains because L is multiplied by the symmetric tensor $\nabla\nabla$ (a dyadic product). Since the first and second derivatives of $f(\boldsymbol{r})$ may be chosen in an arbitrary way, we can now conclude from (80) that

$$L^{s}(\boldsymbol{B}) = L^{s}(-\boldsymbol{B}), \tag{81}$$

$$\nabla \cdot L(\boldsymbol{B}) = \nabla \cdot \tilde{L}(-\boldsymbol{B}), \tag{82}$$

or, if the first relation is used in the second,

$$L^{s}(\boldsymbol{B}) = L^{s}(-\boldsymbol{B}), \tag{83}$$

$$\nabla \cdot L^{a}(\boldsymbol{B}) = -\nabla \cdot L^{a}(-\boldsymbol{B}), \tag{84}$$

where $L^{a} = \frac{1}{2}(L - \tilde{L})$ is the antisymmetric part of the tensor L. These relations are not equivalent to the reciprocal relations

$$L(\boldsymbol{B}) = \tilde{L}(-\boldsymbol{B}) \tag{85}$$

as given in Chapter IV, § 3, because a nabla-operator appears in (82) and (84). The reason for this is the following. If we split L into its symmetric and antisymmetric part also in equation (74), we obtain

$$\rho \frac{\partial u}{\partial t} = -\operatorname{div} \boldsymbol{J}_q = L^{s} : \nabla\nabla T + \{ \nabla \cdot (L^{s} + L^{a}) \} \cdot \nabla T. \tag{86}$$

Since only the divergence of the heat flow has physical meaning, in macroscopic theory, it is clear from (86) that observable results can be obtained for the symmetric part itself, because L^{s} appears also without differential operator, but only for the divergence of the antisymmetric part of the heat conduction tensor, since L^{a} appears only preceded by a differential operator.

Rewriting (84) in the form

$$\nabla \cdot \{ L^{a}(\boldsymbol{B}) + L^{a}(-\boldsymbol{B}) \} = 0 \tag{87}$$

it is seen that the divergence of the part of $L^{a}(\boldsymbol{B})$, which is an even function of the magnetic field $\boldsymbol{B}$, vanishes. In view of the preceding statements it follows then that the even part of $L^{a}(\boldsymbol{B})$ cannot be related to an observable quantity. One may for convenience therefore put it equal to zero so that (83) and (84) together are for all practical purposes identical with

$$L^s(\boldsymbol{B}) = L^s(-\boldsymbol{B}), \tag{88}$$

$$L^a(\boldsymbol{B}) = -L^a(-\boldsymbol{B}), \tag{89}$$

or in other words, with (85). We note that the relations (88) and (89) express the fact that the symmetric part of L is an even function and the antisymmetric part an odd function of the magnetic field strength.

(ii) *Heat conduction and diffusion in an isotropic fluid*

We shall treat one other example of coupled vectorial phenomena, *viz.* heat conduction, diffusion and cross-effects in an n-component isotropic fluid. It will be supposed that no external forces and no magnetic field act on the system. The phenomena of viscous flow and chemical reactions are not considered. It will furthermore be assumed that the barycentric velocity vanishes and hence that changes in the total density ρ are small and negligible in the course of time.

The total entropy production in this system, which we suppose to be adiabatically insulated is

$$\frac{\mathrm{d}S}{\mathrm{d}t} = \rho \int^{V} \frac{\partial s}{\partial t} \mathrm{d}V$$

$$= \rho \int^{V} \left(\frac{1}{T} \frac{\partial u}{\partial t} - \sum_{k=1}^{n-1} \frac{\mu_k - \mu_n}{T} \frac{\partial c_k}{\partial t} \right) \mathrm{d}V, \tag{90}$$

where Gibbs' relation (III.16), adapted to the case under consideration, has been used.

Since the energy and the masses of the various components are conserved we have

$$\int \frac{\partial u}{\partial t} \mathrm{d}V = 0, \tag{91}$$

$$\int \frac{\partial c_k}{\partial t} \mathrm{d}V = 0. \tag{92}$$

With these relations (90) may be rewritten in the torm

$$\frac{\mathrm{d}S}{\mathrm{d}t} = \rho \int^{V} \left(\Delta \frac{1}{T} \frac{\partial u}{\partial t} - \sum_{k=1}^{n-1} \Delta \frac{\mu_k - \mu_n}{T} \frac{\partial c_k}{\partial t} \right) \mathrm{d}V. \tag{93}$$

Here the symbol Δ indicates the difference of a quantity and its value at an arbitrary point $\boldsymbol{r}_0$. We have thus obtained, as in the previous case, an expression for the entropy production [of the form (IV.53)]. This expression is bilinear in the "fluxes" $\partial u/\partial t$ and $\partial c_k/\partial t$ and the thermodynamic forces ΔT^{-1} and $-\Delta\{(\mu_k - \mu_n)/T\}$.

The phenomenological equations for the present case are

$$\frac{\partial u}{\partial t} = \rho \int \left\{ K_{qq}(\boldsymbol{r}, \boldsymbol{r}') \, \Delta \frac{1}{T'} - \sum_{k=1}^{n-1} K_{qk}(\boldsymbol{r}, \boldsymbol{r}') \, \Delta \frac{\mu_k' - \mu_n'}{T'} \right\} \mathrm{d}\boldsymbol{r}' , \tag{94}$$

$$\frac{\partial c_i}{\partial t} = \rho \int \left\{ K_{iq}(\boldsymbol{r}, \boldsymbol{r}') \, \Delta \frac{1}{T'} - \sum_{k=1}^{n-1} K_{ik}(\boldsymbol{r}, \boldsymbol{r}') \, \Delta \frac{\mu_k' - \mu_n'}{T'} \right\} \mathrm{d}\boldsymbol{r}', \quad (i = 1, 2, \ldots, n-1) , \tag{95}$$

with the Onsager relations

$$K_{qq}(\boldsymbol{r}, \boldsymbol{r}') = K_{qq}(\boldsymbol{r}', \boldsymbol{r}) , \tag{96}$$

$$K_{iq}(\boldsymbol{r}, \boldsymbol{r}') = K_{qi}(\boldsymbol{r}', \boldsymbol{r}) , \quad (i = 1, 2, \ldots, n-1) , \tag{97}$$

$$K_{ik}(\boldsymbol{r}, \boldsymbol{r}') = K_{ki}(\boldsymbol{r}', \boldsymbol{r}) , \quad (i, k = 1, 2, \ldots, n-1) , \tag{98}$$

valid at all points $\boldsymbol{r}$ and $\boldsymbol{r}'$ except $\boldsymbol{r} = \boldsymbol{r}_0$ and $\boldsymbol{r}' = \boldsymbol{r}_0$.

On the other hand the system is also characterized by a local entropy source strength [*cf.* (V.25) without chemical terms]

$$\sigma = \boldsymbol{J}_q \cdot \operatorname{grad} \frac{1}{T} - \sum_{k=1}^{n-1} \boldsymbol{J}_k \cdot \operatorname{grad} \frac{\mu_k - \mu_n}{T} \tag{99}$$

and the local linear phenomenological equations [*cf.* (V.26) and (V.27)]

$$\boldsymbol{J}_q = L_{qq} \operatorname{grad} \frac{1}{T} - \sum_{k=1}^{n-1} L_{qk} \operatorname{grad} \frac{\mu_k - \mu_n}{T} , \tag{100}$$

$$\boldsymbol{J}_i = L_{iq} \operatorname{grad} \frac{1}{T} - \sum_{k=1}^{n-1} L_{ik} \operatorname{grad} \frac{\mu_k - \mu_n}{T} , \quad (i = 1, 2, \ldots, n-1) . \tag{101}$$

These local linear laws may be combined with the conservation laws, [*cf.* (V.20) without chemical terms and (V.22)],

$$\rho \frac{\partial u}{\partial t} = - \operatorname{div} \boldsymbol{J}_q , \tag{102}$$

$$\rho \frac{\partial c_k}{\partial t} = - \operatorname{div} \boldsymbol{J}_k , \quad (k = 1, 2, \ldots, n-1) \tag{103}$$

and yield the partial differential equations

$$\rho \frac{\partial u}{\partial t} = - \operatorname{div} \left(L_{qq} \operatorname{grad} \Delta \frac{1}{T} - \sum_{k=1}^{n-1} L_{qk} \operatorname{grad} \Delta \frac{\mu_k - \mu_n}{T} \right) , \tag{104}$$

$$\rho \frac{\partial c_i}{\partial t} = - \operatorname{div} \left(L_{iq} \operatorname{grad} \Delta \frac{1}{T} - \sum_{k=1}^{n-1} L_{ik} \operatorname{grad} \Delta \frac{\mu_k - \mu_n}{T} \right) ,$$

$$(i = 1, 2, \ldots, n-1) . \tag{105}$$

In order to derive the consequences of the Onsager relations (96)–(98) for the coefficients L_{qq}, L_{qk}, L_{iq} and L_{ik} $(i, k = 1, 2, \ldots, n-1)$, we shall transform the differential equations (104) and (105) into integro-differential equations of the form (94) and (95). Using the same procedure as in example (i), *viz.* by introducing δ-functions and integrating by parts, we obtain

$$\rho \frac{\partial u}{\partial t} = - \int \left[\Delta \frac{1}{T'} \boldsymbol{\nabla}' \cdot \{ L'_{qq} \boldsymbol{\nabla}' \, \delta(\boldsymbol{r} - \boldsymbol{r}') \} \right.$$

$$\left. - \sum_{k=1}^{n-1} \Delta \frac{\mu'_k - \mu'_n}{T'} \boldsymbol{\nabla}' \cdot \{ L'_{qk} \boldsymbol{\nabla}' \, \delta(\boldsymbol{r} - \boldsymbol{r}') \} \right] \mathrm{d}\boldsymbol{r}' , \tag{106}$$

$$\rho \frac{\partial c_i}{\partial t} = - \int \left[\Delta \frac{1}{T'} \boldsymbol{\nabla}' \cdot \{ L'_{iq} \boldsymbol{\nabla}' \, \delta(\boldsymbol{r} - \boldsymbol{r}') \} \right.$$

$$\left. - \sum_{k=1}^{n-1} \Delta \frac{\mu'_k - \mu'_n}{T'} \boldsymbol{\nabla}' \cdot \{ L'_{ik} \boldsymbol{\nabla}' \, \delta(\boldsymbol{r} - \boldsymbol{r}') \} \right] \mathrm{d}\boldsymbol{r}' , \; (i = 1, 2, \ldots, n-1) . \tag{107}$$

These equations hold everywhere except at the arbitrarily chosen point $\boldsymbol{r}_0$. Comparison of these equations with (94) and (95) permits one to identify all the coefficients K. Then the Onsager relations (96)–(98) turn out to be

$$\nabla' \cdot \{ L'_{qq} \nabla' \delta(\boldsymbol{r} - \boldsymbol{r}') \} = \nabla \cdot \{ L_{qq} \nabla \delta(\boldsymbol{r} - \boldsymbol{r}') \}, \tag{108}$$

$$\nabla' \cdot \{ L'_{iq} \nabla' \delta(\boldsymbol{r} - \boldsymbol{r}') \} = \nabla \cdot \{ L_{qi} \nabla \delta(\boldsymbol{r} - \boldsymbol{r}') \}, \quad (i = 1, 2, \ldots, n-1), \tag{109}$$

$$\nabla' \cdot \{ L'_{ik} \nabla' \delta(\boldsymbol{r} - \boldsymbol{r}') \} = \nabla \cdot \{ L_{ki} \nabla \delta(\boldsymbol{r} - \boldsymbol{r}') \}, \quad (i, k = 1, 2, \ldots, n-1). \tag{110}$$

This result holds for all points $\boldsymbol{r}$ and $\boldsymbol{r}'$, since the choice of the reference point $\boldsymbol{r}_0$ is arbitrary.

The same mathematical procedure as used in the preceding example shows that the first of these relations is an identity, whereas the others lead to the reciprocal relations

$$L_{iq} = L_{qi}, \quad (i = 1, 2, \ldots, n-1), \tag{111}$$

$$L_{ik} = L_{ki}, \quad (i, k = 1, 2, \ldots, n-1). \tag{112}$$

For all other processes of vectorial or tensorial character, *e.g.* viscous phenomena in anisotropic systems or electromagnetic phenomena, the proof of the reciprocal relations can be given along the lines of the method by which the two preceding examples were treated, or by slightly different but mathematically equivalent techniques*.

We have thus shown that the phenomenological coefficients occurring in the linear laws of vectorial or tensorial character established between the fluxes and thermodynamic forces of the *local entropy production* do indeed obey Onsager reciprocal relations, as stated in Chapter IV, § 3.

§ 5. *Transformation Properties of the Onsager Relations*

Let us finally study in this chapter the influence of linear transformations of the fluxes and thermodynamic forces on the validity of the Onsager relations**.

* P. Mazur and S. R. de Groot, Physica **19** (1953) 961.
S. R. de Groot and P. Mazur, Phys. Rev. **94** (1954) 218.
P. Mazur and S. R. de Groot, Phys. Rev. **94** (1954) 224.
R. Fieschi, S. R. de Groot and P. Mazur, Physica **20** (1954) 67.
R. Fieschi, S. R. de Groot, P. Mazur and J. Vlieger, Physica **20** (1954) 245.
S. R. de Groot and N. G. van Kampen, Physica **21** (1955) 39.
** J. Meixner, Ann. Phys. **43** (1943) 244.

We consider an adiabatically insulated system, the state of which is described by a number of n independent variables. The deviations of these variables from their equilibrium values are denoted by α_1, $\alpha_2, \ldots, \alpha_n$. The deviation of the entropy from its equilibrium value is in first approximation given by [*cf.* (IV.26)]

$$\Delta S = -\tfrac{1}{2} \sum_{i,j=1}^{n} g_{ij}\alpha_j\alpha_i \,, \tag{113}$$

or in matrix notation (*cf.* Appendix I)

$$\Delta S = -\tfrac{1}{2}\mathsf{g} : \boldsymbol{\alpha\alpha} = \tfrac{1}{2}\boldsymbol{\alpha}\cdot\boldsymbol{X} \,, \tag{114}$$

with

$$\boldsymbol{X} = -\mathsf{g}\cdot\boldsymbol{\alpha} \,. \tag{115}$$

The entropy production is

$$\dot{\Delta S} = \dot{\boldsymbol{\alpha}}\cdot\boldsymbol{X} = \boldsymbol{J}\cdot\boldsymbol{X} \,, \tag{116}$$

where the fluxes $\boldsymbol{J}$ are defined by

$$\boldsymbol{J} \equiv \dot{\boldsymbol{\alpha}} \,. \tag{117}$$

These fluxes obey the phenomenological equations

$$\boldsymbol{J} = \mathsf{L}\cdot\boldsymbol{X} \,. \tag{118}$$

The matrix L of phenomenological coefficients satisfies, if we only take the case of $\boldsymbol{\alpha}$-variables (*cf.* Chapter IV, § 3), the Onsager relations of the form

$$\mathsf{L}(\boldsymbol{B}) = \tilde{\mathsf{L}}(-\boldsymbol{B}) \,, \tag{119}$$

where $\boldsymbol{B}$ is the external magnetic field (and may stand also for a possible angular velocity of the system).

Let us now introduce new independent fluxes $\boldsymbol{J}'$, which are linear combinations of the old fluxes $\boldsymbol{J}$ and new independent thermodynamic forces $\boldsymbol{X}'$ which are linear functions of the old thermodynamic forces $\boldsymbol{X}$:

$$\boldsymbol{J}' = \mathsf{P}\cdot\boldsymbol{J} \,, \tag{120}$$

$$\boldsymbol{X}' = \mathsf{Q}\cdot\boldsymbol{X} \,. \tag{121}$$

We shall require that these transformations leave the entropy production invariant

$$\Delta \dot{S} = \boldsymbol{J}' \cdot \boldsymbol{X}' = \boldsymbol{J} \cdot \tilde{\mathsf{P}} \cdot \mathsf{Q} \cdot \boldsymbol{X} = \boldsymbol{J} \cdot \boldsymbol{X}, \tag{122}$$

where $\tilde{\mathsf{P}}$ is the transposed matrix of P. It follows from this condition that

$$\tilde{\mathsf{P}} \cdot \mathsf{Q} = \mathsf{U} + \mathsf{A} \cdot \mathsf{L}, \tag{123}$$

where U is the unit matrix and A an arbitrary antisymmetric matrix:

$$\tilde{\mathsf{A}} = - \mathsf{A}. \tag{124}$$

Indeed when (123) is inserted into (122) the term which contains A drops out, because from (118) we have

$$\boldsymbol{J} \cdot \mathsf{A} \cdot \mathsf{L} \cdot \boldsymbol{X} = \boldsymbol{J} \cdot \mathsf{A} \cdot \boldsymbol{J}, \tag{125}$$

and the second member vanishes identically since A is antisymmetric. Relation (123) gives a connexion between the three matrices P, Q and A which characterize the transformation. Only two of these matrices may be chosen independently.

With the help of (120) and (121) we can transform the phenomenological relations (118) into

$$\boldsymbol{J}' = \mathsf{L}' \cdot \boldsymbol{X}' \quad \text{with} \quad \mathsf{L}' = \mathsf{P} \cdot \mathsf{L} \cdot \mathsf{Q}^{-1}. \tag{126}$$

The new phenomenological coefficient matrix L', which appears here, can be written in a different way if we eliminate Q with the help of (123) and therefore consider P and A as matrices to be chosen independently. We then obtain

$$\mathsf{L}' = \mathsf{P} \cdot (\mathsf{L}^{-1} + \mathsf{A})^{-1} \cdot \tilde{\mathsf{P}}. \tag{127}$$

We shall now show that for this new coefficient scheme L' the Onsager relations remain valid if the antisymmetric matrix A is an odd function of the external magnetic field $\boldsymbol{B}$:

$$\mathsf{A}(-\boldsymbol{B}) = - \mathsf{A}(\boldsymbol{B}). \tag{128}$$

(The matrix P in transformations of fluxes and thermodynamic forces is always chosen independent of $\boldsymbol{B}$.)

Using the conditions (124) and (128) we find indeed from the Onsager relations (119) for the matrix (127):

$$
\begin{aligned}
\mathsf{L}'(\boldsymbol{B}) &= \mathsf{P}\cdot\{\mathsf{L}^{-1}(\boldsymbol{B}) + \mathsf{A}(\boldsymbol{B})\}^{-1}\cdot\tilde{\mathsf{P}} \\
&= \mathsf{P}\cdot\{\tilde{\mathsf{L}}^{-1}(-\boldsymbol{B}) - \tilde{\mathsf{A}}(\boldsymbol{B})\}^{-1}\cdot\tilde{\mathsf{P}} \\
&= \mathsf{P}\cdot\{\tilde{\mathsf{L}}^{-1}(-\boldsymbol{B}) + \tilde{\mathsf{A}}(-\boldsymbol{B})\}^{-1}\cdot\tilde{\mathsf{P}} \\
&= \mathsf{P}\cdot\{\mathsf{L}^{-1}(-\boldsymbol{B}) + \mathsf{A}(-\boldsymbol{B})\}^{-1}\cdot\tilde{\mathsf{P}} \\
&= \tilde{\mathsf{L}}'(-\boldsymbol{B})\,. \qquad (129)
\end{aligned}
$$

These reciprocal relations have the same form as the Onsager relations (119) for the old coefficients.

Let us consider two cases of the transformation described here.

(i) We choose

$$\mathsf{P} = \mathsf{U} \qquad (130)$$

and in agreement with (123)

$$\mathsf{Q} = \mathsf{U} + \mathsf{A}\cdot\mathsf{L}\,. \qquad (131)$$

Equations (120) and (121) become then with (118)

$$\boldsymbol{J}' = \boldsymbol{J}\,, \qquad (132)$$

$$\boldsymbol{X}' = \boldsymbol{X} + \mathsf{A}\cdot\boldsymbol{J}\,. \qquad (133)$$

A practical example of a transformation of the type described here arises for systems in the presence of a magnetic field, where the fluxes and thermodynamic forces are Cartesian vectors. Then the equations (132) and (133) can be written, if we drop the n-dimensional matrix notation,

$$\boldsymbol{J}_i' = \boldsymbol{J}_i\,, \qquad (i = 1, 2, \ldots, m)\,, \qquad (134)$$

$$\boldsymbol{X}_i' = \boldsymbol{X}_i + \sum_{k=1}^{m} \mathsf{A}_{ik}\cdot\boldsymbol{J}_k\,, \quad (i = 1, 2, \ldots, m)\,, \qquad (135)$$

where m is the number of fluxes and thermodynamic forces ($n = 3m$),

$\boldsymbol{J}_i$ and $\boldsymbol{X}_i$ are (Cartesian) vectors and A_{ik} (Cartesian) tensors. Let us now choose in particular

$$\mathsf{A}_{ik} = \mathsf{B}\delta_{ik}\,, \quad (i, k = 1, 2, \ldots, m)\,, \tag{136}$$

where B is the antisymmetric tensor corresponding to the axial vector $\boldsymbol{B}$ (the magnetic field) in the usual way ($B_x \equiv B_{yz} = -B_{zy}$, and cycl.). This choice satisfies the requirements (124) and (128) and leads to a form (135), which can be written as

$$\boldsymbol{X}_i' = \boldsymbol{X}_i + \mathsf{B}\cdot\boldsymbol{J}_i = \boldsymbol{X}_i + \boldsymbol{J}_i \wedge \boldsymbol{B}\,, \tag{137}$$

an expression occurring in practical applications (*cf.* Chapter XIII). We have found that such a transformation does not destroy the validity of the Onsager relations.

(ii) A second, more frequently encountered, case occurs when one has simply

$$\mathsf{A} = 0\,, \tag{138}$$

a condition which satisfies (124) and (128) in a trivial way. Then (123) becomes

$$\mathsf{Q} = \tilde{\mathsf{P}}^{-1}\,, \tag{139}$$

and also (127) and the proof (129) take simple forms. In the absence of a magnetic field this is the only type of transformation under which the reciprocal relations remain valid.

The reason for the invariance of the reciprocal relations can easily be understood in the latter case. From (117) and (120) it follows that the state variables $\boldsymbol{\alpha}$ transform as

$$\boldsymbol{\alpha}' = \mathsf{P}\cdot\boldsymbol{\alpha}\,. \tag{140}$$

From (115), (121), (140) and (139) it follows furthermore that the matrix g of (114) transforms as

$$\mathsf{g}' = \tilde{\mathsf{P}}^{-1}\cdot\mathsf{g}\cdot\mathsf{P}^{-1}\,. \tag{141}$$

It is seen that these transformations leave the deviation of the entropy ΔS invariant:

$$\Delta S = -\tfrac{1}{2}\mathsf{g}' : \boldsymbol{\alpha}'\boldsymbol{\alpha}'\,. \tag{142}$$

We may conclude that the transformations (120) and (121) with (139) correspond to a description of the system by means of a new set of independent state variables. Since the Onsager relations may be derived on the basis of an arbitrary set of independent state variables, the result $L'(\boldsymbol{B}) = \tilde{L}'(-\boldsymbol{B})$ is evident.

It must be stressed that if one requires only the entropy production to be invariant, but does not impose the auxiliary condition (128), there exists the larger class of transformations leading to a phenomenological coefficient scheme (127), for which the reciprocal relations are not valid*. However the transformations performed in physical applications are usually of a form as described above which does conserve the Onsager relations. It is clear that one must convince oneself that the original form of the entropy production, as derived from the various balance equations, is such that the Onsager relations are valid, *i.e.* that fluxes and thermodynamic forces are conjugate according to (115) and (117).

* J. E. Verschaffelt, Bull. Acad. roy. Belg., Cl. Sc. [5] **37** (1951) 853.
R. O. Davies, Physica **18** (1952) 182.
G. J. Hooyman, S. R. de Groot and P. Mazur, Physica **21** (1955) 360.

CHAPTER VII

DISCUSSION OF THE STATISTICAL FOUNDATIONS

§ 1. *Introduction*

In this chapter and the next we shall discuss the foundations of a number of postulates used in the phenomenological macroscopic theory, which was treated in the preceding chapters. In particular this chapter will be devoted to a discussion of the basis of the Onsager reciprocal relations between irreversible processes. From a fundamental point of view these relations as well as irreversible behaviour in general should follow in a straightforward manner from the microscopic laws of motion and the principles of statistical mechanics. However, no microscopic theory of such a general character as exists for equilibrium phenomena is available for non-equilibrium processes. On the other hand it is possible to derive from the microscopic laws of mechanics and the principles of statistical mechanics a number of theorems on the basis of which the reciprocal relations may be obtained with the help of an additional assumption on the behaviour of the processes considered. For the macroscopic description of a system one is not interested in the complete set of mechanical variables, describing its microscopic state, but only in a much more restricted number of variables. These variables could, for instance, be related to the extensive properties of macroscopically infinitesimal regions within the system. These regions are of course of such dimensions that they still contain a large number of the constituent particles of the system. For the average time behaviour of these "coarse grained" variables certain results can be derived on a rigorous statistical mechanical basis. In particular it will be shown that the so-called property of "microscopic reversibility" or "detailed balancing", which these variables obey, is a consequence of the time reversal invariance of the microscopic equations of motion. From this theorem the reciprocal relations follow by postulating that the averages of the "coarse grained" variables obey linear first order differential equations.

Since the coarse grained variables do not depend in a completely definite way on time, that is, since these variables are, what one calls random (or stochastic) variables, the language most appropriate to deal with them is that of the theory of random processes. In order to discuss the processes considered in more detail, we shall, after having derived the Onsager reciprocal relations, make some further, more specific, assumptions. In fact we shall assume that the above mentioned coarse grained variables describe processes which have, what is called, Gaussian Markoff character. In doing so we do not wish to assume that all macroscopic processes considered belong to this specific class of processes. It may, however, be surmised that a number of real phenomena may, within a certain approximation, be adequately described by such Gaussian Markoff processes. The advantage of specifying more precisely the nature of the processes considered is that it enables us to discuss, on the level of the theory of random processes, the behaviour of entropy, which constitutes a central concept of the thermodynamic theory of irreversible processes developed in the previous chapters.

The foregoing consideration will be the subject of more detailed discussions in the following sections.

In Chapter IX some of the problems mentioned above will be considered from a different point of view and discussed on the basis of the kinetic theory of gases.

§ 2. *State Variables and Fluctuations*

Consider an adiabatically insulated system, consisting of N point particles*, N being a very large number. We shall apply to this system the concepts of classical statistical mechanics. The microscopic state of the system is described by a point $(\boldsymbol{r}^N, \boldsymbol{p}^N) = (\boldsymbol{r}_1, \boldsymbol{r}_2, \ldots, \boldsymbol{r}_N; \boldsymbol{p}_1, \boldsymbol{p}_2, \ldots, \boldsymbol{p}_N)$ in the phase space, spanned by the set of canonical coordinates $\boldsymbol{r}^N$ and momenta $\boldsymbol{p}^N$ of the particles.

In statistical mechanics one studies the behaviour of a system by considering a representative ensemble of systems. For an aged system (*i.e.* an adiabatically insulated system which has been allowed to reach statistical equilibrium), with energy between E and $E + \mathrm{d}E$, this ensemble is the "micro-canonical ensemble", which is described by the following probability density:

$$\rho(\boldsymbol{r}^N, \boldsymbol{p}^N) = \left\{ \begin{array}{ll} \rho_0\,, & \text{within the energy shell } (E, E + \mathrm{d}E) \\ 0\,, & \text{elsewhere in phase space} \end{array} \right\}. \qquad (1)$$

* The assumption of point particles is not a necessary one. It is made here only for the sake of simplicity in the presentation.

The constant ρ_0 is determined by the following normalization condition

$$\int \rho(\boldsymbol{r}^N, \boldsymbol{p}^N)\, \mathrm{d}\boldsymbol{r}^N\, \mathrm{d}\boldsymbol{p}^N = 1\,, \tag{2}$$

where the integration extends over the whole of phase space. From equations (1) and (2) we obtain

$$\rho_0^{-1} = \int\limits_{(E,\,E+\mathrm{d}E)} \mathrm{d}\boldsymbol{r}^N\, \mathrm{d}\boldsymbol{p}^N \equiv \Omega\,, \tag{3}$$

where Ω is the volume in phase space of the energy shell defined by the interval $(E, E + \mathrm{d}E)$.

From a macroscopic point of view the state of the system will be described by a set of n extensive variables $A_1, A_2, \ldots, A_n$. One may think, for instance, of the energies, masses, electric charges of small sub-systems. These sub-systems must contain enough particles, so that the concepts of statistical mechanics may be applied to them. Extensive variables have been chosen to describe the states of the system, because of the difficulties encountered if one tries to introduce from the start intensive thermodynamic state variables for non-equilibrium states.

For convenience, a matrix or tensor notation will be introduced (*cf.* Appendix I). We shall consider the quantities A_i with $i = 1,2,\ldots,n$ as the components of a vector $\boldsymbol{A}$. The macroscopic state of a system can then be represented by a point in the so-called $\boldsymbol{A}$-space of which the n Cartesian coordinates are the quantities A_i. The variables $\boldsymbol{A}$ are functions of the dynamical variables $\boldsymbol{r}^N$ and $\boldsymbol{p}^N$ of the system:

$$\boldsymbol{A} = \boldsymbol{A}(\boldsymbol{r}^N, \boldsymbol{p}^N)\,. \tag{4}$$

We introduce the probability $f(\boldsymbol{A})\, \mathrm{d}\boldsymbol{A} = f(A_1, A_2, \ldots, A_n)$ $\mathrm{d}A_1\, \mathrm{d}A_2 \ldots \mathrm{d}A_n$ that the system is in a state for which $\boldsymbol{A}(\boldsymbol{r}^N, \boldsymbol{p}^N)$ lies between $\boldsymbol{A}$ and $\boldsymbol{A} + \mathrm{d}\boldsymbol{A}$. This probability is given by

$$f(\boldsymbol{A})\, \mathrm{d}\boldsymbol{A} = \int\limits_{(\boldsymbol{A},\, \boldsymbol{A}+\mathrm{d}\boldsymbol{A})} \rho(\boldsymbol{r}^N, \boldsymbol{p}^N)\, \mathrm{d}\boldsymbol{r}^N\, \mathrm{d}\boldsymbol{p}^N\,, \tag{5}$$

where the integration extends over the region of phase space corresponding to

$$\boldsymbol{A} \leqslant \boldsymbol{A}(\boldsymbol{r}^N, \boldsymbol{p}^N) \leqslant \boldsymbol{A} + \mathrm{d}\boldsymbol{A}\,. \tag{6}$$

Applying (2) and (3) we obtain for (5)

$$f(\boldsymbol{A})\,\mathrm{d}\boldsymbol{A} = \int_{\substack{(E,E+\mathrm{d}E)\\(\boldsymbol{A},\boldsymbol{A}+\mathrm{d}\boldsymbol{A})}} \rho_0\,\mathrm{d}\boldsymbol{r}^N\,\mathrm{d}\boldsymbol{p}^N = \frac{1}{\Omega}\int_{\substack{(E,E+\mathrm{d}E)\\(\boldsymbol{A},\boldsymbol{A}+\mathrm{d}\boldsymbol{A})}} \mathrm{d}\boldsymbol{r}^N\,\mathrm{d}\boldsymbol{p}^N = \frac{\Omega(\boldsymbol{A})}{\Omega}, \tag{7}$$

where $\Omega(\boldsymbol{A})$ is the volume in phase space containing the points for which the energy lies between E and $E + \mathrm{d}E$ and $\boldsymbol{A}(\boldsymbol{r}^N, \boldsymbol{p}^N)$ between $\boldsymbol{A}$ and $\boldsymbol{A} + \mathrm{d}\boldsymbol{A}$.

We shall assume* that the distribution function $f(\boldsymbol{A})$ is Gaussian:

$$f(\boldsymbol{A})\,\mathrm{d}\boldsymbol{A} \equiv f(A_1, A_2, \ldots, A_n)\,\mathrm{d}A_1\,\mathrm{d}A_2\ldots\mathrm{d}A_n$$

$$= c\exp\{-\tfrac{1}{2}k^{-1}\sum_{i,j} g_{ij}(A_j - \langle A_j\rangle)(A_i - \langle A_i\rangle)\}\,\mathrm{d}A_1\,\mathrm{d}A_2\ldots\mathrm{d}A_n, \tag{8}$$

where g_{ij} are the elements of a symmetric positive definite matrix (which means that the quadratic expression $\sum_{i,j} g_{ij}x_jx_i$, with real x_i, is positive definite), k Boltzmann's constant and $\langle A_i\rangle$ the mean value of A_i:

$$\langle A_i\rangle = \int f(A_1, A_2, \ldots, A_n)\,A_i\,\mathrm{d}A_1\,\mathrm{d}A_2\ldots\mathrm{d}A_n. \tag{9}$$

According to (8) the fluctuations α_i, defined as

$$\alpha_i = A_i - \langle A_i\rangle, \quad (i = 1, 2, \ldots, n), \tag{10}$$

obey the distribution function

$$f(\alpha_1, \alpha_2, \ldots, \alpha_n) = c\exp(-\tfrac{1}{2}k^{-1}\sum_{i,j} g_{ij}\alpha_j\alpha_i). \tag{11}$$

* One could say that the form (8) accepted for the distribution function $f(A_1, A_2, \ldots, A_n)$ defines the class of variables to which the theory given above may be applied. It is on such a distribution function that conventional fluctuation theory is based (see, *e.g.*, R. C. Tolman, The principles of statistical mechanics (Oxford, 1938) § 141; L. Landau and E. Lifshitz, Statistical physics (Oxford, 1938) Ch. VI.

Fluctuations of thermodynamic quantities obey this requirement under a wide range of conditions. From a more fundamental point of view the validity of (8) implies that on a kinetic model the extensive variables A_i are algebraic sums of microscopic variables and that a "central limit theorem" holds (*cf.* Appendix III for the explicit derivation of the Gaussian distribution in a special case).

From the definition (5) and the normalization (2) we get

$$\int f(\alpha_1, \alpha_2, \ldots, \alpha_n)\, d\alpha_1\, d\alpha_2 \ldots d\alpha_n = 1 . \tag{12}$$

The normalization constant c of (11) is found from (12):

$$c = \sqrt{\frac{|g|}{(2\pi k)^n}}, \tag{13}$$

where $|g|$ is the determinant value of the matrix of the quantities g_{ik}.
In matrix notation formula (11) reads

$$f(\boldsymbol{\alpha}) = c \exp(-\tfrac{1}{2}k^{-1}g : \boldsymbol{\alpha}\boldsymbol{\alpha}), \tag{14}$$

where g is a symmetric positive definite matrix with elements g_{ik}, and where $\boldsymbol{\alpha}\boldsymbol{\alpha}$ is a dyadic product (*i.e.* a matrix with elements $\alpha_i\alpha_j$).

A number of average values* will now be calculated with the help of the distribution function (14).

The following quantity will be introduced.

$$\mathbf{X} \equiv k \frac{\partial \ln f}{\partial \boldsymbol{\alpha}}, \quad \left(X_i \equiv k \frac{\partial \ln f}{\partial \alpha_i}, \quad (i = 1, 2, \ldots, n) \right). \tag{15}$$

From this definition and (14) it follows also that

$$\mathbf{X} = -g \cdot \boldsymbol{\alpha}, \quad \left(X_i = -\sum_{k=1}^{n} g_{ik}\alpha_k \right). \tag{16}$$

The average value of the dyadic $\boldsymbol{\alpha}\mathbf{X}$ can now easily be calculated:

$$\langle \boldsymbol{\alpha}\mathbf{X} \rangle = \int \boldsymbol{\alpha}\mathbf{X} f\, d\boldsymbol{\alpha} = k \int \boldsymbol{\alpha} \frac{\partial f}{\partial \boldsymbol{\alpha}}\, d\boldsymbol{\alpha}, \tag{17}$$

where (15) has been used, and where $d\boldsymbol{\alpha}$ stands for $d\alpha_1\, d\alpha_2 \ldots d\alpha_n$.
Partial integration gives

$$\langle \boldsymbol{\alpha}\mathbf{X} \rangle = -k \int f \frac{\partial}{\partial \boldsymbol{\alpha}} \boldsymbol{\alpha}\, d\boldsymbol{\alpha}, \tag{18}$$

* As usual in the theory of random processes, averages can be interpreted either as time averages over a single system, or as averages over an ensemble of systems.

since f vanishes at the integration boundaries. The dyadic differential quotient in the integrand is the unit matrix U with elements δ_{ik} ($\delta_{ik} = 1$ if $i = k$; $\delta_{ik} = 0$ if $i \neq k$), since $\partial\alpha_i/\partial\alpha_k = \delta_{ik}$. In this way one finally obtains, taking into account the normalization (12),

$$\langle \boldsymbol{\alpha}\mathbf{X} \rangle = -\,k\mathsf{U}\,. \tag{19}$$

From this formula we get, by introducing (16), the important result

$$\langle \boldsymbol{\alpha}\boldsymbol{\alpha} \rangle = k\mathsf{g}^{-1}\,, \tag{20}$$

where g^{-1} is the reciprocal matrix of g, (*i.e.* $\mathsf{g}\cdot\mathsf{g}^{-1} = \mathsf{g}^{-1}\cdot\mathsf{g} = \mathsf{U}$). The expressions (20) are called the "variances" of the Gaussian distribution (14).

In general one distinguishes between two types of macroscopic variables. The variables of the first type are even functions of the particle velocities (*e.g.* energy or mass densities). In the following the symbol $\boldsymbol{\alpha}$ will denote more specifically this class of variables. It may, however, occur that also variables are required for the description of the system, which are odd functions of the particle velocities (*e.g.* momentum densities, electric current densities). This second class of variables will be denoted by the symbol $\boldsymbol{\beta} = (\beta_1, \beta_2\,, \ldots, \beta_m)$.

We shall first consider the case that no external magnetic field acts on the system. The distribution function $f(\boldsymbol{\alpha}, \boldsymbol{\beta})$ then satisfies the relation

$$f(\boldsymbol{\alpha}, \boldsymbol{\beta}) = f(\boldsymbol{\alpha}, -\,\boldsymbol{\beta})\,. \tag{21}$$

This relation follows from the expression for $f(\boldsymbol{\alpha}, \boldsymbol{\beta})\, \mathrm{d}\boldsymbol{\alpha}\, \mathrm{d}\boldsymbol{\beta}$ in terms of an integral over a volume in phase space [*cf.* eq. (7)], taking into account the parities of the various quantities involved under a reversal of the particle velocities. (The energy E, occurring in the limits of integration of (7), is an even function of the particle velocities.) The Gaussian distribution (14) is therefore for this case of the form

$$f(\boldsymbol{\alpha}, \boldsymbol{\beta}) = c \exp\{-\tfrac{1}{2}k^{-1}(\mathsf{g} : \boldsymbol{\alpha}\boldsymbol{\alpha} + \mathsf{h} : \boldsymbol{\beta}\boldsymbol{\beta})\}\,, \tag{22}$$

where both g and h are symmetric positive definite matrices and where the normalization constant is

$$c = \sqrt{\frac{|\mathsf{g}|\;|\mathsf{h}|}{(2\pi k)^{n+m}}}\,. \tag{23}$$

No cross-terms between $\boldsymbol{\alpha}$- and $\boldsymbol{\beta}$-type variables occur in (22), in view of the property (21).

With this distribution function we define the two sets of quantities

$$\boldsymbol{X} \equiv k \frac{\partial \ln f}{\partial \boldsymbol{\alpha}} = - g \cdot \boldsymbol{\alpha} , \tag{24}$$

$$\boldsymbol{Y} \equiv k \frac{\partial \ln f}{\partial \boldsymbol{\beta}} = - h \cdot \boldsymbol{\beta} , \tag{25}$$

and obtain in analogy to (19) the formulae

$$\langle \boldsymbol{\alpha X} \rangle = - kU , \tag{26}$$

$$\langle \boldsymbol{\beta Y} \rangle = - kU , \tag{27}$$

$$\langle \boldsymbol{\beta X} \rangle = 0 , \tag{28}$$

$$\langle \boldsymbol{\alpha Y} \rangle = 0 . \tag{29}$$

For the variances of (22) we obtain from (26)–(29) with (24) and (25)

$$\langle \boldsymbol{\alpha\alpha} \rangle = kg^{-1} , \tag{30}$$

$$\langle \boldsymbol{\beta\beta} \rangle = kh^{-1} , \tag{31}$$

$$\langle \boldsymbol{\alpha\beta} \rangle = \langle \boldsymbol{\beta\alpha} \rangle = 0 . \tag{32}$$

If an external magnetic field $\boldsymbol{B}$ acts on the system, the distribution function $f(\boldsymbol{\alpha}, \boldsymbol{\beta}; \boldsymbol{B})$ satisfies instead of (21) the relation

$$f(\boldsymbol{\alpha}, \boldsymbol{\beta}; \boldsymbol{B}) = f(\boldsymbol{\alpha}, -\boldsymbol{\beta}; -\boldsymbol{B}) . \tag{33}$$

This relation follows again from the expression for $f(\boldsymbol{\alpha}, \boldsymbol{\beta}; \boldsymbol{B})$ in terms of an integral over a volume in phase space taking into account, that velocities are reversed if both the momenta and the magnetic field are reversed and that the energy of the system is even in both the momenta and the magnetic field*.

* In classical mechanics the distribution function (33) is independent of $\boldsymbol{B}$: magnetic properties cannot be obtained for classical point particles [*cf.* L. Rosenfeld, Theory of Electrons (Amsterdam, 1951)]. However in quantum mechanics this is not the case; the property (33) is valid in general.

Now the Gaussian distribution has the form

$$f(\boldsymbol{\alpha}, \boldsymbol{\beta}; \boldsymbol{B}) = c \exp \{ -\tfrac{1}{2}k^{-1}(\mathsf{g} : \boldsymbol{\alpha\alpha} + \mathsf{m} : \boldsymbol{\alpha\beta} + \mathsf{n} : \boldsymbol{\beta\alpha} + \mathsf{h} : \boldsymbol{\beta\beta}) \}, \quad (34)$$

where m and n are rectangular matrices, the first of n columns and m rows, the second of m columns and n rows. The matrix m is the transposed matrix of n:

$$\mathsf{m} = \tilde{\mathsf{n}}, \quad (m_{ik} = n_{ki}, \quad (i = 1, 2, \ldots, m; k = 1, 2, \ldots, n)). \quad (35)$$

In view of (33) one has

$$\mathsf{g}(\boldsymbol{B}) = \mathsf{g}(-\boldsymbol{B}), \quad (36)$$

$$\mathsf{m}(\boldsymbol{B}) = -\mathsf{m}(-\boldsymbol{B}), \quad (37)$$

$$\mathsf{n}(\boldsymbol{B}) = -\mathsf{n}(-\boldsymbol{B}), \quad (38)$$

$$\mathsf{h}(\boldsymbol{B}) = \mathsf{h}(-\boldsymbol{B}). \quad (39)$$

From the distribution function (34) we find for the quantities $\boldsymbol{X}$ and $\boldsymbol{Y}$

$$\boldsymbol{X} \equiv k \frac{\partial \ln f}{\partial \boldsymbol{\alpha}} = -\mathsf{g} \cdot \boldsymbol{\alpha} - \mathsf{n} \cdot \boldsymbol{\beta}, \quad (40)$$

$$\boldsymbol{Y} \equiv k \frac{\partial \ln f}{\partial \boldsymbol{\beta}} = -\mathsf{m} \cdot \boldsymbol{\alpha} - \mathsf{h} \cdot \boldsymbol{\beta}, \quad (41)$$

where the relation (35) has been applied. With these definitions one obtains again the results (26)–(29). For the variances of (34) we get from (26)–(29) with (40) and (41)

$$\langle \boldsymbol{\alpha\alpha} \rangle = k(\mathsf{g} - \mathsf{n} \cdot \mathsf{h}^{-1} \cdot \mathsf{m})^{-1}, \quad (42)$$

$$\langle \boldsymbol{\beta\beta} \rangle = k(\mathsf{h} - \mathsf{m} \cdot \mathsf{g}^{-1} \cdot \mathsf{n})^{-1}, \quad (43)$$

$$\langle \boldsymbol{\alpha\beta} \rangle = -k(\mathsf{g} - \mathsf{n} \cdot \mathsf{h}^{-1} \cdot \mathsf{m})^{-1} \cdot \mathsf{n} \cdot \mathsf{h}^{-1} \quad (44)$$

$$\langle \widetilde{\boldsymbol{\alpha\beta}} \rangle = \langle \boldsymbol{\beta\alpha} \rangle = -k(\mathsf{h} - \mathsf{m} \cdot \mathsf{g}^{-1} \cdot \mathsf{n})^{-1} \cdot \mathsf{m} \cdot \mathsf{g}^{-1}. \quad (45)$$

One can check that the right-hand sides of the last two relations are each others transposed matrices, by applying relation (35) and the fact that g and h are symmetric. The parity with respect to reversal of the external magnetic field of the variances follows from (36)–(39). Thus the variances (42) and (43) are even in $\boldsymbol{B}$ and (44) and (45) odd.

According to the Boltzmann entropy postulate the following connection is adopted between the probability of a state and its entropy $S(\boldsymbol{\alpha})$ (for convenience we use again the comprehensive symbol $\boldsymbol{\alpha}$ for both $\boldsymbol{\alpha}$- and $\boldsymbol{\beta}$-type variables):

$$f(\boldsymbol{\alpha}) \sim e^{S(\boldsymbol{\alpha})/k}, \tag{46}$$

or

$$f(\boldsymbol{\alpha}) = f(0)\, e^{\Delta S/k}, \tag{47}$$

where

$$\Delta S = S(\boldsymbol{\alpha}) - S(0). \tag{48}$$

Comparison of (14) and (47) yields

$$\Delta S = -\tfrac{1}{2} g : \boldsymbol{\alpha\alpha}. \tag{49}$$

The entropy $S(0)$ refers to the state of maximum probability $f(0)$, so that (49) may be seen as a Taylor series expansion of the entropy around this state, in which terms of order higher than the second in $\boldsymbol{\alpha}$ have been neglected (first order terms are zero since the expansion was carried out with respect to the maximum value). The matrix $-g$ is then the matrix of the second derivatives of S with respect to $\boldsymbol{\alpha}$. The preceding arguments have frequently been used to derive from (47) and (49) the form (14) of the distribution function $f(\boldsymbol{\alpha})$ (*cf.* footnote p. 86).

The state of maximum probability corresponding to the entropy $S(0)$ is often referred to as the "equilibrium state". On the other hand the word equilibrium may also be used in connexion with a distribution over possible states. Thus one may call (14) the equilibrium distribution of states. Of course thermodynamics does not distinguish between these two concepts of equilibrium, although we know that a system in equilibrium does fluctuate around an average state. That the two concepts of equilibrium are physically almost equivalent may be seen by calculating the average value of $S(\boldsymbol{\alpha})$ with (49) and (20):

$$\langle S(\boldsymbol{\alpha}) \rangle = S(0) - \tfrac{1}{2} g : \langle \boldsymbol{\alpha\alpha} \rangle = S(0) - \tfrac{1}{2} kn, \tag{50}$$

where n is the number of variables $\boldsymbol{\alpha}$. Since $S(0)$ is of the order of Nk where N is the number of particles in the system, the difference between $\langle S(\boldsymbol{\alpha}) \rangle$ and $S(0)$ is virtually negligible. In other words the distribution of states (14) is so sharply peaked that virtually only the "equilibrium state" contributes to the average value of $S(\boldsymbol{\alpha})$. This is in agreement with the fact that no distinction is made in thermodynamics between an "equilibrium state" and an equilibrium distribution.

The variables $\mathbf{X}$ of (15) and (16) can now be defined alternatively as

$$\mathbf{X} = \frac{\partial \Delta S}{\partial \boldsymbol{\alpha}}, \tag{51}$$

and may therefore be interpreted as the intensive thermodynamic state parameters conjugate to the extensive variables $\boldsymbol{\alpha}$*.

§ 3. *Microscopic Reversibility*

In section 2 we have discussed the equilibrium distribution function $f(\boldsymbol{\alpha})$. Let us consider the joint distribution function $f(\boldsymbol{\alpha}, \boldsymbol{\alpha}'; \tau)$. By definition $f(\boldsymbol{\alpha}, \boldsymbol{\alpha}'; \tau)\, \mathrm{d}\boldsymbol{\alpha}\, \mathrm{d}\boldsymbol{\alpha}'$ is the joint probability to find the (aged) system at some initial time in a state for which $\boldsymbol{\alpha} \leqslant \boldsymbol{\alpha}(\boldsymbol{r}^N, \boldsymbol{p}^N) \leqslant \boldsymbol{\alpha} + \mathrm{d}\boldsymbol{\alpha}$ and at a time τ later in a state for which $\boldsymbol{\alpha}' \leqslant \boldsymbol{\alpha}(\boldsymbol{r}^N, \boldsymbol{p}^N) \leqslant \boldsymbol{\alpha}' + \mathrm{d}\boldsymbol{\alpha}'$.

In order to obtain an expression for the second distribution function, we introduce the conditional probability density in phase space $P(\boldsymbol{r}^N, \boldsymbol{p}^N \mid \boldsymbol{r}'^N, \boldsymbol{p}'^N; \tau)$. By definition $P(\boldsymbol{r}^N, \boldsymbol{p}^N \mid \boldsymbol{r}'^N, \boldsymbol{p}'^N; \tau)\, \mathrm{d}\boldsymbol{r}'^N\, \mathrm{d}\boldsymbol{p}'^N$ is the probability to find the system in the range $(\boldsymbol{r}'^N, \boldsymbol{p}'^N; \boldsymbol{r}'^N + \mathrm{d}\boldsymbol{r}'^N, \boldsymbol{p}'^N + \mathrm{d}\boldsymbol{p}'^N)$ at time τ, when initially (at time $\tau = 0$) it was at $(\boldsymbol{r}^N, \boldsymbol{p}^N)$.

If the system is at the point $(\boldsymbol{r}^N, \boldsymbol{p}^N)$ in phase space initially it will be at the point $(\boldsymbol{r}'^N, \boldsymbol{p}'^N)$ at a time τ later. The changes $\Delta \boldsymbol{r}^N = \boldsymbol{r}'^N - \boldsymbol{r}^N$ in coordinates and $\Delta \boldsymbol{p}^N = \boldsymbol{p}'^N - \boldsymbol{p}^N$ in momenta are functions of the initial values $\boldsymbol{r}^N$ and $\boldsymbol{p}^N$ and of the time τ (and follow from the classical equations of motion)

$$\begin{aligned} \boldsymbol{r}'^N - \boldsymbol{r}^N &= \Delta \boldsymbol{r}^N(\boldsymbol{r}^N, \boldsymbol{p}^N; \tau), \quad \text{with} \quad \Delta \boldsymbol{r}^N(\boldsymbol{r}^N, \boldsymbol{p}^N; 0) = 0, \\ \boldsymbol{p}'^N - \boldsymbol{p}^N &= \Delta \boldsymbol{p}^N(\boldsymbol{r}^N, \boldsymbol{p}^N; \tau), \quad \text{with} \quad \Delta \boldsymbol{p}^N(\boldsymbol{r}^N, \boldsymbol{p}^N; 0) = 0. \end{aligned} \tag{52}$$

* Conceptually one must define the values of the entropy and the intensive variables of the sub-systems by determining these quantities when one suddenly isolates the sub-systems in a state $\boldsymbol{\alpha}$ and permits their equilibrium to be reached. We show in Appendix III that such a definition of entropy and intensive "local" thermodynamic variables is in agreement with (46) and (51).

It follows from the definition of P and eq. (52) that

$$P(\boldsymbol{r}^N, \boldsymbol{p}^N \mid \boldsymbol{r}'^N, \boldsymbol{p}'^N ; \tau)$$

$$= \delta\,\{\, \boldsymbol{r}'^N - \boldsymbol{r}^N - \Delta \boldsymbol{r}^N(\boldsymbol{r}^N, \boldsymbol{p}^N ; \tau)\,\} \;\; \delta\,\{\boldsymbol{p}'^N - \boldsymbol{p}^N - \Delta \boldsymbol{p}^N(\boldsymbol{r}^N, \boldsymbol{p}^N ; \tau)\,\}\,, \quad (53)$$

where both factors on the right-hand side are $3N$-dimensional δ-functions. This means that the probability density (53) is zero except for a transition to that point in phase space which should be reached according to the laws of mechanics.

It follows from (53) that the conditional probability density is normalized in such a way that

$$\int P(\boldsymbol{r}^N, \boldsymbol{p}^N \mid \boldsymbol{r}'^N, \boldsymbol{p}'^N ; \tau)\, \mathrm{d}\boldsymbol{r}'^N\, \mathrm{d}\boldsymbol{p}'^N = 1\,. \quad (54)$$

For all points $(\boldsymbol{r}^N, \boldsymbol{p}^N)$ lying in the energy shell $(E, E + \mathrm{d}E)$ we can write this formula alternatively as

$$\int_{(E,E+\mathrm{d}E)} P(\boldsymbol{r}^N, \boldsymbol{p}^N \mid \boldsymbol{r}'^N, \boldsymbol{p}'^N ; \tau)\, \mathrm{d}\boldsymbol{r}'^N\, \mathrm{d}\boldsymbol{p}'^N = 1\,, \quad (55)$$

since all phase trajectories remain inside the energy shell.

The probability density $\rho(\boldsymbol{r}^N, \boldsymbol{p}^N ;\tau)$ may be obtained from the density $\rho(\boldsymbol{r}^N, \boldsymbol{p}^N ;0)$ by means of

$$\rho(\boldsymbol{r}'^N, \boldsymbol{p}'^N ; \tau) = \int \rho(\boldsymbol{r}^N, \boldsymbol{p}^N ; 0)\, P(\boldsymbol{r}^N, \boldsymbol{p}^N \mid \boldsymbol{r}'^N, \boldsymbol{p}'^N ; \tau)\, \mathrm{d}\boldsymbol{r}^N\, \mathrm{d}\boldsymbol{p}^N\,, \quad (56)$$

in view of the meaning of P as defined above. For the stationary micro-canonical ensemble we have

$$\rho(\boldsymbol{r}^N, \boldsymbol{p}^N ; \tau) = \rho(\boldsymbol{r}^N, \boldsymbol{p}^N ; 0) = \rho(\boldsymbol{r}^N, \boldsymbol{p}^N)\,, \quad (57)$$

where $\rho(\boldsymbol{r}^N, \boldsymbol{p}^N)$ is given by (1). From equations (1), (56) and (57) it then follows that

$$\int_{(E,E+\mathrm{d}E)} P(\boldsymbol{r}^N, \boldsymbol{p}^N \mid \boldsymbol{r}'^N, \boldsymbol{p}'^N ; \tau)\, \mathrm{d}\boldsymbol{r}^N\, \mathrm{d}\boldsymbol{p}^N = 1\,, \quad (58)$$

for all $(\boldsymbol{r}'^N, \boldsymbol{p}'^N)$ for which the energy lies between E and $E + \mathrm{d}E$. Again since all phase trajectories remain inside the energy shell, equation (58) may be replaced by

$$\int P(\boldsymbol{r}^N, \boldsymbol{p}^N \mid \boldsymbol{r}'^N, \boldsymbol{p}'^N; \tau)\, \mathrm{d}\boldsymbol{r}^N\, \mathrm{d}\boldsymbol{p}^N = 1\,. \tag{59}$$

In fact the content of this equation is equivalent to that of Liouville's theorem. It is, fundamentally, a consequence of the Hamiltonian form of the equations of motion.

Now the equations of motion of the particles are invariant under "time reversal", *i.e.* under the transformation

$$\left.\begin{array}{ccc} \tau & \rightarrow & -\,\tau \\ \boldsymbol{r}^N & \rightarrow & \boldsymbol{r}^N \\ \boldsymbol{p}^N & \rightarrow & -\,\boldsymbol{p}^N \end{array}\right\}. \tag{60}$$

This means that the function P obeys the relation

$$P(\boldsymbol{r}^N, \boldsymbol{p}^N \mid \boldsymbol{r}'^N, \boldsymbol{p}'^N; \tau) = P(\boldsymbol{r}^N, -\,\boldsymbol{p}^N \mid \boldsymbol{r}'^N, -\,\boldsymbol{p}'^N; -\,\tau)\,. \tag{61}$$

Furthermore, as a consequence of the causal nature of the equations of motion, the "probability" to be at $(\boldsymbol{r}'^N, -\,\boldsymbol{p}'^N)$ a time τ earlier when initially one is at $(\boldsymbol{r}^N, -\,\boldsymbol{p}^N)$ is the same as the "probability" to be at $(\boldsymbol{r}^N, -\,\boldsymbol{p}^N)$ a time τ later when initially one is at $(\boldsymbol{r}'^N, -\,\boldsymbol{p}'^N)$:

$$P(\boldsymbol{r}^N, -\,\boldsymbol{p}^N \mid \boldsymbol{r}'^N, -\,\boldsymbol{p}'^N; -\,\tau) = P(\boldsymbol{r}'^N, -\,\boldsymbol{p}'^N \mid \boldsymbol{r}^N, -\,\boldsymbol{p}^N; \tau)\,. \tag{62}$$

From the two foregoing formulae we obtain

$$P(\boldsymbol{r}^N, \boldsymbol{p}^N \mid \boldsymbol{r}'^N, \boldsymbol{p}'^N; \tau) = P(\boldsymbol{r}'^N, -\,\boldsymbol{p}'^N \mid \boldsymbol{r}^N, -\,\boldsymbol{p}^N; \tau)\,. \tag{63}$$

This relation expresses the fact that if at a certain time we reverse the momenta of the particles, they retrace their former paths with reversed momenta.

The joint probability to find the system initially in the range $(\boldsymbol{r}^N, \boldsymbol{p}^N; \boldsymbol{r}^N + \mathrm{d}\boldsymbol{r}^N, \boldsymbol{p}^N + \mathrm{d}\boldsymbol{p}^N)$ and at a time τ later in the range $(\boldsymbol{r}'^N, \boldsymbol{p}'^N; \boldsymbol{r}'^N + \mathrm{d}\boldsymbol{r}'^N, \boldsymbol{p}'^N + \mathrm{d}\boldsymbol{p}'^N)$ is for the micro-canonical ensemble equal to $\rho(\boldsymbol{r}^N, \boldsymbol{p}^N)\, P(\boldsymbol{r}^N, \boldsymbol{p}^N \mid \boldsymbol{r}'^N, \boldsymbol{p}'^N; \tau)\, \mathrm{d}\boldsymbol{r}^N \mathrm{d}\boldsymbol{p}^N \mathrm{d}\boldsymbol{r}'^N \mathrm{d}\boldsymbol{p}'^N$. Then the joint probability $f(\boldsymbol{\alpha}, \boldsymbol{\alpha}'; \tau)\, \mathrm{d}\boldsymbol{\alpha}\, \mathrm{d}\boldsymbol{\alpha}'$ for the micro-canonical ensemble is

$$f(\boldsymbol{\alpha}, \boldsymbol{\alpha}'; \tau)\, \mathrm{d}\boldsymbol{\alpha}\, \mathrm{d}\boldsymbol{\alpha}'$$
$$= \int\limits_{(\boldsymbol{\alpha}, \boldsymbol{\alpha}+\mathrm{d}\boldsymbol{\alpha})} \int\limits_{(\boldsymbol{\alpha}', \boldsymbol{\alpha}'+\mathrm{d}\boldsymbol{\alpha}')} \rho(\boldsymbol{r}^N, \boldsymbol{p}^N)\, P(\boldsymbol{r}^N, \boldsymbol{p}^N \mid \boldsymbol{r}'^N, \boldsymbol{p}'^N; \tau)\, \mathrm{d}\boldsymbol{r}^N \mathrm{d}\boldsymbol{p}^N \mathrm{d}\boldsymbol{r}'^N \mathrm{d}\boldsymbol{p}'^N \tag{64}$$

It follows with (1) that

$$f(\boldsymbol{\alpha}, \boldsymbol{\alpha}'; \tau)\, \mathrm{d}\boldsymbol{\alpha}\, \mathrm{d}\boldsymbol{\alpha}'$$
$$= \rho_0 \int\limits_{\substack{(\boldsymbol{\alpha}, \boldsymbol{\alpha}+\mathrm{d}\boldsymbol{\alpha}) \\ (E, E+\mathrm{d}E)}} \int\limits_{(\boldsymbol{\alpha}', \boldsymbol{\alpha}'+\mathrm{d}\boldsymbol{\alpha}')} P(\boldsymbol{r}^N, \boldsymbol{p}^N \mid \boldsymbol{r}'^N, \boldsymbol{p}'^N; \tau)\, \mathrm{d}\boldsymbol{r}^N \mathrm{d}\boldsymbol{p}^N \mathrm{d}\boldsymbol{r}'^N \mathrm{d}\boldsymbol{p}'^N$$
$$= \rho_0 \int\limits_{\substack{(\boldsymbol{\alpha}, \boldsymbol{\alpha}+\mathrm{d}\boldsymbol{\alpha}) \\ (E, E+\mathrm{d}E)}} \int\limits_{\substack{(\boldsymbol{\alpha}', \boldsymbol{\alpha}'+\mathrm{d}\boldsymbol{\alpha}') \\ (E, E+\mathrm{d}E)}} P(\boldsymbol{r}^N, \boldsymbol{p}^N \mid \boldsymbol{r}'^N, \boldsymbol{p}'^N; \tau)\, \mathrm{d}\boldsymbol{r}^N \mathrm{d}\boldsymbol{p}^N \mathrm{d}\boldsymbol{r}'^N \mathrm{d}\boldsymbol{p}'^N, \tag{65}$$

using the fact that all phase trajectories remain inside the energy shell $(E, E + \mathrm{d}E)$. It is immediately apparent from equation (64) that the quantity $f(\boldsymbol{\alpha}, \boldsymbol{\alpha}'; \tau)\, \mathrm{d}\boldsymbol{\alpha}\, \mathrm{d}\boldsymbol{\alpha}'$ is stationary, *i.e.* that it is only a function of the time interval τ and not of some initial time t.

From (65) we get with (55) and (58) and the definition (7) (replacing $\boldsymbol{A}$ by $\boldsymbol{\alpha}$)

$$\int f(\boldsymbol{\alpha}, \boldsymbol{\alpha}'; \tau)\, \mathrm{d}\boldsymbol{\alpha}' = f(\boldsymbol{\alpha}), \tag{66}$$

$$\int f(\boldsymbol{\alpha}, \boldsymbol{\alpha}'; \tau)\, \mathrm{d}\boldsymbol{\alpha} = f(\boldsymbol{\alpha}'). \tag{67}$$

We now define the conditional probability density $P(\boldsymbol{\alpha} \mid \boldsymbol{\alpha}'; \tau)$ for the micro-canonical ensemble:

$$P(\boldsymbol{\alpha} \mid \boldsymbol{\alpha}'; \tau)\, \mathrm{d}\boldsymbol{\alpha}' \equiv \frac{f(\boldsymbol{\alpha}, \boldsymbol{\alpha}'; \tau)\, \mathrm{d}\boldsymbol{\alpha}\, \mathrm{d}\boldsymbol{\alpha}'}{f(\boldsymbol{\alpha})\, \mathrm{d}\boldsymbol{\alpha}}. \tag{68}$$

It follows from (3), (7) and (65) that

$$P(\boldsymbol{\alpha} \mid \boldsymbol{\alpha}' ; \tau)\, \mathrm{d}\boldsymbol{\alpha}' = \frac{f(\boldsymbol{\alpha}, \boldsymbol{\alpha}' ; \tau)\, \mathrm{d}\boldsymbol{\alpha}\, \mathrm{d}\boldsymbol{\alpha}'}{\rho_0\, \Omega(\boldsymbol{\alpha})}$$

$$= \frac{1}{\Omega(\boldsymbol{\alpha})} \int\limits_{\substack{(\boldsymbol{\alpha},\, \boldsymbol{\alpha}+\mathrm{d}\boldsymbol{\alpha}) \\ (E,\, E+\mathrm{d}E)}} \int\limits_{\substack{(\boldsymbol{\alpha}',\, \boldsymbol{\alpha}'+\mathrm{d}\boldsymbol{\alpha}') \\ (E,\, E+\mathrm{d}E)}} P(\boldsymbol{r}^N, \boldsymbol{p}^N \mid \boldsymbol{r}'^N, \boldsymbol{p}'^N ; \tau)\, \mathrm{d}\boldsymbol{r}^N \mathrm{d}\boldsymbol{p}^N \mathrm{d}\boldsymbol{r}'^N \mathrm{d}\boldsymbol{p}'^N . \quad (69)$$

It is apparent from this equation that the quantity $P(\boldsymbol{\alpha} \mid \boldsymbol{\alpha}';\tau)\, \mathrm{d}\boldsymbol{\alpha}'$ is also stationary.

For a non-stationary ensemble with density $\rho(\boldsymbol{r}^N, \boldsymbol{p}^N;t)$ the joint probability $\rho(\boldsymbol{r}^N, \boldsymbol{p}^N;t)\, P(\boldsymbol{r}^N, \boldsymbol{p}^N \mid \boldsymbol{r}'^N, \boldsymbol{p}'^N;\tau)\, \mathrm{d}\boldsymbol{r}^N \mathrm{d}\boldsymbol{p}^N \mathrm{d}\boldsymbol{r}'^N \mathrm{d}\boldsymbol{p}'^N$ is non-stationary, since this quantity depends essentially on the initial time t. Instead of (64) we now have for the joint probability density

$$f(\boldsymbol{\alpha}, t ; \boldsymbol{\alpha}', t + \tau)\, \mathrm{d}\boldsymbol{\alpha}\, \mathrm{d}\boldsymbol{\alpha}'$$

$$= \int\limits_{(\boldsymbol{\alpha},\, \boldsymbol{\alpha}+\mathrm{d}\boldsymbol{\alpha})} \int\limits_{(\boldsymbol{\alpha}',\, \boldsymbol{\alpha}'+\mathrm{d}\boldsymbol{\alpha}')} \rho(\boldsymbol{r}^N, \boldsymbol{p}^N ; t)\, P(\boldsymbol{r}^N, \boldsymbol{p}^N \mid \boldsymbol{r}'^N, \boldsymbol{p}'^N ; \tau)\, \mathrm{d}\boldsymbol{r}^N \mathrm{d}\boldsymbol{p}^N \mathrm{d}\boldsymbol{r}'^N \mathrm{d}\boldsymbol{p}'^N . \quad (70)$$

With (54) and (56) we then get, instead of (66) and (67),

$$\int f(\boldsymbol{\alpha}, t ; \boldsymbol{\alpha}', t + \tau)\, \mathrm{d}\boldsymbol{\alpha}' = f(\boldsymbol{\alpha}, t) , \quad (71)$$

$$\int f(\boldsymbol{\alpha}, t ; \boldsymbol{\alpha}', t + \tau)\, \mathrm{d}\boldsymbol{\alpha} = f(\boldsymbol{\alpha}', t + \tau) , \quad (72)$$

where

$$f(\boldsymbol{\alpha}, t)\, \mathrm{d}\boldsymbol{\alpha} = \int\limits_{(\boldsymbol{\alpha},\, \boldsymbol{\alpha}+\mathrm{d}\boldsymbol{\alpha})} \rho(\boldsymbol{r}^N, \boldsymbol{p}^N ; t)\, \mathrm{d}\boldsymbol{r}^N \mathrm{d}\boldsymbol{p}^N \quad (73)$$

is the non-stationary distribution function.

Corresponding to (68) we can define the conditional probability density

$$P(\boldsymbol{\alpha}, t \mid \boldsymbol{\alpha}', t + \tau)\, \mathrm{d}\boldsymbol{\alpha}' \equiv \frac{f(\boldsymbol{\alpha}, t ; \boldsymbol{\alpha}', t + \tau)\, \mathrm{d}\boldsymbol{\alpha}\, \mathrm{d}\boldsymbol{\alpha}'}{f(\boldsymbol{\alpha}, t)\, \mathrm{d}\boldsymbol{\alpha}} . \quad (74)$$

The joint probability (70) and the conditional probability (74) are non-stationary since they depend essentially on the initial time.

The conditional probability (74) will at some specific time $t = t_0$. reduce to the stationary conditional probability density (68), provided that $\rho(\boldsymbol{r}^N, \boldsymbol{p}^N; t)$ is uniform at $t = t_0$ for the points in phase space for which $\boldsymbol{\alpha} \leqslant \boldsymbol{\alpha}(\boldsymbol{r}^N, \boldsymbol{p}^N) \leqslant \boldsymbol{\alpha} + \mathrm{d}\boldsymbol{\alpha}$ and for which the energy lies in the interval $(E, E + \mathrm{d}E)$. Indeed it follows from (70) and (73) that for that case the right-hand side of (74) is independent of $\rho(\boldsymbol{r}^N, \boldsymbol{p}^N; t_0)$, so that we may as well replace $\rho(\boldsymbol{r}^N, \boldsymbol{p}^N; t_0)$ by the micro-canonical probability density $\rho(\boldsymbol{r}^N, \boldsymbol{p}^N)$, which is also uniform over regions $\boldsymbol{\alpha} \leqslant \boldsymbol{\alpha}(\boldsymbol{r}^N, \boldsymbol{p}^N) \leqslant \boldsymbol{\alpha} + \mathrm{d}\boldsymbol{\alpha}$. Hence

$$\frac{f(\boldsymbol{\alpha}, t_0; \boldsymbol{\alpha}', t_0 + \tau)\, \mathrm{d}\boldsymbol{\alpha}\, \mathrm{d}\boldsymbol{\alpha}'}{f(\boldsymbol{\alpha}, t_0)\, \mathrm{d}\boldsymbol{\alpha}} = \frac{f(\boldsymbol{\alpha}, \boldsymbol{\alpha}'; \tau)\, \mathrm{d}\boldsymbol{\alpha}\, \mathrm{d}\boldsymbol{\alpha}'}{f(\boldsymbol{\alpha})\, \mathrm{d}\boldsymbol{\alpha}}, \tag{75}$$

or with (68) and (74),

$$P(\boldsymbol{\alpha}, t_0 \mid \boldsymbol{\alpha}', t_0 + \tau) = P(\boldsymbol{\alpha} \mid \boldsymbol{\alpha}'; \tau), \tag{76}$$

a relation (true at time $t = t_0$, but not at all times) between two quantities of which one pertains to a non-equilibrium ensemble and the other to an equilibrium ensemble. In particular this relation will be valid if a measurement has indicated that the system is at a given initial time t_0 in a specified state in the range $(\boldsymbol{\alpha}_0, \boldsymbol{\alpha}_0 + \mathrm{d}\boldsymbol{\alpha}_0)$ and is therefore represented in phase space, according to the postulate of equal *a priori* probability, by an ensemble with uniform non-zero density for those points on the energy shell in phase space for which $\boldsymbol{\alpha}_0 \leqslant \boldsymbol{\alpha}(\boldsymbol{r}^N, \boldsymbol{p}^N) \leqslant \boldsymbol{\alpha}_0 + \mathrm{d}\boldsymbol{\alpha}_0$ and with zero density elsewhere.

We can interpret the quantity $P(\boldsymbol{\alpha} \mid \boldsymbol{\alpha}'; \tau)\, \mathrm{d}\boldsymbol{\alpha}'$ in a geometrical way as the ratio of two volumes in phase space, as was done for the quantity $f(\boldsymbol{\alpha})\, \mathrm{d}\boldsymbol{\alpha}$ [*cf.* eq. (7)]. Let $(\boldsymbol{\alpha}, \boldsymbol{\alpha} + \mathrm{d}\boldsymbol{\alpha})$ denote the region of the energy shell where $\boldsymbol{\alpha} \leqslant \boldsymbol{\alpha}(\boldsymbol{r}^N, \boldsymbol{p}^N) \leqslant \boldsymbol{\alpha} + \mathrm{d}\boldsymbol{\alpha}$ and let its volume be $\Omega(\boldsymbol{\alpha})$. Let furthermore $\Omega(\boldsymbol{\alpha}, \boldsymbol{\alpha}'; \tau)$ be the volume in phase space of that sub-region of $(\boldsymbol{\alpha}, \boldsymbol{\alpha} + \mathrm{d}\boldsymbol{\alpha})$ whose points move to the region $(\boldsymbol{\alpha}, \boldsymbol{\alpha}' + \mathrm{d}\boldsymbol{\alpha}')$ in the time τ:

$$\Omega(\boldsymbol{\alpha}, \boldsymbol{\alpha}'; \tau) = \int\limits_{\substack{(\boldsymbol{\alpha},\, \boldsymbol{\alpha}+\mathrm{d}\boldsymbol{\alpha}) \\ (E,\, E+\mathrm{d}E)}} \int\limits_{\substack{(\boldsymbol{\alpha}',\, \boldsymbol{\alpha}'+\mathrm{d}\boldsymbol{\alpha}') \\ (E,\, E+\mathrm{d}E)}} P(\boldsymbol{r}^N, \boldsymbol{p}^N \mid \boldsymbol{r}'^N, \boldsymbol{p}'^N; \tau)\, \mathrm{d}\boldsymbol{r}^N \mathrm{d}\boldsymbol{p}^N \mathrm{d}\boldsymbol{r}'^N \mathrm{d}\boldsymbol{p}'^N. \tag{77}$$

Then equation (69) becomes

$$P(\boldsymbol{\alpha} \mid \boldsymbol{\alpha}'; \tau)\, \mathrm{d}\boldsymbol{\alpha}' = \frac{\Omega(\boldsymbol{\alpha}, \boldsymbol{\alpha}'; \tau)}{\Omega(\boldsymbol{\alpha})}. \tag{78}$$

The following properties of $P(\boldsymbol{\alpha} \mid \boldsymbol{\alpha}';\tau)$ follow from (52), (53) and (66)–(69):

$$P(\boldsymbol{\alpha} \mid \boldsymbol{\alpha}' ; \tau) \geqslant 0 , \tag{79}$$

$$P(\boldsymbol{\alpha} \mid \boldsymbol{\alpha}' ; 0) = \delta(\boldsymbol{\alpha} - \boldsymbol{\alpha}') , \tag{80}$$

$$\int P(\boldsymbol{\alpha} \mid \boldsymbol{\alpha}' ; \tau) \, \mathrm{d}\boldsymbol{\alpha}' = 1 , \tag{81}$$

$$\int f(\boldsymbol{\alpha}) P(\boldsymbol{\alpha} \mid \boldsymbol{\alpha}' ; \tau) \, \mathrm{d}\boldsymbol{\alpha} = f(\boldsymbol{\alpha}') . \tag{82}$$

Furthermore $P(\boldsymbol{\alpha} \mid \boldsymbol{\alpha}';\tau)$ possesses the property:

$$f(\boldsymbol{\alpha}) P(\boldsymbol{\alpha} \mid \boldsymbol{\alpha}' ; \tau) = f(\boldsymbol{\alpha}') P(\boldsymbol{\alpha}' \mid \boldsymbol{\alpha} ; \tau) , \tag{83}$$

provided the $\boldsymbol{\alpha}$'s are even functions of the particle velocities. This is the important property of microscopic reversibility (principle of detailed balance), which can be established in the following way. It follows from (65) and (68) that

$$f(\boldsymbol{\alpha}) P(\boldsymbol{\alpha} \mid \boldsymbol{\alpha}' ; \tau) \, \mathrm{d}\boldsymbol{\alpha} \, \mathrm{d}\boldsymbol{\alpha}'$$
$$= \rho_0 \int\limits_{\substack{(\boldsymbol{\alpha},\, \boldsymbol{\alpha}+\mathrm{d}\boldsymbol{\alpha}) \\ (E,\, E+\mathrm{d}E)}} \int\limits_{\substack{(\boldsymbol{\alpha}',\, \boldsymbol{\alpha}'+\mathrm{d}\boldsymbol{\alpha}') \\ (E,\, E+\mathrm{d}E)}} P(\boldsymbol{r}^N, \boldsymbol{p}^N \mid \boldsymbol{r}'^N, \boldsymbol{p}'^N ; \tau) \, \mathrm{d}\boldsymbol{r}^N \mathrm{d}\boldsymbol{p}^N \mathrm{d}\boldsymbol{r}'^N \mathrm{d}\boldsymbol{p}'^N . \tag{84}$$

Applying (63) we obtain

$$f(\boldsymbol{\alpha}) P(\boldsymbol{\alpha} \mid \boldsymbol{\alpha}' ; \tau) \, \mathrm{d}\boldsymbol{\alpha} \, \mathrm{d}\boldsymbol{\alpha}'$$
$$= \rho_0 \int\limits_{\substack{(\boldsymbol{\alpha},\, \boldsymbol{\alpha}+\mathrm{d}\boldsymbol{\alpha}) \\ (E,\, E+\mathrm{d}E)}} \int\limits_{\substack{(\boldsymbol{\alpha}',\, \boldsymbol{\alpha}'+\mathrm{d}\boldsymbol{\alpha}') \\ (E,\, E+\mathrm{d}E)}} P(\boldsymbol{r}'^N, -\boldsymbol{p}'^N \mid \boldsymbol{r}^N, -\boldsymbol{p}^N ; \tau) \, \mathrm{d}\boldsymbol{r}^N \mathrm{d}\boldsymbol{p}^N \mathrm{d}\boldsymbol{r}'^N \mathrm{d}\boldsymbol{p}'^N . \tag{85}$$

With the transformation of variables $(\boldsymbol{r}^N, \boldsymbol{p}^N) \rightarrow (\boldsymbol{r}^N, -\boldsymbol{p}^N)$ and $(\boldsymbol{r}'^N, \boldsymbol{p}'^N) \rightarrow (\boldsymbol{r}'^N, -\boldsymbol{p}'^N)$ this becomes

$$f(\boldsymbol{\alpha}) P(\boldsymbol{\alpha} \mid \boldsymbol{\alpha}' ; \tau) \, \mathrm{d}\boldsymbol{\alpha} \, \mathrm{d}\boldsymbol{\alpha}'$$
$$= \rho_0 \int\limits_{\substack{(\tilde{\boldsymbol{\alpha}}',\, \tilde{\boldsymbol{\alpha}}'+\mathrm{d}\tilde{\boldsymbol{\alpha}}') \\ (E,\, E+\mathrm{d}E)}} \int\limits_{\substack{(\tilde{\boldsymbol{\alpha}},\, \tilde{\boldsymbol{\alpha}}+\mathrm{d}\tilde{\boldsymbol{\alpha}}) \\ (E,\, E+\mathrm{d}E)}} P(\boldsymbol{r}'^N, \boldsymbol{p}'^N \mid \boldsymbol{r}^N, \boldsymbol{p}^N ; \tau) \, \mathrm{d}\boldsymbol{r}'^N \mathrm{d}\boldsymbol{p}'^N \mathrm{d}\boldsymbol{r}^N \mathrm{d}\boldsymbol{p}^N . \tag{86}$$

where $(\tilde{\boldsymbol{\alpha}}, \tilde{\boldsymbol{\alpha}} + d\tilde{\boldsymbol{\alpha}})$ is that region of phase space into which the region $(\boldsymbol{\alpha}, \boldsymbol{\alpha} + d\boldsymbol{\alpha})$ is transformed by the operation $(\boldsymbol{r}^N, \boldsymbol{p}^N) \to (\boldsymbol{r}^N, -\boldsymbol{p}^N)$. (The energy shell $(E, E + dE)$ remains invariant under this transformation). Since the last member of (86) is equal to $f(\tilde{\boldsymbol{\alpha}}')P(\tilde{\boldsymbol{\alpha}}' \mid \tilde{\boldsymbol{\alpha}}; \tau)\, d\tilde{\boldsymbol{\alpha}}'\, d\tilde{\boldsymbol{\alpha}}$, we obtain the following relation:

$$f(\boldsymbol{\alpha})P(\boldsymbol{\alpha} \mid \boldsymbol{\alpha}'; \tau)\, d\boldsymbol{\alpha}\, d\boldsymbol{\alpha}' = f(\tilde{\boldsymbol{\alpha}}')P(\tilde{\boldsymbol{\alpha}}' \mid \tilde{\boldsymbol{\alpha}}; \tau)\, d\tilde{\boldsymbol{\alpha}}'\, d\tilde{\boldsymbol{\alpha}}\,. \tag{87}$$

The $\boldsymbol{\alpha}$-type variables are, in the absence of an external magnetic field, also even functions of the momenta, *i.e.*

$$\boldsymbol{\alpha}(\boldsymbol{r}^N, \boldsymbol{p}^N) = \boldsymbol{\alpha}(\boldsymbol{r}^N, -\boldsymbol{p}^N) = \tilde{\boldsymbol{\alpha}}(\boldsymbol{r}^N, \boldsymbol{p}^N)\,. \tag{88}$$

Therefore (87) becomes

$$f(\boldsymbol{\alpha})P(\boldsymbol{\alpha} \mid \boldsymbol{\alpha}'; \tau)\, d\boldsymbol{\alpha}\, d\boldsymbol{\alpha}' = f(\boldsymbol{\alpha}')P(\boldsymbol{\alpha}' \mid \boldsymbol{\alpha}; \tau)\, d\boldsymbol{\alpha}'\, d\boldsymbol{\alpha}\,, \tag{89}$$

thus establishing the property (83).

For $\boldsymbol{\beta}$-type variables one has, in the absence of an external magnetic field,

$$\boldsymbol{\beta}(\boldsymbol{r}^N, \boldsymbol{p}^N) = -\boldsymbol{\beta}(\boldsymbol{r}^N, -\boldsymbol{p}^N) = -\tilde{\boldsymbol{\beta}}(\boldsymbol{r}^N, \boldsymbol{p}^N)\,, \tag{90}$$

so that the principle of detailed balance for combined $\boldsymbol{\alpha}$- and $\boldsymbol{\beta}$-type variables can be shown to be expressed by

$$f(\boldsymbol{\alpha}, \boldsymbol{\beta})P(\boldsymbol{\alpha}, \boldsymbol{\beta} \mid \boldsymbol{\alpha}', \boldsymbol{\beta}'; \tau) = f(\boldsymbol{\alpha}', \boldsymbol{\beta}')P(\boldsymbol{\alpha}', -\boldsymbol{\beta}' \mid \boldsymbol{\alpha}, -\boldsymbol{\beta}; \tau)\,, \tag{91}$$

where use has been made of (21).

In the presence of an external magnetic field $\boldsymbol{B}$ the operation of time reversal implies, besides the transformation (60), the reversal of the magnetic field

$$\boldsymbol{B} \to -\boldsymbol{B}\,, \tag{92}$$

because, due to the form of the Lorentz force (which is proportional to the vector product of the velocities and the magnetic field strength), the particles only retrace their former paths if both the momenta and the magnetic field are reversed. A similar situation arises for Coriolis forces where the angular velocity vector $\boldsymbol{\omega}$ must be reversed. One then derives along the same lines the result:

$$f(\boldsymbol{\alpha},\boldsymbol{\beta}\,;\boldsymbol{B})P(\boldsymbol{\alpha},\boldsymbol{\beta}\,|\,\boldsymbol{\alpha}',\boldsymbol{\beta}'\,;\boldsymbol{B}\,;\tau)=f(\boldsymbol{\alpha}',\boldsymbol{\beta}',\boldsymbol{B})P(\boldsymbol{\alpha}',-\boldsymbol{\beta}'\,|\,\boldsymbol{\alpha},-\boldsymbol{\beta}\,;-\boldsymbol{B}\,;\tau), \quad (93)$$

where (33) has been used.

Relations (83), (91) and (93) are the basis for a derivation of the Onsager reciprocal relations, which will be given in the next section.

These results have been obtained here within the framework of classical mechanics. They can also be obtained if the motion of the particles is governed by quantum mechanical laws*.

§ 4. *Derivation of the Onsager Reciprocal Relations*

It is empirically known that for macroscopic values of $\boldsymbol{\alpha}$, *i.e.* for values of the α_i much larger than their root mean square values at equilibrium ($\boldsymbol{\alpha}\boldsymbol{\alpha} \gg k\mathsf{g}^{-1}$), the averages of these quantities frequently obey linear differential equations of first order of the type

$$\frac{\partial\, \overline{\boldsymbol{\alpha}}^{\boldsymbol{\alpha}_0}(t)}{\partial t} = -\,\mathsf{M}\cdot\overline{\boldsymbol{\alpha}}^{\boldsymbol{\alpha}_0}(t)\,, \quad (94)$$

where the (conditional) averages are defined by

$$\overline{\boldsymbol{\alpha}}^{\boldsymbol{\alpha}_0}(t) \equiv \int \boldsymbol{\alpha} P(\boldsymbol{\alpha}_0\,|\,\boldsymbol{\alpha}\,;t)\,\mathrm{d}\boldsymbol{\alpha}\,, \quad (95)$$

and where the $\boldsymbol{\alpha}_0$ are given initial values of $\boldsymbol{\alpha}$. The matrix M of real phenomenological coefficients is independent of time. Equations of the type (94) have been verified experimentally over a wide range of conditions**. The solution of (94) can be given in the form

$$\overline{\boldsymbol{\alpha}}^{\boldsymbol{\alpha}_0}(t) = \mathrm{e}^{-\mathsf{M}t}\cdot\boldsymbol{\alpha}_0\,, \quad (96)$$

where the matrix

* N. G. van Kampen, Physica **20** (1954) 603; Fortschritte der Physik **4** (1956) 405.

J. Vlieger, P. Mazur and S. R. de Groot, Physica **27** (1961) 353, 957, 974.

** In fact the macroscopic equations describing irreversible behaviour are often partial differential equations containing also derivatives of the state variables with respect to space coordinates (cf. Chapter IV, § 4). The reader is referred to Chapter VI, § 4 for the discussion of such a case in relation with the theory of the reciprocal relations.

Linear phenomenological equations may be of a more general type than (94) in still another way: the coefficients M need not be independent of time. We refer to Chapter VIII for a discussion of the consequences of time reversal invariance in this case.

$$e^{-\boldsymbol{M}t} \equiv \sum_{n=0}^{\infty} \frac{(-\boldsymbol{M}t)^n}{n!} \tag{97}$$

operates on $\boldsymbol{\alpha}_0$.

Multiplying (96) on the left with $f(\boldsymbol{\alpha}_0)\boldsymbol{\alpha}_0$ and integrating over $d\boldsymbol{\alpha}_0$, we obtain, using also the definition (95),

$$\int \boldsymbol{\alpha}_0 (e^{-\boldsymbol{M}t} \cdot \boldsymbol{\alpha}_0) f(\boldsymbol{\alpha}_0) \, d\boldsymbol{\alpha}_0 = \int\int \boldsymbol{\alpha}_0 \boldsymbol{\alpha} f(\boldsymbol{\alpha}_0) P(\boldsymbol{\alpha}_0 \mid \boldsymbol{\alpha}; t) \, d\boldsymbol{\alpha} \, d\boldsymbol{\alpha}_0 , \tag{98}$$

where $\boldsymbol{\alpha}_0(e^{-\boldsymbol{M}t} \cdot \boldsymbol{\alpha}_0)$ and $\boldsymbol{\alpha}_0\boldsymbol{\alpha}$ are dyadic products. Interchanging the dummy variables $\boldsymbol{\alpha}$ and $\boldsymbol{\alpha}_0$ on the right-hand side and applying microscopic reversibility (83), formula (98) may also be written as

$$\int \boldsymbol{\alpha}_0 (e^{-\boldsymbol{M}t} \cdot \boldsymbol{\alpha}_0) f(\boldsymbol{\alpha}_0) \, d\boldsymbol{\alpha}_0 = \int\int \boldsymbol{\alpha}\boldsymbol{\alpha}_0 f(\boldsymbol{\alpha}_0) P(\boldsymbol{\alpha}_0 \mid \boldsymbol{\alpha}; t) \, d\boldsymbol{\alpha}_0 \, d\boldsymbol{\alpha} . \tag{99}$$

With the use of (95) and (96) this becomes

$$\int \boldsymbol{\alpha}_0 (e^{-\boldsymbol{M}t} \cdot \boldsymbol{\alpha}_0) f(\boldsymbol{\alpha}_0) \, d\boldsymbol{\alpha}_0 = \int (e^{-\boldsymbol{M}t} \cdot \boldsymbol{\alpha}_0)\boldsymbol{\alpha}_0 f(\boldsymbol{\alpha}_0) \, d\boldsymbol{\alpha}_0 . \tag{100}$$

Employing the fluctuation formula (20) one obtains then

$$\boldsymbol{g}^{-1} \cdot e^{-\tilde{\boldsymbol{M}}t} = e^{-\boldsymbol{M}t} \cdot \boldsymbol{g}^{-1} , \tag{101}$$

where $\tilde{\boldsymbol{M}}$ is the transposed matrix of $\boldsymbol{M}$ ($\tilde{M}_{ik} = M_{ki}$). Since this relation must hold for all times t, we have

$$\boldsymbol{g}^{-1} \cdot \tilde{\boldsymbol{M}} = \boldsymbol{M} \cdot \boldsymbol{g}^{-1} . \tag{102}$$

With the definition

$$\boldsymbol{L} = \boldsymbol{M} \cdot \boldsymbol{g}^{-1} , \tag{103}$$

the relations (102) become

$$\tilde{\boldsymbol{L}} = \boldsymbol{L} , \tag{104}$$

since $\boldsymbol{g}$ is a symmetric matrix ($\boldsymbol{g} = \tilde{\boldsymbol{g}}$). This result, *viz.* that the matrix $\boldsymbol{L}$ is symmetric for the case of $\boldsymbol{\alpha}$-variables, and in the absence of an external magnetic field, is known as Onsager's reciprocity theorem. One calls (104) the reciprocal relations*.

* L. Onsager, Phys. Rev. **37** (1931) 405; **38** (1931) 2265.

If one introduces (103) into the regression equation (94), we obtain

$$\frac{\partial \overline{\boldsymbol{\alpha}}^{\boldsymbol{\alpha}_0}}{\partial t} = L \cdot \overline{\boldsymbol{X}}^{\boldsymbol{\alpha}_0}, \tag{105}$$

where the definition (16) of $\boldsymbol{X}$ has also been used.

Thus if the regression equations are written in the form (105), that is if the time derivatives of $\bar{\boldsymbol{\alpha}}^{\boldsymbol{\alpha}_0}$ are written as linear functions of the $\overline{\boldsymbol{X}^{\boldsymbol{\alpha}_0}}$, the matrix of phenomenological coefficients is symmetric according to (104). The variables $\boldsymbol{X}$ are conjugate to the $\boldsymbol{\alpha}$-variables according to [*cf.* (51)]

$$\boldsymbol{X} = \frac{\partial \Delta S}{\partial \boldsymbol{\alpha}}, \tag{106}$$

which establishes a connexion between the form (105) for which the symmetry (104) is valid, and thermodynamics.

In the course of deriving the reciprocal relations (104) the regression laws (94) or (96) have been assumed to hold also for small values of $\boldsymbol{\alpha}_0$, *i.e.* for initial states lying in the region of average equilibrium fluctuations ($\boldsymbol{\alpha}_0 \boldsymbol{\alpha}_0 \simeq kg^{-1}$). Indeed the main contributions to the averages of (98), (99) and (100) are due to small values of $\boldsymbol{\alpha}_0$. The hypothesis that the equations (94) or (96) are correct for such values of $\boldsymbol{\alpha}_0$ does not seem altogether unreasonable, although the limits of its validity can only be assessed on the basis of a purely microscopic theory*. We note that for small values of $\boldsymbol{\alpha}_0$ the regression equations lose their common meaning, namely that already for a single decay these laws should be verified because of an overwhelming probability to decay along the average path. However, in the above derivation the regression laws have only been used to describe the *average* behaviour of a system. In this connexion it is interesting to remark that Svedberg's and Westgren's experiments** on colloid statistics show that the behaviour of "small" density fluctuations is *on the average* in perfect agreement with the macroscopic law of diffusion.

* See *e.g.* E. P. Wigner, J. chem. Phys. **22** (1954) 1912;
H. B. G. Casimir, Rev. mod. Phys. **17** (1945) 343.

** Th. Svedberg, Z. phys. Chem. **77** (1911) 147;
A. Westgren, Arkiv för Matematik, Astronomi och Fysik **11** (1916) nrs. 8 and 14; **13** (1918) nr. 14.

For the case that also $\boldsymbol{\beta}$-variables must be considered the regression equations can be written in the form

$$\frac{\partial \bar{\boldsymbol{\alpha}}^{\alpha_0, \beta_0}}{\partial t} = \mathsf{L}_{\alpha\alpha} \cdot \overline{\boldsymbol{X}}^{\alpha_0, \beta_0} + \mathsf{L}_{\alpha\beta} \cdot \overline{\boldsymbol{Y}}^{\alpha_0, \beta_0} , \tag{107}$$

$$\frac{\partial \bar{\boldsymbol{\beta}}^{\alpha_0, \beta_0}}{\partial t} = \mathsf{L}_{\beta\alpha} \cdot \overline{\boldsymbol{X}}^{\alpha_0, \beta_0} + \mathsf{L}_{\beta\beta} \cdot \overline{\boldsymbol{Y}}^{\alpha_0, \beta_0} , \tag{108}$$

where $\boldsymbol{X}$ is given by (24) or (106) and $\boldsymbol{Y}$ by (25) or [*cf.* (34)] by

$$\boldsymbol{Y} = k \frac{\partial \ln f}{\partial \boldsymbol{\beta}} = \frac{\partial \Delta S}{\partial \boldsymbol{\beta}} \tag{109}$$

with

$$\Delta S = -\tfrac{1}{2} \mathsf{g} : \boldsymbol{\alpha\alpha} - \tfrac{1}{2} \mathsf{m} : \boldsymbol{\alpha\beta} - \tfrac{1}{2} \mathsf{n} : \boldsymbol{\beta\alpha} - \tfrac{1}{2} \mathsf{h} : \boldsymbol{\beta\beta} . \tag{110}$$

From microscopic reversibility in the form (93) one can derive with the help of (36)–(39), along the same lines as above, the following Onsager–Casimir reciprocal relations* for the phenomenological matrices $\mathsf{L}_{\alpha\alpha}$, $\mathsf{L}_{\alpha\beta}$, $\mathsf{L}_{\beta\alpha}$ and $\mathsf{L}_{\beta\beta}$, valid in the presence of an external magnetic field:

$$\mathsf{L}_{\alpha\alpha}(\boldsymbol{B}) = \tilde{\mathsf{L}}_{\alpha\alpha}(-\boldsymbol{B}) , \tag{111}$$

$$\mathsf{L}_{\alpha\beta}(\boldsymbol{B}) = -\tilde{\mathsf{L}}_{\beta\alpha}(-\boldsymbol{B}) , \tag{112}$$

$$\mathsf{L}_{\beta\beta}(\boldsymbol{B}) = \tilde{\mathsf{L}}_{\beta\beta}(-\boldsymbol{B}) . \tag{113}$$

Returning to formula (98), we note that the matrix of phenomenological coefficients M is given as a statistical average by

$$\mathrm{e}^{-\mathsf{M}t} \cdot \mathsf{g}^{-1} = k^{-1} \iint \boldsymbol{\alpha\alpha}_0 \, f(\boldsymbol{\alpha}_0) P(\boldsymbol{\alpha}_0 \mid \boldsymbol{\alpha} ; t) \, \mathrm{d}\boldsymbol{\alpha} \, \mathrm{d}\boldsymbol{\alpha}_0 , \tag{114}$$

where (20) has been used. From this expression we find for the matrix L of (103), again with (20) and the normalization condition (81),

$$\mathsf{L} = -k^{-1} \lim_{t \to 0} \frac{1}{t} \iint \Delta\boldsymbol{\alpha\alpha}_0 \, f(\boldsymbol{\alpha}_0) P(\boldsymbol{\alpha}_0 \mid \boldsymbol{\alpha} ; t) \, \mathrm{d}\boldsymbol{\alpha} \, \mathrm{d}\boldsymbol{\alpha}_0 , \tag{115}$$

where

$$\Delta\boldsymbol{\alpha} = \boldsymbol{\alpha} - \boldsymbol{\alpha}_0 . \tag{116}$$

* H. B. G. Casimir, loc. cit., p. 102.

Likewise for the case that also $\boldsymbol{\beta}$-variables are present we have for the matrices $L_{\alpha\alpha}$, $L_{\alpha\beta}$, $L_{\beta\alpha}$ and $L_{\beta\beta}$:

$$L_{\alpha\alpha} = -k^{-1} \lim_{t\to 0} \frac{1}{t} \iint \Delta\boldsymbol{\alpha}\boldsymbol{\alpha}_0 f(\boldsymbol{\alpha}_0, \boldsymbol{\beta}_0) P(\boldsymbol{\alpha}_0, \boldsymbol{\beta}_0 \mid \boldsymbol{\alpha}, \boldsymbol{\beta}\,; t)\, \mathrm{d}\boldsymbol{\alpha}\, \mathrm{d}\boldsymbol{\beta}\, \mathrm{d}\boldsymbol{\alpha}_0\, \mathrm{d}\boldsymbol{\beta}_0, \quad (117)$$

$$L_{\alpha\beta} = -k^{-1} \lim_{t\to 0} \frac{1}{t} \iint \Delta\boldsymbol{\alpha}\boldsymbol{\beta}_0 f(\boldsymbol{\alpha}_0, \boldsymbol{\beta}_0) P(\boldsymbol{\alpha}_0, \boldsymbol{\beta}_0 \mid \boldsymbol{\alpha}, \boldsymbol{\beta}\,; t)\, \mathrm{d}\boldsymbol{\alpha}\, \mathrm{d}\boldsymbol{\beta}\, \mathrm{d}\boldsymbol{\alpha}_0\, \mathrm{d}\boldsymbol{\beta}_0, \quad (118)$$

$$L_{\beta\alpha} = -k^{-1} \lim_{t\to 0} \frac{1}{t} \iint \Delta\boldsymbol{\beta}\boldsymbol{\alpha}_0 f(\boldsymbol{\alpha}_0, \boldsymbol{\beta}_0) P(\boldsymbol{\alpha}_0, \boldsymbol{\beta}_0 \mid \boldsymbol{\alpha}, \boldsymbol{\beta}\,; t)\, \mathrm{d}\boldsymbol{\alpha}\, \mathrm{d}\boldsymbol{\beta}\, \mathrm{d}\boldsymbol{\alpha}_0\, \mathrm{d}\boldsymbol{\beta}_0, \quad (119)$$

$$L_{\beta\beta} = -k^{-1} \lim_{t\to 0} \frac{1}{t} \iint \Delta\boldsymbol{\beta}\boldsymbol{\beta}_0 f(\boldsymbol{\alpha}_0, \boldsymbol{\beta}_0) P(\boldsymbol{\alpha}_0, \boldsymbol{\beta}_0 \mid \boldsymbol{\alpha}, \boldsymbol{\beta}\,; t)\, \mathrm{d}\boldsymbol{\alpha}\, \mathrm{d}\boldsymbol{\beta}\, \mathrm{d}\boldsymbol{\alpha}_0\, \mathrm{d}\boldsymbol{\beta}_0. \quad (120)$$

In physical applications the $\boldsymbol{\beta}$-type variables necessary to describe the state of the system can frequently be expressed as time derivatives of $\boldsymbol{\alpha}$-variables. Such variables may arise from the inertia of the system. We shall assume that to every $\boldsymbol{\alpha}$-variable a $\boldsymbol{\beta}$-variable is related according to

$$\frac{\mathrm{d}\boldsymbol{\alpha}}{\mathrm{d}t} = \boldsymbol{\beta}\,. \quad (121)$$

We shall discuss the case that no external magnetic field acts on the system. The increment $\Delta\boldsymbol{\alpha} = \boldsymbol{\alpha} - \boldsymbol{\alpha}_0$ which $\boldsymbol{\alpha}$ suffers during a short time Δt is therefore

$$\Delta\boldsymbol{\alpha} = \boldsymbol{\beta}_0\, \Delta t\,, \quad (122)$$

where $\boldsymbol{\beta}_0$ is the initial value of $\boldsymbol{\beta}$. Taking the (conditional) average of (122), we have

$$\lim_{\Delta t\to 0} \frac{\overline{\Delta\boldsymbol{\alpha}}^{\boldsymbol{\alpha}_0, \boldsymbol{\beta}_0}}{\Delta t} \equiv \lim_{\Delta t\to 0} \frac{1}{\Delta t} \iint \Delta\boldsymbol{\alpha} P(\boldsymbol{\alpha}_0, \boldsymbol{\beta}_0 \mid \boldsymbol{\alpha}, \boldsymbol{\beta}\,; \Delta t)\, \mathrm{d}\boldsymbol{\alpha}\, \mathrm{d}\boldsymbol{\beta} = \boldsymbol{\beta}_0\,. \quad (123)$$

Substituting this equation into (117), we obtain

$$L_{\alpha\alpha} = -k^{-1} \iint \boldsymbol{\beta}_0 \boldsymbol{\alpha}_0 f(\boldsymbol{\alpha}_0, \boldsymbol{\beta}_0)\, \mathrm{d}\boldsymbol{\alpha}_0\, \mathrm{d}\boldsymbol{\beta}_0 = 0\,, \quad (124)$$

because, in the absence of a magnetic field, no correlations exist between $\boldsymbol{\alpha}$- and $\boldsymbol{\beta}$-variables according to formula (32). Similarly substituting (123) into (118), one has

$$L_{\alpha\beta} = -k^{-1} \iint \boldsymbol{\beta}_0 \boldsymbol{\beta}_0 f(\boldsymbol{\alpha}_0, \boldsymbol{\beta}_0)\, \mathrm{d}\boldsymbol{\alpha}_0\, \mathrm{d}\boldsymbol{\beta}_0 = -h^{-1}, \tag{125}$$

according to (31). From (112) it then follows that

$$L_{\beta\alpha} = h^{-1}, \tag{126}$$

since h is symmetric. However, $L_{\beta\beta}$ does not reduce in a similar fashion to zero or an equilibrium quantity; it is the quantity determining the time behaviour of the system considered. With these results the regression equations reduce with (24) and (25) to

$$\frac{\partial \overline{\boldsymbol{\alpha}}^{\alpha_0, \beta_0}}{\partial t} = \overline{\boldsymbol{\beta}}^{\alpha_0, \beta_0}, \tag{127}$$

$$\frac{\partial \overline{\boldsymbol{\beta}}^{\alpha_0, \beta_0}}{\partial t} = -h^{-1} \cdot g \cdot \overline{\boldsymbol{\alpha}}^{\alpha_0, \beta_0} - L_{\beta\beta} \cdot h \cdot \overline{\boldsymbol{\beta}}^{\alpha_0, \beta_0}. \tag{128}$$

These two sets of first order differential equations can be written as a single set of second order differential equations in $\boldsymbol{\alpha}$:

$$h \cdot \frac{\partial^2 \overline{\boldsymbol{\alpha}}^{\alpha_0, \beta_0}}{\partial t^2} + h \cdot L_{\beta\beta} \cdot h \cdot \frac{\partial \overline{\boldsymbol{\alpha}}^{\alpha_0, \beta_0}}{\partial t} + g \cdot \overline{\boldsymbol{\alpha}}^{\alpha_0, \beta_0} = 0. \tag{129}$$

Frequently the first term (the "inertia term") may be neglected after a relatively short time. For a single variable α this is the case if $hL_{\beta\beta} \gg \sqrt{g/h}$ and $t \gg L_{\beta\beta}^{-1}\, h^{-1}$. Formula (129) then reduces to

$$\frac{\partial \overline{\boldsymbol{\alpha}}^{\alpha_0, \beta_0}}{\partial t} = L'_{\alpha\alpha} \cdot \overline{\mathbf{X}}^{\alpha_0, \beta_0}, \tag{130}$$

with

$$L'_{\alpha\alpha} = h^{-1} \cdot L_{\beta\beta}^{-1} \cdot h^{-1} \tag{131}$$

and the reciprocal relations

$$L'_{\alpha\alpha} = \tilde{L}'_{\alpha\alpha}, \tag{132}$$

which follow from (131) with (113) and the symmetric character of the matrix h. This is the previous result (104) for the case of α-variables only. In other words we have seen now that within a certain approximation β-variables are not required for the description of the state and the time behaviour of the system. We have, however, obtained a relation (131) between the coefficient matrix $L'_{\alpha\alpha}$ of the regression laws (130) and the matrix $L_{\beta\beta}$, characteristic for the regression laws (128) of the β-variables.

Depending on the physical situation it may happen that the number of β-variables (which are time derivatives of α-variables) required for the description of the system, is smaller than the number of α-variables. The reduction of the set of relations (111)–(113) can easily be performed for such a case also.

§ 5. *Further Properties of the Matrix of Phenomenological Coefficients*

In addition to the Onsager reciprocal relations, some further results concerning the matrices L and M of the preceding section can be derived. These properties follow from the definitions of these matrices as averages of quantities involving the α-variables.

We shall first prove that

$$\xi \cdot L \cdot \xi \geqslant 0\,, \tag{133}$$

where ξ is an arbitrary real n-dimensional vector and L the matrix defined by (103). The property (133) expresses the fact that the quadratic form in ξ_i $(i = 1, 2, \ldots, n)$ with coefficients L_{ij} $(i,j = 1, 2, \ldots, n)$ is positive definite.

Let us split L into a symmetric and an antisymmetric part L^s and L^a respectively:

$$L = L^s + L^a\,, \tag{134}$$

$$L^s = \tfrac{1}{2}L(+ \tilde{L})\,, \quad \tilde{L}^s = L^s, \tag{135}$$

$$L^a = \tfrac{1}{2}(L - \tilde{L})\,, \quad \tilde{L}^a = - L^a\,. \tag{136}$$

The quadratic form of (133) then reduces to

$$\xi \cdot L \cdot \xi = \xi \cdot L^s \cdot \xi\,, \tag{137}$$

since the expression $\xi \cdot L^a \cdot \xi$ vanishes. With (115) this quadratic form becomes

$$\boldsymbol{\xi}\cdot\mathsf{L}^s\cdot\boldsymbol{\xi} = -k^{-1}\lim_{t\to 0}\frac{1}{2t}\int \boldsymbol{\xi}\cdot(\Delta\boldsymbol{\alpha}\boldsymbol{\alpha}_0 + \boldsymbol{\alpha}_0\Delta\boldsymbol{\alpha})\cdot\boldsymbol{\xi}\, f(\boldsymbol{\alpha}_0)P(\boldsymbol{\alpha}_0\,|\,\boldsymbol{\alpha}\,;t)\,\mathrm{d}\boldsymbol{\alpha}\,\mathrm{d}\boldsymbol{\alpha}_0. \quad (138)$$

Consider now the expressions

$$\iint \boldsymbol{\alpha}_0\boldsymbol{\alpha}_0\, f(\boldsymbol{\alpha}_0)P(\boldsymbol{\alpha}_0\,|\,\boldsymbol{\alpha}\,;t)\,\mathrm{d}\boldsymbol{\alpha}_0\,\mathrm{d}\boldsymbol{\alpha} = \int \boldsymbol{\alpha}_0\boldsymbol{\alpha}_0\, f(\boldsymbol{\alpha}_0)\,\mathrm{d}\boldsymbol{\alpha}_0 = k\mathsf{g}^{-1}, \quad (139)$$

$$\iint \boldsymbol{\alpha}\boldsymbol{\alpha}\, f(\boldsymbol{\alpha}_0)P(\boldsymbol{\alpha}_0\,|\,\boldsymbol{\alpha}\,;t)\,\mathrm{d}\boldsymbol{\alpha}_0\,\mathrm{d}\boldsymbol{\alpha} = \int \boldsymbol{\alpha}\boldsymbol{\alpha}\, f(\boldsymbol{\alpha})\,\mathrm{d}\boldsymbol{\alpha} = k\mathsf{g}^{-1}, \quad (140)$$

where (20), (81) and (82) have been used. These relations allow us to write instead of (138)

$$\boldsymbol{\xi}\cdot\mathsf{L}^s\cdot\boldsymbol{\xi}$$
$$= -k^{-1}\lim_{t\to 0}\frac{1}{2t}\iint \boldsymbol{\xi}\cdot(\Delta\boldsymbol{\alpha}\boldsymbol{\alpha}_0 + \boldsymbol{\alpha}_0\Delta\boldsymbol{\alpha} - \boldsymbol{\alpha}\boldsymbol{\alpha} + \boldsymbol{\alpha}_0\boldsymbol{\alpha}_0)\cdot\boldsymbol{\xi}\, f(\boldsymbol{\alpha}_0)P(\boldsymbol{\alpha}_0|\,\boldsymbol{\alpha}\,;t)\,\mathrm{d}\boldsymbol{\alpha}_0\,\mathrm{d}\boldsymbol{\alpha}, \quad (141)$$

since the difference of the right-hand sides of (141) and (138) vanishes. As $\Delta\boldsymbol{\alpha} = \boldsymbol{\alpha} - \boldsymbol{\alpha}_0$, we obtain from the last relation

$$\boldsymbol{\xi}\cdot\mathsf{L}^s\cdot\boldsymbol{\xi} = k^{-1}\lim_{t\to 0}\frac{1}{2t}\iint (\boldsymbol{\xi}\cdot\Delta\boldsymbol{\alpha})^2\, f(\boldsymbol{\alpha}_0)P(\boldsymbol{\alpha}_0\,|\,\boldsymbol{\alpha}\,;t)\,\mathrm{d}\boldsymbol{\alpha}_0\,\mathrm{d}\boldsymbol{\alpha} \geqslant 0. \quad (142)$$

This inequality, which proves (133), follows from the fact that all factors of the integrand are positive.

The property (142) will prove to be sufficient to ensure that the solutions (96) of the differential equations (94) tend to zero as t approaches infinity:

$$\lim_{t\to\infty}\overline{\boldsymbol{\alpha}}^{\boldsymbol{\alpha}_0}(t) = \lim_{t\to\infty}\mathrm{e}^{-\mathsf{M}t}\cdot\boldsymbol{\alpha}_0 = 0, \quad (143)$$

in agreement with the idea that equations of the type (94) describe the regression of fluctuations.

In order to demonstrate this we first transform, with the help of a real matrix, to new variables $\boldsymbol{\alpha}'$, such that

$$\boldsymbol{\alpha}' = \mathsf{A}\cdot\boldsymbol{\alpha}, \quad \boldsymbol{\alpha} = \mathsf{A}^{-1}\cdot\boldsymbol{\alpha}'. \quad (144)$$

Since the distribution function $f(\boldsymbol{\alpha})\,\mathrm{d}\boldsymbol{\alpha}$ [*cf.* (14)] and therefore the quadratic positive definite form $g{:}\boldsymbol{\alpha\alpha}$ must remain invariant under such a linear transformation we have

$$\boldsymbol{\alpha}\cdot g\cdot\boldsymbol{\alpha} = \boldsymbol{\alpha}'\cdot g'\cdot\boldsymbol{\alpha}' , \tag{145}$$

with

$$g' = \tilde{A}^{-1}\cdot g\cdot A^{-1}, \tag{146}$$

a matrix which is seen to be symmetric and positive definite just as g. The transformation (146) from g to g' is called a congruent transformation. The matrix A can be chosen in such a way that g' becomes the unit matrix

$$g' = U , \quad g = \tilde{A}\cdot A. \tag{147}$$

(This is always possible for a congruent transformation of a symmetric positive definite matrix.)

Under the transformation (144) the regression equations (94) become

$$\frac{\partial\, \overline{\boldsymbol{\alpha}'}^{\boldsymbol{\alpha}'_0}}{\partial t} = - M'\cdot\overline{\boldsymbol{\alpha}'}^{\boldsymbol{\alpha}'_0} , \tag{148}$$

with the solution

$$\overline{\boldsymbol{\alpha}'}^{\boldsymbol{\alpha}'_0} = \mathrm{e}^{-M't}\cdot\boldsymbol{\alpha}'_0 , \tag{149}$$

where

$$M' = A\cdot M\cdot A^{-1} . \tag{150}$$

This transformation of the matrix M is called a similarity transformation. Finally the matrix L', which according to (103) and (147) is equal to M',

$$L' = M'\cdot(g')^{-1} = M'\cdot U = M' , \tag{151}$$

is obtained from L by means of a congruent transformation. This follows with the help of (146) and (150):

$$L' = M'\cdot(g')^{-1} = A\cdot M\cdot A^{-1}\cdot A\cdot g^{-1}\cdot\tilde{A} = A\cdot L\cdot\tilde{A} . \tag{152}$$

This result follows also immediately from (144) and the expression (115) for the matrix L which shows that it must transform as the dyadic $\Delta\boldsymbol{\alpha}\boldsymbol{\alpha}_0$.

Under the congruent transformation (152) (the symmetric part of) the matrix L remains positive definite so that

$$\xi \cdot (\mathsf{L}')^{s} \cdot \xi \geqslant 0 . \tag{153}$$

This can also be concluded at once from (142) which is valid for an arbitrary choice of variables $\boldsymbol{\alpha}$. From (151) and (153) we now find

$$\xi \cdot (\mathsf{M}')^{s} \cdot \xi = \xi \cdot (\mathsf{L}')^{s} \cdot \xi \geqslant 0 . \tag{154}$$

This property of the transformed matrix M' is therefore a direct consequence of the general property (133) of the matrix L.

It is convenient at this point to perform a second linear transformation of variables:

$$\boldsymbol{\alpha}'' = \mathsf{B} \cdot \boldsymbol{\alpha}' , \tag{155}$$

where B is the matrix which diagonalizes M', so that

$$\mathsf{B} \cdot \mathsf{M}' \cdot \mathsf{B}^{-1} = \Lambda ; \quad \Lambda_{ij} = \lambda_i \delta_{ij} . \tag{156}$$

The eigenvalues λ_i which are elements of the diagonal matrix Λ, are the n roots of the equation

$$| \mathsf{M}' - \lambda \mathsf{U} | = 0 . \tag{157}$$

Instead of (148) we have now for the $\boldsymbol{\alpha}''$ the differential equations

$$\frac{\partial \overline{\boldsymbol{\alpha}''}^{\boldsymbol{\alpha}_0''}}{\partial t} = - \Lambda \cdot \overline{\boldsymbol{\alpha}''}^{\boldsymbol{\alpha}_0''} \tag{158}$$

with the solution

$$\overline{\boldsymbol{\alpha}''}^{\boldsymbol{\alpha}_0''} = \mathrm{e}^{-\Lambda t} \cdot \boldsymbol{\alpha}_0'' . \tag{159}$$

or, in components,

$$\overline{\alpha_i''}^{\boldsymbol{\alpha}_0''} = \mathrm{e}^{-\lambda_i t} \, \alpha_{i0}'' , \quad (i = 1, 2, \ldots, n) . \tag{160}$$

It can now be shown, that, due to the inequality (154), the real parts of the eigenvalues λ_i $(i = 1, 2, \ldots, n)$ must be positive:

$$\lambda_i^{r} = \tfrac{1}{2}(\lambda_i + \lambda_i^{*}) > 0 . \tag{161}$$

Here λ_i^{*} is the complex conjugate of λ_i.

Proof of (161). The eigenvalues λ_i are the only values of the parameter λ for which the eigenvalue equation

$$\mathsf{M}' \cdot \boldsymbol{x} = \lambda \boldsymbol{x} \tag{162}$$

has non-vanishing solutions $\boldsymbol{x}$. The vectors $\boldsymbol{x}$ which satisfy (162) are called the eigenvectors of (162).

Multiply (162) on the left with the complex conjugate of $\boldsymbol{x}$:

$$\boldsymbol{x}^* \cdot \mathsf{M}' \cdot \boldsymbol{x} = \lambda \boldsymbol{x}^* \cdot \boldsymbol{x} \, . \tag{163}$$

Taking the complex conjugate of this equation one obtains

$$\boldsymbol{x} \cdot \mathsf{M}' \cdot \boldsymbol{x}^* = \boldsymbol{x}^* \cdot \tilde{\mathsf{M}}' \cdot \boldsymbol{x} = \lambda^* \boldsymbol{x}^* \cdot \boldsymbol{x} \, . \tag{164}$$

Adding the last two equations one has

$$\boldsymbol{x}^* \cdot (\mathsf{M}')^{\mathrm{s}} \cdot \boldsymbol{x} = \lambda^{\mathrm{r}} \boldsymbol{x}^* \cdot \boldsymbol{x} \, , \tag{165}$$

where λ^{r} is the real part of λ. Writing the complex vector $\boldsymbol{x}$ in the form

$$\boldsymbol{x} = \boldsymbol{\xi} + \mathrm{i} \boldsymbol{\eta} \, , \tag{166}$$

where $\boldsymbol{\xi}$ and $\boldsymbol{\eta}$ are real vectors, (165) becomes

$$\boldsymbol{\xi} \cdot (\mathsf{M}')^{\mathrm{s}} \cdot \boldsymbol{\xi} + \boldsymbol{\eta} \cdot (\mathsf{M}')^{\mathrm{s}} \cdot \boldsymbol{\eta} = \lambda^{\mathrm{r}} (\boldsymbol{\xi} \cdot \boldsymbol{\xi} + \boldsymbol{\eta} \cdot \boldsymbol{\eta}) \, . \tag{167}$$

According to (154) the left-hand side of this equation is positive definite, which proves (161), *i.e.* that λ_i^{r} must be a positive number.

With this result it follows that the solutions (159) must approach zero as t tends to infinity*:

* In principle the possibility that the real parts λ_i^{r} of the eigenvalues are zero cannot be excluded: this would occur if in (142) the right-hand side vanishes for all values of $\boldsymbol{\xi}$ (in the limit as $t \to 0$). Then (168) does not hold and the solutions of the phenomenological equations are periodic. This corresponds to reversible harmonic motions. We are, however, not concerned here with such reversible phenomena. Whether a given linear phenomenon will be macroscopically reversible or irreversible must follow from a microscopic theory based on the mechanism responsible for it. The main point of the results derived above is that the real parts λ_i^{r} of the eigenvalues may not be negative and that linear phenomena, for which the $\boldsymbol{\alpha}$-variables become larger and larger as $t \to \infty$, cannot occur.

$$\lim_{t\to\infty} \overline{\boldsymbol{\alpha}''}^{\boldsymbol{\alpha}''_0} = 0 . \tag{168}$$

Since the original variables $\boldsymbol{\alpha}$ are linear combinations of the $\boldsymbol{\alpha}''$, this result implies the property stated in formula (143).

Finally we note that under the transformation (155) the matrix $g' = U$ becomes

$$g'' = \tilde{B}^{-1} \cdot g' \cdot B^{-1} = \tilde{B}^{-1} \cdot B^{-1} , \tag{169}$$

in the same way as g' followed from g according to the transformation (146).

For the special case that the matrix L is symmetric due to Onsager's reciprocal relations, that is if only $\boldsymbol{\alpha}$-type variables are considered and no external magnetic field is present, the eigenvalues λ_i are real and the matrix g'' remains the unit matrix. This can be seen as follows: under the transformation of coordinates (144) the matrix L', into which L transforms, remains symmetric as is clear from (152). Therefore the matrix M', which according to (151) is equal to L', is also symmetric. It then follows from comparison of (163) and (164) that λ must be real. Furthermore, since M' is symmetric, it may be reduced to diagonal form with a real orthogonal matrix, that is with a matrix B which has the property

$$\tilde{B}^{-1} = B , \tag{170}$$

so that according to (169) g'' remains indeed the unit matrix. Thus in this special case the two matrices g and M (and therefore also L) can be diagonalized simultaneously.

The foregoing results are valid both for $\boldsymbol{\alpha}$- and $\boldsymbol{\beta}$-type variables, since they do not depend on the behaviour under time reversal of the variables considered.

§ 6. *Gaussian Markoff Processes*

In the preceding sections only those aspects of the processes $\boldsymbol{\alpha}(t)$ have been considered, which were of use in the derivation of the Onsager relations. In particular, besides some results derived from the principles of statistical mechanics, it was postulated that the empirical laws, valid for the mean regression of large fluctuations, also hold for the mean regression of small fluctuations.

On the other hand one may be interested in different aspects, such as the explicit form of the conditional probability density $P(\boldsymbol{\alpha}_0 \mid \boldsymbol{\alpha};t)$, which describes the processes in more detail*. One could then calculate the conditional average of quantities like the product $\boldsymbol{\alpha}\boldsymbol{\alpha}$, which occur in the definition of the entropy $S(\boldsymbol{\alpha})$ given by formula (49) of § 2. This would enable one to find explicitly how the average entropy changes in the course of time when the system was originally in a specified state $\boldsymbol{\alpha}_0$ (Chapter VII, § 8).

For a more detailed description of this sort it is necessary to make further assumptions about the nature of the process.

We shall first establish an integral relation obeyed by the conditional probability density in $\boldsymbol{\alpha}$-space. We have in the first place [*cf.* (72)]

$$\int f(\boldsymbol{\alpha}', t; \boldsymbol{\alpha}, t+\tau)\, \mathrm{d}\boldsymbol{\alpha}' = f(\boldsymbol{\alpha}, t+\tau) \tag{171}$$

and therefore, by introducing the conditional probability density $P(\boldsymbol{\alpha}', t \mid \boldsymbol{\alpha}, t+\tau)$,

$$f(\boldsymbol{\alpha}, t+\tau) = \int f(\boldsymbol{\alpha}', t) P(\boldsymbol{\alpha}', t \mid \boldsymbol{\alpha}, t+\tau)\, \mathrm{d}\boldsymbol{\alpha}'. \tag{172}$$

In particular,

$$f(\boldsymbol{\alpha}, t+\tau) = \int f(\boldsymbol{\alpha}'', 0) P(\boldsymbol{\alpha}'', 0 \mid \boldsymbol{\alpha}, t+\tau)\, \mathrm{d}\boldsymbol{\alpha}'' \tag{173}$$

and

$$f(\boldsymbol{\alpha}, t) = \int f(\boldsymbol{\alpha}'', 0) P(\boldsymbol{\alpha}'', 0 \mid \boldsymbol{\alpha}, t)\, \mathrm{d}\boldsymbol{\alpha}''. \tag{174}$$

Let us now suppose that

$$f(\boldsymbol{\alpha}, 0) = \delta(\boldsymbol{\alpha} - \boldsymbol{\alpha}_0) \tag{175}$$

and that the density in phase space $\rho(\boldsymbol{r}^N; \boldsymbol{p}^N; 0)$ is uniform in the region $(\boldsymbol{\alpha}_0, \boldsymbol{\alpha}_0 + \mathrm{d}\boldsymbol{\alpha}_0)$. Then equation (173) becomes

$$f(\boldsymbol{\alpha}, t+\tau) = P(\boldsymbol{\alpha}_0, 0 \mid \boldsymbol{\alpha}, t+\tau) = P(\boldsymbol{\alpha}_0 \mid \boldsymbol{\alpha}; t+\tau) \tag{176}$$

and (174)

$$f(\boldsymbol{\alpha}, t) = P(\boldsymbol{\alpha}_0, 0 \mid \boldsymbol{\alpha}, t) = P(\boldsymbol{\alpha}_0 \mid \boldsymbol{\alpha};\ t), \tag{177}$$

* M. S. Green, J. chem. Phys. **20** (1952) 1281.
N. Hashitsume, Progr. theor. Phys. **8** (1952) 461.

where (76) has been used. Substituting these results into (172) we obtain

$$P(\boldsymbol{\alpha}_0 \mid \boldsymbol{\alpha}; t+\tau) = \int P(\boldsymbol{\alpha}_0 \mid \boldsymbol{\alpha}'; t) P(\boldsymbol{\alpha}', t \mid \boldsymbol{\alpha}, t+\tau)\, \mathrm{d}\boldsymbol{\alpha}' . \tag{178}$$

This equation is an exact relation between the stationary and the non-stationary probability densities. The conditional probability density $P(\boldsymbol{\alpha}',t \mid \boldsymbol{\alpha},t+\tau)$ is connected with a distribution $\rho(\boldsymbol{r}^N, \boldsymbol{p}^N;t)$ in phase space which is not uniform over the region $(\boldsymbol{\alpha}', \boldsymbol{\alpha}'+\mathrm{d}\boldsymbol{\alpha}')$ and which has developed from the initial distribution $\rho(\boldsymbol{r}^N, \boldsymbol{p}^N;0)$.

We shall now assume that the non-stationary conditional probability density $P(\boldsymbol{\alpha}', t \mid \boldsymbol{\alpha}, t+\tau)$ in the relation (178) may be replaced by the stationary conditional probability density $P(\boldsymbol{\alpha}' \mid \boldsymbol{\alpha};\tau)$, so that this equation becomes the so-called Smoluchowski equation

$$P(\boldsymbol{\alpha}_0 \mid \boldsymbol{\alpha}; t+\tau) = \int P(\boldsymbol{\alpha}_0 \mid \boldsymbol{\alpha}'; t) P(\boldsymbol{\alpha}' \mid \boldsymbol{\alpha}; \tau)\, \mathrm{d}\boldsymbol{\alpha}' . \tag{179}$$

This means that we suppose that the density $\rho(\boldsymbol{r}^N, \boldsymbol{p}^N; t)$ remains "sufficiently" uniform in the course of time over regions $(\boldsymbol{\alpha}, \boldsymbol{\alpha}+\mathrm{d}\boldsymbol{\alpha})$ in phase space*. Processes, for which (179) is valid, are called Markoff processes.

We shall further assume that in short time intervals the variables $\boldsymbol{\alpha}$ change only by small amounts; more specifically we shall assume that

$$\lim_{\tau \to 0} \frac{\overline{\Delta\boldsymbol{\alpha}}^{\boldsymbol{\alpha}}}{\tau} = \lim_{\tau\to 0} \frac{1}{\tau} \int \Delta\boldsymbol{\alpha} P(\boldsymbol{\alpha} \mid \boldsymbol{\alpha}'; \tau)\, \mathrm{d}\boldsymbol{\alpha}' = -\mathsf{M}\cdot\boldsymbol{\alpha} , \tag{180}$$

$$\lim_{\tau \to 0} \frac{\overline{\Delta\boldsymbol{\alpha}\Delta\boldsymbol{\alpha}}^{\boldsymbol{\alpha}}}{\tau} = \lim_{\tau\to 0} \frac{1}{\tau} \int \Delta\boldsymbol{\alpha}\Delta\boldsymbol{\alpha} P(\boldsymbol{\alpha} \mid \boldsymbol{\alpha}'; \tau)\, \mathrm{d}\boldsymbol{\alpha}' = 2\mathsf{Q} , \tag{181}$$

$$\lim_{\tau \to 0} \frac{\overline{\Delta\boldsymbol{\alpha}\Delta\boldsymbol{\alpha} \ldots \Delta\boldsymbol{\alpha}}^{\boldsymbol{\alpha}}}{\tau} = 0 , \tag{182}$$

* See *e.g.* N. G. van Kampen, Fortschritte der Physik **4** (1956) 405. This assumption can certainly not be rigorously justified for all times. To which approximation this hypothesis is valid could only be assessed from a microscopic statistical treatment. Its validity is assumed here for time intervals which are macroscopically of interest.

where $\Delta\boldsymbol{\alpha} = \boldsymbol{\alpha}' - \boldsymbol{\alpha}$ and where in the last formula $\Delta\boldsymbol{\alpha}\Delta\boldsymbol{\alpha}\ldots\Delta\boldsymbol{\alpha}$ stands for an ordered product of more than two factors [*e.g.* $(\Delta\boldsymbol{\alpha}\Delta\boldsymbol{\alpha}\Delta\boldsymbol{\alpha})_{ijk} \to \Delta\alpha_i\Delta\alpha_j\Delta\alpha_k$]. The left-hand sides of (180)–(182) contain the "moments" of various order $\overline{\Delta\boldsymbol{\alpha}}^{\alpha}$, $\overline{\Delta\boldsymbol{\alpha}\Delta\boldsymbol{\alpha}}^{\alpha}$, ..., etc., of the changes $\Delta\boldsymbol{\alpha}$. These relations express that in the limit $\tau \to 0$ only the first two moments are proportional to τ. Of course (180) is nothing but the previously introduced regression law (94). The factor 2 in front of the constant symmetric matrix Q of (181) is introduced for convenience.

Consider now the integral

$$\int \frac{\partial P(\boldsymbol{\alpha}_0 \mid \boldsymbol{\alpha}\,;t)}{\partial t} R(\boldsymbol{\alpha})\, \mathrm{d}\boldsymbol{\alpha}\,, \tag{183}$$

where $R(\boldsymbol{\alpha})$ is an arbitrary function of the $\boldsymbol{\alpha}$'s, which goes to zero for $\boldsymbol{\alpha}$ tending to $\pm\infty$ sufficiently fast. Writing the time derivative as the limit of a difference quotient and using relation (179) one has

$$\begin{aligned}
&\int \frac{\partial P(\boldsymbol{\alpha}_0 \mid \boldsymbol{\alpha}\,;t)}{\partial t} R(\boldsymbol{\alpha})\, \mathrm{d}\boldsymbol{\alpha} \\
&\quad \equiv \lim_{\tau\to 0}\frac{1}{\tau}\int \{P(\boldsymbol{\alpha}_0 \mid \boldsymbol{\alpha}\,;t+\tau) - P(\boldsymbol{\alpha}_0 \mid \boldsymbol{\alpha}\,;t)\}R(\boldsymbol{\alpha})\,\mathrm{d}\boldsymbol{\alpha} \\
&\quad = \lim_{\tau\to 0}\frac{1}{\tau}\left\{\int\int P(\boldsymbol{\alpha}_0 \mid \boldsymbol{\alpha}\,;t)P(\boldsymbol{\alpha} \mid \boldsymbol{\alpha}'\,;\tau)R(\boldsymbol{\alpha}')\,\mathrm{d}\boldsymbol{\alpha}\,\mathrm{d}\boldsymbol{\alpha}' - \int P(\boldsymbol{\alpha}_0 \mid \boldsymbol{\alpha}\,;t)R(\boldsymbol{\alpha})\,\mathrm{d}\boldsymbol{\alpha}\right\},
\end{aligned} \tag{184}$$

where in the double integral dummy variables $\boldsymbol{\alpha}$ and $\boldsymbol{\alpha}'$ have been interchanged.

Developing $R(\boldsymbol{\alpha}')$ in a Taylor series around $\boldsymbol{\alpha}$ with respect to $\boldsymbol{\alpha}' - \boldsymbol{\alpha} = \Delta\boldsymbol{\alpha}$, we obtain

$$\begin{aligned}
&\int \frac{\partial P(\boldsymbol{\alpha}_0 \mid \boldsymbol{\alpha}\,;t)}{\partial t} R(\boldsymbol{\alpha})\, \mathrm{d}\boldsymbol{\alpha} \\
&\qquad = \int P(\boldsymbol{\alpha}_0 \mid \boldsymbol{\alpha}\,;t)\left\{-(\mathsf{M}\cdot\boldsymbol{\alpha})\cdot\frac{\partial}{\partial\boldsymbol{\alpha}}R(\boldsymbol{\alpha}) + \mathsf{Q} : \frac{\partial^2}{\partial\boldsymbol{\alpha}\partial\boldsymbol{\alpha}}R(\boldsymbol{\alpha})\right\}\mathrm{d}\boldsymbol{\alpha}\,,
\end{aligned} \tag{185}$$

where (180)–(182) and the normalization condition (81) have been used. After partial integration (185) gives

$$\int \left\{ \frac{\partial P}{\partial t} - \frac{\partial}{\partial \boldsymbol{\alpha}} \cdot P(\mathsf{M} \cdot \boldsymbol{\alpha}) - \frac{\partial^2}{\partial \boldsymbol{\alpha} \partial \boldsymbol{\alpha}} : P\mathsf{Q} \right\} R(\boldsymbol{\alpha}) \, \mathrm{d}\boldsymbol{\alpha} = 0 \, . \tag{186}$$

Since $R(\boldsymbol{\alpha})$ is an arbitrary function, we may conclude that

$$\frac{\partial P}{\partial t} = \tilde{\mathsf{M}} : \frac{\partial}{\partial \boldsymbol{\alpha}} P\boldsymbol{\alpha} + \mathsf{Q} : \frac{\partial^2}{\partial \boldsymbol{\alpha} \partial \boldsymbol{\alpha}} P = (\mathsf{M} : \mathsf{U})P + \mathsf{M} : \boldsymbol{\alpha} \frac{\partial P}{\partial \boldsymbol{\alpha}} + \mathsf{Q} : \frac{\partial^2}{\partial \boldsymbol{\alpha} \partial \boldsymbol{\alpha}} P \, , \tag{187}$$

where $\tilde{\mathsf{M}}$ is the transposed matrix of M, and U the unit matrix.

Thus it follows from the assumptions (180)–(182) that P must satisfy the partial differential equation (187), which is the so-called Fokker–Planck equation. This equation must now be solved with the initial condition [*cf.* (80)]

$$P(\boldsymbol{\alpha}_0 \mid \boldsymbol{\alpha} \, ; 0) = \delta(\boldsymbol{\alpha} - \boldsymbol{\alpha}_0) \, . \tag{188}$$

Since according to (82) we have

$$\int f(\boldsymbol{\alpha}_0) P(\boldsymbol{\alpha}_0 \mid \boldsymbol{\alpha} \, ; t) \, \mathrm{d}\boldsymbol{\alpha}_0 = f(\boldsymbol{\alpha}) \, , \tag{189}$$

it follows by multiplying (187) with $f(\boldsymbol{\alpha}_0)$ and integrating over $\boldsymbol{\alpha}_0$, that $f(\boldsymbol{\alpha})$ is a stationary solution of the Fokker–Planck equation:

$$(\mathsf{M} : \mathsf{U}) f + \mathsf{M} : \boldsymbol{\alpha} \frac{\partial f}{\partial \boldsymbol{\alpha}} + \mathsf{Q} : \frac{\partial^2 f}{\partial \boldsymbol{\alpha} \partial \boldsymbol{\alpha}} = 0 \, . \tag{190}$$

Inserting the explicit form (14) of the distribution function $f(\boldsymbol{\alpha})$ into this equation, it becomes

$$(\mathsf{Q} - k\mathsf{M} \cdot g^{-1}) : (k^{-2} g \cdot \boldsymbol{\alpha}\boldsymbol{\alpha} \cdot g - k^{-1} g) = 0 \, . \tag{191}$$

This is a relation which the second moments Q must satisfy. Since (191) must hold for all values of $\boldsymbol{\alpha}$ and since the second factor is a symmetric matrix, it follows that the symmetric part of the first factor must vanish. This gives, because Q is symmetric,

$$\mathsf{Q} = \tfrac{1}{2} k (\mathsf{M} \cdot g^{-1} + g^{-1} \cdot \tilde{\mathsf{M}}) \, , \tag{192}$$

or, with the coefficients L of (103),

$$\mathsf{Q} = \tfrac{1}{2} k (\mathsf{L} + \tilde{\mathsf{L}}) \, . \tag{193}$$

These relations determine the matrix Q uniquely. We have in this way determined the non-vanishing second moments (181).

We shall now solve the differential equation (187) with (188) and the result (192). Let us introduce new variables $\boldsymbol{\alpha}$:

$$\boldsymbol{a} = \mathrm{e}^{\boldsymbol{M}t}\cdot\boldsymbol{\alpha}\,. \tag{194}$$

The Fokker–Planck equation (187) then becomes

$$\frac{\partial P}{\partial t} = (\mathsf{M} : \mathsf{U})P + (\mathrm{e}^{\boldsymbol{M}t}\cdot \mathsf{Q}\cdot \mathrm{e}^{\tilde{\boldsymbol{M}}t}) : \frac{\partial^2 P}{\partial\boldsymbol{a}\partial\boldsymbol{a}}\,, \tag{195}$$

with P now a function of $\boldsymbol{a}_0$, $\boldsymbol{a}$ and t, and the initial condition

$$P(\boldsymbol{a}_0 \mid \boldsymbol{a}\,; 0) = \delta(\boldsymbol{a} - \boldsymbol{a}_0)\,. \tag{196}$$

The Fourier transform of P:

$$A(\boldsymbol{a}_0, \boldsymbol{\omega}\,; t) = \left(\frac{1}{2\pi}\right)^n \int P(\boldsymbol{a}_0 \mid \boldsymbol{a}\,; t)\,\mathrm{e}^{-\mathrm{i}\boldsymbol{\omega}\cdot\boldsymbol{a}}\,\mathrm{d}\boldsymbol{a} \tag{197}$$

then satisfies the equation

$$\frac{\partial A}{\partial t} = \{\, \mathsf{M} : \mathsf{U} - (\mathrm{e}^{\boldsymbol{M}t}\cdot \mathsf{Q}\cdot \mathrm{e}^{\tilde{\boldsymbol{M}}t}) : \boldsymbol{\omega}\boldsymbol{\omega}\,\}\,A\,, \tag{198}$$

with the initial condition

$$A(\boldsymbol{a}_0, \boldsymbol{\omega}\,; 0) = \frac{1}{(2\pi)^n}\,\mathrm{e}^{-\mathrm{i}\boldsymbol{\omega}\cdot\boldsymbol{a}_0}\,. \tag{199}$$

The solution of (198) is

$$A(\boldsymbol{a}_0, \boldsymbol{\omega}\,; t) = \frac{1}{(2\pi)^n}\exp\left\{-\mathrm{i}\boldsymbol{\omega}\cdot\boldsymbol{a}_0 + (\mathsf{M} : \mathsf{U})t - \left(\int_0^t \mathrm{e}^{\boldsymbol{M}\xi}\cdot \mathsf{Q}\cdot \mathrm{e}^{\tilde{\boldsymbol{M}}\xi}\,\mathrm{d}\xi\right) : \boldsymbol{\omega}\boldsymbol{\omega}\right\}. \tag{200}$$

Carrying out the integration in the exponent with the help of (192) we obtain

$$\tfrac{1}{2}k\int_0^t \mathrm{e}^{\boldsymbol{M}\xi}\cdot(\mathsf{M}\cdot g^{-1} + g^{-1}\cdot\tilde{\mathsf{M}})\cdot\mathrm{e}^{\tilde{\boldsymbol{M}}\xi}\mathrm{d}\xi = \tfrac{1}{2}k\int_0^t \frac{\mathrm{d}}{\mathrm{d}\xi}(\mathrm{e}^{\boldsymbol{M}\xi}\cdot g^{-1}\cdot\mathrm{e}^{\tilde{\boldsymbol{M}}\xi})\,\mathrm{d}\xi$$

$$= \tfrac{1}{2}k(\mathrm{e}^{\boldsymbol{M}t}\cdot g^{-1}\cdot\mathrm{e}^{\tilde{\boldsymbol{M}}t} - g^{-1}) \equiv \tfrac{1}{2}kw^{-1}\,. \tag{201}$$

The auxiliary symmetric matrix w^{-1} is defined by the last equality. Introducing this result into (200) we have

$$A(\boldsymbol{a}_0, \boldsymbol{\omega}\,; t) = \frac{1}{(2\pi)^n} \exp\{-i\boldsymbol{\omega}\cdot\boldsymbol{a}_0 + (\mathsf{M}:\mathsf{U})t - \tfrac{1}{2}k\mathsf{w}^{-1}:\boldsymbol{\omega\omega}\}. \tag{202}$$

With Fourier's theorem it then follows from (197) and (202) that

$$P(\boldsymbol{a}_0 \mid \boldsymbol{a}\,; t) = \int A(\boldsymbol{a}_0, \boldsymbol{\omega}\,; t)\, e^{i\boldsymbol{\omega}\cdot\boldsymbol{a}}\, d\boldsymbol{\omega}$$

$$= \frac{1}{(2\pi)^n} e^{(\mathsf{M}:\mathsf{U})t} \int e^{i\boldsymbol{\omega}\cdot(\boldsymbol{a}-\boldsymbol{a}_0) - \frac{1}{2}k\mathsf{w}^{-1}:\boldsymbol{\omega\omega}}\, d\boldsymbol{\omega}$$

$$= \frac{1}{(2\pi)^n} e^{(\mathsf{M}:\mathsf{U})t - \frac{1}{2}k^{-1}\mathsf{w}:(\boldsymbol{a}-\boldsymbol{a}_0)(\boldsymbol{a}-\boldsymbol{a}_0)} \int e^{-\frac{1}{2}k\mathsf{w}^{-1}:\{\boldsymbol{\omega} - ik^{-1}\mathsf{w}\cdot(\boldsymbol{a}-\boldsymbol{a}_0)\}\{\boldsymbol{\omega} - ik^{-1}\mathsf{w}\cdot(\boldsymbol{a}-\boldsymbol{a}_0)\}}\, d\boldsymbol{\omega}$$

$$= \sqrt{\frac{|\mathsf{w}|}{(2\pi k)^n}}\, e^{(\mathsf{M}:\mathsf{U})t}\, e^{-\frac{1}{2}k^{-1}\mathsf{w}:(\boldsymbol{a}-\boldsymbol{a}_0)(\boldsymbol{a}-\boldsymbol{a}_0)}. \tag{203}$$

Transforming back to the original variables $\boldsymbol{\alpha}$ with the help of (194) we find

$$P(\boldsymbol{\alpha}_0 \mid \boldsymbol{\alpha}\,; t) = \sqrt{\frac{|\mathsf{v}|}{(2\pi k)^n}}\, e^{-\frac{1}{2}k^{-1}\mathsf{v}:(\boldsymbol{\alpha} - \overline{\boldsymbol{\alpha}}^{\boldsymbol{\alpha}_0})(\boldsymbol{\alpha} - \overline{\boldsymbol{\alpha}}^{\boldsymbol{\alpha}_0})}, \tag{204}$$

where

$$\overline{\boldsymbol{\alpha}}^{\boldsymbol{\alpha}_0} = e^{-\mathsf{M}t}\cdot\boldsymbol{\alpha}_0, \tag{205}$$

and

$$\mathsf{v}^{-1} \equiv e^{-\mathsf{M}t}\cdot\mathsf{w}^{-1}\cdot e^{-\tilde{\mathsf{M}}t} = \mathsf{g}^{-1} - e^{-\mathsf{M}t}\cdot\mathsf{g}^{-1}\cdot e^{-\tilde{\mathsf{M}}t}. \tag{206}$$

The last member of (206) follows with (201). The factor $\sqrt{|\mathsf{v}|/(2\pi k)^n}$ in (204) can be found from the normalization (81)*.

Thus we have found a Gaussian distribution (204) with "first moments" (205), and "variances"

$$\overline{(\boldsymbol{\alpha} - \overline{\boldsymbol{\alpha}}^{\boldsymbol{\alpha}_0})(\boldsymbol{\alpha} - \overline{\boldsymbol{\alpha}}^{\boldsymbol{\alpha}_0})}^{\boldsymbol{\alpha}_0} = k\mathsf{v}^{-1} = k(\mathsf{g}^{-1} - e^{-\mathsf{M}t}\cdot\mathsf{g}^{-1}\cdot e^{-\tilde{\mathsf{M}}t}), \tag{207}$$

* This factor can also be found directly in the course of the transformation from (203) to (204) using the identity

$$|e^{\mathsf{M}t}| = e^{(\mathsf{M}:\mathsf{U})t}.$$

which follow with (204) in the same way as the variances (20) from the distribution (14).

The foregoing treatment shows that the assumption (179) on the Markoffian character of the process with the conditions (180)–(182) on the moments lead to a Gaussian conditional probability density $P(\boldsymbol{\alpha}_0 \mid \boldsymbol{\alpha};t)$. A process of this kind is called a "Gaussian Markoff process".

It should, at this point, be stressed that real irreversible processes can very well have non-Gaussian character, even though they obey linear regression laws. Such processes do not satisfy all the conditions (179)–(182). On the other hand the Gaussian Markoff processes, based on these conditions, may not only serve as a useful illustration of irreversible behaviour, but could also represent a good approximation to a class of irreversible processes*.

In a description with coordinates $\boldsymbol{\alpha}''$ (*cf.* § 5)

$$\boldsymbol{\alpha}'' = \mathsf{B}\cdot\mathsf{A}\cdot\boldsymbol{\alpha}\,, \tag{208}$$

for which the matrix M is diagonal, $P(\boldsymbol{\alpha}''_0 \mid \boldsymbol{\alpha}'';t)$ is a Gaussian distribution with first moments

$$\overline{\boldsymbol{\alpha}''}^{\boldsymbol{\alpha}''_0} = \mathrm{e}^{-\Lambda t}\cdot\boldsymbol{\alpha}''_0\,, \tag{209}$$

and variances

$$\overline{(\boldsymbol{\alpha}'' - \overline{\boldsymbol{\alpha}''}^{\boldsymbol{\alpha}''_0})\,(\boldsymbol{\alpha}'' - \overline{\boldsymbol{\alpha}''}^{\boldsymbol{\alpha}''_0})}^{\boldsymbol{\alpha}''_0} = k(\mathsf{v}'')^{-1} = k\,\{\,(\mathsf{g}'')^{-1} - \mathrm{e}^{-\Lambda t}\cdot(\mathsf{g}'')^{-1}\cdot\mathrm{e}^{-\tilde{\Lambda} t}\}\,. \tag{210}$$

Since it follows from these formulae and the fact that the real part of the elements of Λ are positive [*cf.* § 5, equation (161)] that

$$\lim_{t\to\infty} \overline{\boldsymbol{\alpha}''}^{\boldsymbol{\alpha}''_0} = 0 \tag{211}$$

and

$$\lim_{t\to\infty} (\mathsf{v}'')^{-1} = (\mathsf{g}'')^{-1}\,, \tag{212}$$

it may be concluded that

$$\lim_{t\to\infty} P(\boldsymbol{\alpha}''_0 \mid \boldsymbol{\alpha}''\,;t) = f\,(\boldsymbol{\alpha}'') \tag{213}$$

* L. Onsager and S. Machlup, Phys. Rev. **91** (1953) 1505.

and in a description with the original coordinates $\boldsymbol{\alpha}$

$$\lim_{t\to\infty} P(\boldsymbol{\alpha}_0 \mid \boldsymbol{\alpha}\,; t) = f(\boldsymbol{\alpha})\,. \tag{214}$$

This result makes again apparent the irreversible nature of the processes described. According to formula (210) the variances, which are zero initially and tend to their equilibrium values, as t tends to infinity, are at all times of the order of magnitude of the latter values. This means that the distribution $P(\boldsymbol{\alpha}_0 \mid \boldsymbol{\alpha};t)$ remains sharply peaked in the course of time. Therefore from the macroscopic point of view, that is if one does not measure any deviations of the order of the equilibrium fluctuations, there is an overwhelming probability for any macroscopic deviation $\boldsymbol{\alpha}_0$ to decay on a single occasion already according to the average linear laws. This is in agreement with the phenomenological laws of macroscopic physics.

The theory discussed in the present section applies to both $\boldsymbol{\alpha}$- and $\boldsymbol{\beta}$-variables.

§ 7. *Gaussian Markoff Processes: Langevin Equations*

The set of assumptions (179)–(182) can be replaced by a different but equivalent set of conditions, which demonstrate more explicitly the connexion between the present theory of random fluctuations and the theory of Brownian motion. Equations, analogous to the Langevin equation used to describe the motion of a Brownian particle,

$$\frac{\mathrm{d}\boldsymbol{\alpha}}{\mathrm{d}t} = -\,\mathsf{M}\cdot\boldsymbol{\alpha} + \boldsymbol{\varepsilon}(t)\,, \tag{215}$$

are then taken as starting points*. Such equations, which in the theory of Brownian motion hold for the velocity of a Brownian particle, are supposed to be obeyed here by the vector $\boldsymbol{\alpha}$ of which the components are the variables α_i $(i = 1, 2, \ldots, n)$. The vector $\boldsymbol{\varepsilon}(t)$ represents a random or stochastic "force" term. Averages of products of the components of $\boldsymbol{\varepsilon}$ must furthermore satisfy the following set of requirements:

$$\overline{\varepsilon_i(t)} = 0\,, \quad (i = 1, 2, \ldots, n)\,, \tag{216}$$

* L. Onsager and S. Machlup, Phys. Rev. **91** (1953) 1505.
S. Machlup and L. Onsager, Phys. Rev. **91** (1953) 1512.

$$\overline{\varepsilon_{i_1}(t_1)\,\varepsilon_{i_2}(t_2)} = 2Q_{i_1 i_2}\,\delta(t_1 - t_2)\,, \quad (i_1, i_2 = 1, 2, \ldots, n)\,, \tag{217}$$

$$\overline{\varepsilon_{i_1}(t_1)\,\varepsilon_{i_2}(t_2)\ldots\varepsilon_{i_{2s-1}}(t_{2s-1})} = 0\,, \quad (i_1, i_2, \ldots, i_{2s-1} = 1, 2, \ldots, n)\,, \tag{218}$$

$$\overline{\varepsilon_{i_1}(t_1)\,\varepsilon_{i_2}(t_2)\ldots\varepsilon_{i_{2s}}(t_{2s})} = \sum_{\text{all pairs}} \overline{\varepsilon_{i_p}(t_p)\,\varepsilon_{i_q}(t_q)}\;\overline{\varepsilon_{i_u}(t_u)\,\varepsilon_{i_v}(t_v)}\ldots,$$
$$(i_1, i_2, \ldots, i_{2s} = 1, 2, \ldots, n)\,, \tag{219}$$

where the sum has to be taken over all different ways in which one can divide the $2s$ time points $t_1, t_2, \ldots, t_{2s}$ into pairs. The matrix Q with elements $Q_{i_1 i_2}$ is seen to be symmetric. The conditions (216)–(219) imply that values of the random vectors $\boldsymbol{\varepsilon}(t)$ at different times are wholly uncorrelated. We shall show that these starting points lead to the same result as previously obtained (§ 6).

The formal solution of (215) is

$$\boldsymbol{\alpha}(t) = \mathrm{e}^{-\boldsymbol{M}t}\cdot\boldsymbol{\alpha}_0 + \mathrm{e}^{-\boldsymbol{M}t}\cdot\int_0^t \mathrm{e}^{\boldsymbol{M}\xi}\cdot\boldsymbol{\varepsilon}(\xi)\,\mathrm{d}\xi\,, \tag{220}$$

as can be verified by substitution. With the help of (216)–(219) it then follows that

$$\overline{\boldsymbol{\alpha}}^{\boldsymbol{\alpha}_0} = \mathrm{e}^{-\boldsymbol{M}t}\cdot\boldsymbol{\alpha}_0\,, \tag{221}$$

$$\overline{(\boldsymbol{\alpha} - \overline{\boldsymbol{\alpha}}^{\boldsymbol{\alpha}_0})\,(\boldsymbol{\alpha} - \overline{\boldsymbol{\alpha}}^{\boldsymbol{\alpha}_0})}^{\boldsymbol{\alpha}_0} = 2\int_0^t \mathrm{e}^{-\boldsymbol{M}(t-\xi)}\cdot\boldsymbol{Q}\cdot\mathrm{e}^{-\tilde{\boldsymbol{M}}(t-\xi)}\,\mathrm{d}\xi\,, \tag{222}$$

$$\overline{(\boldsymbol{\alpha} - \overline{\boldsymbol{\alpha}}^{\boldsymbol{\alpha}_0})_{i_1}\,(\boldsymbol{\alpha} - \overline{\boldsymbol{\alpha}}^{\boldsymbol{\alpha}_0})_{i_2}\ldots(\boldsymbol{\alpha} - \overline{\boldsymbol{\alpha}}^{\boldsymbol{\alpha}_0})_{i_{2s-1}}}^{\boldsymbol{\alpha}_0} = 0\,, \quad (i_1, i_2, \ldots, i_{2s-1} = 1, 2, \ldots, n)\,, \tag{223}$$

$$\overline{(\boldsymbol{\alpha} - \overline{\boldsymbol{\alpha}}^{\boldsymbol{\alpha}_0})_{i_1}\,(\boldsymbol{\alpha} - \overline{\boldsymbol{\alpha}}^{\boldsymbol{\alpha}_0})_{i_2}\ldots(\boldsymbol{\alpha} - \overline{\boldsymbol{\alpha}}^{\boldsymbol{\alpha}_0})_{i_{2s}}}^{\boldsymbol{\alpha}_0}$$
$$= \sum_{\text{all pairs}} \overline{(\boldsymbol{\alpha} - \overline{\boldsymbol{\alpha}}^{\boldsymbol{\alpha}_0})_{i_p}\,(\boldsymbol{\alpha} - \overline{\boldsymbol{\alpha}}^{\boldsymbol{\alpha}_0})_{i_q}}^{\boldsymbol{\alpha}_0}\;\overline{(\boldsymbol{\alpha} - \overline{\boldsymbol{\alpha}}^{\boldsymbol{\alpha}_0})_{i_u}\,(\boldsymbol{\alpha} - \overline{\boldsymbol{\alpha}}^{\boldsymbol{\alpha}_0})_{i_v}}^{\boldsymbol{\alpha}_0}\ldots,$$
$$(i_1, i_2, \ldots, i_{2s} = 1, 2, \ldots, n)\,. \tag{224}$$

In the limit of infinitely short times the results (221)–(224) are equivalent to the assumptions (180)–(182) of § 6. These latter assumptions complemented there the hypothesis on the Markoff character (179), which does not have explicitly to be introduced in the present treatment.

For convenience we introduce a symmetric matrix G defined by

$$2Q = k(M \cdot G^{-1} + G^{-1} \cdot \tilde{M}) . \tag{225}$$

Then the variances (222) become

$$\overline{(\alpha - \overline{\alpha}^{\alpha_0})(\alpha - \overline{\alpha}^{\alpha_0})}^{\alpha_0} = k \int_0^t \frac{d}{d\xi} \{ e^{-M(t-\xi)} \cdot G^{-1} \cdot e^{-\tilde{M}(t-\xi)} \} d\xi$$

$$= k(G^{-1} - e^{-Mt} \cdot G^{-1} \cdot e^{-\tilde{M}t}) \equiv kV^{-1} , \tag{226}$$

where the matrix V is defined by the last equality.

Consider now the integral

$$\int_{-\infty}^{+\infty} e^{i\omega \cdot (\alpha - \overline{\alpha}^{\alpha_0})} P(\alpha_0 \mid \alpha ; t) \, d\alpha = \overline{e^{i\omega \cdot (\alpha - \overline{\alpha}^{\alpha_0})}}^{\alpha_0}$$

$$= \sum_{m=0}^{\infty} \frac{i^m}{m!} \overline{\{ \omega \cdot (\alpha - \overline{\alpha}^{\alpha_0}) \}^m}^{\alpha_0} . \tag{227}$$

Due to (223) all odd powers in (227) vanish. The even powers may be expressed with (223) in terms of the variances. This gives

$$\int_{-\infty}^{+\infty} e^{i\omega \cdot (\alpha - \overline{\alpha}^{\alpha_0})} P(\alpha_0 \mid \alpha ; t) \, d\alpha$$

$$= \sum_{s=0}^{\infty} \frac{i^{2s}}{(2s)!} \frac{(2s)!}{2^s s!} \{ \overline{(\alpha - \overline{\alpha}^{\alpha_0})(\alpha - \overline{\alpha}^{\alpha_0})}^{\alpha_0} : \omega\omega \}^s , \tag{228}$$

where $(2s)!/2^s s!$ is the number of ways in which $2s$ elements can be divided into pairs. With (226) we finally obtain

$$\int_{-\infty}^{+\infty} e^{i\omega \cdot (\alpha - \overline{\alpha}^{\alpha_0})} P(\alpha_0 \mid \alpha ; t) \, d\alpha = \sum_{s=0}^{\infty} \frac{1}{s!} (-\tfrac{1}{2} kV^{-1} : \omega\omega)^s = e^{-\frac{1}{2} kV^{-1} : \omega\omega} . \tag{229}$$

With Fourier's theorem we then find

$$P(\boldsymbol{\alpha}_0 \mid \boldsymbol{\alpha}\,;t) = \frac{1}{(2\pi)^n}\int \mathrm{e}^{-\frac{1}{2}kV^{-1}:\boldsymbol{\omega}\boldsymbol{\omega}-\mathrm{i}(\boldsymbol{\alpha}-\overline{\boldsymbol{\alpha}}^{\boldsymbol{\alpha}_0})\cdot\boldsymbol{\omega}}\,\mathrm{d}\boldsymbol{\omega}$$

$$= \sqrt{\frac{|V|}{(2\pi k)^n}}\,\mathrm{e}^{-\frac{1}{2}k^{-1}V:(\boldsymbol{\alpha}-\overline{\boldsymbol{\alpha}}^{\boldsymbol{\alpha}_0})(\boldsymbol{\alpha}-\overline{\boldsymbol{\alpha}}^{\boldsymbol{\alpha}_0})}\,. \tag{230}$$

With the help of the relation [*cf.* (82)]

$$\int f(\boldsymbol{\alpha}_0)P(\boldsymbol{\alpha}_0 \mid \boldsymbol{\alpha}\,;t)\,\mathrm{d}\boldsymbol{\alpha}_0 = f(\boldsymbol{\alpha})\,, \tag{231}$$

it follows furthermore that

$$G = g\,. \tag{232}$$

Thus the result (230) with (221), (226) and (232) is indeed equivalent with the distribution function (204), with (205) and (206), derived in § 6*.

§ 8. *Entropy and Random Fluctuations*

According to the second law of thermodynamics the entropy of an adiabatically insulated system must increase monotonously until thermodynamic equilibrium is established within the system. In the present section we shall investigate whether this behaviour of entropy may be obtained on the basis of the definition of entropy in terms of the random variables $\boldsymbol{\alpha}$ and of the properties of the Gaussian Markoff processes $\boldsymbol{\alpha}(t)$.

For a state $\boldsymbol{\alpha}$ the deviation of entropy ΔS from its maximum value is according to Boltzmann's entropy postulate [*cf.* § 2, formula (49)]

$$\Delta S = -\tfrac{1}{2}g:\boldsymbol{\alpha}\boldsymbol{\alpha}\,. \tag{233}$$

The conditional average which describes the mean behaviour of this quantity in the course of time, when the specified initial state is $\boldsymbol{\alpha}_0$, is given by

$$\overline{\Delta S}^{\boldsymbol{\alpha}_0} = -\tfrac{1}{2}g:\int \boldsymbol{\alpha}\boldsymbol{\alpha}P(\boldsymbol{\alpha}_0 \mid \boldsymbol{\alpha}\,;t)\,\mathrm{d}\boldsymbol{\alpha}\,. \tag{234}$$

* Within the framework of this section the Markoffian character of the processes $\boldsymbol{\alpha}(t)$ follows directly from the explicit form (230) according to a theorem due to Doob (J. L. Doob, Stochastic processes, Chapter V § 8, New York, 1953).

We shall evaluate this quantity with the expression (204) for the probability density $P(\boldsymbol{\alpha}_0 | \boldsymbol{\alpha}; t)$. We then obtain

$$\overline{\Delta S}^{\alpha_0} = -\tfrac{1}{2}g : \int (\boldsymbol{\alpha} - \bar{\boldsymbol{\alpha}}^{\alpha_0})(\boldsymbol{\alpha} - \bar{\boldsymbol{\alpha}}^{\alpha_0}) P(\boldsymbol{\alpha}_0 | \boldsymbol{\alpha}; t)\, \mathrm{d}\boldsymbol{\alpha} - \tfrac{1}{2}g : \bar{\boldsymbol{\alpha}}^{\alpha_0} \bar{\boldsymbol{\alpha}}^{\alpha_0}$$

$$= -\tfrac{1}{2}g : (k\mathsf{v}^{-1} + \bar{\boldsymbol{\alpha}}^{\alpha_0} \bar{\boldsymbol{\alpha}}^{\alpha_0}), \quad (235)$$

where (207) has been used.

It the explicit form (206) of the matrix v^{-1} of the variances and (205) of the first moments $\bar{\boldsymbol{\alpha}}^{\alpha_0}$ of P is introduced this becomes

$$\overline{\Delta S}^{\alpha_0} = -\tfrac{1}{2}g : \{ kg^{-1} + \mathrm{e}^{-Mt} \cdot (\boldsymbol{\alpha}_0 \boldsymbol{\alpha}_0 - kg^{-1}) \cdot \mathrm{e}^{-\tilde{M}t} \}. \quad (236)$$

We shall restrict the discussion to the case of $\boldsymbol{\alpha}$-type variables, in the absence of an external magnetic field, *i.e.* to the case when the Onsager reciprocal relations hold in the form $L = \tilde{L}$ or $M \cdot g^{-1} = g^{-1} \cdot \tilde{M}$. Using the description with coordinates $\boldsymbol{\alpha}''$ for which M becomes a diagonal matrix Λ with real elements, and g the unit matrix $g'' = U$ (*cf.* § 5), this can, with the help of (209) and (210), alternatively be written as

$$\overline{\Delta S}^{\alpha_0} = -\tfrac{1}{2}g'' : [k(g'')^{-1} + \mathrm{e}^{-\Lambda t} \cdot \{ \boldsymbol{\alpha}_0'' \boldsymbol{\alpha}_0'' - k(g'')^{-1} \} \cdot \mathrm{e}^{-\Lambda t}]$$

$$= -\tfrac{1}{2} \sum_{i=1}^{n} \{ k + (\alpha_{i0}''^2 - k)\, \mathrm{e}^{-2\lambda_i t} \}. \quad (237)$$

One sees that in the limit as $t \to \infty$ one has

$$\lim_{t \to \infty} \overline{\Delta S}^{\alpha_0} = -\tfrac{1}{2}nk, \quad (238)$$

in agreement with (50).

Furthermore for any macroscopic initial state such that $\alpha_{i0}^2 \gg k$

$$\overline{\Delta S}^{\alpha_0} = -\tfrac{1}{2} \sum_{i=1}^{n} (k + \alpha_{i0}''^2\, \mathrm{e}^{-2\lambda_i t}). \quad (239)$$

Since for any macroscopic state, $\overline{\Delta S}^{\alpha_0}$ is initially (for $t = 0$) of the order of Nk, where N is the number of particles in the system, each of the terms between brackets will initially be of the order of $(N/n)k$, with n the number of macroscopic variables. The number n is also of the order of the number of sub-systems for which the original $\boldsymbol{\alpha}$-variables were defined, because only a small number of variables is required to describe the state of each sub-system. Therefore N/n is of the order of the number of particles contained in each sub-system. In the usual macroscopic experiments and under a wide range of physical conditions this number will be at least 10^{12} to 10^{16}. It follows then from inspection of (239) that the terms k may be neglected within an interval of time of the order of several, say ten, times the relaxation time λ_i^{-1}. With this limitation, and using again (209), we find that

$$\overline{\Delta S}^{\alpha_0} = -\tfrac{1}{2}\sum_{i=1}^{n} \alpha_{i0}''^2 \, \mathrm{e}^{-2\lambda_i t} = -\tfrac{1}{2}\,\overline{\boldsymbol{\alpha}''}^{\alpha_0} \cdot \overline{\boldsymbol{\alpha}''}^{\alpha_0} = -\tfrac{1}{2}g : \overline{\boldsymbol{\alpha}}^{\alpha_0}\, \overline{\boldsymbol{\alpha}}^{\alpha_0}, \tag{240}$$

where in the final expression we have transformed back to the original variables $\boldsymbol{\alpha}$.

The time derivative of the average $\overline{\Delta S}^{\alpha_0}$ is obtained from (237)

$$\frac{\partial \overline{\Delta S}^{\alpha_0}}{\partial t} = \sum_{i=1}^{n} \lambda_i (\alpha_{i0}''^2 - k)\, \mathrm{e}^{-2\lambda_i t}. \tag{241}$$

Since the λ_i are all real positive numbers the following conclusions may be drawn from (241):
if $\alpha_{i0}''^2 \geqslant k$ for all i, then

$$\frac{\partial \overline{\Delta S}^{\alpha_0}}{\partial t} \geqslant 0\,; \tag{242}$$

if $\alpha_{i0}''^2 < k$ for all i, then

$$\frac{\partial \overline{\Delta S}^{\alpha_0}}{\partial t} < 0\,. \tag{243}$$

No general conclusion concerning the sign of (241) and valid for all times can be drawn, if the initial conditions are such that for some values of i we have $\alpha_{i0}''^2 \geqslant k$, whereas for the other values $\alpha_{i0}''^2 < k$.

Clearly only (242) has the expected form of the second law of thermodynamics. The possibility (243) also exists as a consequence of the fact that entropy has been defined for a single state according to (233). Indeed, if the initial state is the most probable state $\boldsymbol{\alpha}_0'' = 0$, *i.e.* if $\Delta S = 0$ initially, there will be a finite probability to find states with lower entropy in the course of time, as can be seen from the form of the distribution function (204). Therefore for such initial conditions the average entropy must obviously decrease until the equilibrium value $-\frac{1}{2}nk$ of $\overline{\Delta S}^{\alpha_0}$ has been reached. However, for macroscopic initial conditions ($\alpha''^2_{i0} \gg k$), the inequality (242) which does have the form of the second law of thermodynamics, is valid. It has already been stated in § 2 that from the point of view of thermodynamics no distinction is made between values of the entropy which differ an amount of the order nk. Therefore within a purely macroscopic description (243) has no meaning. In other words the behaviour of entropy found in the present discussion is in complete agreement with the laws of macroscopic theory.

For the case of macroscopic initial conditions, equation (241) may be rewritten in the form

$$\frac{\partial \overline{\Delta S}^{\alpha_0}}{\partial t} \doteq \sum_{i=1}^{n} \lambda_i \alpha''^2_{i0}\, \mathrm{e}^{-2\lambda_i t} = \overline{\boldsymbol{\alpha}''}^{\alpha_0} \cdot \mathsf{\Lambda} \cdot \overline{\boldsymbol{\alpha}''}^{\alpha_0} . \tag{244}$$

Transforming back to the original variables $\boldsymbol{\alpha}$ this becomes

$$\frac{\partial \overline{\Delta S}^{\alpha_0}}{\partial t} = \mathsf{L} : \overline{\mathbf{X}}^{\alpha_0}\, \overline{\mathbf{X}}^{\alpha_0} \geqslant 0 , \tag{245}$$

because [*cf.* § 5 (144), (147), (150), (155), (156) and (170)]

$$\begin{aligned} \boldsymbol{\alpha}'' &= \mathsf{B}\cdot\mathsf{A}\cdot\boldsymbol{\alpha}, & \mathsf{\Lambda} &= \mathsf{B}\cdot\mathsf{A}\cdot\mathsf{M}\cdot\mathsf{A}^{-1}\cdot\mathsf{B}^{-1} ; \\ \mathsf{g} &= \tilde{\mathsf{A}}\cdot\mathsf{A}, & \mathsf{B} &= \tilde{\mathsf{B}}^{-1} ; \\ \mathsf{L} &= \mathsf{M}\cdot\mathsf{g}^{-1}, & \overline{\mathbf{X}}^{\alpha_0} &= -\,\mathsf{g}\cdot\overline{\boldsymbol{\alpha}}^{\alpha_0} . \end{aligned} \tag{246}$$

Of course (245) can also be obtained directly by differentiating (240) with respect to time and making use of the regression equations (105).

The previous discussion was based, with regards to the connexion between entropy and the probability of a state, on Boltzmann's entropy postulate. Entropy was there a random variable, so that only an average of this quantity represented the relevant macroscopic quantity. In statistical mechanics it is also possible to give a different definition of entropy. Entropy is then defined from the outset in terms of a distribution over possible states, according to the so-called "Gibbs entropy postulate", as

$$S = -k \int P(\alpha_0 \mid \alpha ; t) \ln \frac{P(\alpha_0 \mid \alpha ; t)}{f(\alpha)\Omega} \, d\alpha , \tag{247}$$

where $f(\alpha)$ is the equilibrium distribution function (14) and Ω a constant. This constant determines the value of entropy at equilibrium, since in the limit as t approaches infinity, and therefore $P(\alpha_0 \mid \alpha;t)$ the equilibrium distribution $f(\alpha)$ [*cf.* (214)], the expression (247) becomes

$$\lim_{t\to\infty} S = k \ln \Omega . \tag{248}$$

In this connexion the reader is referred to the discussion of "equilibrium" at the end of § 2. It is interesting to re-examine the behaviour of entropy on the basis of the definition (247). It will be seen that from a macroscopic point of view the two definitions of entropy lead to identical results.

From (247) and (248) we find for the deviation ΔS of entropy from its equilibrium value

$$\Delta S = -k \int P(\alpha_0 \mid \alpha ; t) \ln \frac{P(\alpha_0 \mid \alpha ; t)}{f(\alpha)} \, d\alpha . \tag{249}$$

Introducing into this expression the explicit forms (14) and (204) of the distribution functions $f(\alpha)$ and $P(\alpha_0 \mid \alpha;t)$, this becomes

$$\Delta S = -\tfrac{1}{2}k \ln \left(\frac{|\mathsf{v}|}{|g|}\right) - \tfrac{1}{2}g : \{ e^{-Mt} \cdot (\alpha_0\alpha_0 - kg^{-1}) \cdot e^{-\tilde{M}t} \} . \tag{250}$$

The first term on the right-hand side of this equation can be written explicitly with the help of (206) as

$$-\tfrac{1}{2}k \ln \left(\frac{|\mathsf{v}|}{|g|}\right) = \tfrac{1}{2}k \ln (|\mathsf{v}^{-1}| \, |g|) = \tfrac{1}{2}k \ln | g^{-1} - e^{-Mt} \cdot g^{-1} \cdot e^{-\tilde{M}t} | \, |g|$$

$$= \tfrac{1}{2}k \ln | U - g \cdot e^{-Mt} \cdot g^{-1} \cdot e^{-\tilde{M}t} | . \tag{251}$$

Considering again the case of a symmetric L-matrix and using coordinates $\boldsymbol{\alpha}''$ for which g is the unit matrix and M a diagonal matrix Λ with real elements, (250), combined with (251), becomes

$$\Delta S = \tfrac{1}{2} \sum_{i=1}^{n} \{ k \ln (1 - \mathrm{e}^{-2\lambda_i t}) - (\alpha_{i0}''^2 - k)\, \mathrm{e}^{-2\lambda_i t} \} . \tag{252}$$

Let us discuss the relative order of magnitude of the various terms in this expression. For macroscopic initial conditions $\sum_{i=1}^{n} \alpha_{i0}''^2$ will be of the order of Nk, where N is the number of particles in the system, [*cf.* the discussion after formula (239)]. The order of magnitude of any of the terms $\alpha_{i0}''^2$ will therefore be $(N/n)k$, with n the number of macroscopic variables. The number N/n is at least of the order of 10^{12} to 10^{16}, as discussed after formula (239). Now each of the logarithmic terms will become of the order of $10^4 k$ to $10^{10} k$ within times of the order of $10^{-10^4} \lambda_i^{-1}$ to $10^{-10^{10}} \lambda_i^{-1}$, that is within times which are completely irrelevant compared to the macroscopic relaxation times λ_i^{-1}. (We may add that the whole theory of the Gaussian Markoff process loses its meaning for such extremely small time intervals). After such times the logarithmic terms may therefore already be neglected with respect to the terms containing $\alpha_{i0}''^2$, which, lie between $10^{12} k$ and $10^{16} k$. Expression (252) thus reduces to

$$\Delta S \simeq -\tfrac{1}{2} \sum_{i=1} \alpha_{i0}''^2 \mathrm{e}^{-2\lambda_i t} = -\tfrac{1}{2} g : \overline{\boldsymbol{\alpha}}^{\alpha_0} \overline{\boldsymbol{\alpha}}^{\alpha_0} , \tag{253}$$

where in the final expression we have transformed back to the original variables $\boldsymbol{\alpha}$. From the point of view of macroscopic theory, one is not interested in the behaviour of ΔS within the short times mentioned, so that (253) represents the relevant result. The result (253) is identical with (240) found on the basis of Boltzmann's entropy postulate.

The time derivative of ΔS is found from (252) to be equal to

$$\frac{\partial \Delta S}{\partial t} = \sum_{i=1}^{n} \lambda_i \left\{ \frac{k\, \mathrm{e}^{-2\lambda_i t}}{1 - \mathrm{e}^{-2\lambda_i t}} + (\alpha_{i0}''^2 - k)\, \mathrm{e}^{-2\lambda_i t} \right\}$$

$$= \sum_{i=1}^{n} \lambda_i \left(\frac{k\, \mathrm{e}^{-2\lambda_i t}}{1 - \mathrm{e}^{-2\lambda_i t}} + \alpha_{i0}''^2 \right) \mathrm{e}^{-2\lambda_i t} \geqslant 0 . \tag{254}$$

The inequality follows from the fact that all λ_i are real positive numbers, and it expresses the second law of thermodynamics. We note that this inequality is valid, with the present definition of entropy, for any initial condition $\boldsymbol{\alpha}_0$. The formula simplifies for macroscopic initial conditions. If we suppose again that α''^2_{i0} lies between $10^{12}k$ and $10^{16}k$, each of the first terms between brackets of (254) will be negligible, *i.e.* of the order of 10^4k to $10^{10}k$ after times of the order of $10^{-4}\lambda_i^{-1}$ to $10^{-10}\lambda_i^{-1}$. Such time intervals are also irrelevant on a macroscopic scale, so that (254) reduces to

$$\frac{\partial\,\Delta S}{\partial t} = \sum_{i=1}^{n} \lambda_i \alpha''^2_{i0}\, \mathrm{e}^{-2\lambda_i t} = L : \overline{\boldsymbol{X}}^{\boldsymbol{\alpha}_0}\,\overline{\boldsymbol{X}}^{\boldsymbol{\alpha}_0} \geqslant 0\,, \tag{255}$$

where in the last expression we have transformed back to the original variables $\boldsymbol{\alpha}$ in the same way as in the passage from (244) to (245). Again this result is identical with the expression (245), found on the basis of Boltzmann's entropy postulate.

The preceding discussion may be generalized to the case of a non-symmetric L-matrix, and also to the case that $\boldsymbol{\beta}$-variables are needed for the description of the system. Notwithstanding the additional complications arising in these cases, the relevant macroscopic features remain unchanged.

The foregoing consideration shows that the two definitions (233) and (249) of entropy lead, for Gaussian Markoff processes, to identical macroscopically relevant results. In this respect the situation is the same as in equilibrium statistical mechanics where the same two definitions of entropy, but now in terms of the molecular coordinates and momenta, yield identical values for the thermodynamic functions.

Let us finally turn our attention to the point of view taken in macroscopic phenomenological theory (Chapter IV, § 3). No distinction is then made between the behaviour of a single system on a single occasion and the average behaviour of a system. Therefore if the entropy of a non-equilibrium situation is given by [*cf.* equation (49)]

$$\Delta S = -\tfrac{1}{2} g : \boldsymbol{\alpha\alpha}\,, \tag{256}$$

the second law of thermodynamics is expressed by

$$\frac{\mathrm{d}\,\Delta S}{\mathrm{d}t} = -\,g : \boldsymbol{\alpha}\frac{\mathrm{d}\boldsymbol{\alpha}}{\mathrm{d}t} = \boldsymbol{X}\cdot\frac{\mathrm{d}\boldsymbol{\alpha}}{\mathrm{d}t} = L : \boldsymbol{XX} \geqslant 0\,, \tag{257}$$

where it has been assumed, in agreement with the point of view just mentioned, that the α-variables themselves obey the phenomenological laws. It is obvious that (256) and (257) are precisely of the form of the macroscopically relevant results derived in this section.

We may note that according to the phenomenological theory the positive definite character of the coefficient matrix L follows from the expression for the second law of thermodynamics (257). On the other hand in the considerations of this chapter the second law is a consequence of the positive definite character of L, which followed itself from general mechanical properties.

CHAPTER VIII

THE FLUCTUATION DISSIPATION THEOREM

§ 1. *Introduction*

In the previous chapter it has been tacitly assumed that the macroscopic $\boldsymbol{\alpha}$-variables form a "complete set" in the following sense: their conditional mean values obey linear differential equations of the first order in the time with constant coefficients. This assumption, which is frequently justified on the time scale of macroscopic measurements, certainly does not cover all situations of interest. It may indeed happen that the evolution in time of the variables studied is influenced amongst other things by a set of unknown parameters, which also determine the state of the system. In this chapter we shall extend the theory of the previous chapter in order to deal with such situations. To this end we shall first need a number of general theorems which we shall establish in the next two sections.

§ 2. *The Correlation Function of Stationary Processes; the Wiener–Khinchin Theorem*

Let us consider the random (or stochastic) vector process $\boldsymbol{\alpha}(t)$ with components $\alpha_1(t), \alpha_2(t), \ldots, \alpha_n(t)$. Such a process is stationary if the distribution functions $f(\boldsymbol{\alpha}_1, t_1; \boldsymbol{\alpha}_2, t_2; \ldots; \boldsymbol{\alpha}_m, t_m)\, \mathrm{d}\boldsymbol{\alpha}_1\, \mathrm{d}\boldsymbol{\alpha}_2 \ldots \mathrm{d}\boldsymbol{\alpha}_m$, representing the joint probability of finding the values of the vector $\boldsymbol{\alpha}$ in the ranges $(\boldsymbol{\alpha}_1, \boldsymbol{\alpha}_1 + \mathrm{d}\boldsymbol{\alpha}_1), (\boldsymbol{\alpha}_2, \boldsymbol{\alpha}_2 + \mathrm{d}\boldsymbol{\alpha}_2), \ldots, (\boldsymbol{\alpha}_m, \boldsymbol{\alpha}_m + \mathrm{d}\boldsymbol{\alpha}_m)$ at times $t_1, t_2, \ldots, t_m$ respectively, are invariant for a shift of the time axis. In Chapter VII, § 3, we have seen that in the stationary micro-canonical ensemble

$$f(\boldsymbol{\alpha}_1, t_1; \boldsymbol{\alpha}_2, t_2) = f(\boldsymbol{\alpha}_1, t_1 + h; \boldsymbol{\alpha}_2, t_2 + h) = f(\boldsymbol{\alpha}_1, \boldsymbol{\alpha}_2; t_2 - t_1). \tag{1}$$

This property of invariance for a shift of the time axis also holds in the micro-canonical ensemble for a joint distribution function $f(\boldsymbol{\alpha}_1, t_1; \boldsymbol{\alpha}_2, t_2; \ldots; \boldsymbol{\alpha}_m, t_m)$ with m arbitrary times. It thus follows that the micro-canonical ensemble generates a stationary random vector process $\boldsymbol{\alpha}(t)$.

The correlation function matrix $\rho(\tau)$ for this process is defined as

$$\rho(\tau) = \left\{ \begin{array}{l} \langle \boldsymbol{\alpha}(t)\, \boldsymbol{\alpha}(t+\tau) \rangle \equiv \iint \boldsymbol{\alpha}\boldsymbol{\alpha}' f(\boldsymbol{\alpha}, t\,;\, \boldsymbol{\alpha}', t+\tau)\, \mathrm{d}\boldsymbol{\alpha}\, \mathrm{d}\boldsymbol{\alpha}', \quad (\text{all } \tau) \\ \iint \boldsymbol{\alpha}\boldsymbol{\alpha}' f(\boldsymbol{\alpha}, \boldsymbol{\alpha}'\,;\, \tau)\, \mathrm{d}\boldsymbol{\alpha}\, \mathrm{d}\boldsymbol{\alpha}' = \iint \boldsymbol{\alpha}\boldsymbol{\alpha}' f(\boldsymbol{\alpha}) P(\boldsymbol{\alpha} \mid \boldsymbol{\alpha}'\,;\, \tau)\, \mathrm{d}\boldsymbol{\alpha}\, \mathrm{d}\boldsymbol{\alpha}', \ (\tau > 0) \end{array} \right\}, \tag{2}$$

where $P(\boldsymbol{\alpha} \mid \boldsymbol{\alpha}'; \tau)$ is the (stationary) conditional probability density [*cf.* (VII.68)].

The stationarity of the process $\boldsymbol{\alpha}(t)$ also implies that

$$\rho(\tau) = \tilde{\rho}(-\tau)\,, \tag{3}$$

since

$$\begin{aligned} \langle \boldsymbol{\alpha}(t)\, \boldsymbol{\alpha}(t+\tau) \rangle &= \iint \boldsymbol{\alpha}\boldsymbol{\alpha}' f(\boldsymbol{\alpha}, t\,;\, \boldsymbol{\alpha}', t+\tau)\, \mathrm{d}\boldsymbol{\alpha}\, \mathrm{d}\boldsymbol{\alpha}' \\ &= \iint \boldsymbol{\alpha}\boldsymbol{\alpha}' f(\boldsymbol{\alpha}, t-\tau\,;\, \boldsymbol{\alpha}', t)\, \mathrm{d}\boldsymbol{\alpha}\, \mathrm{d}\boldsymbol{\alpha}' \\ &= \langle \boldsymbol{\alpha}(t-\tau)\, \boldsymbol{\alpha}(t) \rangle = \widetilde{\langle \boldsymbol{\alpha}(t)\, \boldsymbol{\alpha}(t-\tau) \rangle}\,, \end{aligned} \tag{4}$$

where the property (1) has been used.

On the other hand the property of microscopic reversibility implies for $\boldsymbol{\alpha}$-type variables that

$$\rho(\tau) = \tilde{\rho}(\tau)\,, \tag{5}$$

since

$$\begin{aligned} \iint \boldsymbol{\alpha}\boldsymbol{\alpha}' f(\boldsymbol{\alpha}) P(\boldsymbol{\alpha} \mid \boldsymbol{\alpha}'\,;\, \tau)\, \mathrm{d}\boldsymbol{\alpha}\, \mathrm{d}\boldsymbol{\alpha}' &= \iint \boldsymbol{\alpha}\boldsymbol{\alpha}' f(\boldsymbol{\alpha}') P(\boldsymbol{\alpha}' \mid \boldsymbol{\alpha}\,;\, \tau)\, \mathrm{d}\boldsymbol{\alpha}\, \mathrm{d}\boldsymbol{\alpha}' \\ &= \iint \boldsymbol{\alpha}'\boldsymbol{\alpha} f(\boldsymbol{\alpha}) P(\boldsymbol{\alpha} \mid \boldsymbol{\alpha}'\,;\, \tau)\, \mathrm{d}\boldsymbol{\alpha}\, \mathrm{d}\boldsymbol{\alpha}'\,, \end{aligned} \tag{6}$$

where the first equality follows from the property of microscopic reversibility (VII.83).

If also $\boldsymbol{\beta}$-type variables are included in the description of the system and if an external magnetic field $\boldsymbol{B}$ is applied, one has, instead of (5),

$$\rho_{\alpha\alpha}(\tau\,;\boldsymbol{B}) = \tilde{\rho}_{\alpha\alpha}(\tau\,;\,-\boldsymbol{B})\,, \tag{7}$$

$$\rho_{\alpha\beta}(\tau\,;\boldsymbol{B}) = -\,\tilde{\rho}_{\beta\alpha}(\tau\,;\,-\boldsymbol{B})\,, \tag{8}$$

$$\rho_{\beta\beta}(\tau\,;\boldsymbol{B}) = \tilde{\rho}_{\beta\beta}(\tau\,;\,-\boldsymbol{B})\,, \tag{9}$$

where

$$\rho_{\alpha\alpha}(\tau\,;\boldsymbol{B}) = \langle\, \alpha(t)\,\alpha(t+\tau)\,\rangle_{\boldsymbol{B}}\,, \tag{10}$$

$$\rho_{\alpha\beta}(\tau\,;\boldsymbol{B}) = \langle\, \alpha(t)\,\beta(t+\tau)\,\rangle_{\boldsymbol{B}}\,, \tag{11}$$

$$\rho_{\beta\beta}(\tau\,;\boldsymbol{B}) = \langle\, \beta(t)\,\beta(t+\tau)\,\rangle_{\boldsymbol{B}}\,. \tag{12}$$

The index $\boldsymbol{B}$ indicates that the averages must be performed with distribution functions, which depend on the external magnetic field. Equations (7), (8) and (9) follow from (VII.93), together with (VII.33).

We further note that, according to (VII.20),

$$\rho(0) = \langle\, \alpha(t)\,\alpha(t)\,\rangle = kg^{-1}\,. \tag{13}$$

For the case of linear regression laws (VII.94) with solutions (VII.96) it follows with (VII.95) that the correlation function matrix is of the form

$$\begin{aligned}\rho(\tau) &= \iint \alpha\alpha'\, f(\alpha) P(\alpha \mid \alpha'\,;\tau)\,\mathrm{d}\alpha\,\mathrm{d}\alpha' \\ &= \int \alpha\,(\mathrm{e}^{-M\tau}\cdot\alpha)\, f(\alpha)\,\mathrm{d}\alpha = kg^{-1}\cdot \mathrm{e}^{-\tilde{M}\tau}\,, \quad (\tau > 0)\,.\end{aligned} \tag{14}$$

Since $\rho(\tau)$ satisfies the stationarity condition (3), we have for positive and negative values of τ

$$\rho(\tau) = \left\{ \begin{array}{ll} kg^{-1}\cdot \mathrm{e}^{-\tilde{M}\tau}\,, & (\tau > 0) \\ k\,\mathrm{e}^{-M|\tau|}\cdot g^{-1}\,, & (\tau < 0) \end{array} \right\}. \tag{15}$$

The coefficient matrix L of the linear regression laws is related to the correlation function matrix in the following way:

$$\tilde{L} = g^{-1}\cdot\tilde{M} = -\,k^{-1} \lim_{\tau\to 0} \frac{\partial\rho(\tau)}{\partial\tau}\,. \tag{16}$$

Now there exists for stationary processes an important theorem, the Wiener–Khinchin theorem, which relates the correlation function to the so-called spectrum of the process. In order to establish this theorem let us define the stochastic vector variable $\boldsymbol{\alpha}(t;T)$ such that

$$\boldsymbol{\alpha}(t\,;T) = \left\{ \begin{array}{lll} \boldsymbol{\alpha}(t)\,, & \text{if} & |\,t\,| < T \\ 0\,, & \text{if} & |\,t\,| > T \end{array} \right\}, \quad \lim_{T\to\infty} \boldsymbol{\alpha}(t\,;T) = \boldsymbol{\alpha}(t)\,. \tag{17}$$

We may then develop $\boldsymbol{\alpha}(t;\,T)$ in a Fourier integral

$$\boldsymbol{\alpha}(t\,;T) = \frac{1}{2\pi} \int_{-\infty}^{\infty} \hat{\boldsymbol{\alpha}}(\omega\,;T)\,\mathrm{e}^{-\mathrm{i}\omega t}\,\mathrm{d}\omega\,. \tag{18}$$

For the Fourier transform $\hat{\boldsymbol{\alpha}}(\omega;\,T)$ of $\boldsymbol{\alpha}(t;\,T)$,

$$\hat{\boldsymbol{\alpha}}(\omega\,;T) = \int_{-\infty}^{\infty} \boldsymbol{\alpha}(t\,;T)\,\mathrm{e}^{\mathrm{i}\omega t}\,\mathrm{d}t\,, \tag{19}$$

one has

$$\hat{\boldsymbol{\alpha}}(\omega\,;T) = \hat{\boldsymbol{\alpha}}^{*}(-\,\omega\,;T) \tag{20}$$

(where the asterisk stands for the complex conjugate), since $\boldsymbol{\alpha}(t;T)$ is a real quantity.

Consider the matrix

$$\mathsf{S}(\omega) \equiv \lim_{T\to\infty} \frac{1}{\pi T}\,\hat{\boldsymbol{\alpha}}^{*}(\omega\,;T)\,\hat{\boldsymbol{\alpha}}(\omega\,;T)\,. \tag{21}$$

With (19) we find that

$$\mathsf{S}(\omega) = \frac{1}{\pi} \int_{-\infty}^{\infty} \mathrm{d}\tau\,\mathrm{e}^{\mathrm{i}\omega\tau} \lim_{T\to\infty} \frac{1}{T} \int_{-\infty}^{\infty} \boldsymbol{\alpha}(t\,;T)\,\boldsymbol{\alpha}(t+\tau\,;T)\,\mathrm{d}t\,. \tag{22}$$

On the other hand, according to the fundamental postulate of statistical mechanics the phase space averages of dynamical functions in the micro-canonical ensemble are equal to time averages. Therefore

$$\rho(\tau) = \langle \boldsymbol{\alpha}(t)\, \boldsymbol{\alpha}(t+\tau) \rangle = \lim_{T\to\infty} \frac{1}{T} \int_{-T}^{T} \boldsymbol{\alpha}(t)\, \boldsymbol{\alpha}(t+\tau)\, \mathrm{d}t$$

$$= \lim_{T\to\infty} \frac{1}{T} \int_{-\infty}^{\infty} \boldsymbol{\alpha}(t\,;T)\, \boldsymbol{\alpha}(t+\tau\,;T)\, \mathrm{d}t\,. \tag{23}$$

Introducing this expression into (22) we find that

$$\mathsf{S}(\omega) = \frac{1}{\pi} \int_{-\infty}^{\infty} \rho(\tau)\, \mathrm{e}^{\mathrm{i}\omega\tau}\, \mathrm{d}\tau \tag{24}$$

and by Fourier inversion

$$\rho(\tau) = \frac{1}{2} \int_{-\infty}^{\infty} \mathsf{S}(\omega)\, \mathrm{e}^{-\mathrm{i}\omega\tau}\, \mathrm{d}\omega\,. \tag{25}$$

Equations (24) and (25) express the Wiener–Khinchin theorem which states that the correlation function and the matrix $\mathsf{S}(\omega)$, the so-called spectral density matrix, are each others Fourier transforms.

The spectral density matrix has the following properties:

1. From the definition (21) it follows that

$$\mathsf{S}(\omega) = \tilde{\mathsf{S}}^{*}(\omega)\,. \tag{26}$$

Thus the matrix $\mathsf{S}(\omega)$ is hermitian. [This property also follows from (24) with the stationarity condition (3).]

2. From the reality condition (20) it follows that

$$\mathsf{S}(\omega) = \mathsf{S}^{*}(-\omega)\,. \tag{27}$$

(This property also follows directly from (24) since $\rho(\tau)$ is real.)

3. From the definition (21) it furthermore follows that

$$\boldsymbol{\xi}\cdot\mathsf{S}\cdot\boldsymbol{\xi}^{*} = \lim_{T\to\infty} \frac{1}{\pi T} \,|\, \boldsymbol{\xi}\cdot\hat{\boldsymbol{\alpha}}(\omega\,;T)\,|^{2} \geqslant 0\,, \tag{28}$$

where $\boldsymbol{\xi} = \boldsymbol{\eta} + \mathrm{i}\boldsymbol{\zeta}$ is an arbitrary n-dimensional complex vector ($\boldsymbol{\eta}$ and $\boldsymbol{\zeta}$ are arbitrary real n-dimensional vectors). Thus the hermitian matrix S is positive definite.

We now write $\mathsf{S}(\omega)$ in the following form

$$\mathsf{S}(\omega) = \mathsf{G}(\omega) + \mathrm{i}\mathsf{H}(\omega) , \tag{29}$$

where G and H are real matrices. From (26) it then follows that

$$\mathsf{G}(\omega) = \tilde{\mathsf{G}}(\omega) , \tag{30}$$

$$\mathsf{H}(\omega) = - \tilde{\mathsf{H}}(\omega) . \tag{31}$$

Therefore the real part of S is symmetric and the imaginary part antisymmetric.

Furthermore we have from (27)

$$\mathsf{G}(\omega) = \mathsf{G}(- \omega) , \tag{32}$$

$$\mathsf{H}(\omega) = - \mathsf{H}(- \omega) , \tag{33}$$

so that the real part of S is an even function and the imaginary part an odd function of ω.

Finally it follows from (28) that

$$\boldsymbol{\eta}\cdot\mathsf{S}\cdot\boldsymbol{\eta} = \boldsymbol{\eta}\cdot\mathsf{G}\cdot\boldsymbol{\eta} \geqslant 0 , \tag{34}$$

since with (31)

$$\boldsymbol{\eta}\cdot\mathsf{H}\cdot\boldsymbol{\eta} = 0 . \tag{35}$$

Thus G is a real, positive definite, symmetric matrix, which is even in ω, whereas H is real, antisymmetric and odd in ω.

Writing again (25), and applying (29), (32) and (33), it follows that

$$\rho(\tau) = \int_0^\infty \mathsf{G}(\omega) \cos \omega\tau \, \mathrm{d}\omega + \int_0^\infty \mathsf{H}(\omega) \sin \omega\tau \, \mathrm{d}\omega . \tag{36}$$

From this expression one gets for $\tau = 0$ with the use of (13)

$$\rho(0) = \langle \boldsymbol{\alpha\alpha} \rangle = k\mathsf{g}^{-1} = \int_0^\infty \mathsf{G}(\omega) \, \mathrm{d}\omega . \tag{37}$$

For a single variable α this reads

$$\rho(0) = \langle \alpha^2 \rangle = kg^{-1} = \int_0^\infty G(\omega)\, \mathrm{d}\omega\,, \tag{38}$$

with $G(\omega)$ a positive quantity [see (34)]. The quantity $G(\omega)\,\mathrm{d}\omega$ represents the contribution to the mean fluctuation of α arising from the components of $\alpha(t)$ having frequencies between ω and $\omega + \mathrm{d}\omega$. Hence the name spectral density, and by extension the name spectral density matrix for the matrix S.

We have not yet considered the repercussion of the principle of microscopic reversibility on the matrix S. In view of (5), this property leads for $\boldsymbol{\alpha}$-type variables to

$$\mathsf{S}(\omega) = \tilde{\mathsf{S}}(\omega)\,, \tag{39}$$

as follows from (24), or, in view of (29)–(31), to

$$\mathsf{H}(\omega) = 0\,. \tag{40}$$

Formula (36) becomes therefore simply

$$\boldsymbol{\rho}(\tau) = \int_0^\infty \mathsf{G}(\omega) \cos \omega\tau \, \mathrm{d}\omega\,. \tag{41}$$

For $\boldsymbol{\alpha}$- and $\boldsymbol{\beta}$-variables and in the presence of a magnetic field one has, instead of (24),

$$\mathsf{S}_{\alpha\alpha}(\omega\,;\boldsymbol{B}) = \frac{1}{\pi}\int \boldsymbol{\rho}_{\alpha\alpha}(\tau\,;\boldsymbol{B})\, \mathrm{e}^{\mathrm{i}\omega\tau}\, \mathrm{d}\tau\,, \tag{42}$$

$$\mathsf{S}_{\alpha\beta}(\omega\,;\boldsymbol{B}) = \frac{1}{\pi}\int \boldsymbol{\rho}_{\alpha\beta}(\tau\,;\boldsymbol{B})\, \mathrm{e}^{\mathrm{i}\omega\tau}\, \mathrm{d}\tau\,, \tag{43}$$

$$\mathsf{S}_{\beta\beta}(\omega\,;\boldsymbol{B}) = \frac{1}{\pi}\int \boldsymbol{\rho}_{\beta\beta}(\tau\,;\boldsymbol{B})\, \mathrm{e}^{\mathrm{i}\omega\tau}\, \mathrm{d}\tau\,. \tag{44}$$

The matrices $\mathsf{S}_{\alpha\alpha}$ and $\mathsf{S}_{\beta\beta}$ satisfy again the conditions (26)–(28) or (30)–(34). For $\mathsf{S}_{\alpha\beta}$ we have with

$$\mathsf{S}_{\alpha\beta}(\omega) = \mathsf{G}_{\alpha\beta}(\omega) + \mathrm{i}\mathsf{H}_{\alpha\beta}(\omega)\,, \tag{45}$$

instead of (26), (30) and (31):

$$S_{\alpha\beta}(\omega) = \tilde{S}^*_{\beta\alpha}(\omega)\,, \tag{46}$$

$$G_{\alpha\beta}(\omega) = \tilde{G}_{\beta\alpha}(\omega)\,, \tag{47}$$

$$H_{\alpha\beta}(\omega) = -\tilde{H}_{\beta\alpha}(\omega)\,, \tag{48}$$

whereas both $S_{\alpha\beta}$ and $S_{\beta\alpha}$ satisfy the reality condition (27).

Microscopic reversibility (7)–(9) now implies that

$$S_{\alpha\alpha}(\omega\,;\boldsymbol{B}) = \tilde{S}_{\alpha\alpha}(\omega\,;-\boldsymbol{B})\,, \tag{49}$$

$$S_{\alpha\beta}(\omega\,;\boldsymbol{B}) = -\tilde{S}_{\beta\alpha}(\omega\,;-\boldsymbol{B})\,, \tag{50}$$

$$S_{\beta\beta}(\omega\,;\boldsymbol{B}) = \tilde{S}_{\beta\beta}(\omega\,;-\boldsymbol{B})\,, \tag{51}$$

or, with (45)–(48) and the corresponding relations for $S_{\alpha\alpha}$, $S_{\beta\beta}$, $G_{\alpha\alpha}$, $G_{\beta\beta}$, $H_{\alpha\alpha}$ and $H_{\beta\beta}$:

$$\left.\begin{aligned} G_{\alpha\alpha}(\omega\,;\boldsymbol{B}) &= G_{\alpha\alpha}(\omega\,;-\boldsymbol{B}) \\ H_{\alpha\alpha}(\omega\,;\boldsymbol{B}) &= -H_{\alpha\alpha}(\omega\,;-\boldsymbol{B})\,, \end{aligned}\right. \tag{52}$$

$$\left.\begin{aligned} G_{\alpha\beta}(\omega\,;\boldsymbol{B}) &= -G_{\alpha\beta}(\omega\,;-\boldsymbol{B}) \\ H_{\alpha\beta}(\omega\,;\boldsymbol{B}) &= H_{\alpha\beta}(\omega\,;-\boldsymbol{B})\,, \end{aligned}\right. \tag{53}$$

$$\left.\begin{aligned} G_{\beta\beta}(\omega\,;\boldsymbol{B}) &= G_{\beta\beta}(\omega\,;-\boldsymbol{B}) \\ H_{\beta\beta}(\omega\,;\boldsymbol{B}) &= -H_{\beta\beta}(\omega\,;-\boldsymbol{B})\,. \end{aligned}\right. \tag{54}$$

Note that for $\boldsymbol{B} = 0$ one has

$$\left.\begin{aligned} H_{\alpha\alpha} &= 0 \\ G_{\alpha\beta} &= 0 \\ H_{\beta\beta} &= 0\,. \end{aligned}\right. \tag{55}$$

If the number of β-variables is the same as the number of α-variables and if the β-variables are time derivatives of the α-variables [*cf.* (VII.121) *seq.*], then one has

$$\rho_{\alpha\beta}(\tau) = \langle\, \alpha(t)\, \beta(t+\tau)\,\rangle = \langle\, \alpha(t)\, \frac{\partial}{\partial\tau}\, \alpha(t+\tau)\,\rangle$$

$$= \frac{\partial}{\partial\tau} \langle\, \alpha(t)\, \alpha(t+\tau)\,\rangle = \frac{\partial}{\partial\tau}\, \rho_{\alpha\alpha}(\tau) = \frac{\partial}{\partial\tau}\, \tilde{\rho}_{\alpha\alpha}(-\tau)\,, \tag{56}$$

$$\rho_{\beta\alpha}(\tau) = \tilde{\rho}_{\alpha\beta}(-\tau)\,, \tag{57}$$

$$\rho_{\beta\beta}(\tau) = \langle\, \beta(t)\, \beta(t+\tau)\,\rangle = \langle\, \beta(t)\, \frac{\partial}{\partial\tau}\, \alpha(t+\tau)\,\rangle$$

$$= \frac{\partial}{\partial\tau} \langle\, \beta(t)\, \alpha(t+\tau)\,\rangle = \frac{\partial}{\partial\tau}\, \rho_{\beta\alpha}(\tau) = -\,\frac{\partial^2}{\partial\tau^2}\, \rho_{\alpha\alpha}(\tau)\,. \tag{58}$$

In the last member of (56) we have applied the stationarity condition $\rho_{\alpha\alpha}(\tau) = \tilde{\rho}_{\alpha\alpha}(-\tau)$ [*cf.* (3)]. Equation (57) is the stationarity condition for the correlation function matrix $\rho_{\beta\alpha}(\tau)$. Finally in (58) we have applied both (57) and (56).

From (42), (43) and (44) it now follows with (46), (56), (58) and (25) that

$$S_{\alpha\beta} = \tilde{S}^*_{\beta\alpha} = -\,\mathrm{i}\omega S_{\alpha\alpha}\,, \tag{59}$$

$$S_{\beta\beta} = \omega^2\, S_{\alpha\alpha}\,, \tag{60}$$

or equivalently

$$\begin{aligned} G_{\alpha\beta} &= \tilde{G}_{\beta\alpha} = \omega H_{\alpha\alpha} \\ H_{\alpha\beta} &= -\,\tilde{H}_{\beta\alpha} = -\,\omega G_{\alpha\alpha}\,, \end{aligned} \tag{61}$$

$$\begin{aligned} G_{\beta\beta} &= \omega^2 G_{\alpha\alpha} \\ H_{\beta\beta} &= \omega^2 H_{\beta\beta}\,. \end{aligned} \tag{62}$$

From (61) we find that the matrices $G_{\alpha\beta}$ and $G_{\beta\alpha}$ are antisymmetric and $H_{\alpha\beta}$ and $H_{\beta\alpha}$ are symmetric since $H_{\alpha\alpha}$ and $G_{\alpha\alpha}$ were antisymmetric and symmetric respectively.

Microscopic reversibility is in this case completely contained in equations (49) or (52). Equations (50), (51), (53) and (54) contain no additional information.

As an application we shall compute the spectral densities of the correlation function in two special cases.

First we consider the case of a single α-variable obeying the regression equation (VII.94). The correlation function is then given by [*cf.* (15)]

$$\rho(\tau) = kg^{-1}\,\mathrm{e}^{-M|\tau|}. \tag{63}$$

From (24) we obtain the spectral density

$$\begin{aligned} S(\omega) &= \frac{k}{\pi g}\int_{-\infty}^{\infty} \mathrm{e}^{-M|\tau| + \mathrm{i}\omega\tau}\,\mathrm{d}\tau \\ &= \frac{k}{\pi g}\int_{0}^{\infty} \{\,\mathrm{e}^{(\mathrm{i}\omega - M)\tau} + \mathrm{e}^{-(\mathrm{i}\omega + M)\tau}\,\}\,\mathrm{d}\tau \\ &= \frac{2}{\pi}\,\frac{kMg^{-1}}{\omega^2 + M^2}. \end{aligned} \tag{64}$$

As a second example we take the case of a system in the absence of an external magnetic field described by one α- and one β-variable with $\beta = \dot{\alpha}$. These variables obey the linear regression equations (VII.127) and (VII.128):

$$\frac{\partial\,\overline{\alpha(\tau)}^{\alpha_0,\beta_0}}{\partial\tau} = \overline{\beta(\tau)}^{\alpha_0,\beta_0} \tag{65}$$

$$\frac{\partial\,\overline{\beta(\tau)}^{\alpha_0,\beta_0}}{\partial\tau} = -\,gh^{-1}\,\overline{\alpha(\tau)}^{\alpha_0,\beta_0} - M\,\overline{\beta(\tau)}^{\alpha_0,\beta_0} \tag{66}$$

where we have written M for $hL_{\beta\beta}$. The correlation function matrix is now of the form

$$\rho(\tau) = \begin{pmatrix} \rho_{\alpha\alpha} & \rho_{\alpha\beta} \\ \rho_{\beta\alpha} & \rho_{\beta\beta} \end{pmatrix} = k\begin{pmatrix} g^{-1} & 0 \\ 0 & h^{-1} \end{pmatrix}\cdot\begin{pmatrix} (\mathrm{e}^{-\tilde{\mathsf{M}}\tau})_{\alpha\alpha} & (\mathrm{e}^{-\tilde{\mathsf{M}}\tau})_{\alpha\beta} \\ (\mathrm{e}^{-\tilde{\mathsf{M}}\tau})_{\beta\alpha} & (\mathrm{e}^{-\tilde{\mathsf{M}}\tau})_{\beta\beta} \end{pmatrix},\ (\tau > 0)\,, \tag{67}$$

where the matrix M is given by

$$\mathsf{M} = \begin{pmatrix} 0 & 1 \\ -\,gh^{-1} & -\,M \end{pmatrix}. \tag{68}$$

Expression (67) follows by writing down explicitly the matrix equation (14) for the present two-dimensional case. [Formula (14) has been obtained without any specific reference to the even ($\boldsymbol{\alpha}$) or odd ($\boldsymbol{\beta}$) character of the variables, and therefore still applies.] Only the two diagonal elements of the $\boldsymbol{g}^{-1}$-matrix exist, since equilibrium correlations of an α- with a β-variable vanish. We have denoted $\langle \alpha^2 \rangle$ by kg^{-1} and $\langle \beta^2 \rangle$ by kh^{-1} in accordance with Chapter VII, § 2 and the use of these symbols in equations (65) and (66). The form (68) of the matrix $\boldsymbol{M}$ follows from inspection of equations (65) and (66). Explicit direct evaluation of the matrix elements of (67) in closed form is rather cumbersome. We shall therefore follow a different method. Substituting (65) into (66) we obtain the second order differential equation

$$\frac{\partial^2 \overline{\alpha(\tau)}^{\alpha_0, \beta_0}}{\partial \tau^2} + M \frac{\partial \overline{\alpha(\tau)}^{\alpha_0, \beta_0}}{\partial \tau} + gh^{-1} \overline{\alpha(\tau)}^{\alpha_0, \beta_0} = 0 , \quad (\tau > 0) . \tag{69}$$

Multiplying this equation with α_0 and averaging over the equilibrium distribution $f(\alpha_0, \beta_0)$ we obtain according to the definition of $\rho_{\alpha\alpha}$ [*cf.* (2)]

$$\rho_{\alpha\alpha}(\tau) = \iint \alpha_0 \alpha f(\alpha_0, \beta_0) P(\alpha_0, \beta_0 \mid \alpha, \beta ; \tau) \, \mathrm{d}\alpha_0 \, \mathrm{d}\beta_0 \, \mathrm{d}\alpha \, \mathrm{d}\beta , \tag{70}$$

the differential equation

$$\frac{\partial^2 \rho_{\alpha\alpha}(\tau)}{\partial \tau^2} + M \frac{\partial \rho_{\alpha\alpha}(\tau)}{\partial \tau} + gh^{-1} \rho_{\alpha\alpha}(\tau) = 0 , \quad (\tau > 0) . \tag{71}$$

This equation, which is essentially the equation of the damped harmonic oscillator, has the general solution

$$\rho_{\alpha\alpha}(\tau) = \mathrm{e}^{-\frac{1}{2}M\tau} (c_1 \cos \omega' \tau + c_2 \sin \omega' \tau) , \quad (\tau > 0) , \tag{72}$$

where c_1 and c_2 are constants, and where

$$\omega' = \sqrt{gh^{-1} - \tfrac{1}{4} M^2} . \tag{73}$$

The formulae are written for the case that ω' is real (the "underdamped case"). If ω' is imaginary ("overdamped case"), put $\omega' = \mathrm{i}\omega_1$

and replace in the above equation $\cos i\omega_1\tau$ by $\cosh \omega_1\tau$ and $\sin i\omega_1\tau$ by $i \sinh \omega_1\tau$. Since according to (13) $\rho_{\alpha\alpha}(0) = kg^{-1}$, we find from (72) that

$$c_1 = kg^{-1} . \tag{74}$$

On the other hand, since

$$\left(\frac{\partial \rho_{\alpha\alpha}}{\partial \tau}\right)_{\tau=0} = \rho_{\alpha\beta}(0) = \langle \alpha\beta \rangle = 0 , \tag{75}$$

it follows from (72) that

$$c_2 = \frac{Mc_1}{2\omega'} . \tag{76}$$

The correlation function $\rho_{\alpha\alpha}(\tau)$ is therefore of the form

$$\begin{aligned} \rho_{\alpha\alpha}(\tau) &= kg^{-1}\, e^{-\frac{1}{2}M\tau} \left(\cos \omega'\tau + \frac{M}{2\omega'} \sin \omega'\tau\right) \\ &= kg^{-1}\, e^{-\frac{1}{2}M\tau} \left(\cosh \omega_1\tau + \frac{M}{2\omega_1} \sinh \omega_1\tau\right), \quad (\tau > 0) . \end{aligned} \tag{77}$$

Since $\rho_{\alpha\alpha}(\tau)$ must be an even function of τ we have for all times

$$\rho_{\alpha\alpha}(\tau) = kg^{-1}\, e^{-\frac{1}{2}M|\tau|} \left(\cosh \omega_1\tau + \frac{M}{2\omega_1} \sinh \omega_1 |\tau|\right). \tag{78}$$

In the strongly overdamped case ($M \gg \sqrt{gh^{-1}}$) and after long times ($\tau \gg M^{-1}$) this function reduces to

$$\rho_{\alpha\alpha}(\tau) = kg^{-1}\, e^{-\frac{g}{hM}|\tau|} \tag{79}$$

This is the correlation function for a single α-variable with a "relaxation time" Mh/g. We have already mentioned in Chapter VII, § 4, that within the above approximation the β-variables may be neglected for the description of the system.

The correlation functions $\rho_{\alpha\beta}$, $\rho_{\beta\alpha}$ and $\rho_{\beta\beta}$ can be obtained from (78), by differentiation with respect to time according to (56)–(58).

From the Wiener–Khinchin theorem (24) we obtain for the present case (78) the spectral density

$$S_{\alpha\alpha}(\omega) = \frac{2}{\pi} \frac{Mkh^{-1}}{(\omega^2 - gh^{-1})^2 + \omega^2 M^2} \,. \tag{80}$$

In the strongly overdamped case $(M \gg \sqrt{gh^{-1}})$ and for $\omega \ll M$, this function reduces to

$$S_{\alpha\alpha}(\omega) = \frac{2}{\pi} \frac{kg^{-1}\,(g/hM)}{\omega^2 + (g/hM)^2} \,. \tag{81}$$

This is according to (64) the spectrum corresponding to (79). The condition $\omega \ll M$ expresses the fact that for long times $\tau \gg M^{-1}$ only the low frequency part of the spectrum contributes to $\rho_{\alpha\alpha}(\tau)$.

The spectral densities $S_{\alpha\beta}$, $S_{\beta\alpha}$ and $S_{\beta\beta}$ can be obtained from (80) by multiplication with $-\mathrm{i}\omega$, $\mathrm{i}\omega$ and ω^2 respectively, as follows from (59) and (60).

Processes which are Gaussian are completely determined by the spectrum of the correlation function. Indeed for a Gaussian process the first or equilibrium distribution function as well as the joint distribution function for values of α at two times are Gaussian, so that these distributions are completely determined by their variance matrix, *i.e.* by their correlation function matrix (*cf.* problems 10 and 11 of Chapter VII). If the correlation function matrix is then of the type (15), the processes are Markoffian as well as Gaussian (*cf.* Chapter VII, §§ 6 and 7 and problem 12).

In the two preceding examples we have used the Wiener–Khinchin theorem in order to compute the spectral densities from the known correlation functions. This was possible because we had at our disposal, in the special form of the mean regression equations, sufficient information to calculate the correlation functions. It may however occur that the relevant correlation functions are not as easily accessible. This may for instance be the case, even if one postulates the existence of a sufficient number of variables (a "complete set") of which the conditional mean values obey linear first order differential equations with constant coefficients, but only observes a limited number of these variables, while one knows neither the number nor the properties of the remaining set. If one could in such a case experimentally establish the form of the spectral density one could then obtain from this spectrum the correlation function. Now there exists the so-called "fluctuation dissipation theorem" which indeed enables one to relate in such a case the absorption spectrum (or equivalently the dispersion spectrum) of the relevant variables to the spectral densities of the correlation functions. (The absorption or dispersion spectra are observed by subjecting the system to an external force which influences the variables under consideration.) At the same time microscopic reversibility which determined certain properties of the correlation function matrix is also reflected in properties of the absorption and dispersion spectra. Before establishing the fluctuation dissipation theorem we shall first study some mathematical properties of the principle of causality, which are needed in the proof of this theorem.

§ 3. *The Principle of Causality; the Kramers–Kronig Relations*

Consider n time dependent external driving forces $F_1(t), F_2(t), \ldots, F_n(t)$ acting on a system. These forces will induce time dependent responses $x_1(t), x_2(t), \ldots, x_n(t)$. [For instance in an elastic medium a mechanical force $F(t)$ induces an elongation $x(t)$.] For sufficiently small driving forces the relation between these forces and the responses is linear and of the form

$$x_i(t) = \sum_{k=1}^{n} \int_{-\infty}^{\infty} \kappa_{ik}(t - t')\, F_k(t')\, \mathrm{d}t'$$

$$= \sum_{k=1}^{n} \int_{-\infty}^{\infty} \kappa_{ik}(\tau)\, F_k(t - \tau)\, \mathrm{d}\tau\,, \tag{82}$$

where the coefficients κ_{ik} are certain finite functions of time, specific for the system. In matrix notation (82) becomes

$$\boldsymbol{x}(t) = \int_{-\infty}^{\infty} \kappa(t - t')\cdot \boldsymbol{F}(t')\, \mathrm{d}t'$$

$$= \int_{-\infty}^{\infty} \kappa(\tau)\cdot \boldsymbol{F}(t - \tau)\, \mathrm{d}\tau\,. \tag{83}$$

Since a response may not precede in time the effect, which causes it, we have for κ the following condition:

$$\kappa(t - t') = 0 \quad \text{for} \quad t < t' , \tag{84}$$

or

$$\kappa(\tau) = 0 \quad \text{for} \quad \tau < 0 . \tag{85}$$

This is the expression of the principle of causality for this case. We shall also require that a constant finite driving force gives rise to a constant finite response. This implies that

$$\int_0^\infty \kappa(\tau)\, \mathrm{d}\tau < \infty , \tag{86}$$

i.e. that the above integrals exist and are finite.

We shall now expand the functions $\boldsymbol{x}(t)$, $\boldsymbol{F}(t)$ and $\kappa(t)$ into Fourier integrals*:

$$\boldsymbol{x}(t) = \frac{1}{2\pi} \int_{-\infty}^{\infty} \hat{\boldsymbol{x}}(\omega)\, \mathrm{e}^{-\mathrm{i}\omega t}\, \mathrm{d}\omega , \tag{87}$$

$$\boldsymbol{F}(t) = \frac{1}{2\pi} \int_{-\infty}^{\infty} \hat{\boldsymbol{F}}(\omega)\, \mathrm{e}^{-\mathrm{i}\omega t}\, \mathrm{d}\omega , \tag{88}$$

$$\kappa(t) = \frac{1}{2\pi} \int_{-\infty}^{\infty} \hat{\kappa}(\omega)\, \mathrm{e}^{-\mathrm{i}\omega t}\, \mathrm{d}\omega , \tag{89}$$

where the Fourier transforms $\hat{\boldsymbol{x}}(\omega)$, $\hat{\boldsymbol{F}}(\omega)$ and $\hat{\kappa}(\omega)$ are given by

$$\hat{\boldsymbol{x}}(\omega) = \int_{-\infty}^{\infty} \boldsymbol{x}(t)\, \mathrm{e}^{\mathrm{i}\omega t}\, \mathrm{d}t , \tag{90}$$

* It is assumed here that the functions $\boldsymbol{x}(t)$, $\boldsymbol{F}(t)$ and $\kappa(t)$ satisfy the requirements needed for an expansion into Fourier integrals. In particular we assume here the square integrability of $\kappa(t)$. This implies that the function $\kappa(t)$ tends to zero as t tends to infinity.

$$\hat{\boldsymbol{F}}(\omega) = \int_{-\infty}^{\infty} \boldsymbol{F}(t)\, e^{i\omega t}\, dt\,, \tag{91}$$

$$\hat{\boldsymbol{\kappa}}(\omega) = \int_{-\infty}^{\infty} \boldsymbol{\kappa}(t)\, e^{i\omega t}\, dt\,. \tag{92}$$

With these relations (83) becomes

$$\hat{\boldsymbol{x}}(\omega) = \hat{\boldsymbol{\kappa}}(\omega)\cdot\hat{\boldsymbol{F}}(\omega)\,, \tag{93}$$

where use has been made of the Fourier integral expansion of the delta function:

$$\delta(y) = \frac{1}{2\pi}\int_{-\infty}^{\infty} e^{iky}\, dk\,. \tag{94}$$

The quantity $\hat{\boldsymbol{\kappa}}(\omega)$ may be called a generalized susceptibility matrix.

The condition (86) implies, according to (92), that $\hat{\boldsymbol{\kappa}}(0)$ is finite, or in other words, that $\hat{\boldsymbol{\kappa}}(\omega)$ has no pole at the origin. We shall impose the further restriction on $\hat{\boldsymbol{\kappa}}(\omega)$ that it has no pole (does not become infinite) for any value of ω*.

Let us now find what effect the causality condition (85) has on the susceptibility matrix. For this purpose we extend the definition of the Fourier integral (92) to complex values $w = \omega + i\nu$ of the argument. Since $\boldsymbol{\kappa}(t)$ vanishes for negative times we can write instead of (92)

$$\hat{\boldsymbol{\kappa}}(w) = \int_{0}^{\infty} \boldsymbol{\kappa}(t)\, e^{iwt}\, dt = \int_{0}^{\infty} \boldsymbol{\kappa}(t)\, e^{i\omega t - \nu t}\, dt\,. \tag{95}$$

For positive values of ν the integrals (95) exist and are finite since the factor $e^{-\nu t}$ only enhances their convergence. Furthermore for $\nu \to +\infty$

* It can be shown that poles of $\hat{\boldsymbol{\kappa}}(\omega)$ on the real axis correspond to non-dissipative reversible contributions to the macroscopic laws, describing the evolution in time of the system. Such contributions are therefore not assumed to exist within the framework of the present theory, dealing with irreversible phenomena.

the expression tends to zero. Thus the statement equivalent to (and a consequence of) (85) is:

The function $\hat{\kappa}(w)$ *with* $w = \omega + \mathrm{i}\nu$ *has no poles (no singular points) in the upper half of the complex plane and tends to zero in the limit as* ν *tends to* $+\infty$ *. (96)

In the lower half of the complex plane the integral (95) diverges. Here the expression $\hat{\kappa}(w)$ may only be defined as the analytic continuation of its expression from the upper half plane and may in general have poles.

From (96) we can derive some further formulations of the principle of causality, by applying Cauchy's theorem, which states that for a closed contour in the complex plane one has

$$\oint f(w)\,\mathrm{d}w = 0\,, \tag{97}$$

if the function $f(w)$ has no poles inside that contour. The integral is taken along the contour in counter clockwise direction. We shall apply this theorem to the function

$$f(w) = \frac{\hat{\kappa}(w)}{w-u}\,, \tag{98}$$

where u is real. The contour we choose extends along the whole real axis avoiding the point $w = u$ with a small semi-circle of radius r in the upper half of the complex plane and is closed by an infinite semi-circle also in the upper half of the complex plane. Inside this contour the function (98) has no poles in view of (96). Furthermore since $\hat{\kappa}(w)$ vanishes as ν tends to infinity, the line integral along the infinite semi-circle also vanishes. Therefore Cauchy's theorem (97) yields for the function (98) and the contours chosen:

$$\int_{-\infty}^{u-r} \frac{\hat{\kappa}(\omega)}{\omega-u}\,\mathrm{d}\omega + \int_{u+r}^{\infty} \frac{\hat{\kappa}(\omega)}{\omega-u}\,\mathrm{d}\omega + \int_{\text{semi-circle}} \frac{\hat{\kappa}(w)}{w-u}\,\mathrm{d}w = 0\,. \tag{99}$$

* If the Fourier integrals (87)–(89) had been defined with the opposite sign in the exponential (as is frequently done) the statement (96) would hold for the lower half of the complex plane instead of for the upper half.

The last integral is taken along the small semi-circle with radius r around the point $w = u$ passed in clock-wise direction. In the limit as r tends to zero the first two integrals together reduce to the so-called principal part of the integral from $-\infty$ to ∞. The integration along the small semi-circle yields in the limit $r \to 0$ a contribution $- \mathrm{i}\pi\, \hat{\kappa}(u)$. Therefore we get from (99)

$$\hat{\kappa}(u) = \frac{1}{\pi \mathrm{i}} \mathscr{P} \int_{-\infty}^{\infty} \frac{\hat{\kappa}(\omega)}{\omega - u} \, \mathrm{d}\omega \,, \tag{100}$$

where the symbol $\mathscr{P}$ denotes the principal value integral. This formula is again an alternative mathematical expression for the principle of causality (85) or (96).

Splitting $\hat{\kappa}(\omega)$ into its real and imaginary parts

$$\hat{\kappa}(\omega) = \hat{\kappa}'(\omega) + \mathrm{i}\hat{\kappa}''(\omega) \,, \tag{101}$$

we get from (100)

$$\hat{\kappa}'(u) = \frac{1}{\pi} \mathscr{P} \int_{-\infty}^{\infty} \frac{\hat{\kappa}''(\omega)}{\omega - u} \, \mathrm{d}\omega \,, \tag{102}$$

$$\hat{\kappa}''(u) = -\frac{1}{\pi} \mathscr{P} \int_{-\infty}^{\infty} \frac{\hat{\kappa}'(\omega)}{\omega - u} \, \mathrm{d}\omega \tag{103}$$

Each of these two formulae are equivalent with (100) (they are each others "Hilbert transforms"). They are known in physics as dispersion relations or Kramers–Kronig relations.

From the reality of the matrix $\kappa(t)$ it follows from (92) that the matrix $\hat{\kappa}(\omega)$ satisfies the condition

$$\hat{\kappa}^{*}(\omega) = \hat{\kappa}(-\omega) \,, \tag{104}$$

or with (101)

$$\hat{\kappa}'(\omega) = \hat{\kappa}'(-\omega) \,, \tag{105}$$

$$\hat{\kappa}''(\omega) = -\hat{\kappa}''(-\omega) \,. \tag{106}$$

With these properties we get an alternative form of the Kramers–Kronig relations:

$$\hat{\kappa}'(u) = \frac{2}{\pi} \mathscr{P} \int_0^\infty \frac{\omega \hat{\kappa}''(\omega)}{\omega^2 - u^2} \, d\omega \,, \tag{107}$$

$$\hat{\kappa}''(u) = -\frac{2}{\pi} \mathscr{P} \int_0^\infty \frac{u \hat{\kappa}'(\omega)}{\omega^2 - u^2} \, d\omega \,. \tag{108}$$

Another corollary of (96) is obtained by considering the function $e^{iwt} \hat{\kappa}(w)$ for positive values of t. This function has all the properties stated in (96) for $\hat{\kappa}(w)$. Therefore application of Cauchy's theorem to the function

$$f(w) = \frac{e^{iwt} \hat{\kappa}(w)}{w - u}, \quad (t \geqslant 0) \,, \tag{109}$$

with u again a real quantity, leads in the same way, as we have obtained formula (100), to

$$e^{iut} \hat{\kappa}(u) = \frac{1}{\pi i} \mathscr{P} \int_{-\infty}^\infty \frac{e^{i\omega t} \hat{\kappa}(\omega)}{\omega - u} \, d\omega \,, \quad (t \geqslant 0) \,. \tag{110}$$

In the same way also one obtains for the function $e^{-iwt} \hat{\kappa}(w)$ for negative values of t:

$$e^{-iut} \hat{\kappa}(u) = \frac{1}{\pi i} \mathscr{P} \int_{-\infty}^\infty \frac{e^{-i\omega t} \hat{\kappa}(\omega)}{\omega - u} \, d\omega \,, \quad (t \leqslant 0) \,. \tag{111}$$

These two formulae are also a consequence of the causality principle. For $t = 0$ they reduce to the previous result (100).

We shall apply the last two formulations of the causality principle to equation (83) with $\boldsymbol{F}(t)$ a special given function of time. We first introduce (93) into (87) and obtain

$$\boldsymbol{x}(t) = \frac{1}{2\pi} \int_{-\infty}^\infty e^{-i\omega t} \hat{\kappa}(\omega) \cdot \hat{\boldsymbol{F}}(\omega) \, d\omega \,. \tag{112}$$

Now choose the function $\boldsymbol{F}(t)$ in such a way that

$$\boldsymbol{F}(t) = \boldsymbol{F} S(-t) \,, \tag{113}$$

where $\boldsymbol{F}$ is a vector with constant components and where

$$S(t) = \begin{Bmatrix} 1\,, & \text{if} \quad t > 0 \\ \\ 0\,, & \text{if} \quad t < 0 \end{Bmatrix}. \tag{114}$$

This corresponds to constant driving forces which are lifted at time $t = 0$. Then from (91) one obtains for the Fourier transform

$$\hat{\boldsymbol{F}}(\omega) = \boldsymbol{F}\left\{\pi\delta(\omega) + \mathscr{P}\frac{1}{\mathrm{i}\omega}\right\}. \tag{115}$$

Here the symbol $\mathscr{P}$ indicates that one has to take the principal part of the integral in which the term $1/\mathrm{i}\omega$ occurs*. Introducing this expression

* Formula (115) is obtained in the following way. From (113) and (91) one gets for the Fourier transforms of $S(t)$ and $S(-t)$

$$\hat{S}_+(\omega) = \int_0^\infty \mathrm{e}^{\mathrm{i}\omega t}\,\mathrm{d}t\,,$$

$$\hat{S}_-(\omega) = \int_{-\infty}^0 \mathrm{e}^{\mathrm{i}\omega t}\,\mathrm{d}t\,.$$

For the sum of these two expressions we have with (94)

$$\hat{S}_+(\omega) + \hat{S}_-(\omega) = 2\pi\delta(\omega)\,,$$

and for their difference, using a convergence factor in the integrals

$$\hat{S}_+(\omega) - \hat{S}_-(\omega) = -\frac{2}{\mathrm{i}}\lim_{\lambda\to 0}\frac{\omega}{\omega^2 + \lambda^2}\,.$$

From the last two equations we find

$$\hat{S}_-(\omega) = \pi\delta(\omega) + \frac{1}{\mathrm{i}}\lim_{\lambda\to 0}\frac{\omega}{\omega^2 + \lambda^2} = \pi\delta(\omega) + \mathscr{P}\frac{1}{\mathrm{i}\omega}\,.$$

The last equality is found by considering the integrals in which $\hat{S}_-(\omega)$ occurs.

into (112) one finds

$$\boldsymbol{x}(t) = \frac{1}{2}\left\{ \hat{\kappa}(0) + \frac{1}{\pi \mathrm{i}} \mathscr{P} \int_{-\infty}^{\infty} \mathrm{e}^{-\mathrm{i}\omega t} \frac{\hat{\kappa}(\omega)}{\omega} \mathrm{d}\omega \right\} \boldsymbol{F}. \tag{116}$$

This relation holds for all times.

For negative times we apply (111) for $u = 0$,

$$\hat{\kappa}(0) = \frac{1}{\pi \mathrm{i}} \mathscr{P} \int_{-\infty}^{\infty} \mathrm{e}^{-\mathrm{i}\omega t} \frac{\hat{\kappa}(\omega)}{\omega} \mathrm{d}\omega, \quad (t \leqslant 0), \tag{117}$$

so that (116) then gets the form

$$\boldsymbol{x}(t) = \hat{\kappa}(0) \cdot \boldsymbol{F}, \quad (t \leqslant 0). \tag{118}$$

This trivial result was to be expected, since according to the principle of causality, constant driving forces will induce a constant response for $t \leqslant 0$, even though the force is lifted at $t = 0$.

For positive times we apply (110) for $u = 0$,

$$\hat{\kappa}(0) = \frac{1}{\pi \mathrm{i}} \mathscr{P} \int_{-\infty}^{\infty} \mathrm{e}^{\mathrm{i}\omega t} \frac{\hat{\kappa}(\omega)}{\omega} \mathrm{d}\omega, \quad (t \geqslant 0) \tag{119}$$

and obtain for (116) in this case

$$\boldsymbol{x}(t) = \frac{1}{\pi \mathrm{i}} \mathscr{P} \int_{-\infty}^{\infty} \cos \omega t \frac{\hat{\kappa}(\omega)}{\omega} \mathrm{d}\omega \cdot \boldsymbol{F}, \quad (t \geqslant 0). \tag{120}$$

This formula based on the causality principle will enable us to derive the fluctuation dissipation theorem.

§ 4. *Derivation of the Fluctuation Dissipation Theorem*

We are now in a position to derive the fluctuation dissipation theorem, due to Callen and Greene*.

* H. B. Callen and R. F. Greene, Phys. Rev. **86** (1952) 702; R. F. Greene and H. B. Callen, Phys. Rev. **88** (1952) 1387.

Let us consider again the random variables $\boldsymbol{\alpha}(t)$ discussed in § 2 of this chapter. Under the influence of sufficiently small external driving forces $\boldsymbol{F}(t)$ the mean values of these variables will obey relations of the type (83), *viz.*

$$\bar{\boldsymbol{\alpha}}(t) = \int_{-\infty}^{\infty} \kappa(\tau) \cdot \boldsymbol{F}(t - \tau) \, \mathrm{d}\tau \,, \tag{121}$$

where $\kappa(\tau)$ satisfies the causality condition (85). The mean value $\bar{\boldsymbol{\alpha}}$ in (121) is to be obtained with the help of an appropriate distribution function $f(\boldsymbol{\alpha}, t)$, corresponding to the initial (stationary) conditions imposed on the system at time $t = -\infty$,

$$\bar{\boldsymbol{\alpha}}(t) = \int \boldsymbol{\alpha} f(\boldsymbol{\alpha}, t) \, \mathrm{d}\boldsymbol{\alpha} \,. \tag{122}$$

Due to the time dependent driving forces $\boldsymbol{F}(t)$ the distribution function $f(\boldsymbol{\alpha}, t)$ does not remain stationary.

As in the previous section we choose $\boldsymbol{F}(t)$ in such a way that

$$\boldsymbol{F}(t) = \boldsymbol{F} S(-t) \,, \tag{123}$$

where the function $S(t)$ is given by (114). Equation (121) now becomes, according to (118) and (120),

$$\bar{\boldsymbol{\alpha}}(t) = \int \boldsymbol{\alpha} f(\boldsymbol{\alpha}, t) \, \mathrm{d}\boldsymbol{\alpha} = \mathsf{K}(t) \cdot \boldsymbol{F} \,, \tag{124}$$

with

$$\mathsf{K}(t) = \begin{Bmatrix} \hat{\kappa}(0) \,, & \text{if} \quad t \leqslant 0 \\ \dfrac{1}{\pi \mathrm{i}} \mathscr{P} \displaystyle\int_{-\infty}^{\infty} \cos \omega t \, \frac{\hat{\kappa}(\omega)}{\omega} \, \mathrm{d}\omega \,, & \text{if} \quad t \geqslant 0 \end{Bmatrix} , \tag{125}$$

since all results derived for $\boldsymbol{x}(t)$ in the previous section are valid for the mean values $\bar{\boldsymbol{\alpha}}(t)$.

The distribution function $f(\boldsymbol{\alpha}, t)$ can formally be obtained through the relation

$$f(\boldsymbol{\alpha}', t) = \int f(\boldsymbol{\alpha}, 0 \,; \boldsymbol{\alpha}', t) \, \mathrm{d}\boldsymbol{\alpha} \,, \tag{126}$$

where $f(\boldsymbol{\alpha}, 0; \boldsymbol{\alpha}', t)$ is the joint distribution function of $\boldsymbol{\alpha}$ at time 0 and t. This relation may also be written for $t > 0$ as

$$f(\boldsymbol{\alpha}', t) = \int f(\boldsymbol{\alpha}, 0) P(\boldsymbol{\alpha}, 0 \mid \boldsymbol{\alpha}', t)\, \mathrm{d}\boldsymbol{\alpha}\,, \tag{127}$$

with the conditional distribution function

$$P(\boldsymbol{\alpha}, 0 \mid \boldsymbol{\alpha}', t) \equiv \frac{f(\boldsymbol{\alpha}, 0; \boldsymbol{\alpha}', t)}{f(\boldsymbol{\alpha}, 0)}\,. \tag{128}$$

Now, for the driving forces (123), the evolution in time of $\bar{\boldsymbol{\alpha}}(t)$ is determined by the Hamiltonian (or the phase space transformation function [*cf.* (VII.53)]) of the system in the absence of driving forces. Furthermore, since $f(\boldsymbol{\alpha}, 0) = f(\boldsymbol{\alpha}, -\infty)$ corresponds to a stationary (micro-canonical, or equivalently, canonical) distribution in phase space in the presence of constant driving forces $\boldsymbol{F}$, *i.e.* a distribution which is uniform over domains $(\boldsymbol{\alpha}, \boldsymbol{\alpha} + \mathrm{d}\boldsymbol{\alpha})$ in each energy shell $(E, E + \mathrm{d}E)$, one can show* that [*cf.* (VII.76)]

$$P(\boldsymbol{\alpha}, 0 \mid \boldsymbol{\alpha}', t) = P(\boldsymbol{\alpha} \mid \boldsymbol{\alpha}'; t)\,, \tag{129}$$

where $P(\boldsymbol{\alpha} \mid \boldsymbol{\alpha}'; t)$ is the stationary conditional probability density of the stationary process $\boldsymbol{\alpha}(t)$ in the absence of driving forces. We therefore have from (124), (127) and (129)

$$\int f(\boldsymbol{\alpha}, 0) \left[\int \boldsymbol{\alpha}' P(\boldsymbol{\alpha} \mid \boldsymbol{\alpha}'; t)\, \mathrm{d}\boldsymbol{\alpha}' \right] \mathrm{d}\boldsymbol{\alpha} = \mathsf{K}(t) \cdot \boldsymbol{F}\,, \quad (t \geqslant 0)\,. \tag{130}$$

The integral between brackets represents the conditional average value of $\boldsymbol{\alpha}$ in the stationary ensemble. In view of the assumed linearity of the process we may write

$$\bar{\boldsymbol{\alpha}}^{\boldsymbol{\alpha}(0)}(t) \equiv \int \boldsymbol{\alpha}' P(\boldsymbol{\alpha} \mid \boldsymbol{\alpha}'; t)\, \mathrm{d}\boldsymbol{\alpha}' = \mathsf{n}(t) \cdot \boldsymbol{\alpha}\,, \quad (t \geqslant 0)\,. \tag{131}$$

* Strictly speaking in order to prove relation (129) in the present case it is necessary that the potential energy V of the system due to the external forces is constant in regions $(\boldsymbol{\alpha}, \boldsymbol{\alpha} + \mathrm{d}\boldsymbol{\alpha})$, or in other words, that V is a function of $\boldsymbol{r}^N$ and $\boldsymbol{p}^N$ only through $\boldsymbol{\alpha}(\boldsymbol{r}^N, \boldsymbol{p}^N)$ [*e.g.* $V(\boldsymbol{r}^N, \boldsymbol{p}^N) = \boldsymbol{\alpha}(\boldsymbol{r}^N, \boldsymbol{p}^N) \cdot \boldsymbol{F}$].

The function $\mathsf{n}(t)$ is related to the correlation function $\rho(t)$. One has indeed with (2)

$$\rho(t) = \iint \boldsymbol{\alpha}\boldsymbol{\alpha}' f(\boldsymbol{\alpha}) P(\boldsymbol{\alpha} \mid \boldsymbol{\alpha}' ; t)\, \mathrm{d}\boldsymbol{\alpha}\, \mathrm{d}\boldsymbol{\alpha}'$$

$$= \int \boldsymbol{\alpha} \{ \mathsf{n}(t) \cdot \boldsymbol{\alpha} \} f(\boldsymbol{\alpha})\, \mathrm{d}\boldsymbol{\alpha} = k g^{-1} \cdot \tilde{\mathsf{n}}(t)\,, \quad (t \geqslant 0)\,. \qquad (132)$$

Here $f(\boldsymbol{\alpha})$ is the stationary distribution function of $\boldsymbol{\alpha}$ in the absence of driving forces.

On the other hand, introducing (131) into (130) and using (124), we find that

$$\mathsf{n}(t) \cdot \hat{\kappa}(0) = \mathsf{K}(t)\,, \quad (t \geqslant 0)\,. \qquad (133)$$

Finally with (132),

$$\tilde{\rho}(t) = \mathsf{K}(t) \cdot \hat{\kappa}^{-1}(0) \cdot k g^{-1}\,, \quad (t \geqslant 0)\,. \qquad (134)$$

This relation represents essentially the fluctuation dissipation theorem: it relates the correlation function of spontaneous fluctuations for the stationary process $\boldsymbol{\alpha}(t)$ to the "relaxation function" $\mathsf{K}(t)$ (which contains the susceptibility matrix) and thus, as we shall see in the next section, to the dissipation (or entropy production) of the system under the influence of time dependent driving forces.

We shall rewrite the fluctuation dissipation theorem in a form which is valid for positive as well as for negative time t. To this end we shall use the relation

$$\hat{\kappa}(0) = \frac{1}{T} g^{-1}\,, \qquad (135)$$

which follows from thermodynamics if the forces $\boldsymbol{F}$ are thermodynamically conjugate to the variables $\boldsymbol{\alpha}$ and which will be established in the next section. With this expression and (125) we get for (134)

$$\tilde{\rho}(t) = \frac{kT}{\pi \mathrm{i}} \mathscr{P} \int_{-\infty}^{\infty} \cos \omega t\, \frac{\hat{\kappa}(\omega)}{\omega}\, \mathrm{d}\omega\,, \quad (t \geqslant 0)\,. \qquad (136)$$

Writing for $\hat{\kappa}(\omega)$

$$\hat{\kappa} = \hat{\kappa}^{\mathrm{s}} + \hat{\kappa}^{\mathrm{a}}\,, \qquad (137)$$

where $\hat{\kappa}^s$ and $\hat{\kappa}^a$ are the symmetric and the antisymmetric parts respectively of the susceptibility matrix, we have from (119)

$$\mathscr{P} \int_{-\infty}^{\infty} e^{i\omega t} \frac{\hat{\kappa}^a(\omega)}{\omega} d\omega = 0 , \quad (t \geqslant 0) , \tag{138}$$

since $\hat{\kappa}(0) = g^{-1}/T$ is a symmetric matrix. The above equation also implies that

$$\mathscr{P} \int_{-\infty}^{\infty} \cos \omega t \frac{\hat{\kappa}^a(\omega)}{\omega} d\omega = - i \mathscr{P} \int_{-\infty}^{\infty} \sin \omega t \frac{\hat{\kappa}^a(\omega)}{\omega} d\omega . \tag{139}$$

(This relation is also a corollary of the causality principle.) Therefore (136) may be rewritten in the form

$$\tilde{\rho}(t) = \frac{kT}{\pi i} \mathscr{P} \int_{-\infty}^{\infty} \cos \omega t \frac{\hat{\kappa}^s(\omega)}{\omega} d\omega - \frac{kT}{\pi} \mathscr{P} \int_{-\infty}^{\infty} \sin \omega t \frac{\hat{\kappa}^a(\omega)}{\omega} d\omega$$

$$= \frac{kT}{\pi} \mathscr{P} \int_{-\infty}^{\infty} \cos \omega t \frac{\{ \hat{\kappa}''(\omega) \}^s}{\omega} d\omega - \frac{kT}{\pi} \mathscr{P} \int_{-\infty}^{\infty} \sin \omega t \frac{\{ \hat{\kappa}'(\omega) \}^a}{\omega} d\omega , \tag{140}$$

or

$$\rho(t) = \frac{2kT}{\pi} \mathscr{P} \left\{ \int_{0}^{\infty} \cos \omega t \frac{\{ \hat{\kappa}''(\omega) \}^s}{\omega} d\omega + \int_{0}^{\infty} \sin \omega t \frac{\{ \hat{\kappa}'(\omega) \}^a}{\omega} d\omega \right\} . \tag{141}$$

We have used here the even and odd character of $\hat{\kappa}'(\omega)$ and $\hat{\kappa}''(\omega)$ respectively [*cf.* (105) and (106)].

Now the correlation function must satisfy the stationarity condition (5)

$$\tilde{\rho}(- t) = \rho(t) , \tag{142}$$

which implies that the symmetric part of $\rho(t)$ is an even function of time, and the antisymmetric part an odd function. The form (141) satisfies this condition and therefore represents the correlation function for both positive and negative times.

Comparison with the Wiener–Khinchin theorem in the form (36) shows that G and H, the real and imaginary parts of the spectral density matrix S, are given by*

$$\mathsf{G} = \frac{2kT}{\pi} \frac{\{\hat{\kappa}''(\omega)\}^{s}}{\omega}, \tag{143}$$

$$\mathsf{H} = \frac{2kT}{\pi} \frac{\{\hat{\kappa}'(\omega)\}^{a}}{\omega}. \tag{144}$$

The fluctuation dissipation theorem is completely contained in these two relations: knowledge of the susceptibility matrix completely determines (through the spectral density matrix) the correlation function matrix**.

In the case of $\boldsymbol{\alpha}$-type variables (even variables) microscopic reversibility, which had as a consequence the relations (52) for G and H, implies in view of (143) and (144)

$$\{\hat{\kappa}''(\omega\,;\boldsymbol{B})\}^{s} = \{\hat{\kappa}''(\omega\,; -\boldsymbol{B})\}^{s}, \tag{145}$$

$$\{\hat{\kappa}'(\omega\,;\boldsymbol{B})\}^{a} = -\{\hat{\kappa}'(\omega\,; -\boldsymbol{B})\}^{a}. \tag{146}$$

From the Kramers–Kronig relations (102) and (103) we then find that $\{\hat{\kappa}'(\omega;\boldsymbol{B})\}^{s}$ also satisfies a relation similar to (145) and $\{\hat{\kappa}''(\omega;\boldsymbol{B})\}^{a}$ a relation like (146). Therefore one has the two equalities:

$$\hat{\kappa}^{s}(\omega\,;\boldsymbol{B}) = \hat{\kappa}^{s}(\omega\,; -\boldsymbol{B}), \tag{147}$$

$$\hat{\kappa}^{a}(\omega\,;\boldsymbol{B}) = -\hat{\kappa}^{a}(\omega\,; -\boldsymbol{B}), \tag{148}$$

which may be summarized as

$$\hat{\kappa}(\omega\,;\boldsymbol{B}) = \tilde{\hat{\kappa}}(\omega\,; -\boldsymbol{B}). \tag{149}$$

* If these functions occur in integrals, principal values have to be taken.

** In the theory of electric noise this relation had been established previously by Nyquist [H. Nyquist, Phys. Rev. **32** (1928) 110]. The fluctuation dissipation theorem can in fact be considered as a generalization of the Nyquist relation.

In statistical mechanics of transport processes one can establish relations between transport coefficients and correlation functions which are very similar to the fluctuation dissipation theorem. Such relations are known in statistical mechanics as Kubo relations [R. Kubo, J. Phys. Soc. (Japan) **12** (1957) 570; H. B. Callen, M. L. Barasch and J. L. Jackson, Phys. Rev. **88** (1952) 1382; M. S. Green, J. chem. Phys. **19** (1951) 1036].

The corresponding reciprocal relations for $\boldsymbol{\alpha}$- and $\boldsymbol{\beta}$-type variables may be easily found from (53) and (54).

These relations are an extension of the Onsager reciprocal relations to the coefficients occurring in the macroscopic linear response laws (121)*.

§ 5. *The Entropy Production in a System subjected to External Driving Forces*

We shall first write down the total differential of the entropy, when external driving forces act on the system under consideration. This differential is given by

$$\mathrm{d}S = \frac{1}{T}\,\mathrm{d}U + \boldsymbol{X}\cdot\mathrm{d}\boldsymbol{\alpha}\,. \tag{150}$$

At constant energy this expression is equivalent to (VII.49) and to [*cf.* (VII.16) and (VII.51)]

$$\boldsymbol{X} = -\,\mathsf{g}\cdot\boldsymbol{\alpha} = \left(\frac{\partial S}{\partial \boldsymbol{\alpha}}\right)_U. \tag{151}$$

The change $\mathrm{d}U$ is given by the first law of thermodynamics

$$\mathrm{d}U = \boldsymbol{F}\cdot\mathrm{d}\boldsymbol{\alpha}\,. \tag{152}$$

Here we have assumed that the system is insulated against heat exchange with its surroundings. The term $\boldsymbol{F}\cdot\mathrm{d}\boldsymbol{\alpha}$ represents the infinitesimal work performed on the system. The forces $\boldsymbol{F}$ are thermodynamically conjugate to the variables $\boldsymbol{\alpha}$:

$$\boldsymbol{F} = \frac{\partial U}{\partial \boldsymbol{\alpha}}. \tag{153}$$

If we introduce (152) into (150) we obtain

$$\mathrm{d}S = \left(\frac{\boldsymbol{F}}{T} + \boldsymbol{X}\right)\cdot\mathrm{d}\boldsymbol{\alpha} \tag{154}$$

* We have used throughout the last two sections the generalized susceptibility matrix $\hat{\mathsf{K}}(\omega)$. Frequently one uses instead the generalized admittance matrix $\mathsf{Y}(\omega) = -\,\mathrm{i}\omega\hat{\mathsf{K}}(\omega)$, or its inverse $\mathsf{Z}(\omega) = \mathrm{i}\omega^{-1}\,\hat{\mathsf{K}}^{-1}(\omega)$, called the generalized impedance matrix, which occur in the relation between the Fourier transforms of $\dot{\boldsymbol{\alpha}}(t)$ and $\boldsymbol{F}(t)$.

and
$$\frac{\boldsymbol{F}}{T} + \boldsymbol{X} = \frac{\partial S}{\partial \boldsymbol{\alpha}}. \tag{155}$$

For driving forces which are constant in time, we have seen that the mean values of the random variables $\boldsymbol{\alpha}(t)$ are given by

$$\bar{\boldsymbol{\alpha}}(t) = \hat{\kappa}(0) \cdot \boldsymbol{F}, \tag{156}$$

[*cf.* (121) with (92)]. This mean value may be identified with the most probable value of $\boldsymbol{\alpha}$ in the representative stationary ensemble. According to Boltzmann's entropy postulate the entropy must therefore have a maximum for the value (156) of $\boldsymbol{\alpha}$:

$$\left(\frac{\partial S}{\partial \boldsymbol{\alpha}}\right)_{\boldsymbol{\alpha} = \hat{\kappa}(0) \cdot \boldsymbol{F}} = 0. \tag{157}$$

This condition leads with (155) and (151) to

$$\hat{\kappa}(0) = \frac{1}{T} g^{-1}. \tag{158}$$

This is the relation (135) used in the previous section: *the form* (141) *of the fluctuation dissipation theorem and also* (143) *and* (144) *as well as the reciprocal relations* (145)–(149) *hold if the driving forces* $\boldsymbol{F}(t)$ *are conjugate to the values* $\boldsymbol{\alpha}$ *according to* (153).

For the rate of change of entropy with time we have from (154)

$$\frac{\mathrm{d}S}{\mathrm{d}t} = \left(\frac{\boldsymbol{F}(t)}{T} + \boldsymbol{X}\right) \cdot \frac{\mathrm{d}\boldsymbol{\alpha}}{\mathrm{d}t}. \tag{159}$$

In § 8 of Chapter VII we have calculated for a system without external driving forces the time derivative of the conditional average value of entropy considered as a random variable. We could also have computed the conditional average value of (159) with $\boldsymbol{F} = 0$ considering $\mathrm{d}S/\mathrm{d}t$ to be a random variable. Both methods give the same result for the entropy production, since the two operations of taking a time derivative and of averaging commute. In this section we use the latter method.

We first note that for time independent driving forces (159) vanishes if averaged with the corresponding stationary distribution function for $\boldsymbol{\alpha}$.

For time dependent driving forces the average of (159) is

$$\overline{\frac{\mathrm{d}S}{\mathrm{d}t}} = \frac{\boldsymbol{F}(t)}{T} \cdot \overline{\frac{\mathrm{d}\boldsymbol{\alpha}}{\mathrm{d}t}} + \overline{\boldsymbol{X} \cdot \frac{\mathrm{d}\boldsymbol{\alpha}}{\mathrm{d}t}}$$

$$= \frac{\boldsymbol{F}(t)}{T} \cdot \frac{\mathrm{d}\overline{\boldsymbol{\alpha}(t)}}{\mathrm{d}t} - \mathsf{g} : \overline{\boldsymbol{\alpha}(t)} \frac{\mathrm{d}\overline{\boldsymbol{\alpha}(t)}}{\mathrm{d}t} - \tfrac{1}{2}\mathsf{g} : \frac{\mathrm{d}}{\mathrm{d}t} \overline{(\boldsymbol{\alpha} - \bar{\boldsymbol{\alpha}})(\boldsymbol{\alpha} - \bar{\boldsymbol{\alpha}})} . \quad (160)$$

Here we have used formula (151); the mean value $\bar{\boldsymbol{\alpha}}(t)$ is given by (121). In terms of the Fourier transforms $\hat{\bar{\boldsymbol{\alpha}}}(\omega)$ and $\hat{\boldsymbol{F}}(\omega)$, (160) becomes

$$\overline{\frac{\mathrm{d}S}{\mathrm{d}t}} = \frac{\mathrm{i}}{4\pi^2 T} \int\!\!\int_{-\infty}^{\infty} \omega \hat{\boldsymbol{F}}(\omega') \cdot \hat{\kappa}^*(\omega) \cdot \hat{\boldsymbol{F}}^*(\omega)\, \mathrm{e}^{\mathrm{i}(\omega-\omega')t}\, \mathrm{d}\omega\, \mathrm{d}\omega'$$

$$- \frac{\mathrm{i}}{4\pi^2} \int\!\!\int_{-\infty}^{\infty} \omega \mathsf{g} : \{ \hat{\kappa}^*(\omega) \cdot \hat{\boldsymbol{F}}^*(\omega) \hat{\boldsymbol{F}}(\omega') \cdot \tilde{\hat{\kappa}}(\omega') \}\, \mathrm{e}^{\mathrm{i}(\omega-\omega')t}\, \mathrm{d}\omega\, \mathrm{d}\omega'$$

$$- \tfrac{1}{2}\mathsf{g} : \frac{\mathrm{d}\mathsf{q}(t)}{\mathrm{d}t} . \quad (161)$$

Here we have used the relation between $\hat{\bar{\boldsymbol{\alpha}}}(\omega)$ and $\hat{\boldsymbol{F}}(\omega)$,

$$\hat{\bar{\boldsymbol{\alpha}}}(\omega) = \hat{\kappa}(\omega) \cdot \hat{\boldsymbol{F}}(\omega) , \quad (162)$$

which follows through Fourier transformation of (121) [*cf.* also (93)]. Furthermore we have introduced the abbreviation

$$\mathsf{q}(t) = \overline{(\boldsymbol{\alpha} - \bar{\boldsymbol{\alpha}})(\boldsymbol{\alpha} - \bar{\boldsymbol{\alpha}})} . \quad (163)$$

For a stationary ensemble in the absence of driving forces $\mathsf{q}(t)$ is equal to $k\mathsf{g}^{-1}$ [$\bar{\boldsymbol{\alpha}}$ then vanishes and $\mathsf{q}(t) = \overline{\boldsymbol{\alpha}\boldsymbol{\alpha}} = \langle \boldsymbol{\alpha}\boldsymbol{\alpha} \rangle$, *i.e.* the average taken with the equilibrium distribution function $f(\boldsymbol{\alpha})$].

Let us now consider the time integral of (161), assuming that the driving forces only act on the system for a finite time interval. In this case the tensor $q(t)$ is equal to kg^{-1} at $t = -\infty$ (before switching on the forces), and at time $t = +\infty$, when the system has again relaxed to equilibrium. We therefore get, using (94):

$$\int_{-\infty}^{\infty} \overline{\frac{\mathrm{d}S}{\mathrm{d}t}}\,\mathrm{d}t = \frac{\mathrm{i}}{2\pi T}\int_{-\infty}^{\infty} \omega\hat{\boldsymbol{F}}(\omega)\cdot\hat{\kappa}^*(\omega)\cdot\hat{\boldsymbol{F}}^*(\omega)\,\mathrm{d}\omega$$
$$-\frac{\mathrm{i}}{2\pi}\int_{-\infty}^{\infty} \omega g : \{\,\hat{\kappa}^*(\omega)\cdot\hat{\boldsymbol{F}}^*(\omega)\hat{\boldsymbol{F}}(\omega)\cdot\tilde{\hat{\kappa}}(\omega)\,\}\,\mathrm{d}\omega\,. \tag{164}$$

The second integral vanishes since the integrand is an odd function of ω:

$$g : \{\,\hat{\kappa}^*(\omega)\cdot\hat{\boldsymbol{F}}^*(\omega)\hat{\boldsymbol{F}}(\omega)\cdot\tilde{\hat{\kappa}}(\omega)\,\} = g : \{\,\hat{\kappa}(-\omega)\cdot\hat{\boldsymbol{F}}(-\omega)\hat{\boldsymbol{F}}^*(-\omega)\cdot\tilde{\hat{\kappa}}^*(-\omega)\,\}$$
$$= g : \{\,\hat{\kappa}^*(-\omega)\cdot\hat{\boldsymbol{F}}^*(-\omega)\hat{\boldsymbol{F}}(-\omega)\cdot\tilde{\hat{\kappa}}(-\omega)\,\}\,. \tag{165}$$

We have used here the condition (104) and a similar one for $\hat{\boldsymbol{F}}(\omega)$, as well as the matrix equality $\mathsf{A} : \mathsf{B} = \tilde{\mathsf{A}} : \tilde{\mathsf{B}}$. As a result we have for (164) [*cf.* (160) and (152)]

$$\int_{-\infty}^{\infty} \overline{\frac{\mathrm{d}S}{\mathrm{d}t}}\,\mathrm{d}t = \int_{-\infty}^{\infty} \frac{\boldsymbol{F}}{T}\cdot\frac{\mathrm{d}\bar{\boldsymbol{\alpha}}}{\mathrm{d}t}\,\mathrm{d}t = \frac{1}{T}\int_{-\infty}^{\infty} \overline{\frac{\mathrm{d}U}{\mathrm{d}t}}\,\mathrm{d}t$$
$$= \frac{\mathrm{i}}{2\pi T}\int_{-\infty}^{\infty} \omega\hat{\boldsymbol{F}}(\omega)\cdot\hat{\kappa}^*(\omega)\cdot\hat{\boldsymbol{F}}^*(\omega)\,\mathrm{d}\omega\,. \tag{166}$$

Thus the time integral of the entropy production is equal to the time integral of the rate of change of the energy divided by the overall equilibrium temperature of the system. We expect this quantity to be positive definite. This fact can indeed be established in the following way. With the condition (104) and a similar one for $\hat{\boldsymbol{F}}(\omega)$ one has

$$\hat{\boldsymbol{F}}(\omega)\cdot\hat{\kappa}^*(\omega)\cdot\hat{\boldsymbol{F}}^*(\omega) = \hat{\boldsymbol{F}}^*(-\omega)\cdot\hat{\kappa}(-\omega)\cdot\hat{\boldsymbol{F}}\,(-\omega)$$
$$= \hat{\boldsymbol{F}}(-\omega)\cdot\tilde{\hat{\kappa}}(-\omega)\cdot\hat{\boldsymbol{F}}^*(-\omega)\,. \tag{167}$$

With this relation we find for (166), changing the variable of integration from ω to $-\omega$,

$$\frac{1}{T}\int_{-\infty}^{\infty}\overline{\frac{\mathrm{d}U}{\mathrm{d}t}}\,\mathrm{d}t = -\frac{\mathrm{i}}{2\pi T}\int_{-\infty}^{\infty}\omega\hat{\boldsymbol{F}}(\omega)\cdot\tilde{\hat{\kappa}}(\omega)\cdot\hat{\boldsymbol{F}}^{*}(\omega)\,\mathrm{d}\omega\,. \tag{168}$$

Taking half of the sum of (166) and (168) we obtain

$$\int_{-\infty}^{\infty}\overline{\frac{\mathrm{d}U}{\mathrm{d}t}}\,\mathrm{d}t = \frac{\mathrm{i}}{4\pi}\int_{-\infty}^{\infty}\omega\hat{\boldsymbol{F}}\cdot(\hat{\kappa}^{*}-\tilde{\hat{\kappa}})\cdot\hat{\boldsymbol{F}}^{*}\,\mathrm{d}\omega\,. \tag{169}$$

With the relation

$$\mathsf{S}(\omega) = \frac{\mathrm{i}kT}{\pi\omega}\left\{\hat{\kappa}^{*}(\omega)-\tilde{\hat{\kappa}}(\omega)\right\}, \tag{170}$$

which follows from the fluctuation dissipation theorem (143) and (144), equation (169) becomes

$$\int_{-\infty}^{\infty}\overline{\frac{\mathrm{d}U}{\mathrm{d}t}}\,\mathrm{d}t = \frac{1}{4kT}\int_{-\infty}^{\infty}\omega^{2}\hat{\boldsymbol{F}}\cdot\mathsf{S}\cdot\hat{\boldsymbol{F}}^{*}\,\mathrm{d}\omega \geqslant 0\,. \tag{171}$$

This inequality follows from the property (28) of the matrix S. We have thus established that the time integral of the energy dissipation is a non-negative quantity.

From a macroscopic point of view the positive character of the energy dissipation written in the form (166) or (168) follows simply from the requirement that the entropy production $\mathrm{d}S/\mathrm{d}t$ must be positive. In the macroscopic theory the rate of change of entropy is then given by

$$\frac{\mathrm{d}S}{\mathrm{d}t} = \frac{\boldsymbol{F}}{T}\frac{\mathrm{d}\boldsymbol{\alpha}}{\mathrm{d}t} + \boldsymbol{X}\cdot\frac{\mathrm{d}\boldsymbol{\alpha}}{\mathrm{d}t}\,, \tag{172}$$

where one assumes that the variables $\boldsymbol{\alpha}$ obey themselves the macroscopic relations (121), and not only their mean values. This leads to an expression which differs from (161) only in that it does not contain the last term $-\tfrac{1}{2}\boldsymbol{g} : \mathrm{d}\boldsymbol{q}(t)/\mathrm{d}t$ which may be neglected on a macroscopic scale (*cf.* the discussion in Chapter VII, § 8).

We also wish to study the energy dissipation for the case of "monochromatic" driving forces, oscillating with a frequency ω_0,

$$\boldsymbol{F}(t) = \tfrac{1}{2}\boldsymbol{F}\,\mathrm{e}^{-\mathrm{i}\omega_0 t} + \tfrac{1}{2}\boldsymbol{F}^*\,\mathrm{e}^{\mathrm{i}\omega_0 t}. \tag{173}$$

The Fourier components of these forces are given by

$$\hat{\boldsymbol{F}}(\omega) = \pi\,\{\,\boldsymbol{F}\delta(\omega - \omega_0) + \boldsymbol{F}^*\delta(\omega + \omega_0)\,\}. \tag{174}$$

Introducing this expression into (161) we get

$$\begin{aligned}\overline{\frac{\mathrm{d}S}{\mathrm{d}t}} &= \frac{\mathrm{i}}{4T}\,\omega_0\,\{\,\hat{\kappa}^*(\omega_0) : (\boldsymbol{F}^*\boldsymbol{F} + \boldsymbol{F}^*\boldsymbol{F}^*\,\mathrm{e}^{2\mathrm{i}\omega_0 t})\\ &\quad - \tilde{\hat{\kappa}}(\omega_0) : (\boldsymbol{F}^*\boldsymbol{F} + \boldsymbol{F}\boldsymbol{F}\,\mathrm{e}^{-2\mathrm{i}\omega_0 t})\,\}\\ &\quad - \frac{\mathrm{i}}{4}\,\omega_0\tilde{\hat{\kappa}}(\omega_0)\cdot g\cdot\hat{\kappa}^*(\omega_0) : (\boldsymbol{F}^*\boldsymbol{F}^*\,\mathrm{e}^{2\mathrm{i}\omega_0 t} - \boldsymbol{F}\boldsymbol{F}\,\mathrm{e}^{-2\mathrm{i}\omega_0 t})\\ &\quad - \tfrac{1}{2}g : \frac{\mathrm{d}q(t)}{\mathrm{d}t}.\end{aligned} \tag{175}$$

The integral over time of this expression from $-\infty$ to ∞ diverges since the dissipation during every period $2\pi/\omega_0$ is finite. During one such period the total entropy production is

$$\int_{-\pi/\omega_0}^{\pi/\omega_0} \overline{\frac{\mathrm{d}S}{\mathrm{d}t}}\,\mathrm{d}t = \frac{\mathrm{i}\pi}{2T}\,\boldsymbol{F}\cdot\{\,\hat{\kappa}^*(\omega_0) - \tilde{\hat{\kappa}}(\omega_0)\,\}\cdot\boldsymbol{F}^*, \tag{176}$$

where we have chosen $\omega_0 > 0$. This expression is equal to the energy dissipation during one period divided by the absolute temperature. We therefore also have

$$\begin{aligned}\int_{-\pi/\omega_0}^{\pi/\omega_0} \overline{\frac{\mathrm{d}U}{\mathrm{d}t}}\,\mathrm{d}t &= \frac{\mathrm{i}\pi}{2}\,\boldsymbol{F}\cdot\{\,\hat{\kappa}^*(\omega_0) - \tilde{\hat{\kappa}}(\omega_0)\,\}\cdot\boldsymbol{F}^*\\ &= \frac{\pi^2\omega_0}{2kT}\,\boldsymbol{F}\cdot S(\omega_0)\cdot\boldsymbol{F}^* \geqslant 0.\end{aligned} \tag{177}$$

Here we have again applied the fluctuation dissipation theorem (170). The inequality follows from the property (28). Equation (177) demonstrates that a measurement of the energy dissipation during one period under the influence of "monochromatic" forces of chosen amplitudes and phases enables one to determine the elements of the spectral density matrix $\mathsf{S}(\omega_0)$ for the particular frequency ω_0. By measuring at all frequencies one can thus determine the complete spectral density matrix and by means of the Wiener–Khinchin theorem the correlation function matrix $\rho(t)$. The property of microscopic reversibility may either be tested experimentally or else be used as a means to reduce the number of quantities to be determined experimentally.

The preceding analysis clarifies the physical meaning of the fluctuation dissipation theorem.

It is also useful to write the expression (177) in terms of the real and imaginary parts of the susceptibility matrix. This gives

$$\int_{-\pi/\omega_0}^{\pi/\omega_0} \overline{\frac{\mathrm{d}U}{\mathrm{d}t}}\,\mathrm{d}t = \pi\,\{\,\hat{\kappa}''(\omega_0)\,\}^{\mathrm{s}} : \mathrm{Re}\,\boldsymbol{F}\boldsymbol{F}^* + \pi\,\{\,\hat{\kappa}'(\omega_0)\,\}^{\mathrm{a}} : \mathrm{Im}\,\boldsymbol{F}\boldsymbol{F}^*\,. \tag{178}$$

This formula shows that the dissipation is determined by the imaginary part of the symmetric contribution and the real part of the antisymmetric contribution to the susceptibility matrix.

The Kramers–Kronig relations (102) and (103) or (107) and (108) permit one to measure instead of the dissipation itself, *i.e.* the quantities $\{\,\hat{\kappa}''(\omega)\,\}^{\mathrm{s}}$ and $\{\,\hat{\kappa}'(\omega)\,\}^{\mathrm{a}}$, the complementary quantities $\{\,\hat{\kappa}'(\omega)\,\}^{\mathrm{s}}$ and $\{\,\hat{\kappa}''(\omega)\,\}^{\mathrm{a}}$. This corresponds, whenever possible, to a direct measurement of the dispersion.

Once the form of the spectrum $\mathsf{S}(\omega)$ has been determined, it is often convenient to attempt to analyse this spectrum in terms of a set of relaxation times. We shall encounter an example of such a description in terms of a set of relaxation times in Chapter X.

We may finally remark that the macroscopic requirement that the energy dissipation be a non-negative quantity is sufficient to impose the causality condition (85) on the function $\kappa(\tau)$, which occurs in linear laws of the type (83)*.

* D. C. Youla, L. J. Castriota and H. J. Carlin, I.R.E. Transactions on Circuit Theory (1959) 102; Res. Rep. Microwave Res. Inst., Polytechnic Inst. Brooklyn (1957).
J. Meixner and H. König, Rheol. Acta **1** (1958) 190.
J. Meixner, Zeitschr. Phys. **156** (1959) 200.

CHAPTER IX

DISCUSSION OF FOUNDATIONS BY MEANS OF KINETIC THEORY

§ 1. *Introduction*

The considerations of the preceding two chapters were based on a number of general statistical mechanical properties of systems having many degrees of freedom, while the formalism of the theory of stochastic processes was used extensively.

There is, however, an alternative approach to the same body of problems, which is based on the kinetic theory of gases. A justification of the principles of non-equilibrium thermodynamics by this method is in some respects more limited, since it applies only to irreversible processes that occur in dilute gases (or in metals), even though one uses a formalism which enables one to perform an explicit calculation of transport coefficients in such systems in terms of molecular interactions.

In the following sections first a summary is given of the kinetic theory of gases, containing the elements needed for a discussion of the principles of non-equilibrium thermodynamics. One may then show, as was first done by Prigogine, that the macroscopic thermodynamic expression for the entropy source strength can be derived from the basic equations of the kinetic theory*. In a separate section it is shown

* It must be stressed that the irreversibility itself is already "built" into the fundamental equation, the Boltzmann integro-differential equation, of the theory. (Similarly in Chapters VII and VIII the macroscopic irreversible behaviour as such had been postulated). In the present state of theoretical development a derivation of this equation lies outside the scope of this book. For the derivation of irreversible equations from first principles, see *e.g.*:
L. van Hove, Physica **21** (1955) 517; **23** (1957) 441;
R. Brout, Physica **22** (1956) 509;
I. Prigogine and F. Henin, J. math. Phys. **1** (1960) 349;
I. Prigogine and R. Balescu, Physica **25** (1959) 281, 302;
E. W. Montroll and J. Ward, Physica **25** (1959) 423;
W. Kohn and J. M. Luttinger, Phys. Rev. **108** (1957) 590;
P. Mazur and E. W. Montroll, J. math. Phys. **1** (1960) 70;
P. C. Hemmer, Dynamic and stochastic types of motion in the linear chain, Thesis, Trondheim, Norway (1959).

that these equations do also lead to the Onsager reciprocal relations.

The last section of this chapter contains an analysis of the local entropy production for the case of Brownian motion. Since the fundamental (or kinetic) equation for Brownian motion, the Fokker–Planck equation, has a structure somewhat similar to that of the Boltzmann equation, the derivation of an expression for the entropy source strength from this equation is closely related to the above mentioned derivation for the kinetic theory of gases.

§ 2. *The Boltzmann Equation*

In the kinetic theory of dilute gases the microscopic state of a chemically non-reacting multi-component mixture is specified by the numbers of molecules of each molecular species i, $f_i(\boldsymbol{r}, \boldsymbol{u}_i; t)\mathrm{d}\boldsymbol{r}\,\mathrm{d}\boldsymbol{u}_i$, with positions in the interval $(\boldsymbol{r}, \boldsymbol{r} + \mathrm{d}\boldsymbol{r})$ and velocities in the interval $(\boldsymbol{u}_i, \boldsymbol{u}_i + \mathrm{d}\boldsymbol{u}_i)$ at time t. It is assumed here that the molecules do not have any internal degrees of freedom and exert short-range central forces on each other.

The time behaviour of the system is described by the basic integro-differential equation of Boltzmann*:

$$\frac{\partial f_i}{\partial t} = - \boldsymbol{u}_i \cdot \frac{\partial f_i}{\partial \boldsymbol{r}} - \boldsymbol{F}_i \cdot \frac{\partial f_i}{\partial \boldsymbol{u}_i} + \sum_j C(f_i, f_j) \,. \tag{1}$$

The first two terms on the right-hand side of (1) represent the rate of change of f_i resulting from molecular motion and the acceleration due to an external force $\boldsymbol{F}_i$ (per unit mass) respectively. The terms $C(f_i, f_j)$ represent the rate of change of f_i resulting from binary collisions with particles of the same and of all other molecular species j. The explicit form of these collision terms is

$$C(f_i, f_j) = \iint \{ f_i(\boldsymbol{r}, \boldsymbol{u}_i'; t) f_j(\boldsymbol{r}, \boldsymbol{u}_j'; t) - f_i(\boldsymbol{r}, \boldsymbol{u}_i; t) f_j(\boldsymbol{r}, \boldsymbol{u}_j; t) \} g_{ij} W(\boldsymbol{k}_{ij} \mid \boldsymbol{k}_{ij}'; g_{ij})\, \mathrm{d}\boldsymbol{k}_{ij}'\, \mathrm{d}\boldsymbol{u}_j \,. \tag{2}$$

The primed velocities $\boldsymbol{u}_i'$ and $\boldsymbol{u}_j'$ are the velocities of the particles after the direct collision (or before the restituting collision). The quantity

* For the standard derivation of this equation based on the assumption of molecular chaos, the reader is referred to the textbooks on the kinetic theory of gases, *e.g.*, S. Chapman and T. G. Cowling, The mathematical theory of non-uniform gases (Cambridge, 1952).

$g_{ij} = |\boldsymbol{u}_{ij}|$ is the absolute value of the relative velocity $\boldsymbol{u}_{ij} = \boldsymbol{u}_i - \boldsymbol{u}_j$. The vectors $\boldsymbol{k}_{ij}$ and $\boldsymbol{k}'_{ij}$ are unit vectors in the direction of the relative velocities before and after the collision respectively. Finally the quantity $g_{ij}W(\boldsymbol{k}_{ij} \mid \boldsymbol{k}'_{ij}; g_{ij})\, \mathrm{d}\boldsymbol{k}'_{ij}$ is the conditional probability (per unit time) that the unit vector in the direction of the relative velocity lies in the interval $(\boldsymbol{k}'_{ij}, \boldsymbol{k}'_{ij} + \mathrm{d}\boldsymbol{k}'_{ij})$ after the collision, if its direction was $\boldsymbol{k}_{ij}$ before the collision. One may also say that $W(\boldsymbol{k}_{ij} \mid \boldsymbol{k}'_{ij}; g_{ij})$, which is a function of g_{ij}, is the "cross-section" for changing the direction of $\boldsymbol{u}_{ij}$ from $\boldsymbol{k}_{ij}$ to $\boldsymbol{k}'_{ij}$.

The cross-section $W(\boldsymbol{k}_{ij} \mid \boldsymbol{k}'_{ij}; g_{ij})$ has an important symmetry property, which is due to the time reversal invariance of the microscopic equations of motion. Before stating this property, we first note that the absolute value of the relative velocity of two colliding particles must be the same before and after the collision:

$$g_{ij} = g'_{ij} \,. \tag{3}$$

This equality follows from the fact that the total momentum and the total energy of a pair of particles are conserved in a collision. Now the time reversal invariance of the particle equations of motion (which, as stated in Chapter VII, § 3, is an invariance under a reversal of the motion of the particles) implies that

$$W(\boldsymbol{k}_{ij} \mid \boldsymbol{k}'_{ij}\,; g_{ij}) = W(-\,\boldsymbol{k}'_{ij} \mid -\,\boldsymbol{k}_{ij}\,; g_{ij}.) \tag{4}$$

This means that for any value of g_{ij}, the cross-section for changing the direction of the relative velocity from $\boldsymbol{k}_{ij}$ to $\boldsymbol{k}'_{ij}$ is equal to the cross-section for changing the relative velocity from $-\,\boldsymbol{k}'_{ij}$ to $-\,\boldsymbol{k}_{ij}$. On the other hand, due to the central character (spherical symmetry) of the interaction forces, the cross-section must also be invariant under the transformation $-\,\boldsymbol{k}'_{ij} \to \boldsymbol{k}'_{ij}$, $-\,\boldsymbol{k}_{ij} \to \boldsymbol{k}_{ij}$ (*i.e.* under an inversion of the coordinates), so that

$$W(-\,\boldsymbol{k}'_{ij} \mid -\,\boldsymbol{k}_{ij}\,; g_{ij}) = W(\boldsymbol{k}'_{ij} \mid \boldsymbol{k}_{ij}\,; g_{ij}) \,. \tag{5}$$

Combining (4) and (5) one then obtains

$$W(\boldsymbol{k}_{ij} \mid \boldsymbol{k}'_{ij}\,; g_{ij}) = W(\boldsymbol{k}'_{ij} \mid \boldsymbol{k}_{ij}\,; g_{ij}) \,. \tag{6}$$

Relation (6) is, within the framework of the kinetic theory of dilute gases, the expression for the property of microscopic reversibility (detailed balance).

With the help of (6) one can show that the collision integrals (2) have the following property:

$$\sum_{i,j} \int \psi_i C(f_i, f_j) \, \mathrm{d}\boldsymbol{u}_i$$
$$= \sum_{i,j} \iiint \psi_i (f_i' f_j' - f_i f_j) g_{ij} W(\boldsymbol{k}_{ij} \mid \boldsymbol{k}_{ij}' \,; g_{ij}) \, \mathrm{d}\boldsymbol{k}_{ij}' \, \mathrm{d}\boldsymbol{u}_i \, \mathrm{d}\boldsymbol{u}_j = 0 \,, \tag{7}$$

if ψ_i is a so-called "summational invariant", *i.e.*, if ψ_i is the mass m_i of a molecule of species i, its momentum $m_i \boldsymbol{u}_i$, its kinetic energy $\frac{1}{2} m_i \boldsymbol{u}_i^2$ or a linear combination of these quantities, (f_i and f_i' are abbreviations for $f_i(\boldsymbol{r}, \boldsymbol{u}_i; t)$ and $f_i(\boldsymbol{r}, \boldsymbol{u}_i'; t)$ respectively). In order to prove this statement we first transform to new variables:

$$\boldsymbol{u}_{(ij)} = \frac{m_i \boldsymbol{u}_i + m_j \boldsymbol{u}_j}{m_i + m_j} \,, \tag{8}$$

$$\boldsymbol{u}_{ij} = \boldsymbol{u}_i - \boldsymbol{u}_j \,. \tag{9}$$

The velocity $\boldsymbol{u}_{(ij)}$ is the center of mass velocity of the two particles with velocities $\boldsymbol{u}_i$ and $\boldsymbol{u}_j$. Since the total momentum is conserved in a binary collision we have

$$\boldsymbol{u}_{(ij)} = \boldsymbol{u}_{(ij)}' \,. \tag{10}$$

With the transformation of variables (8) and (9) one has

$$\mathrm{d}\boldsymbol{u}_i \, \mathrm{d}\boldsymbol{u}_j = \mathrm{d}\boldsymbol{u}_{ij} \, \mathrm{d}\boldsymbol{u}_{(ij)} = g_{ij}^2 \, \mathrm{d}g_{ij} \, \mathrm{d}\boldsymbol{k}_{ij} \, \mathrm{d}\boldsymbol{u}_{(ij)} \,, \tag{11}$$

and the integral (7) may be written as

$$\sum_{i,j} \int \psi_i C(f_i, f_j) \, \mathrm{d}\boldsymbol{u}_i$$
$$= \sum_{i,j} \iiiint \psi_i (f_i' f_j' - f_i f_j) g_{ij}^3 W(\boldsymbol{k}_{ij} \mid \boldsymbol{k}_{ij}' \,; g_{ij}) \, \mathrm{d}\boldsymbol{k}_{ij}' \, \mathrm{d}\boldsymbol{k}_{ij} \, \mathrm{d}g_{ij} \, \mathrm{d}\boldsymbol{u}_{(ij)} \,. \tag{12}$$

We also note that

$$\mathrm{d}g_{ij} \, \mathrm{d}\boldsymbol{u}_{(ij)} = \mathrm{d}g_{ij}' \, \mathrm{d}\boldsymbol{u}_{(ij)}' \,, \tag{13}$$

because of (3) and (10). We then obtain, on interchanging in (12) primed and unprimed quantities, the equivalent expression

$$\sum_{i,j} \int \psi_i C(f_i, f_j)\, \mathrm{d}\boldsymbol{u}_i$$
$$= \sum_{i\,j} \iiiint \psi'_i(f_i f_j - f'_i f'_j) g^3_{ij} W(\boldsymbol{k}'_{ij} \mid \boldsymbol{k}_{ij}\,;\, g_{ij})\, \mathrm{d}\boldsymbol{k}'_{ij}\, \mathrm{d}\boldsymbol{k}_{ij}\, \mathrm{d}g_{ij}\, \mathrm{d}\boldsymbol{u}_{(ij)}\,. \quad (14)$$

We now apply the property of microscopic reversibility (6) to (14) and then take one half of the sum of the two equal expressions (12) and (14):

$$\sum_{i,j} \int \psi_i C(f_i, f_j)\, \mathrm{d}\boldsymbol{u}_i$$
$$= \tfrac{1}{2} \sum_{i,j} \iiiint (\psi_i - \psi'_i)\,(f'_i f'_j - f_i f_j) g^3_{ij} W(\boldsymbol{k}_{ij} \mid \boldsymbol{k}'_{ij}\,;\, g_{ij})\, \mathrm{d}\boldsymbol{k}'_{ij}\, \mathrm{d}\boldsymbol{k}_{ij}\, \mathrm{d}g_{ij}\, \mathrm{d}\boldsymbol{u}_{(ij)}\,. \quad (15)$$

Repeating this symmetrizing operation with respect to the dummy indices i and j, one finally has:

$$\sum_{i,j} \int \psi_i C(f_i, f_j)\, \mathrm{d}\boldsymbol{u}_i$$
$$= \tfrac{1}{4} \sum_{i,j} \iiiint (\psi_i + \psi_j - \psi'_i - \psi'_j)\,(f'_i f'_j - f_i f_j) g^3_{ij} W(\boldsymbol{k}_{ij} \mid \boldsymbol{k}'_{ij}\,;\, g_{ij})\, \mathrm{d}\boldsymbol{k}'_{ij}\, \mathrm{d}\boldsymbol{k}_{ij}\, \mathrm{d}g_{ij}\, \mathrm{d}\boldsymbol{u}_{(ij)}\,. \quad (16)$$

Now the expression $\psi_i + \psi_j - \psi'_i - \psi'_j$ vanishes if ψ_i is a summational invariant, *i.e.* if ψ_i is equal to m_i, $m_i\boldsymbol{u}_i$ or $\frac{1}{2}m_i\boldsymbol{u}_i^2$. Consequently equation (7) holds if ψ_i is equal to any of these quantities or to a linear combination thereof. Note that for the quantity m_i each of the integrals of (7) vanishes separately and not only the sum over both i and j of these integrals. This is best seen in expression (15): if ψ_i is equal to m_i, then $\psi_i - \psi'_i = 0$, since the mass of a particle is not changed during an encounter.

§ 3. *The Hydrodynamic Equations*

The hydrodynamic equations or general conservation laws may be derived from the Boltzmann equation. We shall need for this purpose the property (7) of the collision integrals established in the previous section.

Let us first write the statistical expressions for the mass densities ρ_i of the various components:

$$\rho_i = n_i m_i = m_i \int f_i \, \mathrm{d}\boldsymbol{u}_i \,. \tag{17}$$

The quantity n_i is the number density of molecules of species i. The total mass density $\rho = \sum_i \rho_i$ is given by

$$\rho = \sum_i m_i \int f_i \, \mathrm{d}\boldsymbol{u}_i \,. \tag{18}$$

If one differentiates (17) with respect to time one obtains, with the Boltzmann equation (1) and the property (7) of the collision integrals, the law of conservation of mass—the equation of continuity—for the ith component. One has indeed

$$\frac{\partial \rho_i}{\partial t} = \int m_i \frac{\partial f_i}{\partial t} \mathrm{d}\boldsymbol{u}_i = -\int \left\{ m_i \boldsymbol{u}_i \cdot \frac{\partial f_i}{\partial \boldsymbol{r}} + m_i \boldsymbol{F}_i \cdot \frac{\partial f_i}{\partial \boldsymbol{u}_i} - \sum_j m_i C(f_i, f_j) \right\} \mathrm{d}\boldsymbol{u}_i \,. \tag{19}$$

The second term of the last member of (19) vanishes because f_i is assumed to tend to zero sufficiently rapidly for large $\boldsymbol{u}_i$; the last term vanishes due to (7). (As stated in the previous section this property holds without summation over i and j if $\psi_i = m_i$.) One therefore obtains

$$\frac{\partial \rho_i}{\partial t} = -\operatorname{div} \rho_i \boldsymbol{v}_i \,, \tag{20}$$

where

$$\boldsymbol{v}_i = \rho_i^{-1} m_i \int \boldsymbol{u}_i f_i \, \mathrm{d}\boldsymbol{u}_i \tag{21}$$

is the average velocity of component i. Equation (20) is the equation of continuity for component i. If we also introduce the barycentric velocity $\boldsymbol{v}$ by means of

$$\rho \boldsymbol{v} = \sum_i \rho_i \boldsymbol{v}_i = \sum_i m_i \int \boldsymbol{u}_i f_i \, \mathrm{d}\boldsymbol{u}_i \,, \tag{22}$$

equation (20) may be rewritten in the alternative form

$$\frac{\partial \rho_i}{\partial t} = -\operatorname{div} \rho_i \boldsymbol{v} - \operatorname{div} \boldsymbol{J}_i \,, \tag{23}$$

where the diffusion flux $\boldsymbol{J}_i$ is defined as

$$\boldsymbol{J}_i = m_i \int (\boldsymbol{u}_i - \boldsymbol{v}) f_i \, \mathrm{d}\boldsymbol{u}_i \,. \tag{24}$$

The equation of motion, or momentum balance equation can be found in a similar way by differentiating (22) with respect to time. One obtains

$$\frac{\partial \rho \boldsymbol{v}}{\partial t} = - \operatorname{Div} (\rho \boldsymbol{v}\boldsymbol{v} + \mathsf{P}) + \sum_i \rho_i \boldsymbol{F}_i \,, \tag{25}$$

with the following definition of the pressure tensor P:

$$\mathsf{P} = \sum_i m_i \int (\boldsymbol{u}_i - \boldsymbol{v}) \, (\boldsymbol{u}_i - \boldsymbol{v}) \, f_i \, \mathrm{d}\boldsymbol{u}_i \,. \tag{26}$$

It follows from this expression that the pressure tensor of our dilute gas mixture is symmetric.

Let us finally consider the energy balance equation. The internal energy density ρu of the gas is given by

$$\rho u = \tfrac{1}{2} \sum_i m_i \int (\boldsymbol{u}_i - \boldsymbol{v})^2 \, f_i \, \mathrm{d}\boldsymbol{u}_i \,. \tag{27}$$

This relation may be used to define the "kinetic" temperature T

$$\tfrac{3}{2} n k T = \rho u = \tfrac{1}{2} \sum_i m_i \int (\boldsymbol{u}_i - \boldsymbol{v})^2 \, f_i \, \mathrm{d}\boldsymbol{u}_i \,. \tag{28}$$

Here $n = \sum_i n_i$ is the total number density. For an equilibrium system this definition of temperature agrees with the thermodynamic definition (*cf.* also § 4). Differentiating (27) with respect to time, using (1) and applying (7), one obtains

$$\frac{\partial \rho u}{\partial t} = - \operatorname{div} (\rho u \boldsymbol{v} + \boldsymbol{J}_q) - \mathsf{P} : \operatorname{Grad} \boldsymbol{v} + \sum_i \boldsymbol{J}_i \cdot \boldsymbol{F}_i \,, \tag{29}$$

where the heat flux $\boldsymbol{J}_q$ is given by

$$\boldsymbol{J}_q = \tfrac{1}{2} \sum_i m_i \int (\boldsymbol{u}_i - \boldsymbol{v})^2 \, (\boldsymbol{u}_i - \boldsymbol{v}) \, f_i \, \mathrm{d}\boldsymbol{u}_i \,. \tag{30}$$

By solving the Boltzmann equation, one can evaluate the fluxes $\boldsymbol{J}_i$, $\boldsymbol{J}_q$ and P occurring in the hydrodynamic equations (23), (25) and (29) and thus determine the macroscopic behaviour of the system. The method of obtaining solutions of the Boltzmann equation will be outlined in § 5. We shall be especially interested in the approximate solution corresponding to the linear phenomenological laws.

§ 4. *The Entropy Balance Equation, Boltzmann's H-theorem*

After having derived the hydrodynamic equations and obtained the kinetic expressions for the fluxes which occur in these, we shall now establish in an analogous way a balance equation for the entropy and obtain the statistical, kinetic expressions for the entropy flux and the entropy source strength.

In kinetic theory the entropy density ρs is defined as

$$\rho s = -k \sum_i \int f_i (\ln f_i - 1)\, d\boldsymbol{u}_i\,, \tag{31}$$

where k is Boltzmann's constant. Taking the time derivative of (31), one obtains with (1)

$$\begin{aligned}\frac{\partial \rho s}{\partial t} &= -k \sum_i \int \ln f_i \frac{\partial f_i}{\partial t}\, d\boldsymbol{u}_i \\ &= -k \sum_i \int \ln f_i \left\{ -\boldsymbol{u}_i \cdot \frac{\partial f_i}{\partial \boldsymbol{r}} - \boldsymbol{F}_i \cdot \frac{\partial f_i}{\partial \boldsymbol{u}_i} + \sum_j C(f_i, f_j) \right\} d\boldsymbol{u}_i\,. \end{aligned} \tag{32}$$

After partial integration, one has

$$\begin{aligned}\frac{\partial \rho s}{\partial t} &= \frac{\partial}{\partial \boldsymbol{r}} \cdot k \sum_i \int \boldsymbol{u}_i f_i (\ln f_i - 1)\, d\boldsymbol{u}_i - k \sum_{i,j} \int C(f_i, f_j) \ln f_i\, d\boldsymbol{u}_i \\ &= \frac{\partial}{\partial \boldsymbol{r}} \cdot k\boldsymbol{v} \sum_i \int f_i (\ln f_i - 1)\, d\boldsymbol{u}_i + \frac{\partial}{\partial \boldsymbol{r}} \cdot k \sum_i \int (\boldsymbol{u}_i - \boldsymbol{v}) f_i (\ln f_i - 1)\, d\boldsymbol{u}_i \\ &\qquad - k \sum_{i,j} \int C(f_i, f_j) \ln f_i\, d\boldsymbol{u}_i\,. \end{aligned} \tag{33}$$

This equation is of the form

$$\frac{\partial \rho s}{\partial t} = -\operatorname{div} (\rho s \boldsymbol{v} + \boldsymbol{J}_s) + \sigma\,, \tag{34}$$

where

$$J_s = -k \sum_i \int (\boldsymbol{u}_i - \boldsymbol{v}) f_i (\ln f_i - 1) \, \mathrm{d}\boldsymbol{u}_i \tag{35}$$

represents the entropy flux, and

$$\sigma = -k \sum_{i,j} \int C(f_i, f_j) \ln f_i \, \mathrm{d}\boldsymbol{u}_i \tag{36}$$

the entropy source strength.

Let us now rewrite explicitly the statistical expression (36) for the entropy source strength. With (2) this expression becomes

$$\sigma = -k \sum_{i,j} \int\int\int (\ln f_i)(f_i' f_j' - f_i f_j) \, g_{ij} W(\boldsymbol{k}_{ij} \mid \boldsymbol{k}_{ij}' ; g_{ij}) \, \mathrm{d}\boldsymbol{k}_{ij}' \, \mathrm{d}\boldsymbol{u}_j \, \mathrm{d}\boldsymbol{u}_i \, . \tag{37}$$

The integral on the right-hand side of this expression may be symmetrized by the method outlined in § 2 [*cf.* the transformation of expression (7) which yields the symmetrized expression (16)]. This gives

$$\sigma = \tfrac{1}{4} k \sum_{i,j} \int\int\int\int \left(\ln \frac{f_i' f_j'}{f_i f_j} \right) (f_i' f_j' - f_i f_j) \, g_{ij}^3 W(\boldsymbol{k}_{ij} \mid \boldsymbol{k}_{ij}' ; g_{ij}) \, \mathrm{d}\boldsymbol{k}_{ij}' \, \mathrm{d}\boldsymbol{k}_{ij} \, \mathrm{d}g_{ij} \, \mathrm{d}\boldsymbol{u}_{(ij)} \, . \tag{38}$$

In each of the integrands the positive factor $g_{ij}^3 W(\boldsymbol{k}_{ij} \mid \boldsymbol{k}_{ij}'; g_{ij})$ is multiplied by a factor of the form $(x - y) \ln (x/y)$ with $x = f_i' f_j'$ and $y = f_i f_j$. If $x > y$ both $(x - y)$ and $\ln (x/y)$ are positive and if $x < y$ both $(x - y)$ and $\ln (x/y)$ are negative. Therefore the integrand of each of the integrals of (38) is positive or zero so that

$$\sigma \geqslant 0 \, . \tag{39}$$

This inequality*, which expresses the fact that the entropy source strength must be positive or zero, constitutes, within the framework of the kinetic theory of gases, a derivation of the second law of thermodynamics. The inequality (39) is known in the kinetic theory as Boltzmann's H-theorem.

If one integrates (38) over the whole volume occupied by the system, the H-theorem expresses the fact that the entropy of a closed system can only increase in the course of time and in fact must approach a

* For the proof of the H-theorem, when microscopic reversibility does not hold in the form (6), see: E. C. G. Stückelberg, Helv. Phys. Act. **25** (1952) 577; W. Heitler, Anu. Inst. H. Poincaré **15** (1956) 67.

limit as the time t tends to infinity. In this limit the integrals on the right-hand side of (38) must vanish. This can only happen if

$$f'_i f'_j = f_i f_j \,, \tag{40}$$

or, equivalently,

$$\ln f'_i + \ln f'_j = \ln f_i + \ln f_j \,. \tag{41}$$

The H-theorem thus states that (40) is the necessary and sufficient condition for equilibrium ($\sigma = 0$); at equilibrium the number of direct and reverse collisions exactly balance each other in any particular velocity range (principle of detailed balance). The equivalent condition (41) states that at equilibrium the logarithms of the distribution functions f_i are summational invariants. They must therefore be equal to a linear combination of the three quantities m_i, $m_i \boldsymbol{u}_i$ and $\frac{1}{2} m_i \boldsymbol{u}_i^2$:

$$\ln f_i^{\text{eq}} = a_i m_i + m_i \boldsymbol{b} \cdot \boldsymbol{u}_i + \tfrac{1}{2} c m_i \boldsymbol{u}_i^2 \,. \tag{42}$$

The constant a_i may be different for each molecular species i, since m_i itself is conserved in a molecular collision. The constants $\boldsymbol{b}$ and c on the other hand must be the same for every molecular species since only the total momentum and energy of a pair of colliding particles is conserved in an encounter. The constants a_i, $\boldsymbol{b}$, and c are related to the number density n_i of molecules of i, the barycentric velocity $\boldsymbol{v}$, and the temperature T of the system by means of (17), (22) and (28). In terms of these last quantities, the equilibrium or Maxwell distribution function is

$$f_i^{\text{eq}} = n_i \left(\frac{m_i}{2\pi kT} \right)^{\frac{3}{2}} \exp \left\{ - \frac{m_i (\boldsymbol{u}_i - \boldsymbol{v})^2}{2kT} \right\} . \tag{43}$$

If we introduce the thermodynamic potential per unit mass of a component i of an ideal gas mixture*:

$$\mu_i = \frac{kT}{m_i} \left(\ln n_i - \tfrac{3}{2} \ln \frac{2\pi kT}{m_i} \right) , \tag{44}$$

we may also write instead of (43)

* See *e.g.* R. H. Fowler and E. A. Guggenheim, Statistical Thermodynamics (Cambridge, 1932) Ch. III.

$$f_i^{\text{eq}} = \exp\left\{m_i \frac{\mu_i - \frac{1}{2}(\boldsymbol{u}_i - \boldsymbol{v})^2}{kT}\right\}. \tag{45}$$

When the equilibrium solution of the Boltzmann equation is substituted into the definitions (24), (26) and (30) of the fluxes $\boldsymbol{J}_i$, P and $\boldsymbol{J}_q$, it is seen that the diffusion fluxes $\boldsymbol{J}_i$ and the heat flux $\boldsymbol{J}_q$ vanish whereas the pressure tensor reduces to the scalar hydrostatic pressure p:

$$\mathsf{P} = \sum_i \int m_i(\boldsymbol{u}_i - \boldsymbol{v})(\boldsymbol{u}_i - \boldsymbol{v}) f_i^{\text{eq}} \, \mathrm{d}\boldsymbol{u}_i = \tfrac{1}{3}\sum_i \int m_i(\boldsymbol{u}_i - \boldsymbol{v})^2 f_i^{\text{eq}} \, \mathrm{d}\boldsymbol{u}_i \, \mathsf{U} = p\mathsf{U}\,, \tag{46}$$

since all non-diagonal elements of the tensor $(\boldsymbol{u}_i - \boldsymbol{v})(\boldsymbol{u}_i - \boldsymbol{v})$ vanish when averaged over the Maxwell–Boltzmann distribution function. [In (46) U is the three-dimensional unit tensor]. The pressure p can be explicitly evaluated to give

$$p = nkT\,, \tag{47}$$

i.e. the ideal gas law.

We can thus conclude that $\boldsymbol{J}_i$, $\boldsymbol{J}_q$ and the non-diagonal elements of P can only exist if σ does not vanish, that is if irreversible changes occur in the system. Similarly it is seen that the entropy flux $\boldsymbol{J}_s$ vanishes at equilibrium and is therefore also related to irreversible changes.

On substitution of (45) into (31) we obtain the equilibrium value of the entropy density

$$\rho s = -k\sum_i \int f_i^{\text{eq}} \left\{\frac{m_i\mu_i - \frac{1}{2}m_i(\boldsymbol{u}_i - \boldsymbol{v})^2}{kT} - 1\right\} \mathrm{d}\boldsymbol{u}_i = -\frac{1}{T}\left(\sum_i \rho_i\mu_i - \rho u - p\right), \tag{48}$$

where the definitions (17), (27) and the ideal gas law (47) have been applied. We have obtained here the correct thermodynamic relation between the entropy, the thermodynamic potentials, the internal energy and the pressure; in other words the kinetic definition of entropy (31) is equivalent, in equilibrium, with the familiar thermodynamic definition of this quantity.

We shall investigate in the following sections under which conditions the kinetic entropy flux (35) and entropy source strength (36) are identical with the expressions for these quantities, obtained from macroscopic theory by applying the Gibbs relation

$$\frac{\mathrm{d}s}{\mathrm{d}t} = \frac{1}{T}\frac{\mathrm{d}u}{\mathrm{d}t} + \frac{p}{T}\frac{\mathrm{d}v}{\mathrm{d}t} - \sum_i \frac{\mu_i}{T}\frac{\mathrm{d}c_i}{\mathrm{d}t} \tag{49}$$

to non-equilibrium situations. In other words we shall investigate the conditions under which relation (49) is valid outside equilibrium. The macroscopic expressions for the entropy flux and the entropy source strength are [*cf.* (III.25) and (III.26)]

$$\boldsymbol{J}_s = \frac{\boldsymbol{J}'_q}{T} + \sum_i s_i \boldsymbol{J}_i \,, \tag{50}$$

$$\sigma = -\frac{1}{T^2}\boldsymbol{J}'_q \cdot \operatorname{grad} T - \frac{1}{T}\sum_i \boldsymbol{J}_i \cdot \{(\operatorname{grad}\mu_i)_T - \boldsymbol{F}_i\} - \frac{1}{T}\boldsymbol{\Pi} : \operatorname{Grad}\boldsymbol{v} \,, \tag{51}$$

where the "heat flux" $\boldsymbol{J}'_q$ is defined by [*cf.* (III.24)]

$$\boldsymbol{J}'_q = \boldsymbol{J}_q - \sum_i h_i \boldsymbol{J}_i \,, \tag{52}$$

where h_i is the partial specific enthalpy of component i and where $s_i = -(\mu_i - h_i)/T$ is the partial specific entropy of component i. In (51) the viscous pressure tensor $\boldsymbol{\Pi}$ is defined as [*cf.* (II.35)]

$$\boldsymbol{\Pi} = \boldsymbol{P} - p\boldsymbol{U} \,. \tag{53}$$

It will turn out that (35) and (36) are of the form (50) and (51) for the approximate solution of the Boltzmann equation corresponding to the linear phenomenological laws.

§ 5. *The Enskog Method of Solution of the Boltzmann Equation*

Let us introduce the Maxwell distribution function $f_i^{(0)}$ corresponding to the local values of the density, the average velocity and the kinetic temperature:

$$f_i^{(0)} = \exp\left\{ m_i \frac{\mu_i - \frac{1}{2}(\boldsymbol{u}_i - \boldsymbol{v})^2}{kT} \right\} . \tag{54}$$

The macroscopic functions μ_i, $\boldsymbol{v}$ and T(or ρ_i, $\boldsymbol{v}$ and T) which are functions of the space coordinates and of time then satisfy the following set of conditions [*cf.* (17), (22) and (28)]:

$$\rho_i = m_i \int f_i \, \mathrm{d}\boldsymbol{u}_i = m_i \int f_i^{(0)} \, \mathrm{d}\boldsymbol{u}_i \,, \tag{55}$$

$$\rho \boldsymbol{v} = \sum_i m_i \int \boldsymbol{u}_i f \, d\boldsymbol{u}_i = \sum_i m_i \int \boldsymbol{u}_i f_i^{(0)} \, d\boldsymbol{u}_i \,, \tag{56}$$

$$\tfrac{3}{2} nkT = \tfrac{1}{2} \sum_i m_i \int (\boldsymbol{u}_i - \boldsymbol{v})^2 f_i \, d\boldsymbol{u}_i = \tfrac{1}{2} \sum_i m_i \int (\boldsymbol{u}_i - \boldsymbol{v})^2 f_i^{(0)} \, d\boldsymbol{u}_i \,. \tag{57}$$

The function $f_i^{(0)}$ thus represents the equilibrium distribution corresponding to the local time and space dependent values of the density, the average velocity and the kinetic temperature.

We now write the distribution function f_i in the form

$$f_i = f_i^{(0)} (1 + \phi_i) \,. \tag{58}$$

The "perturbation" function $\phi_i(\boldsymbol{r}, \boldsymbol{u}_i; t)$ is a measure for the deviation of the actual distribution function f_i from the local equilibrium distribution function $f_i^{(0)}$.

According to equations (55)–(57) the perturbation function ϕ_i must satisfy the conditions

$$\int f_i^{(0)} \phi_i \, d\boldsymbol{u}_i = 0 \,, \tag{59}$$

$$\sum_i m_i \int \boldsymbol{u}_i f_i^{(0)} \phi_i \, d\boldsymbol{u}_i = 0 \,, \tag{60}$$

$$\tfrac{1}{2} \sum_i m_i \int (\boldsymbol{u}_i - \boldsymbol{v})^2 f_i^{(0)} \phi_i \, d\boldsymbol{u}_i = 0 \,. \tag{61}$$

In order to obtain a series solution of the Boltzmann equation

$$\frac{\partial f_i}{\partial t} + \boldsymbol{u}_i \cdot \frac{\partial f_i}{\partial \boldsymbol{r}} + \boldsymbol{F}_i \cdot \frac{\partial f_i}{\partial \boldsymbol{u}_i} = \sum_j C(f_i, f_j) \,, \tag{62}$$

we substitute the known zero order function $f_i^{(0)}$ into the left-hand side of this equation and linearize the right-hand side with respect to the perturbations ϕ_i and ϕ_j. We get in this way the following set of integral equations for the first approximations $\phi_i^{(1)}$ to the perturbation functions ϕ_i:

$$\frac{\partial f_i^{(0)}}{\partial t} + \boldsymbol{u}_i \cdot \frac{\partial f_i^{(0)}}{\partial \boldsymbol{r}} + \boldsymbol{F}_i \cdot \frac{\partial f_i^{(0)}}{\partial \boldsymbol{u}_i}$$

$$= \sum_j \int\int f_i^{(0)} f_j^{(0)} (\phi_i'^{(1)} + \phi_j'^{(1)} - \phi_i^{(1)} - \phi_j^{(1)}) g_{ij} W(\boldsymbol{k}_{ij} \mid \boldsymbol{k}'_{ij} \,; g_{ij}) \, d\boldsymbol{k}'_{ij} \, d\boldsymbol{u}_j \,. \tag{63}$$

Use has been made here of the fact that the zero order functions $f_i^{(0)}$ satisfy the equality (40). The integral equation (63), together with the auxiliary conditions (59)–(61), specifies uniquely the functions $\phi_i^{(1)}$.

This procedure may be repeated by substituting into the left-hand side of (62) the approximate distribution function $f_i^{(0)}(1 + \phi_i^{(1)})$ and retaining a higher order contribution of the perturbation ϕ_i to the collision integrals. By iteration one obtains in this way the Enskog series solution* of the Boltzmann equation:

$$f_i = f_i^{(0)} + f_i^{(1)} + f_i^{(2)} + \dots$$
$$= f_i^{(0)} (1 + \phi_i^{(1)} + \phi_i^{(2)} + \dots) . \tag{64}$$

We shall be concerned here mainly with the first approximation characterized by $\phi_i^{(1)}$. The differentiations of the function $f_i^{(0)}$ occurring in (63) may be carried out. The resulting expressions involve space and time derivatives of the functions μ_i, $\boldsymbol{v}$ and T (or ρ_i, $\boldsymbol{v}$ and u). In order to eliminate the time derivatives we use the hydrodynamic equations (23), (25) and (29) of § 3. However, for consistency with the order of approximation of (63), we replace f_i by $f_i^{(0)}$ in the integrals for the fluxes $\boldsymbol{J}_i$, $\boldsymbol{J}_q$ and the pressure tensor P which occur in the hydrodynamic equations. This means that we use for the elimination of the time derivatives the "reversible" or Euler hydrodynamic equations with $\boldsymbol{J}_i = 0$, $\boldsymbol{J}_q = 0$ and $\mathsf{P} = p\mathsf{U}$ (*cf.* the discussion of § 4). After these various operations have been carried out one obtains the following equations for the functions $\phi_i^{(1)}$:

$$(kT)^{-1} f_i^{(0)} [m_i \{\tfrac{1}{2}(\boldsymbol{u}_i - \boldsymbol{v})^2 - h_i\} (\boldsymbol{u}_i - \boldsymbol{v}) \cdot (\text{grad}\, T)/T$$
$$+ m_i (\boldsymbol{u}_i - \boldsymbol{v}) \cdot \{(\text{grad}\, \mu_i)_T - \boldsymbol{F}_i - \rho^{-1} \text{grad}\, p + \rho^{-1} \sum_j \rho_j \boldsymbol{F}_j\}$$
$$+ m_i \{(\boldsymbol{u}_i - \boldsymbol{v})(\boldsymbol{u}_i - \boldsymbol{v}) - \tfrac{1}{3}(\boldsymbol{u}_i - \boldsymbol{v})^2 \mathsf{U}\} : \text{Grad}\, \boldsymbol{v}]$$
$$= \sum_j \int\!\!\int f_i^{(0)} f_j^{(0)} (\phi_i'^{(1)} + \phi_j'^{(1)} - \phi_i^{(1)} - \phi_j^{(1)})\, g_{ij} W(\boldsymbol{k}_{ij} \,|\, \boldsymbol{k}'_{ij}\,;\, g_{ij})\, \mathrm{d}\boldsymbol{k}'_{ij}\, \mathrm{d}\boldsymbol{u}_j . \tag{65}$$

In this equation h_i is the enthalpy per unit mass of component i, given by

$$m_i h_i = \tfrac{5}{2} kT . \tag{66}$$

* D. Enskog, Thesis, Uppsala (1917); S. Chapman and T. G. Cowling, The mathematical theory of non-uniform gases (Cambridge, 1952).

It is seen from (65) that the perturbation function $\phi_i^{(1)}$ depends upon space and time only through the quantities μ_i, $\boldsymbol{v}$ and T. It is also clear from the form of the integral equation that $\phi_i^{(1)}$ is linear in the gradients of these quantities. Therefore, if one evaluates the fluxes occurring in the hydrodynamic equations by retaining in f_i the first order perturbation $\phi_i^{(1)}$, the diffusion flux $\boldsymbol{J}_i$, the heat flux $\boldsymbol{J}_q$ and the off-diagonal elements of the pressure tensor P become linear functions of the gradients of the macroscopic functions μ_i, $\boldsymbol{v}$ and T: the first approximation of Enskog corresponds to the linear laws of the phenomenological macroscopic theory and thus covers those transport phenomena in dilute gases which under a wide range of physical conditions are adequately described by these laws. In § 7, in which the Onsager relations will be established on the basis of the kinetic theory, we shall carry the analysis of (65) somewhat further. For our purpose it will not be necessary to obtain an explicit solution for $\phi_i^{(1)}$. Such an explicit solution is required if one wishes to calculate the values of the transport coefficients (the coefficients in the linear laws) in terms of the known molecular interactions. Although one of the most important aspects of the kinetic theory of gases is precisely the fact that it provides us with a tool to achieve such a calculation, the details connected with this problem lie outside the scope of the present monograph and will not be dealt with.

Equation (65) will be of use to perform our program of comparing the statistical expression for the entropy production with the expression obtained from the macroscopic theory.

By an analysis of the higher approximations of Enskog one can show that these involve higher space derivatives of the macroscopic functions μ_i, $\boldsymbol{v}$ and T and higher powers of lower space derivatives. In fact the Enskog solution provides a complete description of the future of a gas in terms of a knowledge at a particular time of the macroscopic functions μ_i, $\boldsymbol{v}$ and T at each point in space. It is clear that the Enskog method only enables one to calculate such deviations from the local Maxwellian distribution function which are entirely related to a spatial non-uniformity of the system. This is not the most general solution of the Boltzmann equation for arbitrary initial deviations from equilibrium. It may, however, be shown that homogeneous perturbations of the Maxwell distribution (*i.e.* perturbations which are spatially uniform) have a relaxation time of the order of a few collision times, (the collision time, or mean time between collisions is of the order of

10^{-8} sec). This provides a justification of the Enskog solution in the sense that after such macroscopically irrelevant times the true solution of the Boltzmann equation may be approximated by the Enskog solution*.

§ 6. *The Entropy Balance Equation in the First Approximation of Enskog*

We shall now compare the statistical expressions for the entropy flux and the entropy source strength with the corresponding expressions of the macroscopic theory.

Let us first consider the entropy flux. If one substitutes into the right-hand side of (35) the Enskog series (64) for the distribution function and expands the logarithm in powers of the perturbation functions, one obtains the following series for $\boldsymbol{J}_s$**:

$$\boldsymbol{J}_s = \boldsymbol{J}_s^{(0)} + \boldsymbol{J}_s^{(1)} + \boldsymbol{J}_s^{(2)} + \dots , \tag{67}$$

with

$$\boldsymbol{J}_s^{(0)} = -k \sum_i \int (\boldsymbol{u}_i - \boldsymbol{v})\, f_i^{(0)} (\ln f_i^{(0)} - 1)\, \mathrm{d}\boldsymbol{u}_i = 0 , \tag{68}$$

$$\begin{aligned}\boldsymbol{J}_s^{(1)} &= -k \sum_i \int (\boldsymbol{u}_i - \boldsymbol{v})\, f_i^{(0)} \phi_i^{(1)} \ln f_i^{(0)}\, \mathrm{d}\boldsymbol{u}_i \\ &= T^{-1} \sum_i m_i \int (\boldsymbol{u}_i - \boldsymbol{v})\, \{\tfrac{1}{2}(\boldsymbol{u}_i - \boldsymbol{v})^2 - \mu_i\}\, f_i^{(0)} \phi_i^{(0)}\, \mathrm{d}\boldsymbol{u}_i ,\end{aligned} \tag{69}$$

$$\begin{aligned}\boldsymbol{J}_s^{(2)} &= -k \sum_i \int (\boldsymbol{u}_i - \boldsymbol{v})\, f_i^{(0)} \phi_i^{(2)} \ln f_i^{(0)} \mathrm{d}\boldsymbol{u}_i - \tfrac{1}{2}k \sum_i \int (\boldsymbol{u}_i - \boldsymbol{v})\, f_i^{(0)} (\phi_i^{(1)})^2 \mathrm{d}\boldsymbol{u}_i \\ &= T^{-1} \sum_i m_i \int (\boldsymbol{u}_i - \boldsymbol{v})\, \{\tfrac{1}{2}(\boldsymbol{u}_i - \boldsymbol{v})^2 - \mu_i\}\, f_i^{(0)} \phi_i^{(2)}\, \mathrm{d}\boldsymbol{u}_i \\ &\qquad - \tfrac{1}{2}k \sum_i \int (\boldsymbol{u}_i - \boldsymbol{v})\, f_i^{(0)} (\phi_i^{(1)})^2\, \mathrm{d}\boldsymbol{u}_i ,\end{aligned} \tag{70}$$

where (45) and the conditions (59)–(61) have been used.

* See also H. Grad, Communications on pure and applied Mathematics **2** (1949) 331.

** The quadratic term in $\phi_i^{(1)}$ is of the same order as the term linear in $\phi_i^{(2)}$, since both contain terms quadratic in the macroscopic gradients.

We may on the other hand expand the statistical expressions for the diffusion flux $\boldsymbol{J}_i$ and the heat flux $\boldsymbol{J}_q$ in an analogous way:

$$\boldsymbol{J}_i = \boldsymbol{J}_i^{(0)} + \boldsymbol{J}_i^{(1)} + \boldsymbol{J}_i^{(2)} + \ldots, \tag{71}$$

$$\boldsymbol{J}_q = \boldsymbol{J}_q^{(0)} + \boldsymbol{J}_q^{(1)} + \boldsymbol{J}_q^{(2)} + \ldots, \tag{72}$$

with

$$\boldsymbol{J}_i^{(0)} = m_i \int (\boldsymbol{u}_i - \boldsymbol{v})\, f_i^{(0)}\, \mathrm{d}\boldsymbol{u}_i = 0\,, \tag{73}$$

$$\boldsymbol{J}_q^{(0)} = \tfrac{1}{2} \sum_i m_i \int (\boldsymbol{u}_i - \boldsymbol{v})^2\, (\boldsymbol{u}_i - \boldsymbol{v})\, f_i^{(0)}\, \mathrm{d}\boldsymbol{u}_i = 0\,, \tag{74}$$

$$\boldsymbol{J}_i^{(1)} = m_i \int (\boldsymbol{u}_i - \boldsymbol{v})\, f_i^{(0)} \phi_i^{(1)}\, \mathrm{d}\boldsymbol{u}_i\,, \tag{75}$$

$$\boldsymbol{J}_q^{(1)} = \tfrac{1}{2} \sum_i m_i \int (\boldsymbol{u}_i - \boldsymbol{v})^2\, (\boldsymbol{u}_i - \boldsymbol{v})\, f_i^{(0)} \phi_i^{(1)}\, \mathrm{d}\boldsymbol{u}_i\,, \tag{76}$$

$$\boldsymbol{J}_i^{(2)} = m_i \int (\boldsymbol{u}_i - \boldsymbol{v})\, f_i^{(0)} \phi_i^{(2)}\, \mathrm{d}\boldsymbol{u}_i\,, \tag{77}$$

$$\boldsymbol{J}_q^{(2)} = \tfrac{1}{2} \sum_i m_i \int (\boldsymbol{u}_i - \boldsymbol{v})^2\, (\boldsymbol{u}_i - \boldsymbol{v})\, f_i^{(0)} \phi_i^{(2)}\, \mathrm{d}\boldsymbol{u}_i\,. \tag{78}$$

From (69), (75) and (76) it then follows that

$$\boldsymbol{J}_s^{(1)} = \frac{1}{T} (\boldsymbol{J}_q^{(1)} - \sum_i \mu_i \boldsymbol{J}_i^{(1)}) = \frac{\boldsymbol{J}_q'^{(1)}}{T} + \sum_i s_i \boldsymbol{J}_i^{(1)}\,. \tag{79}$$

where (52) has been employed.

Comparing (79) with (50) we note that in the first approximation of Enskog the statistical expression for the entropy flux is identical with the expression derived on the basis of the macroscopic formalism of Chapter III.

However, from (70), (77) and (78), it is seen that

$$\boldsymbol{J}_s^{(2)} = \frac{1}{T} (\boldsymbol{J}_q^{(2)} - \sum_i \mu_i \boldsymbol{J}_i^{(2)}) - \tfrac{1}{2} k \sum_i \int (\boldsymbol{u}_i - \boldsymbol{v})\, f_i^{(0)}\, (\phi_i^{(1)})^2\, \mathrm{d}\boldsymbol{u}_i\,. \tag{80}$$

This means that the statistical and the macroscopic expression for the entropy flux do not coincide if second order terms are retained in the evaluation of the fluxes $\boldsymbol{J}_s$, $\boldsymbol{J}_q$ and $\boldsymbol{J}_i$. We may thus conclude on the basis of the kinetic theory of gases, that the macroscopic result (50) for the entropy flux only holds if the first approximation of Enskog is sufficient for an adequate description of the transport phenomena occurring in the system.

This limitation of the macroscopic theory is consistently also found if we compare the kinetic and the macroscopic expressions for the entropy source strength. We can demonstrate this as follows. By introducing the Enskog series (64) into the collision integrals $C(f_i, f_j)$ one obtains the series

$$C(f_i, f_j) = C^{(0)}(f_i, f_j) + C^{(1)}(f_i, f_j) + C^{(2)}(f_i, f_j) + \ldots, \tag{81}$$

with

$$C^{(\nu)}(f_i, f_j) = \sum_{\nu'=0}^{\nu} C(f_i^{(\nu')}, f_j^{(\nu-\nu')}) . \tag{82}$$

The first two terms of the series (81) can explicitly be written as

$$C^{(0)}(f_i, f_j) = \iint (f_i'^{(0)} f_j'^{(0)} - f_i^{(0)} f_j^{(0)})\, g_{ij} W(\boldsymbol{k}_{ij} \mid \boldsymbol{k}'_{ij}\, ; g_{ij})\, \mathrm{d}\boldsymbol{k}'_{ij}\, \mathrm{d}\boldsymbol{u}_j = 0\, , \tag{83}$$

$$C^{(1)}(f_i, f_j) = \iint f_i^{(0)} f_j^{(0)} (\phi_i'^{(1)} + \phi_j'^{(1)} - \phi_i^{(1)} - \phi_j^{(1)})\, g_{ij} W(\boldsymbol{k}_{ij} \mid \boldsymbol{k}'_{ij}\, ; g_{ij})\, \mathrm{d}\boldsymbol{k}'_{ij}\, \mathrm{d}\boldsymbol{u}_j, \tag{84}$$

since the functions $f_i^{(0)}$ satisfy the equality (40). It should be noted that the property (7) holds for any term of the series for $C(f_i, f_j)$ so that

$$\sum_{i,j} \int C^{(\nu)}(f_i, f_j) \ln f_i^{(0)}\, \mathrm{d}\boldsymbol{u}_i = 0\, , \tag{85}$$

because $\ln f_i^{(0)}$ is a linear combination of summational invariants. In view of (83) and (85) we may rewrite the kinetic expression (36) for the entropy source strength as

$$\sigma = -k \sum_{i,j} \int \{C(f_i, f_j) - C^{(0)}(f_i, f_j)\} \ln \frac{f_i}{f_i^{(0)}}\, \mathrm{d}\boldsymbol{u}_i\, . \tag{86}$$

If the series (81) for $C(f_i, f_j)$ and the series (64) for f_i are introduced into (86) one obtains for σ the series

$$\sigma = \sigma^{(1)} + \sigma^{(2)} + \dots, \tag{87}$$

with

$$\sigma^{(1)} = -k \sum_{i,j} \int \phi_i^{(1)} C^{(1)}(f_i, f_j) \, \mathrm{d}\boldsymbol{u}_i, \tag{88}$$

$$\sigma^{(2)} = -k \sum_{i,j} \int \phi_i^{(2)} C^{(1)}(f_i, f_j) \, \mathrm{d}\boldsymbol{u}_i + \tfrac{1}{2}k \sum_{i,j} (\phi_i^{(1)})^2 C^{(1)}(f_i, f_j) \, \mathrm{d}\boldsymbol{u}_i$$
$$- k \sum_{i,j} \int \phi_i^{(1)} C^{(2)}(f_i, f_j) \, \mathrm{d}\boldsymbol{u}_i . \tag{89}$$

In (88) and (89) the right-hand sides would seem to be of the second and third order respectively. It must, however, be borne in mind that the various orders $C^{(\nu)}$ for the collision integrals [or rather the sums $\sum_j C^{(\nu)} (f_i, f_j)$] can be expressed in terms of space derivatives of functions $f_i^{(\nu-1)}$ of order $\nu - 1$, by means of the Boltzmann equation of various orders [see for the case $\nu = 1$, equation (65)]. Thus $\sigma^{(1)}$ and $\sigma^{(2)}$ given by (88) and (89) are in fact of the first and the second order respectively. (Similar considerations hold for any order of approximation to σ.) Indeed with the help of (84) and the integral equation (65) for $\phi_i^{(1)}$, we may rewrite $\sigma^{(1)}$ in the form

$$\sigma^{(1)} = -\frac{1}{T^2} \sum_i m_i \int \{\tfrac{1}{2}(\boldsymbol{u}_i - \boldsymbol{v})^2 - h_i\} (\boldsymbol{u}_i - \boldsymbol{v}) f_i^{(0)} \phi_i^{(1)} \, \mathrm{d}\boldsymbol{u}_i \cdot (\operatorname{grad} T)$$
$$-\frac{1}{T} \sum_i m_i \int (\boldsymbol{u}_i - \boldsymbol{v}) f_i^{(0)} \phi_i^{(1)} \, \mathrm{d}\boldsymbol{u}_i \cdot \{(\operatorname{grad} \mu_i)_T - \boldsymbol{F}_i - \rho^{-1} \operatorname{grad} p + \rho^{-1} \sum_j \rho_j \boldsymbol{F}_j\}$$
$$-\frac{1}{T} \sum_i m_i \int \{(\boldsymbol{u}_i - \boldsymbol{v})(\boldsymbol{u}_i - \boldsymbol{v}) - \tfrac{1}{3}(\boldsymbol{u}_i - \boldsymbol{v})^2 \mathsf{U}\} f_i^{(0)} \phi_i^{(1)} \, \mathrm{d}\boldsymbol{u}_i : (\operatorname{Grad} \boldsymbol{v}) . \tag{90}$$

If we now apply the definitions (75) and (76) of $\boldsymbol{J}_i^{(1)}$ and $\boldsymbol{J}_q^{(1)}$, and also (52), expression (90) for $\sigma^{(1)}$ becomes, with $\sum_i \boldsymbol{J}_i^{(1)} = 0$,

$$\sigma^{(1)} = -\frac{1}{T^2} \boldsymbol{J}_q'^{(1)} \cdot \operatorname{grad} T - \frac{1}{T} \sum_i \boldsymbol{J}_i^{(1)} \cdot \{(\operatorname{grad} \mu_i)_T - \boldsymbol{F}_i\} - \frac{1}{T} \Pi^{(1)} : \operatorname{Grad} \boldsymbol{v} . \tag{91}$$

Here $\Pi^{(1)}$ is the first non-vanishing term of the series for the viscous pressure tensor Π [*cf.* (53) and the kinetic expressions (26) and (46)]:

$$\Pi = \Pi^{(0)} + \Pi^{(1)} + \Pi^{(2)} + \ldots, \tag{92}$$

with

$$\Pi^{(0)} = \sum_i m_i \int \{ (\boldsymbol{u}_i - \boldsymbol{v})(\boldsymbol{u}_i - \boldsymbol{v}) - \tfrac{1}{3}(\boldsymbol{u}_i - \boldsymbol{v})^2 U \} f_i^{(0)} \, d\boldsymbol{u}_i = 0, \tag{93}$$

$$\Pi^{(1)} = \sum_i m_i \int (\boldsymbol{u}_i - \boldsymbol{v})(\boldsymbol{u}_i - \boldsymbol{v}) - \tfrac{1}{3}(\boldsymbol{u}_i - \boldsymbol{v})^2 U \} f_i^{(0)} \phi_i^{(1)} d\boldsymbol{u}_i, \tag{94}$$

$$\Pi^{(2)} = \sum_i m_i \int \{ (\boldsymbol{u}_i - \boldsymbol{v})(\boldsymbol{u}_i - \boldsymbol{v}) - \tfrac{1}{3}(\boldsymbol{u}_i - \boldsymbol{v})^2 U \} f_i^{(0)} \phi_i^{(2)} d\boldsymbol{u}_i. \tag{95}$$

By comparing (91) with (51) we see again that the kinetic and the macroscopically obtained expressions for the entropy source strength are identical in the first approximation of Enskog, *i.e.* as long as linear phenomenological laws hold.

A similar analysis of (89) shows that this is no longer true if second order terms must be retained in the evaluation of the irreversible fluxes. Indeed the first integral on the right-hand side of (89) is identical with the corresponding approximation of (51) and the other (non-vanishing) integrals therefore demonstrate that the macroscopic and kinetic expressions for the entropy source strength are incompatible in this order.

It is useful for a deeper insight into the origin of the limitations of thermodynamics of irreversible processes to consider also the kinetic expression (31) for the entropy density. With the Enskog series, this quantity may be expanded as follows:

$$\rho s = \rho s^{(0)} + \rho s^{(1)} + \rho s^{(2)} + \ldots, \tag{96}$$

with

$$\rho s^{(0)} = -k \sum_i \int f_i^{(0)} (\ln f_i^{(0)} - 1) \, d\boldsymbol{u}_i, \tag{97}$$

$$\rho s^{(1)} = -k \sum_i \int f_i^{(0)} \phi_i^{(1)} \ln f_i^{(0)} \, d\boldsymbol{u}_i = 0, \tag{98}$$

$$\rho s^{(2)} = -\tfrac{1}{2} k \sum_i \int f_i^{(0)} (\phi_i^{(1)})^2 \, d\boldsymbol{u}_i. \tag{99}$$

The contribution $\rho s^{(1)}$ to the entropy density vanishes due to conditions (59)–(61). These conditions have also been used in (99). Thus if only terms linear in $\phi_i^{(1)}$ are retained for the evaluation of the entropy density, this quantity is still the same function of the local kinetic temperature T and of the densities ρ_i as in equilibrium.

Consequently the Gibbs formula (49) may still be expected to hold. In fact up to this order of approximation the entropy depends on space and time only implicitly through its dependence on $T(\boldsymbol{r}; t)$ and $\rho_i(\boldsymbol{r}; t)$. The rate of change with time of the latter quantities may be evaluated, in this order of approximation, with the help of the first order expressions for the irreversible fluxes occurring in the hydrodynamic equations. This justifies the use of the Gibbs relation for departures from equilibrium such that the transport phenomena may be described by linear phenomenological laws. The previous analysis in this section of the entropy flows and the entropy source strength has established this fact explicitly. On the other hand, if terms quadratic in $\phi_i^{(1)}$ (or linear in $\phi_i^{(2)}$) are retained in the expansion (96) the entropy density will, according to (99) become a function of the macroscopic gradients and the Gibbs formula (49) will cease to be satisfied.

The present discussion* has been made on the assumption that the molecules do not have any internal degrees of freedom. It may, however, be extended to include the case of molecules which do have internal degrees of freedom**. If, locally, equipartition of energy may be assumed to be established between the internal and the translational degrees of freedom, it can be shown that the conclusions reached in this section remain valid. If the equipartition of energy is not established within times in which the local Maxwell–Boltzmann distribution for each degree of freedom separately is reached, one obtains a further contribution to the entropy source strength due to a relaxation phenomenon, related to the exchange of energy between translational and internal degrees of freedom. Such a term is also found in macroscopic theory if the entropy is considered to be a function not only of the variables u, ρ and c_i, but also of some internal parameter (*cf.* Chapter X). It has furthermore been shown*** that the macroscopic theory also holds for chemical reactions which

* I. Prigogine, Physica **15** (1949) 271.
S. Ono, Scientific papers of the college of education, University of Tokyo **5** (1955) 87.

** H. G. Reik, Z. Physik **148** (1957) 156; 333.

*** J. Ross and P. Mazur, J. chem. Phys. **35** (1961) 19; see also Chapter X.

do not disturb appreciably the Maxwell–Boltzmann distribution of each separate reacting component.

§ 7. *The Onsager Relations*

In order to complete the present investigation of the validity of thermodynamics of irreversible processes by means of the kinetic theory of gases, we shall also give a derivation, within the framework of this theory, of the Onsager reciprocal relations.

Let us consider again the integral equation (65) which together with the conditions (59)–(61) determines uniquely the perturbation functions $\phi_i^{(1)}$. We have already stated that the form of this equation is such that $\phi_i^{(1)}$ must be a linear function of the gradients of the macroscopic quantities μ_i, $\boldsymbol{v}$ and T:

$$\phi_i^{(1)} = -\boldsymbol{A}_i \cdot \frac{\operatorname{grad} T}{T} - \sum_k \boldsymbol{B}_{ik} \cdot \{(\operatorname{grad} \mu_k)_T - \boldsymbol{F}_k - \rho^{-1} \operatorname{grad} p + \rho^{-1} \sum_j \rho_j \boldsymbol{F}_j\} - \mathsf{C}_i : \operatorname{Grad} \boldsymbol{v}, \tag{100}$$

where the vectors $\boldsymbol{A}_i$ and $\boldsymbol{B}_{ik}$ and the tensor C_i are functions of the velocity $\boldsymbol{u}_i - \boldsymbol{v}$, the local temperature and the local composition. For simplicity's sake we shall now consider the case that the temperature T and the average velocity $\boldsymbol{v}$ are uniform. The function $\phi_i^{(1)}$ may then be written as:

$$\phi_i^{(1)} = -\sum_k \boldsymbol{B}_{ik} \cdot \{(\operatorname{grad} \mu_k)_T - \boldsymbol{F}_k - \rho^{-1} \operatorname{grad} p + \rho^{-1} \sum_j \rho_j \boldsymbol{F}_j\}. \tag{101}$$

However, in view of the Gibbs–Duhem relation

$$\sum_k \rho_k (\operatorname{grad} \mu_k)_T = \operatorname{grad} p, \tag{102}$$

the coefficients $\boldsymbol{B}_{ik}$ are not uniquely determined and may still contain an arbitrary additive term of the form $d_i \rho_k$, where d_i is a constant. This enables us to make the convention

$$\sum_k \boldsymbol{B}_{ik} = 0. \tag{103}$$

For an n-component mixture we may then eliminate by means of this convention the nth term of the sum on the right-hand side of (101) and write

$$\phi_i^{(1)} = -\sum_{k=1}^{n-1} \boldsymbol{B}_{ik} \cdot [\{\text{grad}\,(\mu_k - \mu_n)\}_T + \boldsymbol{F}_n - \boldsymbol{F}_k], \quad (i = 1, 2, \ldots, n) . \quad (104)$$

This form for $\phi_i^{(1)}$ may now be introduced into (65), also assuming T and $\boldsymbol{v}$ uniform. Using again (102) in order to eliminate the term grad p on the left-hand side of this equation, and equating the coefficients of the same (independent) gradients we obtain for the coefficients $\boldsymbol{B}_{ik}$ the integral equations

$$(kT)^{-1} f_i^{(0)} m_i (\boldsymbol{u}_i - \boldsymbol{v}) \left(\delta_{ik} - \frac{\rho_k}{\rho}\right)$$

$$= -\sum_{j=1}^{n} \iint f_i^{(0)} f_j^{(0)} (\boldsymbol{B}'_{ik} + \boldsymbol{B}'_{jk} - \boldsymbol{B}_{ik} - \boldsymbol{B}_{jk})\, g_{ij} W(\boldsymbol{k}_{ij} \mid \boldsymbol{k}'_{ij} ; g_{ij})\, \mathrm{d}\boldsymbol{k}'_{ij}\, \mathrm{d}\boldsymbol{u}_j ,$$

$$(i = 1, 2, \ldots n; k = 1, 2, \ldots, n-1) . \quad (105)$$

According to (59) and (60) the coefficients $\boldsymbol{B}_{ik}$ of (104) must also satisfy the auxiliary conditions

$$\sum_{i=1}^{n} m_i \int (\boldsymbol{u}_i - \boldsymbol{v})\, \boldsymbol{B}_{ik} f_i^{(0)}\, \mathrm{d}\boldsymbol{u}_i = 0 , \quad (k = 1, 2, \ldots, n-1) . \quad (106)$$

Let us now rewrite the statistical expression (75) for the diffusion fluxes $\boldsymbol{J}_k^{(1)}$, in the first approximation of Enskog, by introducing into these the form (104) for $\phi_k^{(1)}$:

$$\boldsymbol{J}_k^{(1)} = -\sum_{l=1}^{n-1} \int m_k (\boldsymbol{u}_k - \boldsymbol{v})\, \boldsymbol{B}_{kl} f_k^{(0)}\, \mathrm{d}\boldsymbol{u}_k \cdot \{\text{grad}\,(\mu_l - \mu_n)_T + \boldsymbol{F}_n - \boldsymbol{F}_l\}$$

$$= -\sum_{l=1}^{n-1} \mathsf{L}_{kl} \cdot [\{\text{grad}\,(\mu_l - \mu_n)\}_T + \boldsymbol{F}_n - \boldsymbol{F}_l] , \quad (k = 1, 2, \ldots, n) . \quad (107)$$

These equations (for $k = 1, 2, \ldots, n-1$) have precisely the form of macroscopic linear phenomenological laws established, for a multi-component system with uniform temperature and uniform average velocity, between the independent fluxes and thermodynamic forces occurring in the entropy production (91) (*cf.* also Chapter IV). The (independent) coefficients L_{kl} in these equations are given by*

$$\mathsf{L}_{kl} = m_k \int (\boldsymbol{u}_k - \boldsymbol{v})\, \boldsymbol{B}_{kl}\, f_k^{(0)}\, \mathrm{d}\boldsymbol{u}_k = L_{kl} \mathsf{U}\,, \tag{108}$$

or

$$L_{kl} = \tfrac{1}{3} m_k \int (\boldsymbol{u}_k - \boldsymbol{v}) \cdot \boldsymbol{B}_{kl}\, f_k^{(0)}\, \mathrm{d}\boldsymbol{u}_k\,, \quad (k, l = 1, 2, \ldots, n-1)\,. \tag{109}$$

On the basis of Onsager's reciprocity theorem (Chapter IV, § 3) and of the developments of Chapter VII, we expect the matrix of coefficients L_{kl} to be symmetric. In the present treatment the proof of this symmetry is straightforward. We multiply both members of equation (105) by $\boldsymbol{B}_{il}$, then integrate with respect to $\boldsymbol{u}_i$ and sum both sides of the resulting equation over the index i (from $i = 1$ to n). With the auxiliary condition (106) and the definition (109) of L_{kl} we then obtain

$$L_{kl} = -\frac{kT}{3} \sum_{i,j=1}^{n} \iiint f_i^{(0)} f_j^{(0)} (\boldsymbol{B}'_{ik} + \boldsymbol{B}'_{jk} - \boldsymbol{B}_{ik} - \boldsymbol{B}_{jk})$$

$$\times \cdot\, \boldsymbol{B}_{il}\, g_{ij} W(\boldsymbol{k}_{ij} \mid \boldsymbol{k}'_{ij}\, ;\, g_{ij})\, \mathrm{d}\boldsymbol{k}'_{ij}\, \mathrm{d}\boldsymbol{u}_i\, \mathrm{d}\boldsymbol{u}_j\,, \quad (k, l = 1, 2, \ldots, n-1)\,. \tag{110}$$

By making use of the property (6) of microscopic reversibility,

$$W(\boldsymbol{k}_{ij} \mid \boldsymbol{k}'_{ij}\, ;\, g_{ij}) = W(\boldsymbol{k}'_{ij} \mid \boldsymbol{k}_{ij}\, ;\, g_{ij})\,, \tag{111}$$

the right-hand side of (110) may be symmetrized in the manner outlined in § 2 [*cf.* (7)–(16)] so that

$$L_{kl} = \frac{kT}{3} \sum_{i,j=1}^{n} \iiiint f_i^{(0)} f_j^{(0)} (\boldsymbol{B}'_{ik} + \boldsymbol{B}'_{jk} - \boldsymbol{B}_{ik} - \boldsymbol{B}_{jk})$$

$$\times (\boldsymbol{B}'_{il} + \boldsymbol{B}'_{jl} - \boldsymbol{B}_{il} - \boldsymbol{B}_{jl})\, g_{ij}^3 W(\boldsymbol{k}_{ij} \mid \boldsymbol{k}'_{ij}\, ;\, g_{ij})\, \mathrm{d}\boldsymbol{k}_{ij}\, \mathrm{d}\boldsymbol{k}'_{ij}\, \mathrm{d}g_{ij}\, \mathrm{d}\boldsymbol{u}_{(ij)}\,. \tag{112}$$

* We note that any vector function $\boldsymbol{B}_{kl}$ of the vector $\boldsymbol{u}_k - \boldsymbol{v}$, the only vector variable of $\boldsymbol{B}_{kl}$ [see (104) and (105)], is equal to the vector $\boldsymbol{u}_k - \boldsymbol{v}$ itself, multiplied by a scalar function of the absolute value of the vector $\boldsymbol{u}_k - \boldsymbol{v}$. Consequently in (108) all diagonal elements of the tensors L_{kl} are equal and all off-diagonal elements vanish.

This expression is symmetric with respect to the indices k and l, and therefore the Onsager reciprocal relations

$$L_{kl} = L_{lk}, \quad (k, l = 1, 2, \ldots, n-1) \tag{113}$$

hold. It should be noted that the property of microscopic reversibility is essential also in the present derivation of the Onsager reciprocal relations.

§ 8. *Brownian Motion*

In the preceding sections of this chapter we studied the entropy balance equation (and the Onsager relations) within the framework of the kinetic theory of gases, which is based on the Boltzmann equation. In the present section we wish to consider the properties of the entropy production, using the kinetic approach of the theory of Brownian motion*. This theory is based on the use of a Fokker–Planck equation, a partial differential equation for the distribution function of the Brownian particles, which has a certain similarity to the Boltzmann integro-differential equation. The main purpose of the formalism of this section is again to establish a *local* form for the entropy source strength and to discuss the time evolution of its various contributions, especially in view of the fact that in Brownian motion we encounter an example where besides the familiar contribution arising from diffusion, inertia terms, due to the kinetic energy, can play a role (*cf.* Chapter III, § 4).

Let us suppose that N (non-interacting) Brownian particles of mass m are immersed in a fluid of volume V. The microscopic state of this ensemble of Brownian particles is specified by a distribution function $f(\boldsymbol{r}, \boldsymbol{u}; t)$ in such a way that $f(\boldsymbol{r}, \boldsymbol{u}; t)\, \mathrm{d}\boldsymbol{r}\, \mathrm{d}\boldsymbol{u}$ gives the number of particles with positions in the interval $(\boldsymbol{r}, \boldsymbol{r} + \mathrm{d}\boldsymbol{r})$ and velocities in the interval $(\boldsymbol{u}, \boldsymbol{u} + \mathrm{d}\boldsymbol{u})$ at the time t. From statistical mechanics one knows that the equilibrium distribution is Maxwellian in the velocities and has the form

$$f^{\mathrm{eq}}(\boldsymbol{u}) = \frac{N}{V}\left(\frac{m}{2\pi kT^{\mathrm{eq}}}\right)^{\frac{3}{2}} \exp\left(-\frac{mu^2}{2kT^{\mathrm{eq}}}\right), \tag{114}$$

* G. E. Uhlenbeck and L. S. Ornstein, Phys. Rev. **36** (1930) 823;
Ming Chen Wang and G. E. Uhlenbeck, Rev. mod. Phys. **17** (1945) 323;
S. Chandrasekhar, Rev. mod. Phys. **15** (1943) 1; all three articles are reprinted in N. Wax, Noise and stochastic processes (Dover, New York, 1954).

where T^{eq} is the temperature of the fluid and k Boltzmann's constant. In the theory of Brownian motion the distribution function $f(\boldsymbol{r}, \boldsymbol{u}; t)$ is assumed to obey a Fokker–Planck equation (*cf.* Chapter VII, § 6) of the following form:

$$\frac{\partial f}{\partial t} + \boldsymbol{u} \cdot \frac{\partial f}{\partial \boldsymbol{r}} = \beta \left(\frac{\partial}{\partial \boldsymbol{u}} \cdot f\boldsymbol{u} + \frac{kT^{eq}}{m} \frac{\partial}{\partial \boldsymbol{u}} \cdot \frac{\partial}{\partial \boldsymbol{u}} f \right) . \tag{115}$$

Here the positive quantity β is the "friction constant", characterizing the mean deceleration which the Brownian particles suffer in the fluid, according to the relation

$$\lim_{\Delta t \to 0} \frac{\overline{\Delta \boldsymbol{u}}^{\boldsymbol{u}_0}}{\Delta t} = - \beta \boldsymbol{u}_0 , \tag{116}$$

where the average is taken for an ensemble of Brownian particles with constant initial velocities $\boldsymbol{u}_0$. The equation (115) is based on the Markoffian character of the Brownian motion. The right-hand side of this equation represents the rate of change of f resulting from the interaction with the fluid. It replaces for the case of Brownian particles the collision term (2) of the Boltzmann integro-differential equation (1).

Since the equilibrium distribution function (114) satisfies the Fokker–Planck equation, the latter may alternatively be written as

$$\frac{\partial f}{\partial t} = - \boldsymbol{u} \cdot \frac{\partial f}{\partial \boldsymbol{r}} + \beta \frac{kT^{eq}}{m} \frac{\partial}{\partial \boldsymbol{u}} \cdot \left(f \frac{\partial}{\partial \boldsymbol{u}} \ln \frac{f}{f^{eq}} \right) . \tag{117}$$

The statistical expression for the mass density $\rho(\boldsymbol{r}; t)$ of the Brownian particles is

$$\rho(\boldsymbol{r}; t) = mn(\boldsymbol{r}; t) = m \int f(\boldsymbol{r}, \boldsymbol{u}; t) \, \mathrm{d}\boldsymbol{u} , \tag{118}$$

where $n(\boldsymbol{r}; t)$ is the number density. From this expression and (114) one finds that at equilibrium

$$\rho^{eq} = mn^{eq} = mN/V . \tag{119}$$

Differentiation of (118) with respect to time, using (117) and assuming that f tends to zero sufficiently rapidly for large $\boldsymbol{u}$, gives

$$\frac{\partial \rho}{\partial t} = - \operatorname{div} \rho \boldsymbol{v} , \tag{120}$$

where the average velocity $\boldsymbol{v}(\boldsymbol{r}; t)$ is defined by

$$\rho \boldsymbol{v}(\boldsymbol{r}\,; t) = m \int \boldsymbol{u} f \,\mathrm{d}\boldsymbol{u}\,. \tag{121}$$

In equilibrium, where f is given by (114), the velocity $\boldsymbol{v}$ vanishes ($\boldsymbol{v}^{\mathrm{eq}} = 0$).

The equation of motion follows now with (117), if the expression (121) is differentiated with respect to time and if (114) is also used:

$$\frac{\partial \rho \boldsymbol{v}}{\partial t} = - \operatorname{Div} (\rho \boldsymbol{v}\boldsymbol{v} + \mathsf{P}) - \beta \rho \boldsymbol{v}\,. \tag{122}$$

Here the pressure tensor P is defined as

$$\mathsf{P}(\boldsymbol{r}\,; t) = m \int (\boldsymbol{u} - \boldsymbol{v})\,(\boldsymbol{u} - \boldsymbol{v})\,f \,\mathrm{d}\boldsymbol{u}\,. \tag{123}$$

Its value at equilibrium, which is obtained with (114), is given by

$$\mathsf{P}^{\mathrm{eq}} = \rho^{\mathrm{eq}} \frac{kT^{\mathrm{eq}}}{m} \mathsf{U} = \frac{N}{V} kT^{\mathrm{eq}}\, \mathsf{U}\,, \tag{124}$$

where U is the unit matrix.

An equivalent form of (122) is

$$\rho \frac{\mathrm{d}\boldsymbol{v}}{\mathrm{d}t} = - \operatorname{Div} \mathsf{P} - \beta \rho \boldsymbol{v}\,, \tag{125}$$

where (120) has been employed.

Let us finally define the internal energy density ρu and the local temperature $T(\boldsymbol{r}; t)$ of the Brownian particles by

$$\rho u(\boldsymbol{r}; t) = \tfrac{1}{2} m \int (\boldsymbol{u} - \boldsymbol{v})^2 f \,\mathrm{d}\boldsymbol{u} \equiv \tfrac{3}{2} n k T(\boldsymbol{r}; t)\,. \tag{126}$$

The equilibrium value of u is, according to (114), given by

$$u^{\mathrm{eq}} = \frac{3}{2} \frac{kT^{\mathrm{eq}}}{m}\,. \tag{127}$$

The balance equation for $u(\boldsymbol{r};t)$ follows with (117), using also (114),

$$\frac{\partial \rho u}{\partial t} = -\operatorname{div}(\rho u \boldsymbol{v} + \boldsymbol{J}_q) - \mathsf{P} : \operatorname{Grad} \boldsymbol{v} - 2\beta\rho\,(u - u^{\mathrm{eq}})\,, \tag{128}$$

where the energy flow is given by

$$\boldsymbol{J}_q(\boldsymbol{r};t) = \tfrac{1}{2}m \int (\boldsymbol{u} - \boldsymbol{v})^2\,(\boldsymbol{u} - \boldsymbol{v}) f \,\mathrm{d}\boldsymbol{u}\,. \tag{129}$$

According to (114) its equilibrium value vanishes ($\boldsymbol{J}_q^{\mathrm{eq}} = 0$). Again with (120), one gets the alternative form of (128)

$$\rho \frac{\mathrm{d}u}{\mathrm{d}t} = -\operatorname{div} \boldsymbol{J}_q - \mathsf{P} : \operatorname{Grad} \boldsymbol{v} - 2\beta\rho(u - u^{\mathrm{eq}})\,. \tag{130}$$

The hydrodynamic equations (120), (122) and (128) are similar in structure to the hydrodynamic equations derived from the Boltzmann equation. However, (122) and (128) contain in addition source terms due to the transfer of momentum and energy from the Brownian particles to the fluid.

Let us now define the entropy density ρs as

$$\rho s(\boldsymbol{r};t) = -k \int f \ln \frac{f}{f^{\mathrm{eq}}\Omega}\,\mathrm{d}\boldsymbol{u}\,, \tag{131}$$

which is the Gibbs postulate, introduced also in Chapter VII and in formula (31) of this chapter*. The quantity Ω determines the value of the entropy of the total system (Brownian particles and fluid) at equilibrium, as follows from (131) with (114):

$$s^{\mathrm{eq}} = \frac{k}{m} \ln \Omega\,. \tag{132}$$

* In the definition (131) a term $k\int f \ln f^{\mathrm{eq}}\,\Omega\,\mathrm{d}\boldsymbol{u}$ occurs, whereas (31), if written for convenience for a single species, contains instead a term $k\int f\,\mathrm{d}\boldsymbol{u} = kn$. The reason for this difference in form can be easily understood. If the form (114) for the equilibrium distribution function f^{eq} is inserted into the first term mentioned above, we obtain with (118), (126) and (132) the expression

$$nm\,T^{\mathrm{eq}-1}\,(\mu^{\mathrm{eq}} - u^{\mathrm{eq}} + T^{\mathrm{eq}}s^{\mathrm{eq}}),$$

where the equilibrium entropy s^{eq} refers to the total system of Brownian particles and fluid. In the case of the Boltzmann theory however μ^{eq}, u^{eq} and s^{eq} all refer to the perfect gas. Therefore the last expression becomes, according to the Euler relation, equal to $pT^{\mathrm{eq}-1}$ or, with Boyle–Gay Lussac's law, equal to kn. This is just the term used in (31), as stated above.

From (131) one has with (117)

$$\frac{\partial \rho s}{\partial t} = -k \int \left(\ln \frac{f}{f^{\mathrm{eq}}\Omega} + 1 \right) \frac{\partial f}{\partial t} \, \mathrm{d}\boldsymbol{u} = -\operatorname{div}(\rho s \boldsymbol{v} + \boldsymbol{J}_s) + \sigma , \quad (133)$$

where

$$\boldsymbol{J}_s(\boldsymbol{r}\,;t) = -k \int \left(\ln \frac{f}{f^{\mathrm{eq}}\Omega} \right) (\boldsymbol{u} - \boldsymbol{v}) f \, \mathrm{d}\boldsymbol{u} \quad (134)$$

represents the entropy flow, and

$$\sigma(\boldsymbol{r}\,;t) = \beta \frac{k^2 T^{\mathrm{eq}}}{m} \int f \left(\frac{\partial}{\partial \boldsymbol{u}} \ln \frac{f}{f^{\mathrm{eq}}} \right)^2 \mathrm{d}\boldsymbol{u} \geqslant 0 , \quad (135)$$

the entropy source strength. Both the entropy flow and the entropy source strength vanish at equilibrium ($f = f^{\mathrm{eq}}$, $\boldsymbol{J}_s^{\mathrm{eq}} = 0$, $\sigma^{\mathrm{eq}} = 0$). The inequality of (135), which expresses the fact that the entropy source strength must be positive or zero, constitutes the second law (or the H-theorem) for Brownian motion.

Let us now choose as a special form of the distribution function f at a *certain initial moment* $t = 0$ the following expression:

$$f(\boldsymbol{r}, \boldsymbol{u}\,; 0) = n(\boldsymbol{r}\,; 0) \left\{ \frac{m}{2\pi k T(\boldsymbol{r}\,; 0)} \right\}^{\frac{3}{2}} \exp \left[- \frac{m \{ \boldsymbol{u} - \boldsymbol{v}(\boldsymbol{r}\,; 0) \}^2}{2kT(\boldsymbol{r}\,; 0)} \right] . \quad (136)$$

This initial distribution is the Gaussian distribution which corresponds to three given initial local "moments" of the velocity: the spatial distribution of the density which is $n(\boldsymbol{r}; 0)$ according to (118), the average velocity field which is $\boldsymbol{v}(\boldsymbol{r}; 0)$ as follows with (121) and the temperature field $T(\boldsymbol{r}; 0)$ which is defined by (126).

The form (136) is the same as the equilibrium distribution function (114) but with local density $n(\boldsymbol{r}; 0)$, local average velocity $\boldsymbol{v}(\boldsymbol{r}; 0)$ and local temperature $T(\boldsymbol{r}; 0)$ instead of the corresponding equilibrium quantities n^{eq}, $\boldsymbol{v}^{\mathrm{eq}} = 0$ and T^{eq}. Physically such a distribution corresponds to an ensemble of Brownian particles which at each point of space is in equilibrium in a fictive fluid of temperature $T(\boldsymbol{r}; 0)$ moving with velocity $\boldsymbol{v}(\boldsymbol{r}; 0)$. One may for instance think of an initial distribution with $T(\boldsymbol{r}; 0) = T^{\mathrm{eq}}$, $\boldsymbol{v}(\boldsymbol{r}; 0) = \boldsymbol{v}^{\mathrm{eq}} = 0$, and $n(\boldsymbol{r}; 0)$ determined by an arbitrary external field of force. The external field is then switched off at $t = 0$, and the distribution starts to change in time according to the Fokker–Planck equation (117). Instead of an external field which determines $n(\boldsymbol{r}; 0)$ one could also have a partition which restricts the Brownian particles to part of the space occupied by the fluid and which subsequently is removed.

It can be shown that the solution of the Fokker–Planck equation with our initial condition is simply

$$f(\boldsymbol{r}, \boldsymbol{u}\,; t) = n(\boldsymbol{r}\,; t) \left\{\frac{m}{2\pi kT(\boldsymbol{r}\,; t)}\right\}^{\frac{3}{2}} \exp\left[-\frac{m\,\{\boldsymbol{u} - \boldsymbol{v}(\boldsymbol{r}\,; t)\}^2}{2kT(\boldsymbol{r}\,; t)}\right], \tag{137}$$

i.e. of the same form as (136) but now for arbitrary times, provided that certain subsidiary conditions are fulfilled. In fact if (137) is inserted into the Fokker–Planck equation (117) it satisfies this equation at all times, if

$$\text{Grad}\, \boldsymbol{v} = \tfrac{1}{3}\,(\text{div}\, \boldsymbol{v})\, \mathsf{U}\,, \tag{138}$$

$$\text{grad}\, T = 0\,, \tag{139}$$

as follows when (120), (122) and (128) are also used. The distribution function (137) can be said to represent a "local equilibrium distribution", depending on the space coordinates and on time only through the non-equilibrium quantities $n(\boldsymbol{r}; t)$, $\boldsymbol{v}(\boldsymbol{r}; t)$ and $T(\boldsymbol{r}; t)$, which satisfy the conditions (138) and (139).

A useful alternative form of (137) is

$$f(\boldsymbol{r}, \boldsymbol{u}\,; t) = \exp\left\{\frac{m[\mu(\boldsymbol{r}\,; t) - \frac{1}{2}\{\boldsymbol{u} - \boldsymbol{v}(\boldsymbol{r}\,; t)\}^2]}{kT(\boldsymbol{r}\,; t)}\right\}, \tag{140}$$

where we have introduced the thermodynamic potential per unit mass of the ideally diluted Brownian particles:

$$\mu(\boldsymbol{r}\,; t) \equiv \frac{kT}{m}\left(\ln n - \tfrac{3}{2}\ln\frac{2\pi kT}{m}\right). \tag{141}$$

In equilibrium (140) reduces to

$$f^{\text{eq}}(\boldsymbol{u}) = \exp\left\{\frac{m(\mu^{\text{eq}} - \frac{1}{2}\boldsymbol{u}^2)}{kT^{\text{eq}}}\right\}, \tag{142}$$

because then $\mu = \mu^{\text{eq}}$, $\boldsymbol{v}^{\text{eq}} = 0$ and $T = T^{\text{eq}}$.

The equation of motion and the energy law now get simple forms, because with (137) one has from (123), (126) and (129):

$$P = nkTU\,,\quad u = \frac{3}{2}\frac{kT}{m}\,,\quad J_q = 0\,, \tag{143}$$

which leads with (139) and (141) to

$$\frac{\mathrm{d}\boldsymbol{v}}{\mathrm{d}t} = -\operatorname{grad}\mu - \beta\boldsymbol{v}\,, \tag{144}$$

$$\frac{3}{2}\frac{\partial T}{\partial t} = -T\operatorname{div}\boldsymbol{v} - 3\beta(T - T^{\mathrm{eq}})\,, \tag{145}$$

for the equation of motion (125) and the energy law (130) respectively

The entropy source strength (135) becomes with the use of the distribution function (137):

$$\sigma(\boldsymbol{r}\,;t) = \beta\rho\left\{\frac{\boldsymbol{v}^2}{T^{\mathrm{eq}}} + 3\frac{k}{m}\frac{(T - T^{\mathrm{eq}})^2}{TT^{\mathrm{eq}}}\right\} = \sigma_1 + \sigma_2\,. \tag{146}$$

The first contribution σ_1 to this expression can alternatively be written with the help of (144) as

$$\sigma_1 = -\frac{\rho\boldsymbol{v}}{T^{\mathrm{eq}}}\cdot\left(\frac{\mathrm{d}\boldsymbol{v}}{\mathrm{d}t} + \operatorname{grad}\mu\right), \tag{147}$$

which contains an inertia term, involving the acceleration, and a term with the gradient of the thermodynamic potential, in the same way as the entropy production (III.39) of a diffusing system, in which the kinetic energy of diffusion was taken into account*. Here however we still have a second contribution σ_2, which is caused by the deviation of the local temperature $T(\boldsymbol{r}; t)$ from the equilibrium temperature T^{eq}.

The relative magnitudes of σ_1 (and its two parts) and σ_2 may be analysed most easily (*v.* the problems of this chapter), if for the

* An inertia term in the entropy production has not been found in the previous sections from the Boltzmann equation. The reason is that the Enskog solution is only valid *after* an initial characteristic relaxation time, during which the inertia term has already been damped out.

distribution function (137) a form is chosen, which is Gaussian in both the coordinates $\boldsymbol{r}$ and the velocities $\boldsymbol{u}$ and satisfies the Fokker–Planck equation at all times. To fix the ideas let us suppose that the friction constant β is of the order of 10^9 sec^{-1}, which is roughly the value for Brownian particles of diameter 10^{-5} cm and mass 10^{-15} gramme immersed in water.

The temperature T^{eq} is supposed to be about 300° K.

One finds then that always if t is much larger than β^{-1} the following inequality is valid:

$$\sigma_1 \gg \sigma_2 \,. \tag{148}$$

Thus one has in this case only the entropy production (147). The same is true for $t \ll \beta^{-1}$, if initially the temperature T is equal to T^{eq}. As for the relative order of magnitude of the two contributions to σ_1, it can be shown that after a time much larger than β^{-1} the inertia term becomes negligibly small, so that only the grad μ term plays a role. One then gets from (144):

$$\boldsymbol{v} = -\beta^{-1} \operatorname{grad} \mu \,, \tag{149}$$

or, with (141): and using the fact that $T = T^{\text{eq}}$ for $t \gg \beta^{-1}$

$$\boldsymbol{v} = -\frac{kT^{\text{eq}}}{m\beta\rho} \operatorname{grad} \rho \,. \tag{150}$$

Now if this expression is inserted into the conservation of mass law (120), we obtain the diffusion law (Smoluchowski equation)

$$\frac{\partial \rho}{\partial t} = D\Delta\rho \,, \tag{151}$$

with the diffusion coefficient

$$D \equiv \frac{kT^{\text{eq}}}{m\beta} \,. \tag{152}$$

It may be stressed again that the law (151) was found for the approximation in which the acceleration $\mathrm{d}\boldsymbol{v}/\mathrm{d}t$ may be neglected but a non-vanishing average velocity $\boldsymbol{v}$ subsists.

The situation where

$$\sigma_1 \ll \sigma_2 \tag{153}$$

can arise for times smaller than β^{-1}, if the initial velocity distribution is sharply peaked around a certain fixed velocity $\boldsymbol{u}_0$ smaller than the final thermal velocity of the Brownian particles.

PART B

APPLICATIONS

CHAPTER X

CHEMICAL REACTIONS AND RELAXATION PHENOMENA

§ 1. *Introduction*

In this first of the chapters dealing more specifically with applications of the theory we shall consider all those irreversible processes which may take place in uniform systems. This group of phenomena is described by phenomenological equations relating scalar fluxes to scalar thermodynamic forces.

Chemical reactions belong to this class of processes. They will be discussed in detail in the following sections. Closely related are the processes which are due to the relaxation of internal variables describing the state of the system.

Finally a special formalism may be set up to deal with systems described by infinite sequences of internal variables, *i.e.* by internal coordinates. This formalism, which permits to derive a number of well-known "kinetic" equations, will be developed in § 6.

§ 2. *Chemical Reactions*

In a uniform multi-component system without bulk viscosity in which one single chemical reaction can take place the general expression (III.21) for the entropy production derived in Chapter III reduces to

$$\sigma = -\frac{JA}{T}, \tag{1}$$

with J the rate of the chemical reaction defined by means of the laws of conservation of mass [*cf.* (II.13)]

$$\rho \frac{\partial c_i}{\partial t} = \nu_i J, \tag{2}$$

and A its affinity [*cf.* (III.18)]

$$A = \sum_{i=1}^{n} \nu_i \mu_i . \tag{3}$$

The coefficients ν_i divided by the molecular masses M_i are proportional to the stoichiometric coefficients of species i in the chemical reaction. They are defined to be positive when species i appears in the second (*i.e.* when it is a "product" in the chemical reaction), negative when it appears in the first member of the reaction equation (*i.e.* when species i is a "reactant" of the chemical reaction). It is convenient to normalize the coefficients ν_i in such a way that

$$\sum_{i=q+1}^{n} \nu_i = 1 . \tag{4}$$

Here the species $i = q + 1, q + 2, \ldots, n$ are all the products of the chemical reaction.

With the law of conservation of mass [*cf.* (II.4)]

$$\sum_{i=1}^{n} \nu_i = 0 \tag{5}$$

it then follows for the reactants that

$$\sum_{i=1}^{q} \nu_i = -1 . \tag{6}$$

The linear phenomenological equation of the chemical reaction is [*cf.* (IV.18)]

$$J = -l \frac{A}{T} . \tag{7}$$

It must be remarked here that the linear approximation of thermodynamics of irreversible processes which is adequate to describe the transport phenomena under most experimental conditions, is not a very satisfactory one for chemical reactions. Indeed large deviations from linear behaviour in general are observed for chemical reactions. Very close to equilibrium however there always exists a region where (7) holds. Before discussing the validity of (7) in more detail and also an extension of this law, let us first analyse the consequences of the linear approximation.

It follows from (2) that

$$J = \frac{\rho}{\nu_1} \frac{\partial c_1}{\partial t} = \frac{\rho}{\nu_2} \frac{\partial c_2}{\partial t} = \ldots = \frac{\rho}{\nu_n} \frac{\partial c_n}{\partial t} . \tag{8}$$

Let us now introduce a parameter ξ satisfying the relation

$$J = \rho \frac{\partial \xi}{\partial t} . \tag{9}$$

or

$$\frac{\partial \xi}{\partial t} = \frac{1}{\nu_1} \frac{\partial c_1}{\partial t} = \frac{1}{\nu_2} \frac{\partial c_2}{\partial t} = \ldots = \frac{1}{\nu_n} \frac{\partial c_n}{\partial t} . \tag{10}$$

Integration of (10) leads to

$$\xi(t) = \frac{c_i(t) - c_i^0}{\nu_i} , \tag{11}$$

where the c_i^0 are the initial mass fractions of the n components. Equation (11) shows that for given c_i^0 specification of the value of ξ at some given time t is equivalent to the specification of the value of all the mass fractions c_i at time t. (If a component k does not take part in the reaction, *i.e.* if $\nu_k = 0$, equation (11) simply tells us that $c_k(t) = c_k^0$). Thus the parameter ξ which is called the progress variable (degree of advancement) of the chemical reaction is a true state variable: together with two other independent thermodynamic variables (*e.g.* the pressure p and the temperature T) it defines completely the state of the system for given initial mass fractions.

The interval in which ξ may vary follows from (11) together with the inequalities

$$0 \leqslant c_i \leqslant 1 . \tag{12}$$

Indeed we have from (11) and (12) for the reactants (with negative ν_i)

$$\frac{(1 - c_i^0)}{\nu_i} \leqslant \xi \leqslant - \frac{c_i^0}{\nu_i} , \quad (i = 1, 2, \ldots, q) \tag{13}$$

and for the products (with positive ν_i)

$$- \frac{c_j^0}{\nu_j} \leqslant \xi \leqslant \frac{(1 - c_j^0)}{\nu_j} , \quad (j = q + 1, q + 2, \ldots, n) . \tag{14}$$

The variable ξ may therefore vary in an interval determined by n inequalities (13) and (14) ($n - 1$ of these inequalities are independent). In particular if the initial mass fractions c_i^0 ($i = q + 1, q + 2, \ldots, n$)

of the products are zero ξ may vary according to (13) and (14) in the interval determined by the inequality

$$0 \leqslant \xi \leqslant \min. \left(- \frac{c_i^0}{\nu_i} \right), \quad (i = 1, 2, \dots, q) . \tag{15}$$

If furthermore one has initially for the reactants $c_i^0 = - \nu_i$, the allowed interval for ξ becomes

$$0 \leqslant \xi \leqslant 1 . \tag{16}$$

The value 0 of ξ corresponds to the initial situation (when with the given initial c_i^0 only reactants are present in the system). The value 1 of ξ would correspond to the complete transformation of reactants into reaction products. This complete transformation is in principle possible with our special choice of c_i^0). Whether complete transformation is actually achieved in a chemical reaction, depends of course on the equilibrium condition for the chemical reaction, or more specifically on the functional dependence of the affinity on the state variables.

If one introduces (10) into the Gibbs equation [*cf.* (III.16)]

$$\frac{\partial s}{\partial t} = \frac{1}{T} \frac{\partial u}{\partial t} + \frac{p}{T} \frac{\partial v}{\partial t} - \sum_i \frac{\mu_i}{T} \frac{\partial c_i}{\partial t}, \tag{17}$$

one obtains with (3)

$$\frac{\partial s}{\partial t} = \frac{1}{T} \frac{\partial u}{\partial t} + \frac{p}{T} \frac{\partial v}{\partial t} - \frac{A}{T} \frac{\partial \xi}{\partial t} . \tag{18}$$

It thus follows that

$$A = - T \left(\frac{\partial s}{\partial \xi} \right)_{u, v} = \left(\frac{\partial u}{\partial \xi} \right)_{s, v} \tag{19}$$

Also, using the specific enthalpy $h = u + pv$, the Helmholtz free energy $f = u - Ts$, and the Gibbs free energy $g = u - Ts + pv$, we have

$$A = \left(\frac{\partial h}{\partial \xi} \right)_{s, p} = \left(\frac{\partial f}{\partial \xi} \right)_{v, T} = \left(\frac{\partial g}{\partial \xi} \right)_{p, T} \tag{20}$$

At equilibrium one has

$$A = 0 . \tag{21}$$

From this condition one can in principle find the equilibrium value

ξ^{eq} of ξ as a function of the two other independent thermodynamic state parameters. Thus for instance

$$\xi^{eq} = \xi^{eq}(v, T) . \tag{22}$$

For "small" deviations from equilibrium and assuming the reaction to proceed at constant volume and temperature we may then write for the affinity A

$$A = \left(\frac{\partial A}{\partial \xi}\right)^{eq}_{v,T} \Delta\xi = \left(\frac{\partial^2 f}{\partial \xi^2}\right)^{eq}_{v,T} \Delta\xi , \tag{23}$$

neglecting terms quadratic in $\Delta\xi \equiv \xi - \xi^{eq}$ and using (20).

Using (9) and (23) we may now rewrite the phenomenological equation (7) in the form

$$\frac{\partial \xi}{\partial t} = -\frac{l}{\rho T}\left(\frac{\partial^2 f}{\partial \xi^2}\right)^{eq}_{v,T} \Delta\xi = -\frac{1}{\tau_{v,T}} \Delta\xi , \tag{24}$$

with $\tau_{v,T}$, the "relaxation time" at constant volume and temperature, given by

$$\tau_{v,T} = \frac{\rho T}{l(\partial^2 f/\partial \xi^2)^{eq}_{v,T}} > 0 . \tag{25}$$

The quantity $\tau_{v,T}$ is positive since both the coefficient l and the second derivative $(\partial^2 f/\partial \xi^2)^{eq}_{v,T}$ are positive: indeed l must be positive as a consequence of the positive definite character of the entropy production (*cf.* Ch. IV, § 1) and $(\partial^2 f/\partial \xi^2)^{eq}_{v,T}$ is positive because the free energy at constant volume and temperature has a minimum at equilibrium (the inequality $(\partial^2 f/\partial \xi^2)^{eq}_{v,T} > 0$ represents in fact the thermodynamic stability condition of the system with respect to the chemical reaction at constant v and T). Integrating (24) one gets, close to equilibrium and at constant T and v, a simple exponential decay of ξ to its equilibrium value

$$\Delta\xi = \Delta\xi(0)\, e^{-t/\tau_{v,T}} . \tag{26}$$

One could also study the establishment of equilibrium under different conditions, for instance at constant pressure and temperature. One then obtains an equation similar to (26) but with $\tau_{v,T}$ replaced by $\tau_{p,T}$, the relaxation time at constant pressure and temperature. Indeed one must then expand A at constant pressure and temperature [*cf.* (23)] and carry out the various steps leading to equation (26). The relaxation time $\tau_{p,T}$ is given by

$$\tau_{p,T} = \frac{\rho T}{l(\partial^2 g/\partial \xi^2)^{\text{eq}}_{p,T}}. \tag{27}$$

The positive character of this quantity follows from the thermodynamic stability condition $(\partial^2 g/\partial \xi^2)^{\text{eq}}_{p,T} > 0$. (Note that in (27) ρ should be considered to represent the density at given p and T and the equilibrium value ξ^{eq} of ξ, since in deriving the exponential decay law one only retains terms linear in $\Delta\xi$.)

Again if one studies the reaction at constant energy and volume one obtains in the linear approximation an exponential decay for $\Delta\xi$ with a relaxation time

$$\tau_{u,v} = -\frac{\rho}{l(\partial^2 s/\partial \xi^2)^{\text{eq}}_{u,v}}, \quad \text{with} \quad \left(\frac{\partial^2 s}{\partial \xi^2}\right)^{\text{eq}}_{u,v} < 0. \tag{28}$$

The various relaxation times introduced are connected through purely thermodynamic relations. In practice the ratio of, say, $\tau_{v,T}$ and $\tau_{p,T}$ is close to 1, so that any of the three quantities $\tau_{v,T}$, $\tau_{p,T}$ or $\tau_{u,v}$ may be used indistinctedly to characterize a chemical reaction.

We have analysed above the consequences of the linear approximation of thermodynamics of irreversible processes for a chemical reaction. In conformity with the use throughout this monograph we discussed the chemical reaction in terms of the changes in *mass fractions* of the various components. It is customary, however, to discuss a chemical reaction in terms of the changes in the *total number of moles* of the components in a reacting mixture, or, if the reaction proceeds at constant volume, in terms of the *molar concentrations* N_i. For simplicity's sake let us consider then reactions which proceed at constant volume. Since ρ, the mass density of the system, is now constant in time, we may rewrite (2) as

$$\frac{\partial N_i}{\partial t} = \bar{\nu}_i \bar{J}, \tag{29}$$

with

$$N_i = \frac{\rho_i}{M_i} = \frac{\rho c_i}{M_i}, \tag{30}$$

$$\bar{\nu}_i = \frac{a\nu_i}{M_i} \tag{31}$$

and

$$\bar{J} = \frac{J}{a}. \tag{32}$$

In the above equations M_i is the molar mass of species i, N_i its molar concentration, $\bar{\nu}_i$ its true stoichiometric coefficient in the chemical reaction and $\bar{J}$ the rate of the chemical reaction according to the conventional description using molar concentrations and true stoichiometric coefficients. The constant a may be determined from (31) and the normalization conditions (4) and (6). One has

$$a = \sum_{i=q+1}^{n} \bar{\nu}_i M_i = - \sum_{i=1}^{q} \bar{\nu}_i M_i . \tag{33}$$

The linear law (7) may be rewritten with (32) as

$$\bar{J} = - \bar{l} \frac{\bar{A}}{T}, \tag{34}$$

with

$$\bar{A} = aA \tag{35}$$

and the transformed phenomenological coefficient

$$\bar{l} = \frac{l}{a^2}. \tag{36}$$

The quantity $\bar{A}$ represents the affinity of the chemical reaction in the conventional chemical description. Indeed with (31)

$$\bar{A} = a \sum_{i=1}^{n} \nu_i \mu_i = \sum_{i=1}^{n} \bar{\nu}_i \bar{\mu}_i , \tag{37}$$

where $\bar{\mu}_i = M_i \mu_i$ is the chemical potential per mole of species i. Note that the above transformation of the flux and the thermodynamic force leaves the entropy production invariant. Indeed with (32) and (35)

$$\sigma = - J \frac{A}{T} = - \bar{J} \frac{\bar{A}}{T}. \tag{38}$$

Now for ideal systems (*i.e.* for reactions in perfect gases or in dilute solutions, when the solvent does not take part in the reaction) one assumes in chemical kinetics that the law of mass action holds, *i.e.* that

$$\frac{\partial N_i}{\partial t} = \bar{\nu}_i \left(k_f \prod_{j=1}^{q} N_j^{-\bar{\nu}_j} - k_r \prod_{j=q+1}^{n} N_j^{\bar{\nu}_j} \right), \tag{39}$$

or

$$\bar{J} = k_f \prod_{j=1}^{q} N_j^{-\bar{\nu}_j} - k_r \prod_{j=q+1}^{n} N_j^{\bar{\nu}_j}, \tag{40}$$

where k_f is the (positive) rate constant for the forward reaction (yielding reaction products) and k_r the (positive) rate constant for the reverse reaction (yielding reactants). At chemical equilibrium $\bar{J}$ vanishes so that (40) becomes

$$k_f \prod_{j=1}^{q} (N_j^{\text{eq}})^{-\bar{\nu}_j} = k_r \prod_{j=q+1}^{n} (N_j^{\text{eq}})^{\bar{\nu}_j}. \tag{41}$$

The rates of the forward and the reverse reaction are then equal: this is called the principle of detailed balance. The law of mass action may be written in a different way by introducing the chemical potentials $\bar{\mu}_i$ which for ideal systems are of the form

$$\bar{\mu}_i = RT \ln n_i + \zeta_i(p, T), \tag{42}$$

with $n_i = N_i / \sum_{j=1}^{n} N_j = N_i/N$. ($N$ is the total number of moles per unit volume, n_i the mole fraction of species i.) We now rewrite (40) as

$$\begin{aligned} \bar{J} &= k_r \prod_{j=q+1}^{n} N_j^{\bar{\nu}_j} \left(\frac{k_f}{k_r} \prod_{j=1}^{n} N_j^{-\bar{\nu}_j} - 1 \right) \\ &= k_r \prod_{j=q+1}^{n} N_j^{\bar{\nu}_j} \left\{ \frac{k_f}{k_r} N^{-\sum_{j=1}^{n} \bar{\nu}_j} e^{(\sum_{j=1}^{n} \bar{\nu}_j \zeta_j - \bar{A})/RT} - 1 \right\}, \end{aligned} \tag{43}$$

using (42) and the definition (37) of the affinity $\bar{A}$. Since at equilibrium both $\bar{J}$ and $\bar{A}$ vanish (this follows from purely thermodynamic considerations), we have furthermore

$$K \equiv \frac{k_f}{k_r} = e^{-\sum_{j=1}^{n} \bar{\nu}_j \zeta_j / RT} N^{\sum_{j=1}^{n} \bar{\nu}_j}. \tag{44}$$

Here K is per definition the equilibrium constant of the reaction. Equation (43) now becomes

$$\bar{J} = k_r \prod_{j=q+1}^{n} N_j^{\nu_j} (e^{-\bar{A}/RT} - 1) . \tag{45}$$

It may be noted that since for perfect gases

$$\zeta_i(p, T) = RT \ln N + C_i(T) , \tag{46}$$

the equilibrium constant K is then given by

$$K(T) = e^{-\Sigma_{j=1}^{n} \nu_j C_j/RT} \tag{47}$$

and is a function of temperature only.

Close to equilibrium or more specifically if $\bar{A} \ll RT$ we may expand the right-hand side of (45) in powers of $\bar{A}$ retaining only the linear term. In this order of approximation the molar concentrations N_j appearing explicitly in (45) must be replaced by their equilibrium values N_j^{eq}. One then obtains

$$\bar{J} = - k_r \prod_{j=q+1}^{n} (N_j^{eq})^{\nu_j} \frac{\bar{A}}{RT} . \tag{48}$$

Comparing (34) and (48) we then obtain with (41) a relation between the phenomenological coefficient $\bar{l}$ and the rate constants k_r or k_f:

$$\bar{l} = k_r \prod_{j=q+1}^{n} \frac{(N_j^{eq})^{\nu_j}}{R} = k_f \prod_{i=1}^{q} \frac{(N_j^{eq})^{-\nu_j}}{R} . \tag{49}$$

It thus turns out that the linear relations of thermodynamics of irreversible processes hold for chemical reactions when the condition

$$A \ll RT \tag{50}$$

is satisfied*. In general this condition is only satisfied in the very last stage of a reaction. We may not conclude from this fact that thermodynamics of irreversible processes is inadequate as a whole to describe

* I. Prigogine, P. Outer and Cl. Herbo, J. Phys. and Colloid Chem. **52** (1948) 321.

chemical reactions. Indeed an analysis based on the kinetic theory of gases (*cf.* Ch. IX) indicates that the expression (1) or (38) for the entropy production holds even when (50) is not satisfied. In fact it turns out that (1) is valid for those reactions for which the law of mass action (45) holds*. This is the case for (gas)reactions which do not disturb appreciably the Maxwell velocity distribution of the chemical components. Therefore only the *linear* approximation of the phenomenological laws breaks down in general for chemical reactions.

For transport processes on the other hand we have seen (Ch. IX) that the thermodynamic form of the entropy production is valid only as long as the linear phenomenological laws hold. In this connexion it is interesting to note that the law of mass action can be obtained from a linear law in some internal coordinate space of the reacting mixture (*cf.* § 6).

§ 3. *Coupled Chemical Reactions*

Let us now consider the case that several, say r, reactions take place between the n components of our (uniform) system. The entropy production is then given by

$$\sigma = - \sum_{j=1}^{r} J_j \frac{A_j}{T}, \tag{51}$$

where the rates J_j are defined by

$$\rho \frac{\partial c_i}{\partial t} = \sum_{j=1}^{r} \nu_{ij} J_j \tag{52}$$

and the affinities A_j by

$$A_j = \sum_{i=1}^{n} \nu_{ij} \mu_i . \tag{53}$$

The coefficient ν_{ij} divided by the molecular mass M_i is proportional to the stoichiometric coefficient of species i in reaction j. Conservation of mass implies [*cf.* (4)]

$$\sum_{i=1}^{n} \nu_{ij} = 0 , \quad (j = 1, 2, \ldots, r) . \tag{54}$$

As in the case of a single reaction we choose for the reactants of each

* J. Ross and P. Mazur, J. chem. Physics **35** (1961) 19.

reaction j ($i = 1, 2, \ldots, q_j$) the normalization

$$\sum_{i=1}^{q_j} \nu_{ij} = -1, \quad (j = 1, 2, \ldots, r) \tag{55}$$

and in view of (54), for the products ($i = q_j + 1, \ q_j + 2, \ldots, n$)

$$\sum_{i=q_j+1}^{n} \nu_{ij} = 1, \quad (j = 1, 2, \ldots, r). \tag{56}$$

We must now distinguish between two cases: (i) the number of reactions r assumed to occur between the n components is smaller or equal to $n - 1$; (ii) the number of these reactions is larger than $n - 1$.

(i) Let us first consider the case

$$r \leqslant n - 1. \tag{57}$$

Let us then define r progress variables ξ_j by the relations

$$\rho \frac{\partial \xi_j}{\partial t} = J_j, \quad (j = 1, 2, \ldots, r). \tag{58}$$

From (52) we have

$$\frac{\partial c_i}{\partial t} = \sum_{j=1}^{r} \nu_{ij} \frac{\partial \xi_j}{\partial t}, \quad (i = 1, 2, \ldots, n). \tag{59}$$

Integration of these relations leads to

$$c_i(t) - c_i^0 = \sum_{j=1}^{r} \nu_{ij} \xi_j(t), \quad (i = 1, 2, \ldots, n). \tag{60}$$

Since $\sum_{i=1}^{n} c_i = 1$, at most $n - 1$ of these equations are independent. Therefore at most $n - 1$ parameters ξ_j can be solved unambiguously from (60). In a system of n reacting species at most $n - 1$ independent proper variables ξ_j exist and thus at most $n - 1$ independent chemical reactions, whose values at some given time t together with the initial values c_i^0 determine the composition c_i of the system at time t: the independent parameters ξ_j can be considered to be thermodynamic state variables. The intervals in which the ξ_j may vary follow from (60) and from the inequalities

$$0 \leqslant c_i \leqslant 1\,, \tag{61}$$

as in the case of a single chemical reaction.

We can now easily generalize most of the considerations of the previous section concerning a single chemical reaction. If for $r < n-1$ the chemical reactions of (51) are independent, the linear relations corresponding to (7) are

$$J_j = -\sum_{m=1}^{r} l_{jm} \frac{A_m}{T}\,, \tag{62}$$

with the Onsager relations

$$l_{jm} = l_{mj}\,. \tag{63}$$

The matrix of coefficients l_{jm} is positive definite since σ, the entropy production, is positive definite.

At thermodynamic equilibrium we have

$$A_j = 0\,, \quad (j = 1, 2, \ldots, r)\,. \tag{64}$$

From these conditions we can in principle determine the equilibrium values ξ_j^{eq} of the ξ_j as functions of for instance the temperature and the density of the system

$$\xi_j^{\mathrm{eq}} = \xi_j(v, T)\,. \tag{65}$$

For reactions proceeding at constant volume and temperature, we have up to first order in $\Delta\xi_j \equiv \xi_j - \xi_j^{\mathrm{eq}}$ [*cf.* (23)]

$$A_j = \sum_{m=1}^{r} \left(\frac{\partial A_j}{\partial \xi_m}\right)^{\mathrm{eq}} \Delta\xi_m\,. \tag{66}$$

Straightforward generalization of (18) leads to

$$A_j = \left(\frac{\partial f}{\partial \xi_j}\right)_{v,\,T}, \tag{67}$$

so that (66) becomes

$$A_j = \sum_{m=1}^{r} \left(\frac{\partial^2 f}{\partial \xi_j \partial \xi_m}\right)_{v,\,T}^{\mathrm{eq}} \Delta\xi_m = \sum_{m=1}^{r} g_{jm} \Delta\xi_m\,, \tag{68}$$

with
$$g_{jm} = \left(\frac{\partial^2 f}{\partial \xi_j \partial \xi_m} \right)^{\text{eq}}_{v,\,T} . \tag{69}$$

The matrix with elements g_{jm} is positive definite: this is the thermodynamic stability condition of the system with respect to the chemical reactions at constant v and T.

Introduction of (62) with (68) into (58) yields

$$\frac{\partial \xi_j}{\partial t} = - \frac{1}{\rho T} \sum_{m,\,k=1}^{r} l_{jm} g_{mk} \Delta \xi_k = - \sum_{k=1}^{r} m_{jk} \Delta \xi_k . \tag{70}$$

The solutions of this set of simultaneous first order differential equations may be written in the form

$$\Delta \xi_j(t) = \sum_{k,\,l=1}^{r} c_{jk}^{-1}\, c_{kl} \Delta \xi_l(0)\ \mathrm{e}^{-t/\tau_k} . \tag{71}$$

Here the r relaxation times τ_k are the reciprocal eigenvalues of the matrix with elements m_{jk}. The matrix with elements c_{jk} is that matrix which diagonalizes the matrix m_{jk} through a similarity transformation. Since both matrices l_{jm} and g_{mk} are symmetric and positive definite, it follows that the relaxation times τ_k are real and positive. The proof of this statement has been given in Chapter VII, § 5, where we discussed the linear regression equations of arbitrary fluctuations. The transcription of the proof to suit the case under consideration is trivial and will not be repeated here. Equation (71) thus demonstrates that the time behaviour of ξ_j follows from superposition of simple exponential decay laws. The relaxation times τ_k introduced here refer to reactions proceeding at constant v and T. As in the previous section it is, however, also possible to introduce other sets of relaxation times, *e.g.* for reactions proceeding at constant energy and density. The explicit time behaviour of the mass fractions c_i is obtained by introducing (71) into (60):

$$\Delta c_i(t) = \sum_{j,\,k,\,l=1}^{r} \nu_{ij} c_{jk}^{-1} c_{kl} \Delta \xi_l(0)\ \mathrm{e}^{-t/\tau_k} , \tag{72}$$

where $\Delta c_i(t) \equiv c_i(t) - c_i^{\text{eq}}$ is the deviation of $c_i(t)$ from its equilibrium value.

The conventional chemical description [*cf.* (29)–(37)] of coupled chemical reactions proceeding at constant volume is as follows. Instead of (52) we have

$$\frac{\partial N_i}{\partial t} = \sum_{j=1}^{r} \bar{\nu}_{ij} \bar{J}_j , \tag{73}$$

with the true stoichiometric coefficients $\bar{\nu}_{ij}$

$$\bar{\nu}_{ij} = \frac{a_j \nu_{ij}}{M_i} \tag{74}$$

and

$$\bar{J}_j = \frac{J_j}{a_j} . \tag{75}$$

The constants a_j follow from the normalization conditions (55) and (56)

$$a_j = \sum_{i=q_j+1}^{n} \bar{\nu}_{ij} M_i = - \sum_{i=1}^{q_j} \bar{\nu}_{ij} M_i . \tag{76}$$

The linear laws (62) become

$$\bar{J}_j = - \sum_{m=1}^{r} \bar{l}_{jm} \frac{\bar{A}_m}{T} , \tag{77}$$

with the affinities

$$\bar{A}_m = a_m A_m = a_m \sum_{i=1}^{n} \nu_{im} \mu_i = \sum_{i=1}^{n} \bar{\nu}_{im} \bar{\mu}_i \tag{78}$$

and the phenomenological coefficients

$$\bar{l}_{jm} = \frac{l_{jm}}{a_j a_m} . \tag{79}$$

The Onsager relations (63) become according to (79)

$$\bar{l}_{jm} = \bar{l}_{mj} . \tag{80}$$

Thus the reciprocal relations still hold as they should for the transformation (75) and (78) of fluxes and thermodynamic forces (*cf.* Chapter VI, § 5).

Suppose now that not all the chemical reactions ($r \leqslant n - 1$) are independent, in particular let us assume that only $r - 1$ reactions are independent. Then the stoichiometric coefficients of the rth reaction can be obtained from those of the first $r - 1$ reactions according to

$$\bar{v}_{ir} = \sum_{j=1}^{r-1} \bar{v}_{ij} b_{jr} , \tag{81}$$

where the coefficients b_{jr} are integers. In fact (81) defines a dependent reaction. As a consequence of (81) the *affinities* (not the rates!) of the r reactions become dependent. Indeed

$$\bar{A}_r = \sum_{i=1}^{n} \bar{\mu}_i \bar{v}_{ir} = \sum_{j=1}^{r-1} \sum_{i=1}^{n} \bar{\mu}_i \bar{v}_{ij} b_{jr} = \sum_{j=1}^{r-1} \bar{A}_j b_{jr} . \tag{82}$$

Although the linear relations (77) with the Onsager relations (79) may be written down when either the fluxes only or both the fluxes and the thermodynamic forces are dependent, (*cf.* Ch. VI, § 3), this is not the case when only the affinities are dependent. This can easily be understood on the basis of the discussion in Chapter VI: if in the case of dependent fluxes only, one would try to invert the linear laws and write the affinities as linear functions of the fluxes this could not be achieved since as a consequence of the dependence amongst the fluxes the determinant value of the matrix of phenomenological coefficients is zero (the inverse matrix does not exist).

We must therefore first eliminate the dependent affinity from the entropy production [*cf.* (51), (75) and (78)]

$$\sigma = \sum_{j=1}^{r} \bar{J}_j \frac{\bar{A}_j}{T} . \tag{83}$$

We then get with (82)

$$\sigma = \sum_{j=1}^{r-1} (\bar{J}_j + b_{jr} \bar{J}_r) \frac{\bar{A}_j}{T} = \sum_{j=1}^{r-1} \bar{J}_r' \frac{\bar{A}_r}{T} , \tag{84}$$

with

$$\bar{J}_r' \equiv \bar{J}_j + b_{jr} \bar{J}_r , \quad (j = 1, 2, \ldots, r - 1) . \tag{85}$$

For the $r - 1$ new rates and affinities in (84) we may then write down linear equations of the form (77). The coefficient matrix of these relations satisfies the reciprocal relations. According to (85), the r rates $\bar{J}_j$ are not unambiguously determined by the affinities. Indeed $\bar{J}_j$

is only determined up to a constant c_j which satisfies the relation

$$c_j = -b_{jr}c_r, \quad (j = 1, 2, \ldots, r-1), \tag{86}$$

with c_r a completely arbitrary additive constant in $\bar{J}_r$. We note that if (81) is introduced into (73) we get

$$\frac{\partial N_i}{\partial t} = \sum_{j=1}^{r-1} \bar{\nu}_{ij}(\bar{J}_j + b_{jr}\bar{J}_r) = \sum_{j=1}^{r-1} \bar{\nu}_{ij}\bar{J}'_j. \tag{87}$$

The changes in composition are thus completely determined by the new rates $\bar{J}'_j$ in agreement with the conclusions obtained above. From (87) $r-1$ independent progress variables could be determined unambiguously. Mathematically this is expressed by saying that in view of (81) the rank of the matrix $\| \bar{\nu}_{ij} \|$ or $\| \nu_{ij} \|$ is $r-1$ and not r.

(ii) Let us now consider the case that the number of chemical reactions r assumed to occur between the n species is such that

$$r > n - 1. \tag{88}$$

We have already seen that in an n-component reacting system all changes in composition can be accounted for in terms of at most $n-1$ independent progress variables, or, equivalently in terms of at most $n-1$ independent chemical reactions. We have, however, now assumed that $r > n-1$ reactions actually occur. If this assumption has any meaning at all, it implies that at least $r-n+1$ of the reactions are dependent. Let us assume that indeed the maximum number $n-1$ of the r reactions is independent. Then obviously one has $r-n+1$ relations of the type (81):

$$\bar{\nu}_{il} = \sum_{j=1}^{n-1} \bar{\nu}_{ij} b_{jl}, \quad (l = n, n+1, \ldots, r) \tag{89}$$

and consequently $r-n+1$ dependent affinities

$$\bar{A}_l = \sum_{i=1}^{n} \bar{\mu}_i \bar{\nu}_{il} = \sum_{j=1}^{n-1} \sum_{i=1}^{n} \bar{\mu}_i \bar{\nu}_{ij} b_{jl}$$

$$= \sum_{j=1}^{n-1} \bar{A}_j b_{jl}, \quad (l = n, n+1, \ldots, r). \tag{90}$$

Eliminating these dependent affinities from the entropy production (83) one obtains

$$\sigma = \sum_{j=1}^{n-1} \left(\bar{J}_j + \sum_{l=n}^{r} b_{jl}\bar{J}_l\right)\frac{\bar{A}_j}{T} = \sum_{j=1}^{n-1} \bar{J}'_j \frac{\bar{A}_j}{T}, \tag{91}$$

with the new rates

$$\bar{J}'_j \equiv \bar{J}_j + \sum_{l=n}^{r} b_{jl}\bar{J}_l, \quad (j = 1, 2, \ldots, n-1). \tag{92}$$

Equations (73) become with (89) and (92)

$$\frac{\partial N_i}{\partial t} = \sum_{j=1}^{n-1} \bar{\nu}_{ij}\bar{J}'_j. \tag{93}$$

We have thus reduced the problem of r chemical reactions for $r > n - 1$, to the problem of $n - 1$ independent chemical reactions considered previously.

Since from a macroscopic point of view it is sufficient to assume that at most $n - 1$ independent reactions occur in the system, one might inquire what the physical meaning is of the introduction of more reaction schemes. The answer to this question is the following: there is no meaning to it unless one wishes to indicate that all the mechanisms of transformation corresponding to the r proposed reaction schemes actually exist. This, however, is an assumption of chemical kinetics and not one of macroscopic physics. We shall come back to this point in the next section.

§ 4. *Unimolecular Reactions; the Principle of Detailed Balance*

In this section we shall study in more detail a special class of chemical reactions. We consider a system of n components, all having equal molecular masses. The n species could for instance correspond to n internal states of the same molecule. We shall assume that all possible reactions of the type

$$B_i \leftrightarrows B_k, \quad (i, k = 1, 2, \ldots, n) \tag{94}$$

take place between the n components. Here B_i and B_k represent a molecule of species i and k respectively. We then have $\frac{1}{2}n(n-1)$ of these possible reactions of which only $n - 1$ are independent. If the reaction takes place at constant volume we have [*cf.* (73)]

$$\frac{\partial N_i}{\partial t} = \sum_{j=1}^{\frac{1}{2}n(n-1)} \bar{\nu}_{ij} \bar{J}_j \,. \tag{95}$$

Since each component participates in $(n - 1)$ reactions only, we may rewrite this in the form

$$\frac{\partial N_i}{\partial t} = \sum_{\substack{k=1 \\ k \neq i}}^{n} \bar{\nu}_{ik} \bar{J}_{ik} \,, \tag{96}$$

where $\bar{\nu}_{ik}$ is the stoichiometric coefficient of species i in its transformation into species k and $\bar{J}_{ik} \equiv \bar{J}_{ki}$ the rate of this reaction. Conservation of mass in a reaction implies that

$$\bar{\nu}_{ik} = - \bar{\nu}_{ki} \,, \quad (i, k = 1, 2, \ldots, n) \,. \tag{97}$$

Note that according to (94) and (97) $|\,\bar{\nu}_{ik}\,| = 1 - \delta_{ik}$.

In the notation adopted by us, the entropy production (83) may be written as

$$\sigma = - \tfrac{1}{2} \sum_{\substack{i,k=1 \\ i \neq k}}^{n} \bar{J}_{ik} \frac{\bar{A}_{ik}}{T} \,, \tag{98}$$

with the affinities

$$\bar{A}_{ik} = \bar{\nu}_{ik} \bar{\mu}_i + \bar{\nu}_{ki} \bar{\mu}_k = \bar{\nu}_{ik} (\bar{\mu}_i - \bar{\mu}_k) = \bar{A}_{ki} \,. \tag{99}$$

From this definition of $\bar{A}_{ik}$ it follows that

$$\bar{\nu}_{ji} \bar{\nu}_{ik} \bar{A}_{jk} = \bar{\nu}_{jk} \bar{\nu}_{ik} \bar{A}_{ji} + \bar{\nu}_{jk} \bar{\nu}_{ji} \bar{A}_{ik} \,. \tag{100}$$

We can now choose as independent affinities those of the $n - 1$ reactions between species 1 and all other species. The affinities of all other reactions are related to this independent set of affinities according to (100). Elimination of the dependent set from (98) leads after some simple transformations, using (97), to

$$\sigma = - \sum_{k=2}^{n} \left(\bar{J}_{1k} + \sum_{i=2}^{n} \frac{\bar{\nu}_{ik}}{\bar{\nu}_{1k}} \bar{J}_{ik} \right) \frac{\bar{A}_{1k}}{T} \,, \tag{101}$$

or

$$\sigma = - \sum_{k=2}^{n} \bar{J}'_{1k} \frac{\bar{A}_{1k}}{T}, \tag{102}$$

with the new rates of the $n - 1$ independent reactions

$$\bar{J}'_{1k} = \bar{J}_{1k} + \sum_{i=2}^{n} \frac{\bar{\nu}_{ik}}{\bar{\nu}_{1k}} \bar{J}_{ik} = \sum_{i=1}^{n} \frac{\bar{\nu}_{ik}}{\bar{\nu}_{1k}} \bar{J}_{ik} . \tag{103}$$

In terms of these rates, we have with (97) instead of (96):

$$\frac{\partial N_1}{\partial t} = \sum_{k=2}^{n} \bar{\nu}_{1k} \bar{J}'_{1k} , \tag{104}$$

$$\frac{\partial N_i}{\partial t} = \bar{\nu}_{i1} \bar{J}'_{1i} , \quad (i = 2, 3, \ldots, n) . \tag{105}$$

The linear relations between independent rates and affinities are

$$\bar{J}'_{1k} = - \sum_{j=2}^{n} \bar{l}_{kj} \frac{\bar{A}_{1j}}{T}, \quad (k = 2, 3, \ldots, n) , \tag{106}$$

with the reciprocal relations

$$\bar{l}_{kj} = \bar{l}_{jk} , \quad (k, j = 2, 3, \ldots, n) . \tag{107}$$

Now (106) may be rewritten in the following way

$$\begin{aligned} \bar{J}'_{1k} &= - \bar{l}_{kk} \frac{\bar{A}_{1k}}{T} - \sum_{\substack{j=2 \\ j \neq k}}^{n} \bar{l}_{kj} \frac{\bar{A}_{1j}}{T} \\ &= - \left(\bar{l}_{kk} + \sum_{\substack{j=2 \\ j \neq k}}^{n} \frac{\bar{\nu}_{1j}}{\bar{\nu}_{1k}} \bar{l}_{kj} \right) \frac{\bar{A}_{1k}}{T} - \sum_{\substack{j=2 \\ j \neq k}}^{n} \frac{\bar{\nu}_{1j}}{\bar{\nu}_{kj}} \bar{l}_{kj} \frac{\bar{A}_{kj}}{T} \\ &= - \bar{l}_{1k} \frac{\bar{A}_{1k}}{T} - \sum_{\substack{j=2 \\ j \neq k}}^{n} \frac{\bar{\nu}_{1j}}{\bar{\nu}_{kj}} \bar{l}_{kj} \frac{\bar{A}_{kj}}{T} , \end{aligned} \tag{108}$$

where we have used (100) and where

$$\bar{l}_{1k} \equiv \bar{l}_{kk} + \sum_{\substack{j=2 \\ j \neq k}}^{n} \frac{\bar{\nu}_{1j}}{\bar{\nu}_{1k}} \bar{l}_{kj} = \sum_{j=2}^{n} \frac{\bar{\nu}_{1j}}{\bar{\nu}_{1k}} \bar{l}_{kj} . \tag{109}$$

With (103) we finally get

$$\bar{J}_{1k} + \sum_{i=2}^{n} \frac{\bar{\nu}_{ik}}{\bar{\nu}_{1k}} \bar{J}_{ik} = - \bar{l}_{1k} \frac{\bar{A}_{1k}}{T} - \sum_{i=2}^{n} \frac{\bar{\nu}_{1i}}{\bar{\nu}_{ki}} \bar{l}_{ki} \frac{\bar{A}_{ki}}{T} . \tag{110}$$

Putting arbitrary constants in the definition of $\bar{J}_{ik}$ equal to zero and taking into account the Onsager relations (107), we may thus make the following associations:

$$\bar{J}_{1k} = - \bar{l}_{1k} \frac{\bar{A}_{1k}}{T}, \quad (k = 2, 3, \ldots, n) , \tag{111}$$

$$\bar{J}_{ik} \equiv \bar{J}_{ki} = - \frac{\bar{\nu}_{1i}\bar{\nu}_{1k}}{\bar{\nu}_{ik}\bar{\nu}_{ki}} \bar{l}_{ki} \frac{\bar{A}_{kl}}{T}, \quad (i, k = 2, 3, \ldots, n; i \neq k) . \tag{112}$$

The meaning of these associations is the following: due to the validity of the Onsager relations the set (111) and (112) of $\frac{1}{2}n(n - 1)$ *uncoupled, but dependent*, chemical reactions is *equivalent* with the set (106) of $n - 1$ *coupled, but independent* chemical reactions in the sense that all macroscopically observable results are obviously the same for both sets. Indeed both sets when introduced into (96) or (104) and (105) lead to identical differential equations

$$\frac{\partial N_1}{\partial t} = - \sum_{k=2}^{n} \bar{\nu}_{1k} \bar{l}_{1k} \frac{A_{1k}}{T} , \tag{113}$$

$$\frac{\partial N_i}{\partial t} = - \bar{\nu}_{i1} \bar{l}_{1i} \frac{\bar{A}_{1i}}{T} - \sum_{\substack{k=2 \\ k \neq i}}^{n} \frac{\bar{\nu}_{1i}\bar{\nu}_{1k}}{\bar{\nu}_{ki}} \bar{l}_{ki} \frac{\bar{A}_{ki}}{T}, \quad (i = 2, 3, \ldots, n) . \tag{114}$$

Take the simplest case when $n = 3$. The two equations (106) are then

$$\bar{J}'_{12} = - \bar{l}_{22} \frac{\bar{A}_{12}}{T} - \bar{l}_{23} \frac{\bar{A}_{13}}{T} , \tag{115}$$

$$\bar{J}'_{13} = - \bar{l}_{32} \frac{\bar{A}_{12}}{T} - \bar{l}_{33} \frac{\bar{A}_{13}}{T} , \tag{116}$$

with $\bar{l}_{23} = \bar{l}_{32}$. For the equivalent set of three uncoupled reactions we have according to (111), (112) and (109)

$$\bar{J}_{12} = -\left(\bar{l}_{22} + \frac{\bar{\nu}_{13}}{\bar{\nu}_{12}}\bar{l}_{23}\right)\frac{\bar{A}_{12}}{T}, \tag{117}$$

$$\bar{J}_{13} = -\left(\bar{l}_{33} + \frac{\bar{\nu}_{12}}{\bar{\nu}_{13}}\bar{l}_{32}\right)\frac{\bar{A}_{13}}{T}, \tag{118}$$

$$\bar{J}_{23} = -\frac{\bar{\nu}_{12}\bar{\nu}_{13}}{\bar{\nu}_{23}\bar{\nu}_{32}}\bar{l}_{23}\frac{\bar{A}_{23}}{T}. \tag{119}$$

Assume now that we have either

$$\bar{\nu}_{12}\bar{l}_{22} + \bar{\nu}_{13}\bar{l}_{23} = 0, \tag{120}$$

or

$$\bar{\nu}_{13}\bar{l}_{33} + \bar{\nu}_{12}\bar{l}_{32} = 0 \tag{121}$$

for all values of the density and the temperature. Then the set of two coupled independent reactions $1 \rightleftarrows 2$ and $1 \rightleftarrows 3$ is equivalent with the two uncoupled reactions $1 \rightleftarrows 2$ and $2 \rightleftarrows 3$ or $1 \rightleftarrows 3$ and $2 \rightleftarrows 3$. If conditions of the type (120) or (121) are satisfied, it is possible to transform from a set of r independent coupled chemical reactions to an equivalent set of r uncoupled reactions through a linear transformation of rates and affinities involving only integer coefficients*. Whereas it is always possible to diagonalize the matrix of phenomenological coefficients $\bar{l}_{ik}$, the transformation that achieves this diagonalization will in general not involve coefficients of such a simple type.

To summarize the preceding analysis we may state again that due to the Onsager reciprocal relations we can always *interpret* all macroscopic results for reactions of the type (94) in terms of a set of $\frac{1}{2}n(n-1)$ uncoupled (but dependent through their affinities) chemical reactions.

We shall now show that together with a kinetic assumption (which does not contain microscopic reversibility) the Onsager relations lead to the principle of detailed balance.

Let us make the following kinetic assumptions: the mechanism of the "transitions" (94) is unimolecular. Then the number of molecules of species i making a "transition" to species k per unit volume and in unit time is given by $\lambda_{ik}N_i$, where λ_{ik} is the (time independent) transition probability from i to k. The transition probabilities are normalized

* I. Prigogine, Bull. Acad. Roy. Belg., Cl. Sc. [5] **32** (1946) 30.

in such a way that

$$\sum_{k=1}^{n} \lambda_{ik} = 1 . \tag{122}$$

The total rate of change of N_i is then given by

$$\frac{\partial N_i}{\partial t} = - \sum_{\substack{k=1 \\ k \neq i}}^{n} \lambda_{ik} N_i + \sum_{\substack{k=1 \\ k \neq i}}^{n} \lambda_{ki} N_k . \tag{123}$$

The first sum represents the "loss" of N_i due to transitions to "states" k, the second sum the "gain" due to transitions from states k. Rate equations of this type are called "master equations". We shall not make the customary assumption that at equilibrium $\lambda_{ik} N_i^{\text{eq}} = \lambda_{ki} N_k^{\text{eq}}$, which expresses the principle of detailed balance, but derive this property from the reciprocal relations. Our kinetic assumption determines the rates $\bar{J}_{ik}$ of the $\frac{1}{2}n(n-1)$ "transitions" or chemical reactions unambiguously as

$$\bar{\nu}_{ik} \bar{J}_{ik} = - \lambda_{ik} N_i + \lambda_{ki} N_k . \tag{124}$$

We note that in order that stationary solutions of (123) may exist, the transition probabilities must also satisfy the relations

$$\sum_{k=1}^{n} \lambda_{ki} N_k^{\text{eq}} = N_i^{\text{eq}} , \tag{125}$$

where (122) has been employed. Using (122) the master equation (123) may also be written in the form

$$\frac{\partial N_i}{\partial t} = \sum_{k=1}^{n} (\lambda_{ki} - \delta_{ki}) N_k , \tag{126}$$

or alternatively, applying (125)

$$\frac{\partial N_i}{\partial t} = \sum_{k=1}^{n-1} (\lambda_{ki} - \delta_{ki}) \left(N_k - \frac{N_k^{\text{eq}}}{N_n^{\text{eq}}} N_n \right) . \tag{127}$$

We now formulate the thermodynamical theory of the reactions (94) in a manner slightly different from the one developed previously [equations (94)–(114)]: instead of (98) we may also write for the entropy production, using (95), (97) and (99),

$$\sigma = \sum_{i=1}^{n} \frac{\partial N_i}{\partial t} \frac{\bar{\mu}_i}{T}. \tag{128}$$

At chemical equilibrium the affinities $\bar{A}_{ik} = \bar{\nu}_{ik}(\bar{\mu}_i - \bar{\mu}_k)$ vanish, so that

$$\bar{\mu}_i^{eq} = \bar{\mu}_k^{eq}, \quad (i, k = 1, 2, \ldots, n). \tag{129}$$

Furthermore since $\sum_i N_i = N$ must remain constant for unimolecular reactions we also have

$$\sum_i \frac{\partial N_i}{\partial t} = 0, \tag{130}$$

so that we may write instead of (128)

$$\sigma = \sum_{i=1}^{n-1} \frac{\partial N_i}{\partial t} \left(\frac{\bar{\mu}_i - \bar{\mu}_n}{T} \right). \tag{131}$$

In (131) the quantities $\partial N_i / \partial t$ are to be considered as fluxes and the quantities $\bar{\mu}_i - \bar{\mu}_n$ as thermodynamic forces. The linear relations corresponding to this formalism are

$$\frac{\partial N_i}{\partial t} = \sum_{k=1}^{n-1} L_{ik} \frac{(\bar{\mu}_k - \bar{\mu}_n)}{T}, \quad (i = 1, 2, \ldots, n-1), \tag{132}$$

with the reciprocal relations

$$L_{ik} = L_{ki}, \quad (i, k = 1, 2, \ldots, n-1). \tag{133}$$

Let us assume that the system under consideration is a perfect gas mixture: we then have for the chemical potential per mole of species i [*cf.* (42) and (46)]

$$\bar{\mu}_i = RT \ln N_i + C_i(T) \tag{134}$$

and therefore

$$\bar{\mu}_i - \bar{\mu}_n = RT \ln \left(\frac{N_i}{N_n} \right) + C_i(T) - C_n(T). \tag{135}$$

Using the equilibrium condition (129) we may also write

$$\bar{\mu}_i - \bar{\mu}_n = RT \ln \frac{N_i N_n^{\text{eq}}}{N_i^{\text{eq}} N_n} . \tag{136}$$

If we linearize, for small deviations from equilibrium, the right-hand side with respect to $\Delta N_i \equiv N_i - N_i^{\text{eq}}$ and $\Delta N_n \equiv N_n - N_n^{\text{eq}}$ we get

$$\bar{\mu}_i - \bar{\mu}_n = RT \left(\frac{\Delta N_i}{N_i^{\text{eq}}} - \frac{\Delta N_n}{N_n^{\text{eq}}} \right) = RT \left(\frac{N_i}{N_i^{\text{eq}}} - \frac{N_n}{N_n^{\text{eq}}} \right) . \tag{137}$$

Introducing this expression into (132) one has

$$\frac{\partial N_i}{\partial t} = R \sum_{k=1}^{n-1} L_{ik} \left(\frac{N_k}{N_k^{\text{eq}}} - \frac{N_n}{N_n^{\text{eq}}} \right), \quad (i = 1, 2, \ldots, n-1) . \tag{138}$$

Comparing (138) with (127) and using (122) and (133) we then have

$$\sum_{k=1}^{n-1} (\lambda_{ki} - \delta_{ki}) \left(N_k - \frac{N_k^{\text{eq}}}{N_n^{\text{eq}}} N_n \right) = R \sum_{k=1}^{n-1} L_{ik} (N_k^{\text{eq}})^{-1} \left(N_k - \frac{N_k^{\text{eq}}}{N_n^{\text{eq}}} N_n \right),$$
$$(i = 1, 2, \ldots, n-1) . \tag{139}$$

Since this relation must hold for arbitrary molar concentrations N_k we may conclude that

$$RL_{ik} = (\lambda_{ki} - \delta_{ki}) N_k^{\text{eq}}, \quad (i, k = 1, 2, \ldots, n-1) . \tag{140}$$

In view of the Onsager relations (133), and also of (122) and (125), we thus obtain the following conditions between the transition probabilities λ_{ik}

$$\lambda_{ik} N_i^{\text{eq}} = \lambda_{ki} N_k^{\text{eq}}, \quad (i, k = 1, 2, \ldots, n) . \tag{141}$$

This condition is the expression for the principle of detailed balance.

It is not surprising that we obtain this principle here as a consequence of the reciprocal relations, since it is in fact a special form of the principle of microscopic reversibility (detailed balance in $\boldsymbol{\alpha}$-space: the range of values of α corresponds here to the discrete set of indices i) from which the Onsager relations were derived.

§ 5. *Relaxation Phenomena*

From the point of view of thermodynamics of irreversible processes, the theoretical treatment of relaxation phenomena* is closely related to that of chemical reactions. Consider indeed a uniform system whose state is specified by a set of n external parameters and m internal even (α-type) variables ξ_i ($i = 1, 2, \ldots, m$). The latter could be for instance the progress variables of chemical reactions, but may more generally describe the internal structure of the system or designate some "internal temperature" which characterizes the energy distribution over internal states of the system.

For the total differential of the entropy per unit mass of the system we may write

$$T\,\mathrm{d}s = \sum_{i=1}^{n} Z_i\,\mathrm{d}\zeta_i - \sum_{i=1}^{m} A_i\,\mathrm{d}\xi_i\,, \tag{142}$$

where $\zeta_1, \zeta_2, \ldots, \zeta_n$ are the external parameters characterizing the state of the system; the quantities $Z_i = T\partial s/\partial\zeta_i$ are the parameters thermodynamically conjugate to ζ_i. In tensor notation (142) may also be written as

$$T\,\mathrm{d}s = \boldsymbol{Z}\cdot\mathrm{d}\boldsymbol{\zeta} - \boldsymbol{A}\cdot\mathrm{d}\boldsymbol{\xi}\,. \tag{143}$$

Here $\boldsymbol{Z}$ for instance is an n-dimensional vector with components $Z_1, Z_2, \ldots, Z_n$.

At thermodynamic equilibrium and for given values of the set $\boldsymbol{\xi}$ the entropy must have a maximum so that one has

$$\boldsymbol{A}\cdot\mathrm{d}\boldsymbol{\xi} = 0\,, \quad \text{at equilibrium}\,. \tag{144}$$

As in the case of chemical reactions the affinities vanish therefore at equilibrium

$$\boldsymbol{A} = 0\,. \tag{145}$$

* J. Meixner and H. G. Reik, Encyclopedia of Physics, Volume III/2 (1959) § 23, α and β;
J. Meixner, Kolloid Zeitschr. **134** (1953) 3; Zeitschr. Phys. **139** (1954) 30.

From this condition one then finds the equilibrium values ξ^{eq} of the internal variables

$$\xi^{eq} = \xi^{eq}(\zeta) . \tag{146}$$

Let us now assume that for all transformations of the system the irreversible rate of change of entropy per unit mass is given by

$$\left(\frac{\partial s}{\partial t}\right)_{irr} = -\frac{\boldsymbol{A}}{T} \cdot \frac{\partial \xi}{\partial t} \geqslant 0 . \tag{147}$$

What is implied here is that for instance for all adiabatic transformations one has*

$$\boldsymbol{Z} \cdot d\zeta = 0 . \tag{148}$$

For the "fluxes" $\partial \xi / \partial t$ in (147) we have the following linear relations

$$\frac{\partial \xi}{\partial t} = - \mathsf{L} \cdot \frac{\boldsymbol{A}}{T} , \tag{149}$$

where the matrix of phenomenological coefficients satisfies the Onsager reciprocal relations

$$\mathsf{L} = \tilde{\mathsf{L}} . \tag{150}$$

For sufficiently small deviations from equilibrium and constant values of the external parameters ζ, one has

$$\boldsymbol{A} = \frac{\partial \boldsymbol{A}}{\partial \xi} \cdot \Delta\xi = - T \frac{\partial^2 s}{\partial \xi \partial \xi} \cdot \Delta \xi , \tag{151}$$

with $\Delta\xi = \xi - \xi^{eq}$, the deviations of the ξ-variables from their equilibrium values. The matrix $\partial \boldsymbol{A} / \partial \xi$ has elements $(\partial \boldsymbol{A} / \partial \xi)_{ij} = \partial A_i / \partial \xi_j$. With the abbreviations

$$\frac{\partial^2 s}{\partial \xi \partial \xi} = - \mathsf{g} , \tag{152}$$

where g is a symmetric, non-negative definite matrix, (151) may also be written as

* The restriction imposed here is not a very serious one. One could easily develop the theory of relaxation phenomena in more general cases. We refer in this connexion to the theory of acoustic relaxation in Chapter XII.

$$A = Tg \cdot \Delta\xi \,. \tag{153}$$

The phenomenological relations (149) may therefore also be written in the form

$$\frac{\partial \Delta\xi}{\partial t} = -\, L \cdot g \cdot \Delta\xi = -\, M \cdot \Delta\xi \,, \tag{154}$$

with

$$M = L \cdot g \,. \tag{155}$$

The solutions of this set of equations are identical with the solutions of the rate equations for the progress variables of chemical reactions [*cf.* (71)]. Again one gets for each variable a superposition of simple exponential decay laws. Formally we may write

$$\Delta\xi(t) = \mathrm{e}^{-Mt} \cdot \Delta\xi(0) \,. \tag{156}$$

These "regression laws" describe the "relaxation" of the various internal variables.

It is of special interest to study the influence of relaxation on the behaviour in time of the variables $\boldsymbol{Z}$ when the external parameters ζ vary with time in a prescribed way.

Let us choose a reference state in the system in which the ζ-variables have values ζ^0 and the $\boldsymbol{Z}$-variables corresponding values $\boldsymbol{Z}^0$. Furthermore the reference state is an equilibrium state for which $\boldsymbol{A} = 0$ and $\xi = \xi^{\mathrm{eq}}$. We may now write for small deviations from the reference state

$$\Delta \boldsymbol{Z} = \frac{\partial \boldsymbol{Z}}{\partial \zeta} \cdot \Delta\zeta + \frac{\partial \boldsymbol{Z}}{\partial \xi} \cdot \Delta\xi \,, \tag{157}$$

$$\boldsymbol{A} = \frac{\partial \boldsymbol{A}}{\partial \zeta} \cdot \Delta\zeta + \frac{\partial \boldsymbol{A}}{\partial \xi} \cdot \Delta\xi \,. \tag{158}$$

Introducing the abbreviations

$$a = \frac{\partial \boldsymbol{Z}}{\partial \zeta} = T \frac{\partial^2 s}{\partial \zeta \partial \zeta} \,, \tag{159}$$

$$b = \frac{\partial \boldsymbol{Z}}{\partial \xi} = T \frac{\partial^2 s}{\partial \xi \partial \zeta} = T \widetilde{\frac{\partial^2 s}{\partial \zeta \partial \xi}} = -\widetilde{\frac{\partial \boldsymbol{A}}{\partial \zeta}} \,, \tag{160}$$

$$Tg = \frac{\partial A}{\partial \xi} = -T \frac{\partial^2 s}{\partial \xi \partial \xi}, \tag{161}$$

where a and g are both symmetric matrices, equations (157) and (158) become

$$\Delta Z = a \cdot \Delta \zeta + b \cdot \Delta \xi, \tag{162}$$

$$A = -\tilde{b} \cdot \Delta \zeta + Tg \cdot \Delta \xi. \tag{163}$$

Let us now assume that the external parameters $\Delta\zeta$, and therefore also all other variables, vary harmonically with time with frequencies ω

$$\Delta\zeta(t) \sim e^{-i\omega t}. \tag{164}$$

We then have from (149)

$$i\omega \Delta \xi = L \cdot \frac{A}{T}. \tag{165}$$

If we then eliminate $\Delta\xi$ and A from (162) and (163), we obtain

$$\Delta Z = \kappa(\omega) \cdot \Delta \zeta, \tag{166}$$

where the "generalized susceptibility" $\kappa(\omega)$ is given by

$$\kappa(\omega) = a + T^{-1} b \cdot (g - i\omega L^{-1})^{-1} \cdot \tilde{b}. \tag{167}$$

For frequency $\omega = 0$, this reduces to

$$\kappa(0) = a + T^{-1} b \cdot g^{-1} \cdot \tilde{b}, \tag{168}$$

the thermodynamic coefficients at complete equilibrium when $A = 0$. One shows indeed from (143) and (157)–(161) that

$$a + T^{-1} b \cdot g^{-1} \cdot \tilde{b} = T \left(\frac{\partial^2 s}{\partial \zeta \partial \zeta} \right)_{A=0}. \tag{169}$$

On the other hand in the limit as $\omega \to \infty$, we have

$$\kappa(\omega) = a. \tag{170}$$

These are the thermodynamic coefficients when the internal state is frozen in. It should also be noted that for all frequencies ω, the susceptibility $\kappa(\omega)$ is a symmetric matrix as follows from inspection of (167) with the Onsager reciprocal relations (150).

For sufficiently small frequencies ω, we may, by expanding the matrix $(U + i\omega g^{-1} \cdot L^{-1})^{-1}$ in powers of ω, write the generalized susceptibility (167) as

$$\begin{aligned}\kappa(\omega) &= a + T^{-1} b \cdot g^{-1} \cdot \tilde{b} + i\omega T^{-1} b \cdot g^{-1} \cdot L^{-1} \cdot g^{-1} \cdot \tilde{b} \\ &= \kappa(0) + i\omega T^{-1} b \cdot g^{-1} \cdot L^{-1} \cdot g^{-1} \cdot \tilde{b} \\ &= \kappa(0) + i\omega L' ,\end{aligned} \tag{171}$$

where we have used (168) and introduced the matrix

$$L' \equiv T^{-1} b \cdot g^{-1} \cdot L^{-1} \cdot g^{-1} \cdot \tilde{b} . \tag{172}$$

Equation (166) now becomes

$$\Delta \boldsymbol{Z} - \Delta \boldsymbol{Z}^{\text{eq}} = i\omega L' \cdot \Delta \zeta , \tag{173}$$

where

$$\Delta \boldsymbol{Z}^{\text{eq}} = \kappa(0) \cdot \Delta \zeta , \tag{174}$$

or

$$\Delta \boldsymbol{Z} - \Delta \boldsymbol{Z}^{\text{eq}} = - L' \cdot \frac{\partial \Delta \zeta}{\partial t} . \tag{175}$$

At low frequencies it is therefore as if the quantities $\Delta \boldsymbol{Z} - \Delta \boldsymbol{Z}^{\text{eq}}$ represent fluxes which depend linearly on the thermodynamic forces $\partial \Delta \zeta / \partial t$. The new phenomenological coefficients L' given by (172) satisfy the reciprocal relations as a consequence of the symmetry of the matrix L.

These effective phenomenological coefficients L' can thus be completely understood in terms of internal processes, which, for relatively slow transformations, need not be taken into account explicitly*.

* It is possible to interpret in the same way cross-coefficients which couple external irreversible phenomena (for instance viscous flow) of the type (175) to internal relaxation processes (for instance chemical reactions) in terms of a set of fast internal processes which have not been taken into account explicitly. The coefficients l_{vm} and l_{mv} of equations (IV.17) and (IV.18) are an example of such cross-coefficients [see J. Meixner, Z. Physik **131** (1951) 456].

If one calculates the irreversible rate of entropy production (147) for small ω, one finds with (163), (165) and (172)

$$T\left(\frac{\partial s}{\partial t}\right)_{\text{irr}} = (\mathrm{i}\omega)^2 \mathsf{L}' : \Delta\zeta\Delta\zeta = \mathrm{i}\omega\Delta\zeta\cdot(\Delta \mathbf{Z} - \Delta \mathbf{Z}^{\text{eq}})\,. \tag{176}$$

This expression for the entropy production corresponds to the effective linear law (173).

We shall encounter later, when we consider the theory of acoustic and elastic relaxations (*cf.* Ch. XII) specific examples in which the effect of the relaxation of internal variables may at low frequencies be described, without explicitly taking into account these variables, by a linear phenomenological law for an external variable.

The structure of the susceptibility $\kappa(\omega)$ becomes more transparent, if one chooses the internal variables in such a way that g is the unit matrix, and L is diagonal (with positive diagonal elements $1/\tau_1$, $1/\tau_2$, etc.); (such a choice is always possible since both the matrices g and L are symmetric and positive definite, *cf.* § 3 and Ch. VII, § 5). One then gets for the elements of the matrix $\kappa(\omega)$

$$\kappa_{ik}(\omega) = a_{ik} - T\sum_j \frac{1}{1+\mathrm{i}\omega\tau_j}\, b_{ij}b_{kj}\,. \tag{177}$$

Thus each element of $\kappa(\omega)$ consists of a sum of terms having a simple structure. To the superposition of simple exponential decay laws for the internal variables, correspond susceptibilities which are superpositions of terms of the form $1/(1+\mathrm{i}\omega\tau_j)$. The real and imaginary parts of susceptibilities of this type are called relaxation-type dispersion and absorption spectra. These real and imaginary parts satisfy the Kramers–Kronig or dispersion relations discussed in Ch. VIII, where the general properties of susceptibilities have been studied, as well as their connexion with correlation functions of spontaneous fluctuations (fluctuation dissipation theorem). If in addition to the even ($\boldsymbol{\alpha}$-type) internal variables, one also has odd ($\boldsymbol{\beta}$-type) internal variables, the structure of the corresponding susceptibilites becomes somewhat more complicated than encountered above.

§ 6. *Internal Degrees of Freedom**

We shall now assume that the internal state of a closed uniform

* I. Prigogine and P. Mazur, Physica **19** (1953) 241.

system must be specified by an infinite sequence of internal variables $\xi(y)$, where the parameter y is continuous. The internal variables $\xi(y)$ may represent the density $\rho(y)$ of constituent particles of a system in a state characterized by a given value of the parameter y. The parameter y may represent for instance an angle specifying the orientation of a molecule in space, or a velocity component of a Brownian particle etc., and may be called an internal coordinate. The system is then said to have an internal degree of freedom.

The total differential δs of the entropy per unit mass of the system may be written as

$$\delta s = \frac{1}{T}\,\delta u + \frac{p}{T}\,\delta v - \frac{1}{\rho_{\text{tot}}T}\int \mu(y)\delta\rho(y)\,\mathrm{d}y\,, \tag{178}$$

where $\mu(y)$ is a chemical potential per unit mass for the configuration y, and ρ_{tot} is the total density.

For constant energy and volume the entropy must have a maximum at thermodynamic equilibrium, when the densities $\rho(y)$ have their equilibrium values, so that

$$\delta s = -\frac{1}{\rho_{\text{tot}}T}\int \mu(y)\delta\rho(y)\,\mathrm{d}y = 0\,, \quad \text{at equilibrium}\,. \tag{179}$$

Since the total density is constant, we also have

$$\int \delta\rho(y)\,\mathrm{d}y = 0\,. \tag{180}$$

From this condition, together with (179), it follows that at equilibrium the chemical potential $\mu(y)$ becomes independent of the variable y

$$\mu^{\text{eq}}(y) = a\,. \tag{181}$$

Here a is a quantity independent of y. The irreversible rate of entropy production at constant energy and volume is given by

$$\frac{\partial s}{\partial t} = -\frac{1}{\rho_{\text{tot}}T}\int \mu(y)\,\frac{\partial\rho(y\,;t)}{\partial t}\,\mathrm{d}y \geqslant 0\,. \tag{182}$$

We shall now write the rate of change of the density $\rho(y)$ in the form

$$\frac{\partial \rho(y)}{\partial t} = - \frac{\partial J(y)}{\partial y}. \tag{183}$$

This relation defines the quantity $J(y)$, which formally may be interpreted as a flux in the space of the internal coordinate, up to an arbitrary additive constant. In view of (180) $J(y)$ must have the same value at both limits of the domain in which the variable y has been defined. The arbitrary additive constant in the definition of $J(y)$ may then, without loss of generality, be chosen in such a way that this flux vanishes at these limits. If the continuity equation (183) is introduced into (182), one obtains

$$\frac{\partial s}{\partial t} = \frac{1}{\rho_{\mathrm{tot}} T} \int \mu(y) \frac{\partial J(y)}{\partial y} \, \mathrm{d}y \geqslant 0 , \tag{184}$$

or upon partial integration

$$\frac{\partial s}{\partial t} = - \frac{1}{\rho_{\mathrm{tot}} T} \int J(y) \frac{\partial \mu(y)}{\partial y} \, \mathrm{d}y \geqslant 0 , \tag{185}$$

since the integrated part vanishes.

We can in general write phenomenological equations for the fluxes and thermodynamic forces occurring in (185) as integral equations of the form

$$J(y) = - \int L(y, y') \frac{\partial \mu(y')}{\partial y'} \, \mathrm{d}y' . \tag{186}$$

In many cases, however, one can establish phenomenological equations of a much simpler form for the following reason. In a number of systems, which are described with the help of internal variables, the internal coordinate y changes only in a continuous way, *i.e.*, y does not instantaneously change by a finite amount. This is the case if the elementary interaction mechanisms for the constituent particles (for instance collisions) are such that they produce only infinitesimal changes of the internal coordinate y, whereas finite variations result from the cumulative action of a large number of elementary interactions. Examples of this behaviour are found in the variation of the velocity of a Brownian particle or in the change of orientation of a dipole molecule in an external electric field, in so far as the motion of the dipole can be considered as a rotational Brownian motion. In all these cases (examples of which will be treated below) the continuous flux along the internal coordinate, formally defined by formula (183), can be considered to represent a true physical current which is determined at each point y by the local properties which exist at this point. We shall therefore assume now that not only the integral (185) is positive, but also that the integrand

$$- J(y) \frac{\partial \mu(y)}{\partial y} \geqslant 0 \tag{187}$$

is essentially a positive quantity. The phenomenological equation between the flux and thermodynamic force, occurring in (187), is simply

$$J(y) = - L \frac{\partial \mu(y)}{\partial y}, \tag{188}$$

in conformity with the idea that the flux does not depend in this case on the whole distribution along the coordinate y, as in (186).

We shall show that this equation can serve as a basis for the derivation of various "kinetic" equations* for the motion in an "internal coordinate space". We shall consider (i) a chemical reaction, considered as a diffusion along an internal coordinate, (ii) Brownian motion in velocity space and (iii) the orientation of dipoles. In all three cases we shall study ideal systems, in which the thermodynamic potential $\mu(y)$ is given by

$$\mu(y) = \frac{kT}{m} \ln \rho(y) + C(y), \tag{189}$$

where m is the mass of a constituent particle and where $C(y)$ can depend on external parameters (pressure and temperature for instance) and in principle also on the internal coordinate y.

In equilibrium one finds from (189) with (181)

$$\frac{kT}{m} \ln \rho^{\text{eq}}(y) = - C(y) + a. \tag{190}$$

* J. Meixner, Zeitschr. Physik **149** (1957) 624.

From this equation one can determine C as a function of the variable y if the distribution $\rho^{eq}(y)$ is known.

(i) *The chemical reaction.* Let us study a chemical reaction $1 \rightleftarrows 2$, in which a substance 1 is transformed into a substance 2, and let us restrict ourselves to a transformation which may be considered as a diffusion along an internal coordinate y, the two products 1 and 2 being separated by a potential barrier*. Due to the existence of the potential barrier a quasi-stationary state is set up during the chemical reaction, in which the flux $J(y)$ is almost everywhere constant and equal to J. We then have from (185)

$$\frac{\partial s}{\partial t} = -\frac{1}{\rho_{\text{tot}} T} \int J(y) \frac{\partial \mu(y)}{\partial y} \, \mathrm{d}y \simeq -\frac{J(\mu_2 - \mu_1)}{\rho_{\text{tot}} T}, \tag{191}$$

which contains the product of the reaction rate J and the affinity $A = \mu_2 - \mu_1$ of our chemical reaction, in the usual way, as discussed before (*cf.* this chapter and Chapter III). We wish to show now that the stationarity condition $J(y) = \text{const.}$, together with the phenomenological equation (188) does not necessarily lead to a linear relation between the reaction rate J and the affinity A of the chemical reaction. We can in the first place write (188) with (189) as

$$J(y) = -L\left(\frac{kT}{m\rho}\frac{\partial \rho}{\partial y} + \frac{\partial C}{\partial y}\right), \tag{192}$$

or, defining the "mobility along the internal coordinate" per unit mass**,

$$U \equiv \frac{L}{\rho}, \tag{193}$$

* H. A. Kramers, Physica **7** (1940) 284.
According to (190) we have

$$\rho^{eq} = \text{const. } e^{-mC(y)/kT},$$

so that $C(y)$ represents a potential energy per unit mass along the internal coordinate y. If the two stable products are separated by a sufficiently high potential barrier (much larger than the potential energy difference of the two products 1 and 2), then in equilibrium almost all molecules are in states 1 and 2.

** In first approximation this mobility will be independent of the density and will be assumed to be independent of y.

as

$$J(y) = -U\left(\frac{kT}{m}\frac{\partial \rho}{\partial y} + \rho\frac{\partial C}{\partial y}\right). \tag{194}$$

If we introduce furthermore the "diffusion coefficient along the internal coordinate", defined by

$$D \equiv \frac{kT}{m} U\,, \tag{195}$$

we can transform (194) into

$$\begin{aligned} J(y) &= -D\,\mathrm{e}^{-UC/D}\frac{\partial}{\partial y}\,\mathrm{e}^{UC/D + \ln\rho} \\ &= -D\,\mathrm{e}^{-UC/D}\frac{\partial}{\partial y}\,\mathrm{e}^{m\mu/kT}. \end{aligned} \tag{196}$$

Multiplying both members of this relation by $D^{-1}\exp(UC/D)$, one obtains after integration with the help of the stationarity condition $J(y) = \text{const.}$ the following result

$$J = -\frac{\displaystyle\int_1^2 \frac{\partial}{\partial y}\exp\frac{m\mu}{kT}\,\mathrm{d}y}{\displaystyle\int_1^2 \frac{1}{D}\exp\frac{UC}{D}\,\mathrm{d}y}. \tag{197}$$

With the abbreviation

$$l \equiv \frac{\displaystyle\frac{m}{k}\exp\frac{m\mu_1}{kT}}{\displaystyle\int_1^2 \frac{1}{D}\exp\frac{UC}{D}\,\mathrm{d}y}, \tag{198}$$

one can then write (197) in the form

$$J = \frac{k}{m}\,l(1 - \mathrm{e}^{-mA/kT})\,, \tag{199}$$

which is the relation between reaction rate J and affinity A given by chemical kinetics [*cf.* (45)]. It is then seen that the linear law (192) does not necessarily lead to a linear relationship between the reaction rate and the affinity. As already stated in § 2 non-linear relations of the type (199) for chemical reactions do still lie inside the domain of validity of thermodynamics of irreversible processes.

(ii) *Brownian motion in velocity space.* Another important case where the concept of internal degree of freedom can be used is Brownian motion. We may then consider a (Cartesian) velocity component u_x of the Brownian particles as the internal coordinate y, while the density $\rho(y)$ represents the velocity distribution function $f(u_x)$. In equilibrium this distribution is Maxwellian:

$$f^{\mathrm{eq}}(u_x) = \text{const. } e^{-mu_x^2/2kT}. \tag{200}$$

or

$$\frac{kT}{m} \ln f^{\mathrm{eq}}(u_x) = -\tfrac{1}{2}u_x^2 + \text{const.} \tag{201}$$

Comparing (201) with (190) one finds that for the case considered here, the function $C(u_x)$, appearing in the chemical potential $\mu(u_x)$ [*cf.* (189)], is of the form

$$C(u_x) = \tfrac{1}{2}u_x^2 + \text{const.} \tag{202}$$

The chemical potential $\mu(u_x)$ has therefore the form

$$\mu(u_x) = \tfrac{1}{2}u_x^2 + \frac{kT}{m} \ln f(u_x) + \text{const.} \tag{203}$$

The phenomenological equation (188) becomes then (with $y = u_x$)

$$J(u_x) = -\beta\left(fu_x + \frac{kT}{m}\frac{\partial f}{\partial u_x}\right), \tag{204}$$

where

$$\beta \equiv \frac{L}{f} \tag{205}$$

is the so-called friction constant per unit mass for the Brownian particles (it is the analogue of the mobility U defined by equation (193) in the previous application). If expression (204) is inserted into (183) we get, considering β to be independent of u_x,

$$\frac{\partial f(u_x)}{\partial t} = \beta \left(\frac{\partial}{\partial u_x} f u_x + \frac{kT}{m} \frac{\partial^2 f}{\partial u_x^2} \right), \tag{206}$$

which is the Fokker–Planck equation for the distribution function $f(u_x)$ of the Brownian particles.

By generalizing the present one-dimensional formalism to the three-dimensional velocity space of Brownian particles (*i.e.* to three internal degrees of freedom or three internal coordinates u_x, u_y and u_z) one obtains the complete Fokker–Planck equation for the distribution function $f(u_x, u_y, u_z)$, (*cf.* Ch. IX, § 8). We thus see that a kinetic equation like the Fokker–Planck equation may be obtained from thermodynamics of irreversible processes by making a special assumption concerning the entropy production* [*cf.* (187)].

(iii) *Dipole orientation.* Let us finally apply the scheme of this section to the orientation of dipole molecules in an external electric field. The orientation of the dipoles, with dipole moment p per unit of mass, can be characterized by the polar angle θ, counted with respect to the electric field E. The quantity $\rho(\theta)\,\mathrm{d}\theta$ represents the number of molecules with an orientation between θ and $\theta + \mathrm{d}\theta$. At equilibrium $\rho(\theta)$ is given by the Boltzmann distribution

$$\rho^{\text{eq}}(\theta) = \text{const. } \mathrm{e}^{\frac{mpE\,\cos\theta}{kT}} \sin\theta, \tag{207}$$

where $mpE \cos\theta$ is the potential energy of a dipole making an angle θ with the electric field. If we compare this equation with (190) for $y = \theta$ we find that the function $C(\theta)$ is now of the form

$$C(\theta) = -\, pE \cos\theta - \frac{kT}{m} \ln \sin\theta + \text{const.} \tag{208}$$

The chemical potential $\mu(\theta)$ has therefore the form [*cf.* (189)]

$$\mu(\theta) = \frac{kT}{m} \ln \frac{\rho(\theta)}{\sin\theta} - pE\cos\theta + \text{const.} \tag{209}$$

* Without this special assumption thermodynamics of irreversible processes would have led to an integral equation for the function $f(\boldsymbol{u})$ of the "master equation" type (such an equation has been discussed in § 4 for the case of discrete internal variables). This master equation for $f(\boldsymbol{u})$ corresponds in fact to a linearized Boltzmann equation.

The phenomenological equation (188) becomes for this example (with $y = \theta$)

$$J(\theta) = -\frac{L \sin\theta}{m\rho}\left(kT\frac{\partial \rho \sin^{-1}\theta}{\partial\theta} + \rho m p E\right), \tag{210}$$

or

$$J(\theta) = -\sin\theta\left(D\frac{\partial \rho \sin^{-1}\theta}{\partial\theta} + U\rho p E\right), \tag{211}$$

where we have introduced the rotational mobility

$$U \equiv \frac{L}{\rho} \tag{212}$$

and the rotational diffusion constant

$$D \equiv \frac{kT}{m} U . \tag{213}$$

If one introduces (211) into (183), one gets

$$\frac{\partial \rho}{\partial t} = \frac{\partial}{\partial\theta}\left\{\sin\theta\left(D\frac{\partial \rho \sin^{-1}\theta}{\partial\theta} + U\rho p E\right)\right\}, \tag{214}$$

or

$$\frac{\partial n}{\partial t} = \frac{1}{\sin\theta}\frac{\partial}{\partial\theta}\left\{\sin\theta\left(D\frac{\partial n}{\partial\theta} + UnpE\sin\theta\right)\right\}, \tag{215}$$

where the density $n(\theta)$ is defined as

$$n(\theta) = \frac{\rho(\theta)}{\sin\theta}. \tag{216}$$

The "rotational diffusion equation" (215) has been obtained by Debije in his theory of dielectric relaxation.

Quite generally the formalism of this section leads to Fokker–Planck type partial differential equations in some internal coordinate space.

CHAPTER XI

HEAT CONDUCTION, DIFFUSION AND CROSS-EFFECTS

§ 1. *Heat Conduction*

In the preceding chapter the thermodynamic theory of scalar processes was discussed. In this chapter two important vectorial phenomena will be considered. These phenomena are heat conduction and diffusion, which will be treated separately, as well as in combination with each other or with chemical reactions.

We shall, in this section, consider systems in which only heat conduction takes place. The entropy source strength is then given by

$$\sigma = -\frac{1}{T^2} \boldsymbol{J}_q' \cdot \operatorname{grad} T , \tag{1}$$

as formula (III.25) shows. (We may notice that, since no diffusion phenomena occur, the two heat flows $\boldsymbol{J}_q'$ and $\boldsymbol{J}_q$ discussed in § 3 of Chapter III are identical.) The phenomenological equation is then [*cf.* (IV.22)]

$$\boldsymbol{J}_q' = - \mathsf{L}_{qq} \cdot \frac{\operatorname{grad} T}{T^2} , \tag{2}$$

where in the general case of an anisotropic system the quantity L_{qq} is a tensor, related to the heat conductivity tensor $\boldsymbol{\lambda}$ as

$$\boldsymbol{\lambda} = \frac{\mathsf{L}_{qq}}{T^2} . \tag{3}$$

Using this connexion, the phenomenological equation gets the form

$$\boldsymbol{J}_q' = - \boldsymbol{\lambda} \cdot \operatorname{grad} T , \tag{4}$$

which is Fourier's law. The Onsager relations for λ in the presence of a magnetic field can according to the theory of Chapter VIII, § 4, be written as

$$\lambda(\boldsymbol{B}) = \tilde{\lambda}(-\boldsymbol{B}) , \tag{5}$$

which is also the form (IV.58). Let us split the conduction tensor into its symmetric and antisymmetric parts

$$\lambda^{s} = \tfrac{1}{2}(\lambda + \tilde{\lambda}) , \tag{6}$$

$$\lambda^{a} = \tfrac{1}{2}(\lambda - \tilde{\lambda}) . \tag{7}$$

The Onsager relations (5) imply for these parts

$$\lambda^{s}(\boldsymbol{B}) = \lambda^{s}(-\boldsymbol{B}) , \tag{8}$$

$$\lambda^{a}(\boldsymbol{B}) = -\lambda^{a}(-\boldsymbol{B}) . \tag{9}$$

An alternative representation of the antisymmetric tensor is the axial vector $\boldsymbol{\lambda}^{a}$ with components

$$\lambda_{1}^{a} = -\lambda_{23}^{a} = \lambda_{32}^{a} , \text{ (cycl.)} , \tag{10}$$

which is called the Righi–Leduc vector. The phenomenological equation (4) can then be written as

$$\boldsymbol{J}_{q}' = -\lambda^{s}\cdot \text{grad}\, T - \boldsymbol{\lambda}^{a} \wedge \text{grad}\, T . \tag{11}$$

The Onsager relations (5) or their alternative forms (8) and (9) have not yet been experimentally verified for the case in which a magnetic field is present.

In the absence of a magnetic field one has simply

$$\lambda = \tilde{\lambda} \tag{12}$$

as Onsager relations. This formula, which expresses the fact that the heat conduction tensor is symmetric, has played a certain historical role long before it was recognized by Onsager that its origin was to be found in time reversal invariance. As a matter of fact the relations (12) were found to be satisfied experimentally in crystals of which the spatial symmetries alone would not warrant the symmetric character of the heat conduction tensor. Let us consider, for instance, a crystal belonging to the tetragonal or hexagonal class. Then with a proper choice of coordinate axes the heat conduction tensor must from purely spatial symmetry considerations have the form

$$\lambda = \begin{pmatrix} \lambda_{xx} & \lambda_{xy} & 0 \\ -\lambda_{xy} & \lambda_{xx} & 0 \\ 0 & 0 & \lambda_{zz} \end{pmatrix} \tag{13}$$

Spatial symmetry does not require λ_{xy} to vanish in certain crystals. This means that the heat flow in the x–y-plane will not have the same direction as the temperature gradient. The oldest and very accurate experiments designed to detect the spiral heat flow which would arise if λ_{xy} does not vanish are due to Soret and Voigt*. They observed no spiral motion, *i.e.* the coefficient λ_{xy} turned out to be zero, or in other words (12) was experimentally found to be valid. This result was amongst the data from which Onsager, in 1931, conceived the ideas for the establishment of his symmetry relations based on time reversal invariance.

Let us come back to the case of systems, where a magnetic field $\boldsymbol{B}$ is present, but which would be isotropic in the absence of the magnetic field. The heat conduction tensor has then the form (13) if $\boldsymbol{B}$ is applied parallel to the z-axis, because the phenomenological equation (4) is invariant for rotations around this axis. Furthermore any axis at right angles to the z-axis is a two-fold axis of rotation. Therefore the equations (4) are also invariant for a rotation by an angle π around the x-axis. This leads for the symmetric part of (13) to

$$\lambda_{xx}(\boldsymbol{B}) = \lambda_{xx}(-\boldsymbol{B}), \quad \lambda_{zz}(\boldsymbol{B}) = \lambda_{zz}(-\boldsymbol{B}) \tag{14}$$

and for the antisymmetric part of (13) to

$$\lambda_{xy}(\boldsymbol{B}) = -\lambda_{xy}(-\boldsymbol{B}), \tag{15}$$

* Ch. Soret, Arch. Sc. phys. nat (Genève) **29** (1893) 4;
W. Voigt, Gött. Nachr. (1903) 87.

so that the Onsager relations (8) and (9) give no new information about the system, because they are already satisfied for reasons of spatial symmetry. It is clear from the form (13) how one can measure the relevant phenomenological coefficients. The heat conductivity λ_{xx} for instance is measured if one imposes the condition

$$\frac{\partial T}{\partial y} = 0 \tag{16}$$

on the system, because then (4) with (13) leads to

$$\lambda_{xx} = -\frac{J'_{qx}}{\partial T/\partial x}, \tag{17}$$

of which the right-hand side is a measurable quantity. The form (13) shows furthermore that a temperature gradient in the x-direction gives rise to a heat flow in the y-direction and *vice versa*. This phenomenon is called the Righi–Leduc effect. It is the thermal analogon of the Hall-effect which arises if electric conduction is studied in a magnetic field (see Ch. XIII, § 5). The Righi–Leduc effect can be measured in the situation where one applies a temperature gradient in the x-direction and determines the ensuing temperature gradient in the y-direction, while the heat flow is possible only in the x-direction, *i.e.* the sample is thermally insulated in the y-direction. Then

$$J'_{qy} = 0 \tag{18}$$

and therefore with (4) and (13) one has for the Righi–Leduc effect

$$\frac{\lambda_{xy}}{\lambda_{xx}} = \frac{\partial T/\partial y}{\partial T/\partial x}, \tag{19}$$

where the right-hand side contains the measured quantities. The Righi–Leduc coefficient λ_{xy} follows then if λ_{xx} is known from (17).

Finally it may be remarked that for isotropic and cubic systems the tensor $\boldsymbol{\lambda}$ reduces to a scalar multiple of the unit tensor (*cf.* Ch. IV and VI).

§ 2. *Diffusion. General Remarks*

In this section and the following three the phenomenon of isothermal diffusion in isotropic, non-reacting mixtures in the absence of external forces will be studied. Later we shall remove some of the restrictions mentioned and consider diffusion under the influence of external forces (§ 6 of this chapter, and also Chapter XIII), diffusion in non-isothermal systems (§ 7) and diffusion in reacting systems (§ 8). In none of these cases will it be supposed that an external magnetic field acts on the system.

Before the particular properties of binary, ternary etc. systems are considered in sections 4 and 5, we shall introduce a few general notions concerning the description of diffusion phenomena. Let us first write down the entropy source strength for our n-component system:

$$\sigma = - \sum_{i=1}^{n} \boldsymbol{J}_i \cdot \frac{(\text{grad}\ \mu_i)_T}{T}\,, \tag{20}$$

which form follows from (III.25). Here the diffusion flow is given by (II.9):

$$\boldsymbol{J}_i = \rho_i(\boldsymbol{v}_i - \boldsymbol{v})\,, \quad (i = 1, 2, \ldots, n)\,, \tag{21}$$

where $\boldsymbol{v}$ is the barycentric velocity (II.7). Since in diffusion experiments phenomena due to shear viscosity do not occur and since phenomena due to bulk viscosity are in general negligibly small, we have omitted the corresponding terms in the entropy source strength (20). As explained in § 2 of Chapter V, we can assume that during the diffusion process a system enclosed in a reservoir is practically in mechanical equilibrium. Then in the absence of external forces the pressure is uniform over the system. Furthermore at mechanical equilibrium we can, according to Prigogine's theorem (Ch. V, § 2), replace the diffusion flow $\boldsymbol{J}_i$ occurring in (20) by a diffusion flow

$$\boldsymbol{J}_i^a = \rho_i(\boldsymbol{v}_i - \boldsymbol{v}^a)\,, \quad (i = 1, 2, \ldots, n)\,, \tag{22}$$

where $\boldsymbol{v}^a$ is an arbitrary reference velocity. We then have, instead of (20) for the entropy source strength

$$\sigma = - \sum_{i=1}^{n} \boldsymbol{J}_i^a \cdot \frac{(\text{grad}\ \mu_i)_{T,\,p}}{T}\,. \tag{23}$$

We shall use various reference velocities, which can all be written as weighted averages of the component velocities $\boldsymbol{v}_k$ in the following way

$$\boldsymbol{v}^a = \sum_{i=1}^{n} a_i \boldsymbol{v}_i \,, \quad \left(\sum_{i=1}^{n} a_i = 1 \right), \tag{24}$$

where $a_1, a_2, \ldots, a_n$ are the (normalized) weights. In the following table we list four amongst the most useful choices of weights and corresponding reference velocities and diffusion flows

weights a_i	reference velocity $\boldsymbol{v}^a = \Sigma_i a_i \boldsymbol{v}_i$	diffusion flow $\boldsymbol{J}^a = \rho_i (\boldsymbol{v}_i - \boldsymbol{v}^a)$
mass fractions c_i	barycentric velocity $\boldsymbol{v} = \Sigma_i c_i \boldsymbol{v}_i$	$\boldsymbol{J}_i = \rho_i (\boldsymbol{v}_i - \boldsymbol{v})$
molar fractions n_i	mean molar velocity $\boldsymbol{v}^m = \Sigma_i n_i \boldsymbol{v}_i$	$\boldsymbol{J}_i^m = \rho_i (\boldsymbol{v}_i - \boldsymbol{v}^m)$
$\rho_i v_i$	mean volume velocity $\boldsymbol{v}^0 = \Sigma_i \rho_i v_i \boldsymbol{v}_i$	$\boldsymbol{J}_i^0 = \rho_i (\boldsymbol{v}_i - \boldsymbol{v}^0)$
δ_{in}	n^{th} component velocity $\boldsymbol{v}_n = \Sigma_i \delta_{in} \boldsymbol{v}_i$	$\boldsymbol{J}_i^r = \rho_i (\boldsymbol{v}_i - \boldsymbol{v}_n)$

The first example is based on the use of the mass fractions (II.12)

$$c_i = \frac{\rho_i}{\rho}, \quad \left(\sum_{i=1}^{n} \rho_i = \rho \right) \tag{25}$$

as weights and leads to the hitherto employed barycentric velocity as reference velocity. The second makes use of molar (or molecular) fractions

$$n_i = \frac{N_i}{N}, \quad \left(\sum_{i=1}^{n} N_i = N \right), \tag{26}$$

where N_i is the molar density of component i, related to the mass density ρ_i by means of

$$\rho_i = M_i N_i \,, \tag{27}$$

with M_i the molar mass of i. The third example makes use of weights $\rho_i v_i$, where v_i is the partial specific volume of component i, related to the molar partial specific volume $\bar{v}_i$ as

$$\bar{v}_i = M_i v_i \,, \tag{28}$$

so that the weights $\rho_i v_i$ can alternatively be written as $N_i \bar{v}_i$. The fourth example, finally, leads to the so-called "relative diffusion flows", counted with respect to the nth component.

All diffusion flows $\boldsymbol{J}_i^a$ employed here contain the factor ρ_i. If one introduces N_i instead of ρ_i, one obtains the so-called molar diffusion flows

$$\bar{\boldsymbol{J}}_i^a = N_i(\boldsymbol{v}_i - \boldsymbol{v}^a) = \frac{1}{M_i} \boldsymbol{J}_i^a , \tag{29}$$

used frequently in molecular considerations.

One could use the fluxes and thermodynamic forces occurring in (20) or (23) in order to establish phenomenological equations. This gives simple symmetric forms for these equations, but a number of relations exist between their coefficients, because the fluxes and the thermodynamic forces are not independent. In fact one has from (22) and (24) the following relation between the fluxes

$$\sum_{i=1}^{n} \frac{a_i}{c_i} \boldsymbol{J}_i^a = 0 , \tag{30}$$

and from the Gibbs–Duhem relation (V.6) for the case of uniform pressure

$$\sum_{i=1}^{n} c_i \, (\text{grad}\, \mu_i)_{p,\,T} = 0 , \tag{31}$$

which is a relation between the thermodynamic forces. With the help of the last two relations we can eliminate $\boldsymbol{J}_n^a$ and grad μ_n from the entropy source strength (23). We then obtain an expression which contains independent fluxes and independent forces only:

$$\sigma = \sum_{i=1}^{n-1} \boldsymbol{J}_i^a \cdot \boldsymbol{X}_i^a , \tag{32}$$

with the thermodynamic forces given by

$$\boldsymbol{X}_i^a = - \sum_{k=1}^{n-1} A_{ik}^a \frac{(\text{grad}\, \mu_k)_{p,\,T}}{T} , \quad (k = 1, 2, \ldots, n-1) , \tag{33}$$

where

$$A_{ik}^a = \delta_{ik} + \frac{a_i}{a_n} \frac{c_k}{c_i} , \quad (i, k = 1, 2, \ldots, n-1) . \tag{34}$$

Since pressure and temperature are uniform, the gradients of the chemical potentials depend only on the gradients of the parameters which describe the composition of the system.

It may be noted that formulae analogous to (30)–(34) can be written down for "molar" quantities n_k, $\bar{\boldsymbol{J}}_k^a$, $\bar{\mu}_k = M_k\mu_k$ and $\bar{\boldsymbol{X}}_k^a = M_k\boldsymbol{X}_k^a$ instead of the "mass" quantities c_k, $\boldsymbol{J}_k^a$, μ_k and $\boldsymbol{X}_k^a$. We then obtain instead of (34) the expression

$$\bar{A}_{ik}^a = \delta_{ik} + \frac{a_i}{a_n}\frac{n_k}{n_i} = M_i A_{ik}^a \frac{1}{M_k}. \tag{35}$$

We shall write the phenomenological equations as linear relations between the fluxes and thermodynamic forces of (32)

$$\boldsymbol{J}_i^a = \sum_{k=1}^{n-1} L_{ik}^a \boldsymbol{X}_k^a, \quad (k = 1, 2, \ldots, n-1), \tag{36}$$

where the L_{ik}^a are scalar phenomenological coefficients*. For convenience we introduce a matrix notation in which the sets of ordinary vectors $\boldsymbol{J}_1^a, \boldsymbol{J}_2^a, \ldots, \boldsymbol{J}_{n-1}^a$, $\boldsymbol{X}_1^a, \boldsymbol{X}_2^a, \ldots, \boldsymbol{X}_{n-1}^a$ and grad μ_1, grad $\mu_2, \ldots,$ grad μ_{n-1} are considered as components of $(n-1)$-dimensional vectors $\boldsymbol{J}^a$, $\boldsymbol{X}^a$ and grad μ. Similarly A_{ik}^a and L_{ik}^a (with $i, k = 1, 2, \ldots, n-1$) are components of $(n-1)$-dimensional tensors A^a and L^a. Then formulae (32), (33) and (36) are written as

$$\sigma = \boldsymbol{J}^a \cdot \boldsymbol{X}^a, \tag{37}$$

$$\boldsymbol{X}^a = -\mathsf{A}^a \cdot \frac{\text{grad}\,\mu}{T}, \tag{38}$$

$$\boldsymbol{J}^a = \mathsf{L}^a \cdot \boldsymbol{X}^a, \tag{39}$$

where dots stand for interior products in the $(n-1)$-dimensional space (*cf.* Appendix I). If we introduce the diagonal matrix M of which the elements are

$$M_{ik} = \delta_{ik} M_k, \tag{40}$$

we can write the relation (35) as

$$\bar{\mathsf{A}}^a = \mathsf{M} \cdot \mathsf{A}^a \cdot \mathsf{M}^{-1}. \tag{41}$$

* In principle the L_{ik}^a are tensors in Cartesian space, which reduce to scalar multiples of the unit tensor because the system is isotropic.

If we had chosen to use instead of $\boldsymbol{v}^a$ a different reference velocity

$$\boldsymbol{v}^b = \sum_{k=1}^{n} b_i \boldsymbol{v}_i \tag{42}$$

with weights b_i, we would have had corresponding fluxes and thermodynamic forces $\boldsymbol{J}^b$ and $\boldsymbol{X}^b$ in the entropy source strength

$$\sigma = \boldsymbol{J}^b \cdot \boldsymbol{X}^b . \tag{43}$$

The relation between both kinds of fluxes can be written as

$$\boldsymbol{J}^a = \mathsf{B}^{ab} \cdot \boldsymbol{J}^b , \tag{44}$$

where the elements of the $(n-1)$-dimensional tensor B^{ab} are given by

$$B^{ab}_{ik} = \delta_{ik} + \left(a_n \frac{b_k}{b_n} - a_k \right) \frac{c_i}{c_k}, \quad (i, k = 1, 2, \ldots, n-1) . \tag{45}$$

The proof of this formula is found by expressing both $\boldsymbol{J}^a$ and $\boldsymbol{J}^b$ in formula (44) in terms of the set of independent velocities $\boldsymbol{v}_1, \boldsymbol{v}_2, \ldots, \boldsymbol{v}_n$ (this is done with the help of (22) and (24) and the corresponding formulae with b instead of a) and by identifying the coefficients of these velocities. From (44) it follows that the matrix B^{ba} is equal to the reciprocal of B^{ab}. Its elements are found from those of B^{ab} by interchanging a_i and b_i $(i = 1, 2, \ldots, n)$:

$$B^{ba}_{ik} = (\mathsf{B}^{ab})^{-1}_{ik} = \delta_{ik} + \left(b_n \frac{a_k}{a_n} - b_k \right) \frac{c_i}{c_k} . \tag{46}$$

We shall also need the reciprocal of A^a of which the elements are, as follows from (34),

$$(\mathsf{A}^a)^{-1}_{ik} = \delta_{ik} - \frac{c_k}{c_i} a_i . \tag{47}$$

The transformation of the thermodynamic forces, corresponding to the transformation of fluxes (44), is

$$\boldsymbol{X}^a = \widetilde{\mathsf{B}^{ba}} \cdot \boldsymbol{X}^b , \tag{48}$$

because σ is invariant [see (37) and (43)]. From (38) it is clear that A^a transforms as $\mathbf{X}^a$, so that we have

$$\mathsf{A}^a = \widetilde{\mathsf{B}^{ba}} \cdot \mathsf{A}^b . \tag{49}$$

Finally the transformation of the phenomenological coefficient matrix is given by

$$\mathsf{L}^a = \mathsf{B}^{ab} \cdot \mathsf{L}^b \cdot \widetilde{\mathsf{B}^{ab}} , \tag{50}$$

as follows from (39), (44) and (48). (*Cf.* also Chapter VI, § 5 for the same formulae.)

We shall now introduce a general *definition of diffusion coefficients* by writing the linear laws

$$\boldsymbol{J}^a = - \mathsf{D}^{ax} \cdot \text{grad}\, \boldsymbol{x} , \tag{51}$$

where grad $\boldsymbol{x}$ is an $(n-1)$-dimensional vector with components grad x_1, grad x_2, . . . , grad x_{n-1} and where x_i stands for some quantity which characterizes the composition of the mixture, such as ρ_i, c_i, N_i or n_i. We have then a tensor D^{ax} of $(n-1)^2$ diffusion coefficients, corresponding to the reference velocity $\boldsymbol{v}^a$ and the composition parameters $\boldsymbol{x}$. For a different reference velocity $\boldsymbol{v}^b$ formula (51) is

$$\boldsymbol{J}^b = - \mathsf{D}^{bx} \cdot \text{grad}\, \boldsymbol{x} , \tag{52}$$

which shows that D^{ax} is related to D^{bx} just as $\boldsymbol{J}^a$ is to $\boldsymbol{J}^b$, *i.e.* according to (44)

$$\mathsf{D}^{ax} = \mathsf{B}^{ab} \cdot \mathsf{D}^{bx} , \tag{53}$$

where B^{ab} is by its nature independent of the choice of $\boldsymbol{x}$. Now in practice the actual procedure in describing diffusion is the following. One chooses once for all a certain velocity, which we could call the "basic" velocity $\boldsymbol{v}^b$, and a certain "basic" set of composition parameters $\boldsymbol{x}$ and one calls the corresponding coefficients D^{bx} the diffusion coefficients. This does not mean, however, that one wishes also to fix the choice of the diffusion flows as $\boldsymbol{J}^b$ and the composition parameters as $\boldsymbol{x}$. One still wants to leave the possibility open of using different kinds of reference velocities $\boldsymbol{v}^a$ in the diffusion flow $\boldsymbol{J}^a$ and different kinds of composition parameters $\boldsymbol{y}$, of which the gradients are connected with the gradients of the set $\boldsymbol{x}$ by

$$\text{grad}\,\boldsymbol{x} = \frac{\partial \boldsymbol{x}}{\partial \boldsymbol{y}} \cdot \text{grad}\,\boldsymbol{y}\,, \tag{54}$$

where $\partial\boldsymbol{x}/\partial\boldsymbol{y}$ is a $(n-1)$-dimensional matrix with elements $\partial x_i/\partial y_j$. In other words the diffusion laws used in practice have the general form

$$\boldsymbol{J}^a = -\,\mathsf{B}^{ab}\cdot\mathsf{D}^{bx}\cdot\frac{\partial \boldsymbol{x}}{\partial \boldsymbol{y}} \cdot \text{grad}\,\boldsymbol{y}\,, \tag{55}$$

as follows from (51) with (53) and (54). The important point of these laws is that with fixed $\boldsymbol{v}^b$ and $\boldsymbol{x}$, one has made a definite choice of the diffusion coefficients D^{bx}, independent of the choices of $\boldsymbol{v}^a$ and $\boldsymbol{y}$, which are still open. In practice one makes these choices in such a way that the law (55) has the most convenient form for the problem studied. We shall encounter a number of examples in the following sections.

We have introduced in the above treatment phenomenological coefficients and diffusion coefficients. The relation between these coefficients is found from comparison of (38) and (39) with (51) and (53):

$$\mathsf{D}^{bx} = T^{-1}\mathsf{L}^b\cdot\mathsf{A}^b\cdot\boldsymbol{\mu}^x = T^{-1}\mathsf{B}^{ba}\cdot\mathsf{L}^a\cdot\mathsf{A}^a\cdot\boldsymbol{\mu}^x\,, \tag{56}$$

where $\boldsymbol{\mu}^x$ is the $(n-1)$-dimensional matrix with elements

$$\mu^x_{ik} = (\partial\mu_i/\partial x_k)_{p,\,T,\,x_j}\,,\quad (i, k = 1, 2, \ldots, n-1; j \neq k)\,. \tag{57}$$

It will turn out to be convenient to introduce as an abbreviation the tensor

$$\mathsf{G}^{ax} = \mathsf{A}^a\cdot\boldsymbol{\mu}^x\,. \tag{58}$$

Then (56) gets the form

$$\mathsf{D}^{bx} = T^{-1}\mathsf{L}^b\cdot\mathsf{G}^{bx} = T^{-1}\mathsf{B}^{ba}\cdot\mathsf{L}^a\cdot\mathsf{G}^{ax}\,. \tag{59}$$

The Onsager relations for systems in the absence of a magnetic field read

$$\mathsf{L}^b = \tilde{\mathsf{L}}^b\,. \tag{60}$$

According to (59) these relations have as a consequence the following *Onsager relations for the diffusion coefficients**

$$\mathsf{D}^{bx}\cdot(\mathsf{G}^{bx})^{-1} = \widetilde{(\mathsf{G}^{bx})^{-1}}\cdot\widetilde{\mathsf{D}^{bx}}\,. \tag{61}$$

These are $\frac{1}{2}(n-1)(n-2)$ relations, which reduce the number of $(n-1)^2$ diffusion coefficients in (52) to $\frac{1}{2}n(n-1)$ independent coefficients.

Finally we note that since $\bar{\mu}_i = M_i\mu_i$ we find the following relation between two matrices of the type (57) (using (27) and the definition (40) of the matrix M):

$$\bar{\mu}^N = \mathsf{M}\cdot\mu^p\cdot\mathsf{M}\,, \tag{62}$$

and from this formula with (41) and (58)

$$\bar{\mathsf{G}}^{aN} \equiv \bar{\mathsf{A}}^a\cdot\bar{\mu}^N \equiv \mathsf{M}\cdot\mathsf{A}^a\cdot\mathsf{M}^{-1}\cdot\mathsf{M}\cdot\mu^p\cdot\mathsf{M} = \mathsf{M}\cdot\mathsf{G}^{ap}\cdot\mathsf{M}\,, \tag{63}$$

a connexion that will be used in the following sections.

§ 3. *Thermodynamic Symmetry Relations for Chemical Potentials*

Certain symmetry relations exist between the $(n-1)^2$ elements of the matrix μ^x_{ik}, defined in formula (57). For the case that the composition parameters x_i stand for the mass fractions c_i, this can be seen, for instance, in the following way. Let us write down the total differential of the specific Gibbs function g

$$\begin{aligned} \mathrm{d}g &= -\,s\,\mathrm{d}T - v\,\mathrm{d}p - \sum_{i=1}^{n}\mu_i\,\mathrm{d}c_i \\ &= -\,s\,\mathrm{d}T - v\,\mathrm{d}p - \sum_{i=1}^{n-1}(\mu_i-\mu_n)\,\mathrm{d}c_i\,. \end{aligned} \tag{64}$$

Since this is a total differential we have the relation for the cross-differential quotients

$$\frac{\partial}{\partial c_k}\left(\frac{\partial g}{\partial c_i}\right) = \frac{\partial}{\partial c_i}\left(\frac{\partial g}{\partial c_k}\right), \quad (i, k = 1, 2, \ldots, n-1)\,, \tag{65}$$

* For certain combinations of b_k and x_k, for instance for $b_k = c_k$ and $x_k = c_k$ and also for $b_k = \rho_k v_k$ and $x_k = \rho_k$ the form (61) corresponds to the form $\mathsf{M}\cdot\mathsf{g}^{-1} = \mathsf{g}^{-1}\cdot\mathsf{M}$ of Chapters IV and VII. Then the matrix G is an example of a symmetric matrix g (*cf.* the end of § 3).

which becomes here

$$\mu^c_{ik} - \mu^c_{nk} = \mu^c_{ki} - \mu^c_{ni}, \quad (i, k = 1, 2, \ldots, n-1). \tag{66}$$

If we eliminate μ_n with the help of the Gibbs–Duhem relation (31), these relations become

$$(\mathsf{A}^c \cdot \boldsymbol{\mu}^c)_{ik} = (\mathsf{A}^c \cdot \boldsymbol{\mu}^c)_{ki}, \quad (i, k = 1, 2, \ldots, n-1), \tag{67}$$

where the matrix A^c is given by (34) with $a_i = c_i$, *i.e.*

$$A^c_{ij} = \delta_{ij} + \frac{c_j}{c_n}, \quad (i, j = 1, 2, \ldots, n-1). \tag{68}$$

We can write (67) symbolically, with the notation (58), as

$$\mathsf{G}^{cc} = \widetilde{\mathsf{G}^{cc}}, \tag{69}$$

or, explicitly, with (68), as

$$\mu^c_{ik} + \frac{1}{c_n} \sum_{j=1}^{n-1} c_j \mu^c_{jk} = \mu^c_{ki} + \frac{1}{c_n} \sum_{j=1}^{n-1} c_j \mu^c_{ji}, \quad (i, k = 1, 2, \ldots, n-1). \tag{70}$$

These are $\frac{1}{2}(n-1)(n-2)$ relations between the elements of μ^c_{ik}, which leaves $\frac{1}{2}n(n-1)$ independent elements.

The analogous relations for molar quantities connect elements of the matrix $\bar{\boldsymbol{\mu}}^n$

$$(\bar{\mathsf{A}}^n \cdot \bar{\boldsymbol{\mu}}^n)_{ik} = (\bar{\mathsf{A}}^n \cdot \bar{\boldsymbol{\mu}}^n)_{ki}, \tag{71}$$

or

$$\bar{\mathsf{G}}^{nn} = \widetilde{\bar{\mathsf{G}}^{nn}}, \tag{72}$$

or

$$\bar{\mu}^n_{ik} + \frac{1}{n_n} \sum_{j=1}^{n-1} n_j \bar{\mu}^n_{jk} = \bar{\mu}^n_{ki} + \frac{1}{n_n} \sum_{j=1}^{n-1} n_j \bar{\mu}^n_{ji}, \quad (i, k = 1, 2, \ldots, n-1). \tag{73}$$

This follows from a relation as (64), but containing molar quantities such as $\bar{s} = \sum_{i=1}^n n_i \bar{s}_i$ instead of mass quantities as $s = \sum_{i=1}^n c_i s_i$, etc. [or alternatively directly from (67), (69) and (70)].

If we wish to find relations amongst the elements of the matrix $\boldsymbol{\mu}^\rho$, then we need to know the matrix $\partial \boldsymbol{\rho}/\partial \boldsymbol{c}$ of which the elements are $\partial \rho_i/\partial c_k$ $(i, k = 1, 2, \ldots, n-1)$, because

$$\boldsymbol{\mu}^c = \boldsymbol{\mu}^\rho \cdot \frac{\partial \boldsymbol{\rho}}{\partial \boldsymbol{c}}. \tag{74}$$

The matrix $\partial \boldsymbol{c}/\partial \boldsymbol{\rho}$, which occurs in the connexion

$$\boldsymbol{\mu}^{\rho} = \boldsymbol{\mu}^{c} \cdot \frac{\partial \boldsymbol{c}}{\partial \boldsymbol{\rho}} \tag{75}$$

is the reciprocal of the matrix $\partial \boldsymbol{\rho}/\partial \boldsymbol{c}$, since according to (74)

$$\boldsymbol{\mu}^{\rho} = \boldsymbol{\mu}^{c} \cdot \left(\frac{\partial \boldsymbol{\rho}}{\partial \boldsymbol{c}} \right)^{-1} . \tag{76}$$

An explicit expression for $\partial \rho_i/\partial c_k$ follows immediately from the thermodynamical identity (see Appendix II, formula (15), with $a = v = \rho^{-1}$, the specific volume)

$$\frac{\partial \rho}{\partial c_k} = \rho^2 (v_n - v_k) , \quad (k = 1, 2, \ldots, n-1) \tag{77}$$

(where v_k is the partial specific volume of component k), because with $\rho = \rho_i/c_i$, this formula yields

$$\frac{\partial \rho_i}{\partial c_k} = \rho \left\{ \delta_{ik} + \rho_i (v_n - v_k) \right\} , \quad (i, k = 1, 2, \ldots, n-1) , \tag{78}$$

the required derivatives. As can be checked, the reciprocal elements of this matrix are given by

$$\frac{\partial c_i}{\partial \rho_k} = \frac{1}{\rho} \left\{ \delta_{ik} + c_i \left(\frac{v_k}{v_n} - 1 \right) \right\} , \quad (i, k = 1, 2, \ldots, n-1) . \tag{79}$$

The formulae with molar quantities, analogous to (74), (75), (78) and (79), are

$$\bar{\boldsymbol{\mu}}^{n} = \bar{\boldsymbol{\mu}}^{N} \cdot \frac{\partial \boldsymbol{N}}{\partial \boldsymbol{n}} , \tag{80}$$

$$\bar{\boldsymbol{\mu}}^{N} = \bar{\boldsymbol{\mu}}^{n} \cdot \frac{\partial \boldsymbol{n}}{\partial \boldsymbol{N}} , \tag{81}$$

$$\frac{\partial N_i}{\partial n_k} = N \left\{ \delta_{ik} + N_i (\bar{v}_n - \bar{v}_k) \right\} , \tag{82}$$

$$\frac{\partial n_i}{\partial N_k} = \frac{1}{N} \left\{ \delta_{ik} + n_i \left(\frac{\bar{v}_k}{\bar{v}_n} - 1 \right) \right\} . \tag{83}$$

With the help of the relations $\bar{\mu}_i = M_i \mu_i$, $\rho_i = M_i N_i$, (62), (74), (75) and (78)–(83) one can express the elements of any two of the four matrices $\boldsymbol{\mu}^c$, $\boldsymbol{\mu}^\rho$, $\boldsymbol{\mu}^n$ and $\boldsymbol{\mu}^N$ (or also the same with $\bar{\boldsymbol{\mu}}$) into each other. For instance, from (74) with (78) it follows that

$$\mu_{ik}^c = \rho \left\{ \mu_{ik}^\rho + \left(\sum_{j=1}^{n-1} \mu_{ij}^\rho \rho_j \right) (v_n - v_k) \right\}, \quad (i, k = 1, 2, \ldots, n-1), \tag{84}$$

and from (81) with (83) that

$$\bar{\mu}_{ik}^N = \frac{1}{N} \left\{ \bar{\mu}_{ik}^n + \left(\sum_{j=1}^{n-1} \bar{\mu}_{ij}^n n_j \right) \left(\frac{\bar{v}_k}{\bar{v}_n} - 1 \right) \right\}, \quad (i, k = 1, 2, \ldots, n-1). \tag{85}$$

Both these results will be used in § 5 of this chapter.

We note that the elements of the transformation matrix B^{ab}, given by (45), become for the case $a_i = \rho_i v_i$ and $b_i = c_i$

$$B_{ik}^{0c} = \delta_{ik} + \rho_i (v_n - v_k), \quad (i, k = 1, 2, \ldots, n-1), \tag{86}$$

so that in view of (78) the simple and useful connexion

$$\frac{\partial \boldsymbol{\rho}}{\partial \boldsymbol{c}} = \rho \mathsf{B}^{0c} \tag{87}$$

exists. Taking the reciprocal matrices at both sides of this relation we obtain the equivalent formula

$$\frac{\partial \boldsymbol{c}}{\partial \boldsymbol{\rho}} = \frac{1}{\rho} \mathsf{B}^{0c}, \tag{88}$$

because $\mathsf{B}^{ba} = (\mathsf{B}^{ab})^{-1}$ [*cf.* (46)].

We shall now show that one can still derive another symmetry relation by using the identity (87), because from (67) we have, employing (49) and (74),

$$\widetilde{\mathsf{B}^{0c}} \cdot \mathsf{A}^0 \cdot \boldsymbol{\mu}^\rho \cdot \frac{\partial \boldsymbol{\rho}}{\partial \boldsymbol{c}} = \widetilde{\frac{\partial \boldsymbol{\rho}}{\partial \boldsymbol{c}}} \cdot \widetilde{\boldsymbol{\mu}^\rho} \cdot \widetilde{\mathsf{A}^0} \cdot \mathsf{B}^{0c}. \tag{89}$$

Indeed with (87), this equality gives immediately the symmetry relation

$$\mathsf{A}^0 \cdot \mu^\rho = \widetilde{\mu^\rho} \cdot \widetilde{\mathsf{A}^0}\,, \tag{90}$$

or, with the notation (58),

$$\mathsf{G}^{0\rho} = \widetilde{\mathsf{G}^{0\rho}}\,. \tag{91}$$

The analogous molar relation is

$$\bar{\mathsf{G}}^{0N} = \widetilde{\bar{\mathsf{G}}^{0N}}\,. \tag{92}$$

We found that the matrices G^{cc} and $\mathsf{G}^{0\rho}$ (and also $\bar{\mathsf{G}}^{nn}$ and $\bar{\mathsf{G}}^{0N}$) were symmetric. It may be remarked here that these matrices play the role of the symmetric matrix g, which was used in the discussions of Chapters IV and VII. That this is indeed the case can be understood immediately, because the matrices G^{cc} and $\mathsf{G}^{0\rho}$ are both second derivatives of the specific Gibbs function g. For instance G^{cc}_{ik} is equal to $-\partial^2 g/\partial c_k \partial c_i$ as explained in the beginning of this section. Similarly $G^{0\rho}_{ik}$ is essentially connected to $\partial^2 g/\partial \rho_k \partial \rho_i$.

Written with the help of the symmetric matrices g, the Onsager relation (61) for the diffusion coefficients gets the form $\mathsf{M} \cdot \mathsf{g}^{-1} = \mathsf{g}^{-1} \cdot \tilde{\mathsf{M}}$, discussed in Chapters IV and VII.

§ 4. *Diffusion in Binary Systems*

The most simple case of diffusion is the mixing of two chemical components ($n = 2$). Then the matrices (34) and (45) reduce to ordinary numbers, given by

$$A^{\bar{a}}_{11} = \frac{1}{a_2}\,, \quad B^{ab}_{11} = \frac{a_2}{b_2}\,. \tag{93}$$

The entropy production (32) with (33) contains only one flux and one thermodynamic force

$$\sigma = \boldsymbol{J}^a_1 \cdot \boldsymbol{X}^a_1 = -\frac{1}{a_2 T} \boldsymbol{J}^a_1 \cdot \operatorname{grad} \mu_1\,. \tag{94}$$

The diffusion process is described by the phenomenological equation (55)

$$\boldsymbol{J}^a_1 \,(= M_1 \bar{\boldsymbol{J}}^a_1) = -\frac{a_2}{b_2} D^{bx} \frac{\partial x_1}{\partial y_1} \operatorname{grad} y_1\,, \tag{95}$$

containing one single diffusion coefficient. According to the procedure outlined in the preceding section, we first choose b and x. For the latter we shall take the density ($x_1 = \rho_1$), for the former we shall try all possibilities of the table in § 2, to explore which of these will turn out to be the most convenient for the application to the most important physical situations (see end of this section). Subsequently we have a choice of a for which we also take the same four possibilities. Finally for y_1 we use ρ_1, c_1, N_1 or n_1, but we write only one or two results, which give (95) a simple form. The results are compiled in the following table which gives the values of the factor

$$Y^{ab\rho} \equiv \frac{a_2}{b_2} \frac{\partial \rho_1}{\partial y_1} \operatorname{grad} y_1 , \tag{96}$$

appearing at the right-hand side of (95), with $x_1 = \rho_1$ (see bottom).

The values of $\partial \rho_1 / \partial y_1$ with $y_1 = c_1$, n_1 or N_1 which are needed to find the expressions in this table can be found from the following three formulae

$$\rho_1 = M_1 N_1 , \tag{97}$$

$$\frac{\partial c_1}{\partial n_1} = \frac{c_1 c_2}{n_1 n_2} = M_1 M_2 \frac{N^2}{\rho^2} , \tag{98}$$

$Y^{ab\rho}$	$a_2 = c_2$	$a_2 = n_2$	$a_2 = \rho_2 v_2$	$a_2 = \delta_{22} = 1$
$b_2 = c_2$	$\operatorname{grad} \rho_1$ $= M_1 \operatorname{grad} N_1$	$\frac{\rho}{M_2 N} \operatorname{grad} \rho_1$	$v_2 \rho \operatorname{grad} \rho_1$ $= M_1 v_2 \rho \operatorname{grad} N_1$	$c_2^{-1} \operatorname{grad} \rho_1$
$b_2 = n_2$	$\frac{M_2 N}{\rho} \operatorname{grad} \rho_1$ $= \frac{M_1 M_2 N}{\rho} \operatorname{grad} N_1$	$\operatorname{grad} \rho_1$ $= M_1 \operatorname{grad} N_1$	$M_2 v_2 N \operatorname{grad} \rho_1$ $= M_1 M_2 v_2 N \operatorname{grad} N_1$	$n_2^{-1} \operatorname{grad} \rho_1$
$b_2 = \rho_2 v_2$	$\rho \operatorname{grad} c_1$ $= \frac{N^2}{\rho} M_1 M_2 \operatorname{grad} n_1$	$M_1 N \operatorname{grad} n_1$	$\operatorname{grad} \rho_1$ $= M_1 \operatorname{grad} N_1$	$\frac{\rho}{c_2} \operatorname{grad} c_1$ $= \frac{M_1 N}{n_2} \operatorname{grad} n_1$
$b_2 = \delta_{22} = 1$	$c_2 \operatorname{grad} \rho_1$ $= M_1 c_2 \operatorname{grad} N_1$	$n_2 \operatorname{grad} \rho_1$ $= M_1 n_2 \operatorname{grad} N_1$	$\rho_2 v_2 \operatorname{grad} \rho_1$ $= M_1 \rho_2 v_2 \operatorname{grad} N_1$	$\operatorname{grad} \rho_1$ $= M_1 \operatorname{grad} N_1$

$$\frac{\partial \rho_1}{\partial c_1} = \rho^2 v_2 \,, \left(\text{or } \frac{\partial N_1}{\partial n_1} = N^2 \bar{v}_2 \right) . \tag{99}$$

The first of these formulae is the same as (27). The second follows directly from (25), (26) and (27), which yield

$$c_1 = \frac{\rho_1}{\rho_1 + \rho_2} = \frac{n_1 M_1}{n_1 M_1 + (1 - n_1) M_2} . \tag{100}$$

The third is the special case $n = 2$, $i = 1$ and $k = 1$ of formula (78), when v_1 is eliminated with the help of $\rho_1 v_1 + \rho_2 v_2 = 1$. The formula within brackets (99) is completely analogous to the preceding formula, but written with molar quantities instead of mass quantities.

From the table it is apparent that the choice $b_2 = \rho_2 v_2$ for the weights of the "basic velocity" $\boldsymbol{v}^b$ gives rise to simple forms of the diffusion law. So, if from now on, we stick to this choice, then both parameters b and x are fixed ($b_2 = \rho_2 v_2$, $x_1 = \rho_1$) and we shall call the corresponding coefficient D^{bx} *the* binary diffusion coefficient and denote it simply by D. From (96) and the third row of the table we find then that some useful and simple forms of the diffusion law (95) can be written down, all with D as diffusion coefficient (D has the same dimension as a surface per unit of time). These equations read, with the notations of § 2 [see the table and formula (29)]:

$$\boldsymbol{J}_1 = - \rho D \operatorname{grad} c_1 = - D N^2 M_1 M_2 \rho^{-1} \operatorname{grad} n_1 \,, \quad (a_2 = c_2) \,, \tag{101}$$

$$\bar{\boldsymbol{J}}_1^m = - N D \operatorname{grad} n_1 \,, \quad (a_2 = n_2) \,, \tag{102}$$

$$\boldsymbol{J}_1^0 = - D \operatorname{grad} \rho_1 \,, \quad \text{or} \quad \bar{\boldsymbol{J}}_1^0 = - D \operatorname{grad} N_1 \,, \quad (a_2 = \rho_2 v_2) \,, \tag{103}$$

$$\boldsymbol{J}_1^r = - D \frac{\rho}{c_2} \operatorname{grad} c_1 \,, \quad \text{or} \quad \bar{\boldsymbol{J}}_1^r = - D \frac{N}{n_2} \operatorname{grad} n_1 \,, \quad (a_2 = 1) \,. \tag{104}$$

First of all it may be remarked that (101) has the form of the diffusion law and contains the diffusion coefficients, which are used in the kinetic theory ot gases.

The relations (101)–(104) are useful especially for the establishment of differential equations which describe the behaviour of systems enclosed in reservoirs. Indeed for systems confined to a closed vessel it can often be argued that a certain average velocity $\boldsymbol{v}^a$ vanishes. If that is the case, then the law of conservation of mass can be written, using also the definitions (22) and (29), as

$$\frac{\partial \rho_1}{\partial t} = - \operatorname{div} \rho_1 \boldsymbol{v}_1 = - \operatorname{div} \boldsymbol{J}_1^a , \tag{105}$$

or, alternatively,

$$\frac{\partial N_1}{\partial t} = - \operatorname{div} N_1 \boldsymbol{v}_1 = - \operatorname{div} \bar{\boldsymbol{J}}_1^a . \tag{106}$$

We shall enumerate four special cases with vanishing $\boldsymbol{v}^a$ where the differential equations, which one obtains from (105) or (106) with the help of one of the phenomenological equations (101)–(104), have very simple forms under a few more subsidiary conditions. In the four cases $\boldsymbol{v}^a$ will be chosen as $\boldsymbol{v}$, $\boldsymbol{v}^m$, $\boldsymbol{v}^0$ and $\boldsymbol{v}_2$ respectively.

Before we deal with these special cases it is useful to remark that the assumption that viscous shear phenomena give no contribution to the entropy production (*cf.* § 2), leads to a vanishing rotation of the barycentric velocity $\boldsymbol{v}$. This may be seen in the following way. According to Chapters III and IV the entropy production due to viscous phenomena is equal to

$$-\frac{1}{T} \overset{\circ}{\Pi} : (\overset{\circ}{\mathrm{Grad}}\, \boldsymbol{v})^s - \frac{1}{T} \Pi \operatorname{div} \boldsymbol{v} = 2\frac{\eta}{T} (\overset{\circ}{\mathrm{Grad}}\, \boldsymbol{v})^s : (\overset{\circ}{\mathrm{Grad}}\, \boldsymbol{v})^s + \frac{\eta_v}{T} (\operatorname{div} \boldsymbol{v})^2 , \tag{107}$$

where the two parts are due to shear and bulk viscosity respectively. Since $\eta > 0$, the vanishing of the first part leads to the conditions

$$\frac{\partial v_\alpha}{\partial x_\beta} + \frac{\partial v_\beta}{\partial x_\alpha} = 0 , \ (\alpha \neq \beta) ; \quad \frac{\partial v_1}{\partial x_1} = \frac{\partial v_2}{\partial x_2} = \frac{\partial v_3}{\partial x_3} , \tag{108}$$

where v_1, v_2, v_3 are the Cartesian components of $\boldsymbol{v}$. The general solution* of this set of differential equations is the velocity field

$$\boldsymbol{v}(\boldsymbol{r}\,; t) = \boldsymbol{a} + \boldsymbol{b} \wedge \boldsymbol{r} + c\boldsymbol{r} + 2\boldsymbol{r}(\boldsymbol{d}\cdot\boldsymbol{r}) - \boldsymbol{d}r^2 , \tag{109}$$

where the quantities $\boldsymbol{a}$, $\boldsymbol{b}$, c and $\boldsymbol{d}$ are time functions independent of the space coordinates. The divergence and the rotation of this field are given by

* J. Meixner, Ann. Physik [5] **41** (1942) 409.
J. Meixner and H. G. Reik, Encyclopedia of Physics III/2 (1959) p. 448.

$$\operatorname{div} \boldsymbol{v} = 3c + 6(\boldsymbol{d} \cdot \boldsymbol{r}), \quad \operatorname{rot} \boldsymbol{v} = 2\boldsymbol{b} + 4\boldsymbol{d} \wedge \boldsymbol{r}. \tag{110}$$

The condition of mechanical equilibrium

$$\frac{\mathrm{d}\boldsymbol{v}}{\mathrm{d}t} \equiv \frac{\partial \boldsymbol{v}}{\partial t} + (\boldsymbol{v} \cdot \operatorname{grad})\boldsymbol{v} = 0, \tag{111}$$

supposed to be valid throughout the system, yields with the solution (109):

$$\boldsymbol{b} = 0, \quad \boldsymbol{d} = 0, \tag{112}$$

so that (110) becomes

$$\operatorname{div} \boldsymbol{v} = 3c, \quad \operatorname{rot} \boldsymbol{v} = 0. \tag{113}$$

We wish to emphasize that we only neglected the effect of shear viscosity, but not of volume viscosity [last term in (107)]. The quantity div $\boldsymbol{v}$ does not vanish, as (113) shows.

Let us now consider our four special cases.

(i) In systems where the density ρ can be considered as practically constant in time and uniform, the conservation of mass law implies that the divergence of the barycentric velocity $\boldsymbol{v}$ vanishes. From div $\boldsymbol{v} = 0$, rot $\boldsymbol{v} = 0$ and the boundary condition (the normal component of $\boldsymbol{v}$ vanishes at the boundary of the vessel) it follows then that the barycentric velocity vanishes everywhere. In the physical situation just described, which occurs *e.g.* in diffusion experiments in isomer mixtures or in isotopic mixtures (with not too different molecular masses) and in mass diluted systems (*i.e.* $c_1 \ll 1$ and $c_2 \simeq 1$), we find from (25), (101) and (105)

$$\frac{\partial c_1}{\partial t} = D \triangle c_1, \tag{114}$$

where it has been supposed that D is practically uniform and where $\triangle$ is the Laplacian operator. This differential equation has the form of Fourier's equation, established originally for the description of heat conduction.

(ii) In systems where the molar density N is constant in time and uniform, we have as a consequence of the conservation of mass a vanishing divergence of the mean molecular velocity $\boldsymbol{v}^m$. Making use of the definition $\boldsymbol{v}^m = \sum_i n_i \boldsymbol{v}_i$, the formulae (26) and (29) and the fact that N is uniform, one can write the rotation of $\boldsymbol{v}^m$ as

$$\text{rot } \boldsymbol{v}^m = \frac{1}{N} \sum_{i=1}^{n} \text{rot } \bar{\boldsymbol{J}}_i + \text{rot } \boldsymbol{v} . \tag{115}$$

If we neglect transverse effects (and in particular disregard the Coriolis force), the isotropy of the fluid requires that the phenomenological equations have a form of the type

$$\bar{\boldsymbol{J}}_i = \sum_k L_{ik} \text{ grad } Y_k , \tag{116}$$

where L_{ik} and Y_k are scalar quantities. Then

$$\text{rot } \bar{\boldsymbol{J}}_i = \sum_k (\text{grad } L_{ik}) \wedge (\text{grad } Y_k) . \tag{117}$$

Now the coefficients L_{ik} are functions of the local values of the state parameters, which, in the case of uniform temperature, are the concentrations (and the pressure). This set of state parameters is essentially equivalent to the specification of the quantities Y_k. If, as is generally the case, all grad Y_k are parallel, the right-hand side of (117) vanishes because in each term of the sum the two gradients are parallel.

From (113), (115) and (117) it then follows that rot $\boldsymbol{v}^m = 0$. This result together with div $\boldsymbol{v}^m = 0$ and the boundary condition, yields a vanishing mean molecular velocity. We now find for this case from (26), (102) and (106) with uniform D,

$$\frac{\partial n_1}{\partial t} = D \triangle n_1 , \tag{118}$$

which is again Fourier's differential equation. The condition of constant N is realized in diffusion experiments in perfect gases and, approximately, also in molar diluted systems ($n_1 \ll 1$, $n_2 \simeq 1$).

(iii) In a system where the mean volume velocity $\boldsymbol{v}^0$ can be neglected we have from (105) and (106), supposing D to be practically uniform, if (103) is inserted

$$\frac{\partial \rho_1}{\partial t} = D \triangle \rho_1 \tag{119}$$

and, equivalently since $\rho_1 = M_1 N_1$,

$$\frac{\partial N_1}{\partial t} = D \triangle N_1 \,. \tag{120}$$

We may remark that the condition of vanishing mean volume velocity $\boldsymbol{v}^0$, which we used here, is fulfilled in systems where the partial specific volumes v_i do not appreciably depend on the concentrations (and the pressure) and are therefore practically uniform and constant in time. This is the case in many liquid systems. It is also true in perfect gas mixtures and in mixtures of isomers or isotopes.

We can prove that $\boldsymbol{v}^0$ then vanishes in the following manner. From the identity $\sum_{i=1}^n \rho_i v_i = 1$ one has

$$\sum_{i=1}^{n} \left(v_i \frac{\partial \rho_i}{\partial t} + \rho_i \frac{\partial v_i}{\partial t} \right) = 0 \,. \tag{121}$$

With the help of the conservation of mass law and the definition $\boldsymbol{v}^0 = \sum_i \rho_i v_i \boldsymbol{v}_i$, this identity becomes

$$\operatorname{div} \boldsymbol{v}^0 = \sum_{i=1}^{n} \rho_i \left(\frac{\partial v_i}{\partial t} + \boldsymbol{v}_i \cdot \operatorname{grad} v_i \right) . \tag{122}$$

The expression at the right-hand side vanishes according to the assumption stated above. The rotation of $\boldsymbol{v}^0$ can, according to the definition of $\boldsymbol{v}^0$ and the identity $\sum_i \rho_i v_i = 1$, be written as

$$\operatorname{rot} \boldsymbol{v}^0 = \sum_{i=1}^{n} v_i \operatorname{rot} \boldsymbol{J}_i + \operatorname{rot} \boldsymbol{v} \,, \tag{123}$$

where use has been made of the assumption that v_i is uniform. One can now show in a way completely analogous to the derivation of $\operatorname{rot} \bar{\boldsymbol{J}}_i = 0$ in case (ii) that $\operatorname{rot} \boldsymbol{J}_i$ vanishes. From $\operatorname{rot} \boldsymbol{J}_i = 0$ it follows with (113) and (123) that $\operatorname{rot} \boldsymbol{v}^0 = 0$. This result together with the result $\operatorname{div} \boldsymbol{v}^0 = 0$, found before, and the fact that the normal component of $\boldsymbol{v}^0$ vanishes at the boundary of the vessel, yields the condition of vanishing volume flow $\boldsymbol{v}^0$, which we set out to prove.

(iv) In a mass diluted system ($c_1 \ll 1$, $c_2 \simeq 1$) the velocity $\boldsymbol{v}_2$ will be practically equal to the barycentric velocity $\boldsymbol{v}$. The latter velocity however is negligibly small in a mass diluted system as explained in the first example. Thus we can also neglect $\boldsymbol{v}_2$. From (25), (104) and (105) we obtain for this system the differential equation (114), if again we consider D as practically uniform (for a mass diluted system ρ is practically uniform and constant).

For a molar diluted system ($n_1 \ll 1$, $n_2 \simeq 1$) the velocity $\boldsymbol{v}_2$ will be practically equal to the mean molecular velocity $\boldsymbol{v}^m$ which as argued in the second example can now be neglected. Thus again $\boldsymbol{v}_2$ is negligible. Then we obtain from (26), (104) and (106) as a differential equation formula (118), because for a molar diluted system N is practically uniform and constant.

It is clear that some physical systems can belong to two or more of the four classes enumerated here.

Let us finally write down the connexion between the diffusion coefficient D introduced above and the phenomenological coefficient L^a. From the general formula (56) with (93) and the choices $b_2 = \rho_2 v_2$ and $x_1 = \rho_1$ one has

$$D = L^a \frac{\rho_2 v_2}{a_2^2 T} \frac{\partial \mu_1}{\partial \rho_1}, \tag{124}$$

or with (98) and (99)

$$D = L^a \frac{c_2}{a_2^2 \rho T} \frac{\partial \mu_1}{\partial c_1} = L^a \frac{n_1 n_2}{a_2^2 \rho_1 T} \frac{\partial \mu_1}{\partial n_1}. \tag{125}$$

The chemical potential μ_1 can be written as

$$\mu_1 = \frac{RT}{M_1} \ln f_1 n_1 + \text{const.}, \tag{126}$$

where f_1 is the "activity coefficient", which is equal to unity for ideal mixtures. With this expression for μ_1, formula (125) becomes

$$D = L^a \frac{R n_2}{a_2^2 M_1 \rho_1} \left(1 + \frac{\partial \ln f_1}{\partial \ln n_1}\right), \tag{127}$$

where the last factor is missing for ideal mixtures.

§ 5. *Diffusion in Multi-component Systems*

For the n-component mixture we had established diffusion laws of the form (51) or (55), involving a tensor of $(n-1)^2$ diffusion coefficients. In this section we study the diffusion laws for mixtures with $n > 2$.

The case of ternary mixtures is particularly interesting, because it is the most simple system for which an Onsager relation exists*.

It is convenient to write the diffusion law for the n-component mixture (51) with (53) in the form

$$\boldsymbol{J}^a = - \mathcal{D}^{abx} \cdot \boldsymbol{Y}^{abx}, \tag{128}$$

where the two matrices at the right-hand side are given by

$$\mathcal{D}^{abx} = \mathsf{B}^{ab} \cdot \mathsf{D}^{bx} \cdot (\mathsf{B}^{ab})^{-1}, \tag{129}$$

$$\boldsymbol{Y}^{abx} = \mathsf{B}^{ab} \cdot \text{grad}\, \boldsymbol{x}. \tag{130}$$

It is clear that for binary mixtures, where B^{ab} is an ordinary number, in (129) the factors B^{ab} and $(B^{ab})^{-1}$ drop out, so that one gets back the description of the preceding section. For the general n-component case we shall choose for the basic weights b_k the quantities $\rho_k v_k$ and for the composition parameters x_k the densities ρ_k, because one can show that, just as in the binary mixture case, simple forms of the diffusion laws arise. To simplify the notation we shall omit the indices b and x from the quantities appearing in the preceding formulae, if the choice just mentioned has been made. We thus write

$$\boldsymbol{J}^a = - \mathcal{D}^a \cdot \boldsymbol{Y}^a, \tag{131}$$

$$\mathcal{D}^a = \mathsf{B}^a \cdot \mathsf{D} \cdot (\mathsf{B}^a)^{-1}, \tag{132}$$

$$\boldsymbol{Y}^a = \mathsf{B}^a \cdot \text{grad}\, \boldsymbol{\rho}. \tag{133}$$

For the last formula one obtains with (45), dropping the matrix notation

$$\begin{aligned} \boldsymbol{Y}_i^a &= \text{grad}\, \rho_i + \frac{a_n}{b_n} c_i \sum_{j=1}^{n-1} \frac{b_j}{c_j} \text{grad}\, \rho_j - c_i \sum_{j=1}^{n-1} \frac{a_j}{c_j} \text{grad}\, \rho_j \\ &= \text{grad}\, \rho_i + \frac{a_n}{b_n} c_i \sum_{j=1}^{n} \frac{b_j}{c_j} \text{grad}\, \rho_j - c_i \sum_{j=1}^{n} \frac{a_j}{c_j} \text{grad}\, \rho_j \\ &= \text{grad}\, \rho_i - c_i \sum_{j=1}^{n} \frac{a_j}{c_j} \text{grad}\, \rho_j, \quad (i = 1, 2, \ldots, n-1), \end{aligned} \tag{134}$$

* For an experimental confirmation see: P. J. Dunlop and L. J. Gosting, J. phys. Chem. **63** (1959) 86.

where in the last member $b_k = \rho_k v_k$ has been inserted and the relation $\sum_{j=1}^{n} v_j \delta\rho_j = 0$, valid at constant temperature and pressure, has been used (see Appendix II). For the four conventional choices of a_k one obtains the following table from (134):

	$a_k = c_k$	$a_k = n_k$	$a_k = \rho_k v_k$	$a_k = \delta_{kn}$
$\mathbf{Y}_i^a$	$\rho \operatorname{grad} c_i$	$M_i N \operatorname{grad} n_i$	$\operatorname{grad} \rho_i$ $= M_i \operatorname{grad} N_i$	$\operatorname{grad} \rho_i - \frac{c_i}{c_n} \operatorname{grad} \rho_n$ $= M_i \left\{ \operatorname{grad} N_i - \frac{n_i}{n_n} \operatorname{grad} N_n \right\}$

The expressions in the first two columns are found with the help of (25), (26) and (27). For the choice $a_k = c_k$ we have the usual diffusion flows $\boldsymbol{J}_i$ (counted with respect to the barycentric motion). From (131), (133) and the table we obtain the diffusion laws

$$\boldsymbol{J}_i = - \rho \sum_{k=1}^{n-1} \mathscr{D}_{ik} \operatorname{grad} c_k \,, \quad (i = 1, 2, \ldots, n-1) \,. \tag{135}$$

From comparison of this diffusion law and (51) with the choices $a_k = c_k$ and $x_k = c_k$ we find

$$\mathscr{D}_{ik} = \rho^{-1} D_{ik}^{cc} \,, \tag{136}$$

which shows that the script symbols $\mathscr{D}$ are also essentially examples of the general diffusion coefficients defined by (51).

For the second choice, $a_k = n_k$, one gets from (131) and (133), introducing also the molar diffusion flow (29) and using the second expression in the table,

$$\bar{\boldsymbol{J}}_i^m = - N \sum_{k=1}^{n-1} \frac{M_k}{M_i} \mathscr{D}_{ik}^m \operatorname{grad} n_k \,. \tag{137}$$

This formula contains a coefficient which is also essentially an example of (51), because comparison of (51) and (137) shows that

$$M_k \mathscr{D}_{ik}^m = N^{-1} D_{ik}^{mm} \,, \tag{138}$$

where D^{mm} is the special case with $a_k = n_k$ and $x_k = n_k$ of the general diffusion coefficient D^{ax}.

For the third choice, $a_k = \rho_k v_k$, the set (131) with (133) gives, since B^a is now the unit matrix,

$$J_i^0 = -\sum_{k=1}^{n-1} \mathscr{D}_{ik}^0 \operatorname{grad} \rho_k = -\sum_{k=1}^{n-1} D_{ik} \operatorname{grad} \rho_k\,, \quad (i = 1, 2, \ldots, n-1). \tag{139}$$

The $\mathscr{D}_{ik}^0$ are now (trivially) the diffusion coefficients D_{ik}. With (27) and (29) one finds an alternative form of the last formula

$$\bar{J}_i^0 = -\sum_{k=1}^{n-1} \frac{M_k}{M_i} \mathscr{D}_{ik}^0 \operatorname{grad} N_k = -\sum_{k=1}^{n-1} \frac{M_k}{M_i} D_{ik} \operatorname{grad} N_k\,, \quad (i = 1, 2, \ldots, n-1). \tag{140}$$

The fourth choice gives a somewhat more complicated form of the diffusion laws:

$$J_i^r = -\sum_{k=1}^{n-1} \mathscr{D}_{ik}^r \left(\operatorname{grad} \rho_k - \frac{c_k}{c_n} \operatorname{grad} \rho_n \right), \quad (i = 1, 2, \ldots, n-1), \tag{141}$$

or, equivalently,

$$\bar{J}_i^r = -\sum_{k=1}^{n-1} \frac{M_k}{M_i} \mathscr{D}_{ik}^r \left(\operatorname{grad} N_k - \frac{n_k}{n_n} \operatorname{grad} N_n \right), \quad (i = 1, 2, \ldots, n-1). \tag{142}$$

Just as in the preceding section we can apply these formulae in particular physical situations. We generalize now the four cases, already enumerated for the binary systems, to multi-component systems, enclosed in reservoirs.

(i) Systems of constant density ρ (isotopic mixtures, or mass diluted systems, *i.e.* $c_i \ll 1$ for $i = 1, 2, \ldots, n-1$ and $c_n \simeq 1$). Then $\boldsymbol{v} = 0$ and the conservation of mass law (105) (valid for $i = 1, 2, \ldots, n-1$) becomes with (135), according to the same arguments as in the preceding section:

$$\frac{\partial c_i}{\partial t} = \sum_{k=1}^{n-1} \mathscr{D}_{ik} \triangle c_k\,, \quad (i = 1, 2, \ldots, n-1). \tag{143}$$

(ii) Systems of constant molar density N (perfect gases or molar diluted systems, *i.e.* $n_i \ll 1$ for $i = 1, 2, \ldots, n-1$ and $n_n \simeq 1$).

From $\boldsymbol{v}^m = 0$ and mass conservation (106), but for $i = 1, 2, \ldots, n-1$, with (137) it follows that

$$\frac{\partial n_i}{\partial t} = \sum_{k=1}^{n-1} \frac{M_k}{M_i} \mathscr{D}_{ik}^m \triangle n_k\,, \quad (i = 1, 2, \ldots, n-1)\,. \tag{144}$$

(iii) Systems with negligible mean volume velocity $\boldsymbol{v}^0$ (*i.e.* systems of which the partial specific volumes v_k do not appreciably depend on the concentrations, *e.g.* liquid systems, as explained in the preceding section) one has from (105) with (139)

$$\frac{\partial \rho_i}{\partial t} = \sum_{k=1}^{n-1} D_{ik} \triangle \rho_k\,, \quad (i = 1, 2, \ldots, n-1)\,, \tag{145}$$

or, equivalently, from (106) with (140)

$$\frac{\partial N_i}{\partial t} = \sum_{k=1}^{n-1} \frac{M_k}{M_i} D_{ik} \triangle N_k\,, \quad (i = 1, 2, \ldots, n-1)\,. \tag{146}$$

(iv) In mass diluted systems ($c_i \ll 1$, $i = 1, 2, \ldots, n-1$; $c_n \simeq 1$) one finds from (105) and (141) again the set (143), because $(c_k/c_n)\text{grad}\, \rho_n$ is negligible in such mixtures and $\boldsymbol{v}_n \simeq \boldsymbol{v} = 0$. Similarly for molar diluted systems ($n_i \ll 1$, $i = 1, 2, \ldots, n-1$; $n_n \simeq 1$) from (106) and (142) one obtains the set (144), because now $(n_k/n_n)\text{grad}\, N_n$ is negligible and $\boldsymbol{v}_n \simeq \boldsymbol{v}^m = 0$.

Let us now turn our attention to ternary mixtures ($n = 3$). If we use the diffusion law (139), then one has, explicitly,

$$\boldsymbol{J}_1^0 = -D_{11}\,\text{grad}\, \rho_1 - D_{12}\,\text{grad}\, \rho_2\,, \tag{147}$$

$$\boldsymbol{J}_2^0 = -D_{21}\,\text{grad}\, \rho_1 - D_{22}\,\text{grad}\, \rho_2\,. \tag{148}$$

We have four diffusion coefficients, amongst which one Onsager relation exists. Omitting as usual the indices b and x, which indicates that the choices $b_k = \rho_k v_k$ and $x_k = \rho_k$ are made, this relation is, according to (61) and since $\mathsf{G}^{0\rho} \equiv \mathsf{G}$ is a symmetric matrix [see (91)]:

$$\mathsf{D}\cdot\mathsf{G}^{-1} = \mathsf{G}^{-1}\cdot\tilde{\mathsf{D}}\,, \tag{149}$$

or equivalently

$$G \cdot D = \tilde{D} \cdot G , \tag{150}$$

where [*cf.* (58) and (91)]

$$G = A^0 \cdot \mu^\rho = \tilde{G} . \tag{151}$$

Written explicitly the last two formulae read for the ternary system

$$G_{11} D_{12} + G_{12} D_{22} = D_{11} G_{12} + D_{21} G_{22} , \tag{152}$$

$$G_{ik} = \sum_{m=1}^{2} A^0_{im} \mu^\rho_{mk} = G_{ki} , \quad (i, k = 1, 2) . \tag{153}$$

To verify experimentally the validity of the Onsager relation (152) one must not only measure the four diffusion coefficients D_{11}, D_{12}, D_{21} and D_{22}, but also the equation of state of the mixture, which allows one to compute the thermodynamic quantities μ^ρ_{ik} and v_i which occur in (153).

It may, incidentally, be remarked that only three of the four quantities μ^ρ_{ik} are independent, because from the equality (70), which reads for the ternary mixture ($n = 3$)

$$c_1 \mu^c_{11} - (1 - c_2)\, \mu^c_{12} + (1 - c_1) \mu^c_{21} - c_2 \mu^c_{22} = 0 , \tag{154}$$

and the relation (84), one finds eliminating v_3 with the help of $\sum_{k=1}^{3} \rho_k \, v_k = 1$ the connexion

$$\rho_1 v_2 \mu^\rho_{11} - (1 - \rho_2 v_2) \mu^\rho_{12} + (1 - \rho_1 v_1) \mu^\rho_{21} - \rho_2 v_1 \mu^\rho_{22} = 0 . \tag{155}$$

If we consider ideal mixtures we can specify explicitly the thermodynamic derivatives occurring in the reciprocal relations for the diffusion coefficients. We shall show that these relations are then of a very simple form. As always when an explicit theoretical equation of state is used, it is slightly more convenient to employ molar quantities. We can maintain the form (150) [or, for ternary mixtures (152)], for the Onsager relations, but we write (151), in view of (63), as

$$G = M^{-1} \cdot \bar{G}^{0N} \cdot M^{-1} , \tag{156}$$

with

$$\bar{G}^{0N} = \bar{A} \cdot \bar{\mu}^N , \tag{157}$$

(where, as often, the indices 0 and ρ are omitted, *i.e.* G stands for $G^{0\rho}$ and $\bar{A}$ stands for $\bar{A}^0$). Now for ideal mixtures one has

$$\bar{\mu}_i = RT \ln n_i + \text{const.}, \quad (i = 1, 2, \ldots, n-1) \tag{158}$$

and therefore

$$\bar{\mu}^n_{ik} = RT \frac{\delta_{ik}}{n_i}, \quad (i, k = 1, 2, \ldots, n-1). \tag{159}$$

[These expressions satisfy the identity (73).] With this formula and (85) we obtain

$$\bar{\mu}^N_{ik} = \frac{RT}{N}\left(\frac{\delta_{ik}}{n_i} + \frac{\bar{v}_k}{\bar{v}_n} - 1\right), \quad (i, k = 1, 2, \ldots, n-1). \tag{160}$$

If this result is inserted into (157) one gets, using also (35) with $a_i = N_i \bar{v}_i$,

$$\bar{G}^{0N}_{ik} = \frac{RT}{N}\left\{\frac{\delta_{ik}}{n_i} - \frac{(\bar{v}_i - \bar{v}_n)(\bar{v}_k - \bar{v}_n)}{\bar{v}_n^2} + \frac{\bar{v}_i\bar{v}_k}{n_n\bar{v}_n^2}\right\}, \quad (i, k = 1, 2, \ldots, n-1), \tag{161}$$

so that (156) gives with this expression and with (40)

$$G_{ik} = \frac{RT}{M_i M_k N}\left\{\frac{\delta_{ik}}{n_i} - \frac{(\bar{v}_i - \bar{v}_n)(\bar{v}_k - \bar{v}_n)}{\bar{v}_n^2} + \frac{\bar{v}_i\bar{v}_k}{n_n\bar{v}_n^2}\right\},$$

$$(i, k = 1, 2, \ldots, n-1). \tag{162}$$

These are the quantities which we need in the Onsager relations (150). Note that $\bar{G}^{0N}$ and $G \equiv G^{0\rho}$ are symmetric matrices, as they should be according to (92) and (91).

We can still further specify the ideal mixture by assuming that the system is a mixture of perfect gases. Then we have besides (158) also

$$\bar{v}_1 = \bar{v}_2 = \ldots = \bar{v}_n = N^{-1}, \tag{163}$$

so that (162) simplifies to

$$G_{ik} = \frac{RT}{M_i M_k N}\left(\frac{\delta_{ik}}{n_i} + \frac{1}{n_n}\right), \quad (i, k = 1, 2, \ldots, n-1). \tag{164}$$

If this result is used in the Onsager relations (150), one obtains as connexions between the diffusion coefficients

$$\frac{1}{M_i^2 n_i} D_{ik} - \frac{1}{M_k^2 n_k} D_{ki}$$

$$= \frac{1}{n_n} \sum_{j=1}^{n-1} \frac{1}{M_j} \left(\frac{1}{M_k} D_{ji} - \frac{1}{M_i} D_{jk} \right), \quad (i, k = 1, 2, \ldots, n-1). \tag{165}$$

For the ternary perfect gas mixture ($n = 3$) this gives, with $i = 1$ and $k = 2$, the connexion

$$n_2(1 - n_2)M_2^2 D_{12} - n_1(1 - n_1)M_1^2 D_{21} = n_1 n_2 M_1 M_2 (D_{11} - D_{22}) \tag{166}$$

between the four diffusion coefficients.

§ 6. *Diffusion in Rotating Systems*

If external forces are exerted on a mixture of chemical components, then the diffusion phenomena in such a system are influenced by these forces. An important example is a mixture with charged components under the influence of an electromagnetic field. This case is treated in detail in Chapters XIII and XIV. Here we wish to consider another special case, namely uniformly rotating systems. In such systems the various components are subjected to centrifugal as well as to Coriolis forces, which both arise as a result of the rotation of the system. The force per unit mass on component k in a system, which rotates with an angular frequency ω, is given by

$$\boldsymbol{F}_k = \omega^2 \boldsymbol{r} + 2\boldsymbol{v}_k \wedge \boldsymbol{\omega}, \quad (k = 1, 2, \ldots, n), \tag{167}$$

where $\omega^2 \boldsymbol{r}$ is the centrifugal force ($\boldsymbol{r}$ is the distance from the axis of rotation) and where the vector product $2\, \boldsymbol{v}_k \wedge \boldsymbol{\omega}$ is the Coriolis force, both per unit mass.

If just as in the two preceding sections we consider isothermal, non-reacting n-component mixtures in which viscous phenomena can be neglected, we have for the entropy source strength the expression

$$\sigma = \frac{1}{T} \sum_{k=1}^{n} \boldsymbol{J}_k \cdot \{ \boldsymbol{F}_k - (\text{grad}\, \mu_k)_T \}, \tag{168}$$

as follows from (III.21).

It is to be remarked that the explicit contribution of the forces in this expression vanishes both for the centrifugal and the Coriolis parts. This follows simply from the fact that $\sum_{k=1}^{n} \boldsymbol{J}_k = 0$ (*cf.* Ch. II) for the centrifugal force $\omega^2 \boldsymbol{r}$, which is the same for all components k. For the Coriolis force it can be seen if we write $\boldsymbol{J}_k$ explicitly as $\rho_k(\boldsymbol{v}_k - \boldsymbol{v})$, because then we obtain for the corresponding term in the entropy source strength an expression proportional to

$$\sum_{k=1}^{n} \rho_k(\boldsymbol{v}_k - \boldsymbol{v}) \cdot \boldsymbol{v}_k \wedge \boldsymbol{\omega} = \sum_{k=1}^{n} \rho_k \boldsymbol{v}_k \cdot \boldsymbol{v}_k \wedge \boldsymbol{\omega} - \rho \boldsymbol{v} \cdot \boldsymbol{v} \wedge \boldsymbol{\omega} , \tag{169}$$

where the last term is obtained with the help of the definition $\rho \boldsymbol{v} \equiv \sum_{k=1}^{n} \rho_k \boldsymbol{v}_k$. All terms at the right-hand side vanish, because of the identities

$$\boldsymbol{v}_k \cdot \boldsymbol{v}_k \wedge \boldsymbol{\omega} = 0 , \quad \boldsymbol{v} \cdot \boldsymbol{v} \wedge \boldsymbol{\omega} = 0 . \tag{170}$$

The vanishing of the term $\sum_k \boldsymbol{J}_k \cdot \boldsymbol{F}_k$ corresponds to the "reversible" character which one would expect for motion under the influence of purely mechanical forces. It does not mean that one gets back the same value for the entropy source strength, as in the absence of external forces. At mechanical equilibrium the influence of the external forces is implicitly present in the other term of the entropy source strength, involving $(\text{grad}\ \mu_k)_T$. This gradient will now contain, in contrast with the case of § 2, a term proportional to the pressure gradient which arises itself as a result of the external forces. Before we will show this explicitly, we make use of Prigogine's theorem (Ch. V, § 2) for systems in mechanical equilibrium and write instead of (168)

$$\sigma = \frac{1}{T} \sum_{k=1}^{n} \boldsymbol{J}_k^a \cdot \{ \boldsymbol{F}_k - (\text{grad}\ \mu_k)_T \} , \tag{171}$$

where $\boldsymbol{J}_k^a$ is the diffusion flow (22), counted with respect to the arbitrary velocity $\boldsymbol{v}^a$. It may be stressed here that (171) is obtained from (168) without deleting the vanishing part $\sum_k \boldsymbol{J}_k \cdot \boldsymbol{F}_k$. That this procedure is the correct way of applying Prigogine's theorem follows from its derivation (Ch. V, § 2), which is based on the relation (V.2). This relation contains the complete bracket which appears in (168). Therefore $\boldsymbol{F}_k$ had to be retained in (168) in order to obtain the correct result (171).

The entropy source strength (171) can also be written as

$$\sigma = \frac{1}{T} \sum_{k=1}^{n} \boldsymbol{J}_k^a \cdot \{ \boldsymbol{F}_k - v_k \operatorname{grad} p_k - (\operatorname{grad} \mu_k)_{p,T} \} . \tag{172}$$

We can express the gradient of the pressure p in terms of the mechanical forces because at mechanical equilibrium we have from (V.1)

$$\operatorname{grad} p = \sum_{k=1}^{n} \rho_k \boldsymbol{F}_k = \rho(\omega^2 \boldsymbol{r} + 2\boldsymbol{v} \wedge \boldsymbol{\omega}) , \tag{173}$$

where the last member is obtained with (167) and $\sum_k \rho_k \boldsymbol{v}_k = \rho \boldsymbol{v}$.

Furthermore we have the identity

$$\sum_{k=1}^{n} \boldsymbol{J}_k^a \cdot \boldsymbol{v}_k \wedge \boldsymbol{\omega} = \sum_{k=1}^{n} \boldsymbol{J}_k^a \cdot \boldsymbol{v} \wedge \boldsymbol{\omega} , \tag{174}$$

which follows with $\rho \boldsymbol{v} = \sum_k \rho_k \boldsymbol{v}_k$ from (22) and (170). Now the expression (172) becomes with the help of (167), (173) and (174)

$$\sigma = \frac{1}{T} \sum_{k=1}^{n} \boldsymbol{J}_k^a \cdot \{ (1 - \rho v_k) (\omega^2 \boldsymbol{r} + 2\boldsymbol{v} \wedge \boldsymbol{\omega}) - (\operatorname{grad} \mu_k)_{p,T} \} . \tag{175}$$

Just as in section 2 of this chapter we can eliminate $\boldsymbol{J}_n^a$ and $(\operatorname{grad} \mu_n)_{p,T}$ with the help of (30) and (31). Furthermore we can also eliminate the partial specific volume v_n by means of the identity

$$\sum_{k=1}^{n} c_k(1 - \rho v_k) = 0 , \tag{176}$$

which follows from (25) and $\sum_k \rho_k v_k = 1$. In such a way we obtain an expression for the entropy source strength of the form (32), with the thermodynamic forces

$$\boldsymbol{X}_k^a = \sum_{j=1}^{n-1} A_{kj}^a \{ (1 - \rho v_j) (\omega^2 \boldsymbol{r} + 2\boldsymbol{v} \wedge \boldsymbol{\omega}) - (\operatorname{grad} \mu_j)_{p,T} \} / T ,$$
$$(k = 1, 2, \ldots, n-1) , \tag{177}$$

where the coefficients A_{kj}^a are again given by (34).

Let us first consider the case of thermodynamic *equilibrium*, *i.e.* the simultaneous vanishing of all fluxes and thermodynamic forces. This case is of practical interest for determining molecular masses by means of *sedimentation* in an ultracentrifuge. From the form (177) we may conclude that at the sedimentation equilibrium the bracket expressions vanish because the thermodynamic forces are zero:

$$(1 - \rho v_k)\,(\omega^2 \boldsymbol{r} + 2\boldsymbol{v} \wedge \boldsymbol{\omega}) - (\text{grad}\, \mu_k)_{p,T} = 0\,, \quad (k = 1, 2, \ldots, n-1). \tag{178}$$

For a closed vessel the velocities $\boldsymbol{v}_k$, and consequently also the barycentric velocity $\boldsymbol{v}$, vanish at equilibrium. Therefore (178) becomes

$$(1 - \rho v_k)\omega^2 \boldsymbol{r} - (\text{grad}\, \mu_k)_{p,T} = 0\,, \quad (k = 1, 2, \ldots, n-1)\,. \tag{179}$$

This relation allows us to describe the distribution of the chemical components in sedimentation equilibrium. Let us first take the example of a binary mixture ($n = 2$). Then (179) gives

$$(1 - \rho v_1)\omega^2 \boldsymbol{r} - (\text{grad}\, \mu_1)_{p,T} = 0\,, \tag{180}$$

where the first factor could alternatively be written as $\rho_2(v_2 - v_1)$, involving the difference of partial specific volumes of the two components. The gradient of the chemical potential can be expressed as a gradient of some parameter which characterizes the composition, for instance n_1, c_1, N_1 or ρ_1, by means of

$$(\text{grad}\, \mu_1)_{p,T} = \left(\frac{\partial \mu_1}{\partial n_1}\right)_{p,T} \text{grad}\, n_1 = \left(\frac{\partial \mu_1}{\partial c_1}\right)_{p,T} \text{grad}\, c_1\,, \quad \text{etc.} \tag{181}$$

With the help of these connexions and the formulae (97)–(99) and (126), the equation (180) gives the results

$$\frac{\text{grad}\, n_1}{n_1} = M_1(1 - \rho v_1)\left(1 + \frac{\partial \ln f_1}{\partial \ln n_1}\right)^{-1} \frac{\omega^2 \boldsymbol{r}}{RT}\,, \tag{182}$$

$$\frac{\text{grad}\, c_1}{c_1} = M_1\left\{1 - c_1\left(1 - \frac{M_2}{M_1}\right)\right\}(1 - \rho v_1)\left(1 + \frac{\partial \ln f_1}{\partial \ln n_1}\right)^{-1} \frac{\omega^2 \boldsymbol{r}}{RT}\,, \tag{183}$$

$$\frac{\text{grad}\, \rho_1}{\rho_1} = \frac{\text{grad}\, N_1}{N_1} = M_1\,(1 - \rho v_1)\,(1 - c_1 \rho v_1)\,\frac{1}{n_2}\left(1 + \frac{\partial \ln f_1}{\partial \ln n_1}\right)^{-1} \frac{\omega^2 \boldsymbol{r}}{RT}\,. \tag{184}$$

From these equations the molecular mass M_1 can be calculated, if the other relevant quantities have been measured.

For ideal mixtures in which the activity coefficient $f_1 = 1$, the factor containing this quantity drops out. Then formula (182) reduces to the well-known Svedberg equation* for the sedimentation equilibrium. In practical applications to the determination of the molecular mass M_1 the expressions on the left of (182)–(184) are often taken to be equal. From the above it can be seen which approximations are necessary in order to justify this procedure. In practical cases one has usually $M_2 \ll M_1$. Generally the solution will also be molar diluted, so that $n_2 \simeq 1$. If the solution is so highly diluted that also $c_2 \simeq 1$ the right-hand sides of (183) and (184) reduce to the right-hand side of (182).

For n-component mixtures we have, if we use the molar fractions n_i as composition parameters,

$$\operatorname{grad} \mu_j = \sum_{i=1}^{n-1} \frac{\partial \mu_j}{\partial n_i} \operatorname{grad} n_i \,. \tag{185}$$

The equilibrium condition (179) with (185) gives $n - 1$ equations from which one can solve the $n - 1$ quantities grad n_i, which determine the distribution of matter in the system. In such a way, for ideal mixtures, where one has formula (159) for the derivative occurring in (185), one finds

$$\frac{\operatorname{grad} n_j}{n_j} = M_j(1 - \rho v_j) \frac{\omega^2 \boldsymbol{r}}{RT} \,. \tag{186}$$

With the help of (25), (26) and (27) we can express the gradients of the mass fractions in terms of the gradients of the molar fractions:

$$\frac{\operatorname{grad} c_j}{c_j} = \frac{\operatorname{grad} n_j}{n_j} - \sum_{m=1}^{n-1} c_m \frac{M_m - M_n}{M_m} \frac{\operatorname{grad} n_m}{n_m} \,. \tag{187}$$

If we insert (186) into this formula we obtain the following result:

$$\frac{\operatorname{grad} c_j}{c_j} = \left\{ M_j(1 - \rho v_j) - \sum_{m=1}^{n-1} c_m (M_m - M_n)(1 - \rho v_m) \right\} \frac{\omega^2 \boldsymbol{r}}{RT} \,, \tag{188}$$

* Th. Svedberg und K. O. Pedersen, Die Ultrazentrifuge (Steinkopff, Dresden und Leipzig, 1940).

which gives the mass distribution of an ideal n-component mixture at sedimentation equilibrium.

Another method to determine molecular masses is by studying the *sedimentation velocity*. The actual measurement in the ultracentrifuge consists in following the sedimentation rate of the layer between solution and pure solvent. In this case the fluxes and the thermodynamic forces are both different from zero. We have the phenomenological equations (36) with the thermodynamic forces (177). In practice one can neglect the Coriolis force. Then for an isotropic system the tensors L^a_{ik} reduce to scalars L^a_{ik}. In this way the phenomenological equations become

$$\boldsymbol{J}^a_i = \sum_{k,\,j=1}^{n-1} L^a_{ik} A^a_{kj} \left\{ (1 - \rho v_j)\omega^2 \boldsymbol{r} - \sum_{m=1}^{n-1} \mu^x_{jm} \operatorname{grad} x_m \right\} / T\,, \qquad (i = 1, 2, \ldots, n-1)\,, \tag{189}$$

where the abbreviation (57) has been used. If we introduce diffusion coefficients D^{bx}_{ik} instead of the phenomenological coefficients L^a_{ik} by means of the relationship (56), these phenomenological equations get the form

$$\boldsymbol{J}^a_i = \sum_{k,\,j=1}^{n-1} B^{ab}_{ik} D^{bx}_{kj} \left\{ \sum_{m=1}^{n-1} (\mu^x)^{-1}_{jm} (1 - \rho v_m)\omega^2 \boldsymbol{r} - \operatorname{grad} x_j \right\}, \qquad (i = 1, 2, \ldots, n-1)\,. \tag{190}$$

For the case of a binary mixture ($n = 2$) this becomes

$$\boldsymbol{J}^a_1 = B^{ab}_{11} D^{bx} \{ (\mu^x)^{-1} (1 - \rho v_1)\omega^2 \boldsymbol{r} - \operatorname{grad} x_1 \}\,. \tag{191}$$

With (93) and the choices $x_1 = \rho_1$ and $b_2 = \rho_2 v_2$, which were discussed in § 4 of this chapter, one has

$$\boldsymbol{J}^a_1 = \frac{a_2}{\rho_2 v_2} D \{ (\mu^\rho)^{-1} (1 - \rho v_1)\omega^2 \boldsymbol{r} - \operatorname{grad} \rho_1 \}\,, \tag{192}$$

where D was just called the diffusion coefficient. As long as the system is still far from equilibrium the system is homogeneous over a large region between the boundary (separating solution and pure solvent) and the outer wall, *i.e.* grad $\rho_1 = 0$. Therefore, in this region (192) becomes

$$J_1^a = \frac{a_2}{\rho_2 v_2} D\,(\mu^\rho)^{-1}\,(1 - \rho v_1)\omega^2 \boldsymbol{r}\,. \tag{193}$$

With the help of (22), (25), (98), (99) and (126) this formula can be written as

$$M_1 = \frac{|\boldsymbol{v}_1 - \boldsymbol{v}^a|}{\omega^2 r} \frac{RTn_2}{D(1 - \rho v_1)a_2} \left(1 + \frac{\partial \ln f_1}{\partial \ln n_1}\right), \tag{194}$$

which shows which quantities must be known in order to calculate the molecular mass M_1. In particular one must measure the first factor, which is sometimes called the sedimentation coefficient. Two particular choices of the weights a_i and the corresponding reference velocity $\boldsymbol{v}^a$ give useful results. The first of these is obtained by choosing for $\boldsymbol{v}^a$ the mean molar velocity $\boldsymbol{v}^m$. [The weight a_2 is then n_2 (see § 2).] Expression (194) becomes then

$$M_1 = \frac{|\boldsymbol{v}_1 - \boldsymbol{v}^m|}{\omega^2 r} \frac{RT}{D(1 - \rho v_1)} \left(1 + \frac{\partial \ln f_1}{\partial \ln n_1}\right). \tag{195}$$

This choice is practical for molar diluted systems, because then $\boldsymbol{v}^m$ vanishes (*cf.* § 4).

The other choice of practical interest is $a_2 = \rho_2 v_2$, which means that $\boldsymbol{v}^a$ is now the mean volume velocity (see § 2). Then (194) becomes, if v_2 is eliminated with the help of $\rho_1 v_1 + \rho_2 v_2 = 1$:

$$M_1 = \frac{|\boldsymbol{v}_1 - \boldsymbol{v}^0|}{\omega^2 r} \frac{RTn_2}{D(1 - \rho v_1)(1 - c_1 \rho v_1)} \left(1 + \frac{\partial \ln f_1}{\partial \ln n_1}\right). \tag{196}$$

As explained in § 4 of this chapter we may neglect the mean volume velocity $\boldsymbol{v}^0$ in a number of cases, in particular if the system is a liquid confined to a vessel.

In the same way as after formulae (182)–(184) one can make the approximation $n_2 \simeq 1$, or (if, as usual in centrifugation experiments, $M_1 \gg M_2$) the more restrictive approximation $c_2 \simeq 1$. Then (195) and (196) respectively lead, in the case of an ideal solution ($f_1 = 1$), to the well-known Svedberg equation:

$$M_1 = \frac{|\boldsymbol{v}_1|}{\omega^2 r} \frac{RT}{D(1 - \rho v_1)}\,. \tag{197}$$

The equations (196) and (197) are valid throughout the region where grad c_1 is negligible. In practice one measures $|\boldsymbol{v}_1|$ by observing a part of the boundary layer of which the motion also follows from (196) or (197).

Both the Svedberg equations for the sedimentation equilibrium and the sedimentation velocity have thus been derived and their limits of validity indicated. In particular all quantities pertaining to the non-equilibrium case have been defined just as rigorously as those for the equilibrium case.

The sedimentation process in multi-component mixtures can be studied with the help of equation (190)*. A convenient choice of the weights a_k and b_k is again $\rho_k v_k$. Then the velocity $\boldsymbol{v}^a$ is the mean volume velocity which as discussed above can often be put equal to zero. Then, using the densities ρ_j for the parameter x_j, formula (190) reduces to

$$\rho_i \boldsymbol{v}_i = \sum_{j=1}^{n-1} D_{ij} \left\{ \sum_{m=1}^{n-1} (\mu^\rho)^{-1}_{jm} (1 - \rho v_m)\omega^2 \boldsymbol{r} \right\}, \quad (i = 1, 2, \ldots, n-1) \tag{198}$$

in the region where the gradients of ρ_j can be neglected. These equations permit one to calculate molecular masses in an analogous way as explained above for the binary mixtures, if the relevant quantities which occur are measured.

We have seen in the foregoing how we could use diffusion coefficients instead of the phenomenological coefficients. Quite often still a different kind of quantities, the so-called *mobilities*, are used instead of the phenomenological coefficients, especially for the description of diffusion phenomena in systems on which external forces are exerted. We shall indicate in the following how the mobilities can be defined in general and how these quantities are related to the diffusion coefficients.

The phenomenological equations had the form (189). Now instead of putting these equations in the form (190) we can alternatively write

$$\boldsymbol{J}_i^a = \rho_i M_i \sum_{j=1}^{n-1} U_{ij}^a (1 - \rho v_j)\omega^2 \boldsymbol{r} - \sum_{j,m=1}^{n-1} B_{ij}^{ab} D_{jm}^{bx} \operatorname{grad} x_m, \quad (i = 1, 2, \ldots, n-1), \tag{199}$$

* G. J. Hooyman, H. Holtan Jr., P. Mazur and S. R. de Groot, Physica **19** (1953) 1095.
G. J. Hooyman, P. Mazur and S. R. de Groot, Kolloid Zeitschr. **140** (1955) 165.
G. J. Hooyman, Physica **22** (1956) 761.
O. Lamm, J. phys. Chem. **59** (1955) 1149.

where the coefficients U^a_{ij} are called the *mobilities*. They have the dimension of a velocity per unit force on a mole (or on N molecules). It is clear from comparison of this equation with (189) that the mobilities are related to the phenomenological coefficients in the following way

$$\frac{1}{T}\sum_{k=1}^{n-1} L^a_{ik} A^a_{kj} = \rho_i M_i U^a_{ij}, \quad (i, j = 1, 2, \ldots, n-1). \tag{200}$$

The right-hand member of this relation transforms under changes of the basic weights a_k in the same way as the flux $\boldsymbol{J}^a$ or the diffusion coefficients D^{ax} [(44) and (53)]. It is to be noted that the mobilities U^a_{ij} are referred to the "arbitrary" reference velocities $\boldsymbol{v}^a$, which occur in $\boldsymbol{J}^a_i$, whereas the diffusion coefficients are referred to a "basic" velocity $\boldsymbol{v}^b$ which is chosen once for all. If we eliminate the phenomenological coefficients from (56) and (200) [or compare (199) with (190)], then the connexion between mobilities and diffusion coefficients is established:

$$\rho_i M_i U^a_{ij} = \sum_{k,m=1}^{n-1} B^{ab}_{ik} D^{bx}_{km} (\mu^x)^{-1}_{mj}, \quad (i, j = 1, 2, \ldots, n-1). \tag{201}$$

This connexion is a generalized Fokker–Einstein relationship. If we make the choice $b_k = v_k\rho_k$ and $x_k = \rho_k$, as was also done in the preceding sections, then (201) reads

$$\rho_i M_i U^a_{ij} = \sum_{k,m=1}^{n-1} B^{a0}_{ik} D_{km} (\mu^\rho)^{-1}_{mj}, \quad (i, j = 1, 2, \ldots, n-1). \tag{202}$$

For the binary mixture ($n = 2$) we have $B^{a0}_{11} = a_2/\rho_2 v_2$. Then the relation between the mobility U^a_{11}, which we shall denote as U^a, and the diffusion coefficient becomes

$$\rho_1 M_1 U^a = \frac{a_2}{\rho_2 v_2} D(\mu^\rho)^{-1}. \tag{203}$$

With (25), (98), (99) and (126) we can write this as

$$D = \frac{n_2}{a_2} RTU^a \left(1 + \frac{\partial \ln f_1}{\partial \ln n_1}\right). \tag{204}$$

For an ideal system the activity coefficient $f_1 = 1$, so that the preceding relation becomes

$$D = \frac{n_2}{a_2} RTU^a . \tag{205}$$

Finally if we choose the molar description, where the weight factor $a_2 = n_2$, we have

$$D = RTU^m , \tag{206}$$

the Fokker–Einstein relation. Historically this relation arose for the first time in the study of Brownian motion of heavy particles in a solvent. We then have a case of a molar diluted binary system (mole fraction of the Brownian particles $n_1 \ll 1$), for which the mean molar velocity $\boldsymbol{v}^m$ is negligible, so that the differential equation (118) describes its behaviour.

§ 7. *Thermal Diffusion* (*Soret Effect*) *and Dufour Effect*

In this section we study the phenomena which arise in a mixture if both the concentrations and the temperature are non-uniform over the system. We shall consider isotropic fluids in which viscous phenomena may be neglected. Furthermore no external forces are supposed to be present. Under these conditions the pressure is uniform over the system if we assume that mechanical equilibrium is rapidly established, as is the case for systems confined to closed reservoirs (*cf.* § 2). The entropy production follows then from (IV.13)

$$\sigma = - \boldsymbol{J}'_q \cdot \frac{\text{grad } T}{T^2} - \sum_{k=1}^{n-1} \boldsymbol{J}_k \cdot \frac{\{ \text{grad } (\mu_k - \mu_n) \}_{T,p}}{T} . \tag{207}$$

With the help of the Gibbs–Duhem relation

$$\sum_{k=1}^{n} c_k \delta\mu_k = 0 , \quad (p \text{ and } T \text{ constant}) , \tag{208}$$

we can eliminate μ_n from (207). This gives

$$\sigma = - \boldsymbol{J}'_q \cdot \frac{\text{grad } T}{T^2} - \sum_{k,m=1}^{n-1} \boldsymbol{J}_k \cdot \frac{A_{km}(\text{grad } \mu_m)_{p,T}}{T} , \tag{209}$$

where

$$A_{km} = \delta_{km} + \frac{c_m}{c_n}, \quad (k, m = 1, 2, \ldots, n-1). \tag{210}$$

(This is a result which follows also from (34) if we put $a_k = c_k$ as it should be done for the present case where the barycentric velocity has been chosen as reference velocity.) We can express the gradients of the chemical potentials in terms of the concentration gradients

$$(\text{grad}\, \mu_m)_{p,T} = \sum_{i=1}^{n-1} \mu_{mi}^c \,\text{grad}\, c_i, \quad (m = 1, 2, \ldots, n-1). \tag{211}$$

The matrix elements μ_{mi}^c are abbreviations for the derivatives $(\partial\mu_m/\partial c_i)_{p,T,c_j}$, $(j \neq i)$. This is also the matrix (57) for the choice $x_i = c_i$ as concentration parameter. The phenomenological equations for the fluxes and thermodynamical forces, which appear in (209) with (210) and (211) are now

$$\boldsymbol{J}'_q = -L_{qq}\frac{\text{grad}\, T}{T^2} - \sum_{k,m,j=1}^{n-1} L_{qk}\frac{A_{km}\mu_{mj}^c \,\text{grad}\, c_j}{T}, \tag{212}$$

$$\boldsymbol{J}_i = -L_{iq}\frac{\text{grad}\, T}{T^2} - \sum_{k,m,j=1}^{n-1} L_{ik}\frac{A_{km}\mu_{mj}^c \,\text{grad}\, c_j}{T}, \quad (i = 1, 2, \ldots, n-1), \tag{213}$$

with the Onsager relations

$$L_{iq} = L_{qi}, \quad L_{ik} = L_{ki}, \quad (i, k = 1, 2, \ldots, n-1). \tag{214}$$

The coefficients L_{qq} and L_{ik} are related to the heat conductivity and to the diffusion coefficients in the manner indicated in the preceding sections. The coefficients L_{iq} are characteristic for the phenomenon of *thermal diffusion, i.e.* a flow of matter caused by a temperature gradient, as equation (213) shows. It is usually called the Soret effect in liquids*. A reciprocal phenomenon, *viz.* a heat flow caused by concentration gradients, also exists, as is clear from (212). This effect is called the *Dufour-effect* and its magnitude depends on the coefficients L_{qk}**.

* C. Ludwig, S-B Akad. Wiss., Vienna **20** (1856) 539.
Ch. Soret, Arch. Sci. phys. nat., Geneve **2** (1879) 48; **4** (1880) 209;
Comptes Rendus Acad. Sci., Paris **91** (1880) 279.
S. Chapman and F. W. Dootson, Phil. Mag. **33** (1917) 248.

** L. Dufour, Arch. Sci. phys. nat., Geneve **45** (1872) 9; Ann. Phys. [5] **28** (1873) 490.
K. Clusius and L. Waldmann, Naturw. **30** (1942) 711.

In a closed reservoir with an applied temperature gradient, concentration gradients will build up until in a final, stationary state the diffusion flows (213) vanish. Then the concentration gradients are given by

$$\frac{\operatorname{grad} c_j}{\operatorname{grad} T} = - \frac{\sum_{m,k,i=1}^{n-1} (\mu^c)^{-1}_{jm} A^{-1}_{mk} L^{-1}_{ki} L_{iq}}{T}, \quad (j = 1, 2, \ldots, n-1), \tag{215}$$

where the reciprocal matrix of (210) is given by

$$A^{-1}_{mk} = \delta_{mk} - c_k, \quad (m, k = 1, 2, \ldots, n-1) \tag{216}$$

[which is a special case with $a_k = c_k$ of formula (47)].

Let us now consider in somewhat more detail the case of a *binary mixture* ($n = 2$). Then it follows from (210) that

$$A_{11} = \frac{1}{c_2} \tag{217}$$

and the phenomenological equations (212) and (213) become

$$J'_q = - L_{qq} \frac{\operatorname{grad} T}{T^2} - L_{q1} \frac{\mu^c_{11}}{c_2 T} \operatorname{grad} c_1, \tag{218}$$

$$J_1 = - L_{1q} \frac{\operatorname{grad} T}{T^2} - L_{11} \frac{\mu^c_{11}}{c_2 T} \operatorname{grad} c_1, \tag{219}$$

with the Onsager relation

$$L_{1q} = L_{q1} \tag{220}$$

and the inequalities which follow from the fact that the entropy source strength is positive definite

$$L_{qq} \geqslant 0, \quad L_{11} \geqslant 0, \quad L_{qq} L_{11} - \tfrac{1}{4}(L_{1q} + L_{q1})^2 = L_{qq} L_{11} - (L_{1q})^2 \geqslant 0. \tag{221}$$

Instead of the phenomenological coefficients occurring in (218) and (219) we can introduce the following set of coefficients

$$\lambda = \frac{L_{qq}}{T^2}, \qquad \text{(heat conductivity)}, \tag{222}$$

$$D'' = \frac{L_{q1}}{\rho c_1 c_2 T^2}, \qquad \text{(Dufour coefficient)}, \tag{223}$$

$$D' = \frac{L_{1q}}{\rho c_1 c_2 T^2}, \qquad \text{(thermal diffusion coefficient)}, \tag{224}$$

$$D = \frac{L_{11}\mu_{11}^c}{\rho c_2 T}, \qquad \text{(diffusion coefficient)}. \tag{225}$$

The "direct" effects of heat conductivity and diffusion have already been discussed in the preceding sections, and the definitions (222) and (225) were already given in formulae (3) and (125) with $a_2 = c_2$. The cross-phenomena of thermal diffusion and Dufour effect are now given by the coefficient D' and D''. With the use of the definitions (222)–(225) we can write (218) and (219) in the form

$$\boldsymbol{J}_q' = -\lambda \operatorname{grad} T - \rho_1 \mu_{11}^c T D'' \operatorname{grad} c_1, \tag{226}$$

$$\boldsymbol{J}_1 = -\rho c_1 c_2 D' \operatorname{grad} T - \rho D \operatorname{grad} c_1. \tag{227}$$

Another useful form of these phenomenological equations is obtained by introducing molar quantities. The partial molar chemical potential $\bar{\mu}_1$ is equal to $M_1\mu_1$. Furthermore one has

$$(\operatorname{grad} \mu_1)_{p,T} = \mu_{11}^c \operatorname{grad} c_1 = \mu_{11}^n \operatorname{grad} n_1. \tag{228}$$

Now if we use also (97) in equation (226), and (29) and (98) in equation (227), the following forms are obtained

$$\boldsymbol{J}_q' = -\lambda \operatorname{grad} T - N_1 \bar{\mu}_{11}^n T D'' \operatorname{grad} n_1, \tag{229}$$

$$\bar{\boldsymbol{J}}_1 \frac{\rho}{M_2 N} = -N n_1 n_2 D' \operatorname{grad} T - ND \operatorname{grad} n_1 \left(= \boldsymbol{J}_1 \frac{\rho}{M_1 M_2 N}\right). \tag{230}$$

The right-hand sides of these equations contain "molar" quantities instead of the "mass" quantities in (226) and (227), but the same phenomenological coefficients. It may be noted that $\bar{\boldsymbol{J}}_1$, just as $\boldsymbol{J}_1$, is counted with respect to the barycentric velocity as reference velocity. The forms (229) and (230) are usually employed in the kinetic theory of gases.

The Onsager relation (220) implies for the coefficients (223) and (224)

$$D' = D'' . \tag{231}$$

The inequalities (221) imply for the coefficients (222)–(225)

$$\lambda \geqslant 0 \, , \; D \geqslant 0 \, , \; (D')^2 \leqslant \frac{\lambda D}{T \rho c_1^2 c_2 \mu_{11}^c} . \tag{232}$$

To obtain the second of these inequalities the thermodynamic stability condition $\mu_{11}^c \geqslant 0$ was used. In ideal systems one has formula (159), which gives with (25), (26), (27) and (98)

$$\mu_{11}^c = \frac{RT}{c_1 \{ M_1 - c_1(M_1 - M_2) \}} . \tag{233}$$

The last inequality of (232) can then be written as

$$(D')^2 \leqslant \frac{\lambda D \{ M_1 - c_1(M_1 - M_2) \}}{RT^2 \rho c_1 c_2} . \tag{234}$$

For a system enclosed in a reservoir we may usually neglect convection phenomena. Furthermore if the concentration gradients are not too large we may consider the overall density ρ as roughly uniform. Then the differential equations, which are obtained from the energy law (II.34) and the mass law (II.13) together with the phenomenological equations (226) and (227), have relatively simple forms

$$\rho c_p \frac{\partial T}{\partial t} = - \operatorname{div} \boldsymbol{J}_q' = \operatorname{div} (\lambda \operatorname{grad} T + \rho_1 \mu_{11}^c T D'' \operatorname{grad} c_1) , \tag{235}$$

$$\frac{\partial c_1}{\partial t} = - \operatorname{div} \frac{\boldsymbol{J}_1}{\rho} = \operatorname{div} (c_1 c_2 D' \operatorname{grad} T + D \operatorname{grad} c_1) . \tag{236}$$

Let us now consider the experimental situations in which the cross-coefficients D' and D'' are measured. For measuring the thermal diffusion coefficient D' one fixes a temperature difference between two walls of a closed vessel. The temperature distribution as a function of space coordinates and of time can be found from (235). In practice the term with grad c_1 may be neglected in (235) so that one has to solve the heat conduction differential equation. (In many cases a temperature distribution which is linear in one coordinate direction and independent of the two other coordinates and of the time is a sufficient approximation to the solution of the heat conduction equation.) With this result for T one can then solve (236) and find the concentration as a function of the coordinates and of time*. If at a time $t = 0$ the concentration in the vessel is uniform (grad $c_1 = 0$), then as time goes on a concentration gradient will be built up. Finally a stationary state will be reached in which the diffusion flow (227) or (230) vanishes. Then

$$\frac{D'}{D} = -\frac{1}{c_1 c_2} \frac{\text{grad } c_1}{\text{grad } T} = -\frac{1}{n_1 n_2} \frac{\text{grad } n_1}{\text{grad } T}. \tag{237}$$

This quotient is called the "Soret coefficient" and sometimes denoted as s_T. Other quantities, also characteristic for the separation achieved in thermal diffusion experiments, used in litterature are the "thermal diffusion factor" α and the "thermal diffusion ratio" k_T defined as

$$\alpha = \frac{D'}{D} T, \quad k_T = c_1 c_2 \frac{D'}{D} T. \tag{238}$$

The mathematical analysis of the behaviour of the concentration as a function of space coordinates and of time shows that the final stationary state is reached after a time large compared to a characteristic time Θ, given by

$$\Theta = \frac{a^2}{\pi D}, \tag{239}$$

where a is the distance between the two walls of different temperature. Roughly speaking one can say that the concentration gradient builds up exponentially in time with a characteristic time constant (239),

* S. R. de Groot, Physica **9** (1952) 699.
S. R. de Groot, L'effet Soret, diffusion thermique dans les phases condensées (North-Holland Publishing Company, Amsterdam, 1945).

except in an initial period where a more complicated dependence on time exists.

The values of the Soret coefficient (237) measured, turn out to be of the order of magnitude of 10^{-3} to 10^{-5} reciprocal degrees, both in gaseous and liquid mixtures. With the knowledge of D, the diffusion coefficient, one finds then the value of D', the thermal diffusion coefficient. Since the diffusion coefficients D are of the order of 10^{-5} cm^2/sec in liquids and 10^{-1} cm^2/sec in gases, the thermal diffusion coefficient D' is about 10^{-8} to 10^{-10} cm^2 sec^{-1} degree^{-1} in liquids and 10^{-4} to 10^{-6} cm^2 sec^{-1} degree^{-1} in gases.

The Dufour coefficient D'' can be measured by mixing two different fluids and by determining the ensuing temperature gradient. In the differential equation (236) the term with grad T can be neglected, and the resulting diffusion equation can be solved, so that c_1 is known as a function of the space coordinates and of time. Then with this result the equation (235) can be solved and one finds the temperature field. It turns out that if one allows two different fluid substances, which were at the same temperature initially, to diffuse into each other, a maximal temperature difference ΔT is reached after a time $c_p \rho a^2/\lambda$ with a the linear dimension of the reservoir. This temperature difference corresponds to a concentration difference Δc_1, such that

$$\frac{D''}{\lambda} \simeq -\frac{1}{T\rho_1\mu_{11}^c}\frac{\Delta T}{\Delta c_1} = -\frac{1}{TN_1\bar{\mu}_{11}^n}\frac{\Delta T}{\Delta n_1}. \tag{240}$$

Experimentally one finds in gases temperature differences ΔT of the order of one degree centigrade. These turn out to correspond to values of D'' equal to those of D'. The Onsager relation (231) is thus experimentally confirmed*. In liquids the Dufour effect is much harder to detect. This can be seen from (240), since $D'' = D'$ in liquids is about 10^4 times smaller than in gases, λ is about 10^2 times larger than in gases and the density ρ_1 is about 10^3 times larger than in gases. The other quantities are of the same order of magnitude in liquids and in gases. So according to (240) the effect ΔT will be about 10^3 times smaller in liquids than in gases, that is 10^{-3} degree centigrade, which is indeed difficult to observe.

We can, with the help of the values for the physical constants,

* L. Waldmann, J. Phys. Rad. **7** (1946) 129; Z. Naturf. **4A** (1949) **105**.

check whether the inequalities (232) or (234) are satisfied. It turns out that indeed for gases the left-hand side of (234) is about 10^{-2} or 10^{-3} of the right-hand side, and for diluted solutions this proportion is 10^{-4} to 10^{-5}.

The well-known thermogravitational method for the separation of isotopes, designed by Clusius and Dickel*, is a means to enhance the separation effect, produced by thermal diffusion, through gravitational convection currents. In the theoretical description of this process it is necessary to take into account the viscous flow of the fluid** and also the convection terms in the differential equations.

It is of some interest to consider the *heat conduction* in diffusing mixtures. We shall give expressions for the four heat conduction coefficients defined by

$$\boldsymbol{J}_q' = -\chi' \operatorname{grad} T\,, \quad \boldsymbol{J}_q = -\chi \operatorname{grad} T\,, \quad (\operatorname{grad} c_i = 0)\,, \tag{241}$$

$$\boldsymbol{J}_q' = -\kappa' \operatorname{grad} T\,, \quad \boldsymbol{J}_q = -\kappa \operatorname{grad} T\,, \quad (\boldsymbol{J}_i = 0)\,, \tag{242}$$

where the first two equations refer to a uniformly mixed state ($\operatorname{grad} c_i = 0$), such as occurs in the beginning of the thermal diffusion experiment described above, and the other two to situations in which the diffusion flows $\boldsymbol{J}_i$ vanish (this is the case in the final stationary state of the thermal diffusion experiment). We have introduced coefficients χ' and κ' connected to the reduced heat flow $\boldsymbol{J}_q'$, which plays a role in the kinetic theory, as well as coefficients χ and κ connected to the ordinary heat flow $\boldsymbol{J}_q$ usually employed by experimentalists, since it is directly measured. The connexion between these two heat fluxes following from (III.24) and (II.15) is

$$\boldsymbol{J}_q' = \boldsymbol{J}_q - \sum_{k=1}^{n-1} (h_k - h_n)\boldsymbol{J}_k \tag{243}$$

* K. Clusius and G. Dickel, Nature **26** (1938) 546; Z. physik. Chem. B**44** (1939) 397, 451.

** P. Debije, Ann. Physik **36** (1939) 284.

W. M. Furry, R. Clark Jones and L. Onsager, Phys. Rev. **55** (1939) 1083.

S. R. de Groot, Physica **9** (1942) 801.

S. R. de Groot, W. Hoogenstraaten and C. J. Gorter, Physica **9** (1942) 923

S. R. de Groot, L'effet Soret, diffusion thermique dans les phases condensées (North-Holland Publishing Company, Amsterdam, 1945).

R. Clark Jones and W. H. Furry, Rev. mod. Phys. **18** (1946) 151.

From (212), (213), (241), (242), (243) and (222) one finds for the heat conductivities

$$\chi' = \frac{L_{qq}}{T^2} = \lambda\,, \quad \chi = \frac{L_{qq} + \sum_{k=1}^{n-1} (h_k - h_n) L_{kq}}{T^2}\,, \tag{244}$$

$$\kappa' = \frac{L_{qq} - \sum_{k,i=1}^{n-1} L_{qk} L_{ki}^{-1} L_{iq}}{T^2}\,, \quad \kappa = \kappa'\,. \tag{245}$$

For the binary mixture ($n = 2$) these formulae become

$$\chi' = \frac{L_{qq}}{T^2} = \lambda\,, \quad \chi = \frac{L_{qq} + (h_1 - h_2) L_{1q}}{T^2}\,, \tag{246}$$

$$\kappa' = \frac{L_{qq} - L_{q1} L_{11}^{-1} L_{1q}}{T^2}\,, \quad \kappa = \kappa'\,, \tag{247}$$

or with the set of coefficients (222)–(225) and with (231)

$$\chi' = \lambda\,, \quad \chi = \lambda + D'(h_1 - h_2)\rho c_1 c_2\,, \tag{248}$$

$$\kappa' = \lambda - \frac{(D')^2}{D} \mu_{11}^c \rho c_1^2 c_2 T\,, \quad \kappa = \kappa'\,. \tag{249}$$

We have from (232), (249) and the stability condition $\mu_{11}^c \geqslant 0$: .

$$\kappa' < \lambda\,. \tag{250}$$

Quantitatively it turns out that the difference between these two coefficients is at the highest a few percent only.

The coefficient χ can be greater or smaller than λ, as (248) shows.

Let us finally introduce the various "*quantities of transfer*" which are sometimes used for the description of transfer phenomena in mixtures. The ordinary or relative quantities of transfer are the (reduced) heat of transfer $Q_k'^*$, the heat of transfer Q_k^*, and the entropy of transfer S_k^*, defined by the following relations

$$\boldsymbol{J}_q' = \sum_{k=1}^{n-1} Q_k'^* \boldsymbol{J}_k\,, \quad (\text{grad}\, T = 0)\,, \tag{251}$$

$$J_q = \sum_{k=1}^{n-1} Q_k^* J_k \,, \quad (\text{grad}\, T = 0)\,, \tag{252}$$

$$J_s = \sum_{k=1}^{n-1} S_k^* J_k \,, \quad (\text{grad}\, T = 0)\,, \tag{253}$$

in all three cases in a system with uniform temperature. From (212), (213) and (251) we find that $Q_k'^*$ can be expressed in the phenomenological coefficients

$$Q_i'^* = \sum_{k=1}^{n-1} L_{qk} L_{ki}^{-1}\,, \quad (i = 1, 2, \ldots, n-1)\,. \tag{254}$$

With the Onsager relations (214) this can alternatively be written as

$$Q_i'^* = \sum_{k=1}^{n-1} L_{ik}^{-1} L_{kq}\,, \quad (i = 1, 2, \ldots, n-1)\,. \tag{255}$$

It is seen from (215) that this expression occurs in the result for the concentration gradients in the stationary state of the thermal diffusion experiment. With the help of (254) and (255) one can eliminate the coefficients L_{qk} and L_{kq} from the phenomenological equations (212) and (213). If one introduces furthermore the quantities

$$X_k^* = X_k - Q_k'^* \frac{(\text{grad}\, T)}{T^2}\,, \tag{256}$$

$$J_q^* = J_q' - \sum_{m=1}^{n-1} Q_m'^* J_m\,, \tag{257}$$

which leave the entropy source strength (209) invariant, then the phenomenological equations get the "diagonal" form

$$J_q^* = -\left(L_{qq} - \sum_{m,k=1}^{n-1} Q_m'^* L_{mk} Q_k'^*\right) \frac{\text{grad}\, T}{T^2}\,, \tag{258}$$

$$J_i = \sum_{k=1}^{n-1} L_{ik} X_k^*\,. \tag{259}$$

From (243) and from

$$J_q' = T\left\{ J_s - \sum_{k=1}^{n-1} (s_k - s_n) J_k \right\}, \tag{260}$$

which follows from (II.15) and (III.26), we find the relations between the quantities of transfer, defined by (251)–(253)

$$Q_k^* = Q_k'^* + h_k - h_n , \tag{261}$$

$$S_k^* = \frac{Q_k'^*}{T} + s_k - s_n = \frac{Q_k^* - \mu_k + \mu_n}{T} . \tag{262}$$

The heat of transfer Q_k^* was employed already by Eastman and Wagner*.

A different set of similar quantities, the "absolute quantities of transfer", can be defined if we introduce first the following "absolute flows":

$$\boldsymbol{J}_k^{\text{abs}} \equiv \rho_k \boldsymbol{v}_k = \boldsymbol{J}_k + \rho_k \boldsymbol{v} , \tag{263}$$

$$\boldsymbol{J}_q^{\text{abs}} \equiv \boldsymbol{J}_q + h\rho \boldsymbol{v} = \boldsymbol{J}_q' + \sum_{k=1}^{n} h_k \rho_k \boldsymbol{v}_k , \tag{264}$$

$$\boldsymbol{J}_{s,\,\text{tot}} \equiv \boldsymbol{J}_s + s\rho \boldsymbol{v} = \frac{\boldsymbol{J}_q'}{T} + \sum_{k=1}^{n} s_k \rho_k \boldsymbol{v}_k . \tag{265}$$

The identity signs indicate definitions; the last members are obtained from (II.9), (III.24) and (III.26) and the thermodynamic relations

$$h\rho = \sum_k h_k \rho_k , \quad s\rho = \sum_k s_k \rho_k . \tag{266}$$

From (263)–(265) it follows that one can write

$$\boldsymbol{J}_q' = \boldsymbol{J}_q - \sum_{k=1}^{n} h_k \boldsymbol{J}_k = \boldsymbol{J}_q^{\text{abs}} - \sum_{k=1}^{n} h_k \boldsymbol{J}_k^{\text{abs}} , \tag{267}$$

$$\boldsymbol{J}_q' = T\left(\boldsymbol{J}_s - \sum_{k=1}^{n} s_k \boldsymbol{J}_k\right) = T\left(\boldsymbol{J}_{s,\,\text{tot}} - \sum_{k=1}^{n} s_k \boldsymbol{J}_k^{\text{abs}}\right) , \tag{268}$$

which shows that $\boldsymbol{J}_q'$ is invariant for the change from "relative" to "absolute" flows.

* E. D. Eastman, J. Am. chem. Soc. **48** (1926) 1482; **49** (1927) 794; **50** (1928) 283, 292.
C. Wagner, Ann. Physik [5] **3** (1929) 629; **6** (1930) 370.

We define now the absolute (reduced) heat of transfer $Q'^{*}_{k,\,\mathrm{abs}}$, the absolute heat of transfer $Q^{*}_{k,\,\mathrm{abs}}$ and the absolute entropy of transfer $S^{*}_{k,\,\mathrm{abs}}$ by the equations

$$\boldsymbol{J}'_q = \sum_{k=1}^{n} Q'^{*}_{k,\,\mathrm{abs}} \boldsymbol{J}_k^{\mathrm{abs}}, \quad (\operatorname{grad} T = 0)\,, \tag{269}$$

$$\boldsymbol{J}_q^{\mathrm{abs}} = \sum_{k=1}^{n} Q^{*}_{k,\,\mathrm{abs}} \boldsymbol{J}_k^{\mathrm{abs}}, \quad (\operatorname{grad} T = 0)\,, \tag{270}$$

$$\boldsymbol{J}_{s,\,\mathrm{tot}} = \sum_{k=1}^{n} S^{*}_{k,\,\mathrm{abs}} \boldsymbol{J}_k^{\mathrm{abs}}, \quad (\operatorname{grad} T = 0)\,, \tag{271}$$

all valid in the isothermal state. The sums at the right-hand side are extended over the values of k from 1 to n, in contrast to what was done in (251)–(253).

We shall now show that the following relations hold for the absolute quantities of transfer, which connect them to the relative quantities of transfer and amongst themselves:

$$Q'^{*}_{k,\,\mathrm{abs}} - Q'^{*}_{n,\,\mathrm{abs}} = Q'^{*}_{k}\,, \quad \sum_{k=1}^{n} Q'^{*}_{k,\,\mathrm{abs}}\, \rho_k = 0\,, \tag{272}$$

$$Q^{*}_{k,\,\mathrm{abs}} - Q^{*}_{n,\,\mathrm{abs}} = Q^{*}_{k}\,, \quad \sum_{k=1}^{n} Q^{*}_{k,\,\mathrm{abs}}\, \rho_k = h\rho\,, \tag{273}$$

$$S^{*}_{k,\,\mathrm{abs}} - S^{*}_{n,\,\mathrm{abs}} = S^{*}_{k}\,, \quad \sum_{k=1}^{n} S^{*}_{k,\,\mathrm{abs}}\, \rho_k = \rho s\,. \tag{274}$$

The relations (272) follow when the right-hand sides of (251) and (269) are expressed in the independent vectors $\boldsymbol{J}_k$ ($k = 1, 2, \ldots, n-1$) and $\boldsymbol{v}$ [using (II.15) and (263)] and when the coefficients of those vectors are equated. The relations (273) and (274) follow with the same procedure applied to the last equalities of (267) and (268), if the definitions (252), (253), (270) and (271) are introduced and the formulae (II.15) and (263) are employed. The absolute entropy of transfer is used in the discussion on the Peltier coefficient in Chapter XIII, § 9.

§ 8. *Heat Conduction and Thermal Diffusion in Reacting Systems*

In this section we study systems in which, besides heat conduction, diffusion and their cross-effects, a number of chemical reactions amongst the components of the mixture occur as well. We shall investigate in particular the influence of thermal diffusion and of chemical reactions on the heat conduction and on the distribution of matter in the stationary state of a system with a non-uniform temperature distribution*. Let us assume that the system is a fluid mixture of n components and that r independent chemical reactions may take place. The entropy source strength is now, again in the absence of external forces and neglecting viscous phenomena, according to (III.21)

$$\sigma = \boldsymbol{J}_q \cdot \text{grad} \frac{1}{T} - \sum_{i=1}^{n} \boldsymbol{J}_i \cdot \text{grad} \frac{\mu_i}{T} - \frac{1}{T} \sum_{j=1}^{r} J_j A_j \geqslant 0 . \tag{275}$$

The vectorial fluxes and thermodynamic forces are not independent. In the preceding sections we have always eliminated the flux $\boldsymbol{J}_n$ by means of (II.15). One can however eliminate one flux in such a way that the source strength remains symmetric with respect to the chemical components**. This is done by eliminating grad T^{-1} from σ by means of the following form of the Gibbs–Duhem relation

$$h \, \text{grad} \frac{1}{T} - \sum_{i=1}^{n} c_i \, \text{grad} \frac{\mu_i}{T} = 0 , \quad (\text{grad}\, p = 0) . \tag{276}$$

This formula is obtained from (V.5) with

$$h = \mu + Ts = \sum_{i=1}^{n} c_i \mu_i + Ts \tag{277}$$

for the specific enthalpy. We have supposed here that the system is in mechanic equilibrium (grad $p = 0$). With (276) the entropy source strength (275) becomes

$$\sigma = - \sum_{i=1}^{n} \boldsymbol{K}_i \cdot \text{grad} \frac{\mu_i}{T} - \frac{1}{T} \sum_{j=1}^{r} J_j A_j \geqslant 0 , \tag{278}$$

* J. Meixner, Z. Naturf. 7A (1952) 553.
I. Prigogine and R. Buess, Acad. roy. Belg., Bull. Cl. Sc. **38** (1952) 711, 851.
W. Nernst, Boltzmann Festschrift (1904) 904.
P. A. M. Dirac, Proc. Cambr. Phil. Soc. **22** (1925) 132.
** W. Byers Brown, Trans. Far. Soc. **54** (1958) 772.

where the abbreviation

$$\boldsymbol{K}_i = \boldsymbol{J}_i - \frac{c_i}{h} \boldsymbol{J}_q, \quad (i = 1, 2, \ldots, n) \tag{279}$$

has been used. The $n + 1$ vectorial fluxes $\boldsymbol{J}_1, \boldsymbol{J}_2, \ldots, \boldsymbol{J}_n$ and $\boldsymbol{J}_q$, of which only n are independent because of the relation (II.15), can be expressed in terms of the n independent fluxes $\boldsymbol{K}_i$:

$$\boldsymbol{J}_i = \boldsymbol{K}_i - c_i \sum_{k=1}^{n} \boldsymbol{K}_k, \quad (i = 1, 2, \ldots, n), \tag{280}$$

$$\boldsymbol{J}_q = -h \sum_{i=1}^{n} \boldsymbol{K}_i. \tag{281}$$

The reduced heat flow (III.24) becomes in terms of $\boldsymbol{K}_i$:

$$\boldsymbol{J}'_q = -\sum_{i=1}^{n} h_i \boldsymbol{K}_i. \tag{282}$$

The phenomenological equations for the fluxes and forces of (278) become according to the Curie principle for isotropic systems (*cf.* Ch. IV and VI):

$$\boldsymbol{K}_i = -\sum_{k=1}^{n} a_{ik} \operatorname{grad} \frac{\mu_k}{T}, \quad (i = 1, 2, \ldots, n), \tag{283}$$

$$J_j = -\sum_{j'=1}^{r} l_{jj'} \frac{A_{j'}}{T}, \qquad (j = 1, 2, \ldots, r). \tag{284}$$

The Onsager relations are

$$a_{ik} = a_{ki}, \quad (i, k = 1, 2, \ldots, n), \tag{285}$$

$$l_{jj'} = l_{j'j}, \quad (j, j' = 1, 2, \ldots, r). \tag{286}$$

We can write the equations (284) in an equivalent form, involving n fluxes and n thermodynamic forces, by multiplying them with the stoichiometric coefficients ν_{ij} (Ch. II and X) and summing over $j = 1, 2, \ldots, r$. We then get with the help of (III.18)

$$m_i = \sum_{k=1}^{n} b_{ik} \psi_k, \quad (i = 1, 2, \ldots, n), \tag{287}$$

where the following quantities have been introduced

$$m_i \equiv \sum_{j=1}^{r} \nu_{ij} J_j , \quad (i = 1, 2, \ldots , n) , \tag{288}$$

$$b_{ik} \equiv \sum_{j, j'=1}^{r} \nu_{ij} l_{jj'} \nu_{kj'} , \quad (i, k = 1, 2, \ldots , n) , \tag{289}$$

$$\psi_k \equiv - \frac{\mu_k}{T} . \tag{290}$$

The quantity m_i of (288) is the mass of component i produced by the chemical reactions. It has already been mentioned in Chapter X that the number of independent reactions r is at most $n - 1$. Therefore the matrix b_{ik} (289), which has n rows and columns, is of rank $r \leqslant n - 1$. It is symmetric

$$b_{ik} = b_{ki} , \quad (i, k = 1, 2, \ldots , n) , \tag{291}$$

because of the Onsager relation (286).

The function (290) is often called the Planck potential. It appears also in the phenomenological equations (283), which can be written as

$$\boldsymbol{K}_i = \sum_{k=1}^{n} a_{ik} \operatorname{grad} \psi_k , \quad (i = 1, 2, \ldots , n) . \tag{292}$$

With the help of (III.18), (288) and (290) the entropy source strength (278) becomes

$$\sigma = \sum_{i=1}^{n} (\boldsymbol{K}_i \cdot \operatorname{grad} \psi_i + m_i \psi_i) \geqslant 0 , \tag{293}$$

involving the Planck potential ψ_i in both the terms referring to transport phenomena and in those referring to the chemical transitions. In an n-dimensional matrix notation we can write this entropy production as

$$\sigma = \boldsymbol{K} \cdot \operatorname{grad} \boldsymbol{\psi} + \boldsymbol{m} \cdot \boldsymbol{\psi} \geqslant 0 \tag{294}$$

and the phenomenological equations (292) and (287) as

$$\boldsymbol{K} = \mathsf{a} \cdot \operatorname{grad} \boldsymbol{\psi} , \tag{295}$$

$$\boldsymbol{m} = \mathsf{b} \cdot \boldsymbol{\psi} . \tag{296}$$

With these equations the entropy source strength (294) gets the form

$$\sigma = \mathsf{a} : (\text{grad}\,\boldsymbol{\psi})\,(\text{grad}\,\boldsymbol{\psi}) + \mathsf{b} : \boldsymbol{\psi}\boldsymbol{\psi}\,. \tag{297}$$

The two matrices a and b are symmetric according to (285) and (291)

$$\mathsf{a} = \tilde{\mathsf{a}}\,, \tag{298}$$

$$\mathsf{b} = \tilde{\mathsf{b}}\,. \tag{299}$$

They can therefore be diagonalized simultaneously by means of a congruent transformation:

$$\tilde{\mathsf{Q}}\cdot\mathsf{a}\cdot\mathsf{Q} = \mathsf{U}\,, \tag{300}$$

$$\tilde{\mathsf{Q}}\cdot\mathsf{b}\cdot\mathsf{Q} = \mathsf{\Lambda}\,, \tag{301}$$

where U is the unit matrix and $\mathsf{\Lambda}$ a diagonal matrix with r diagonal elements, say $\Lambda_1, \Lambda_2, \ldots, \Lambda_r \neq 0$ and $n - r$ vanishing elements, say $\Lambda_{r+1} = \Lambda_{r+2} = \ldots = \Lambda_n = 0$. From formulae (300) and (301) it follows that

$$\mathsf{Q}^{-1}\cdot\mathsf{a}^{-1}\cdot\mathsf{b}\cdot\mathsf{Q} = \mathsf{\Lambda}\,. \tag{302}$$

Thus $\mathsf{a}^{-1}\cdot\mathsf{b}$ is diagonalized by a similarity transformation. It follows therefore that the quantities Λ_i $(i = 1, 2, \ldots, n)$ are the roots of the equation

$$|\,\mathsf{a}^{-1}\cdot\mathsf{b} - \Lambda_i\mathsf{U}\,| = 0\,, \tag{303}$$

or of

$$|\,\mathsf{b} - \Lambda_i\mathsf{a}\,| = 0\,, \tag{304}$$

where the vertical bars indicate the determinant value. The Λ_i determine roughly the relative magnitude of the chemical reaction rates as compared to the rates of the transport phenomena. According to the equations (300), (301) and (304) the elements of the matrices Q and $\mathsf{\Lambda}$ depend on the elements of a and b. [We shall later treat a particular case ($n = 2$, $r = 1$) explicitly.] It is convenient to introduce transformed fluxes and thermodynamic forces:

$$\boldsymbol{K}^* = \tilde{\mathsf{Q}}\cdot\boldsymbol{K}\,, \quad \boldsymbol{m}^* = \tilde{\mathsf{Q}}\cdot\boldsymbol{m}\,, \tag{305}$$

$$\boldsymbol{\psi}^* = \mathsf{Q}^{-1}\cdot\boldsymbol{\psi}\,. \tag{306}$$

Using these quantities, the entropy source strength (294) becomes, assuming that the a's and b's and therefore also the elements of Q are independent of the space coordinates,

$$\sigma = \boldsymbol{K}^* \cdot \text{grad}\, \boldsymbol{\psi}^* + \boldsymbol{m}^* \cdot \boldsymbol{\psi}^* \geqslant 0\,, \tag{307}$$

or similar to (297),

$$\sigma = (\text{grad}\, \boldsymbol{\psi}^*) \cdot (\text{grad}\, \boldsymbol{\psi}^*) + \Lambda : \boldsymbol{\psi}^* \boldsymbol{\psi}^* \geqslant 0\,, \tag{308}$$

where (300) and (301) have also been employed.

Written explicitly in components this formula has the form

$$\sigma = \sum_{i=1}^{n} (\text{grad}\, \psi_i^*)^2 + \sum_{i=1}^{r} \Lambda_i (\psi_i^*)^2 \geqslant 0\,, \tag{309}$$

because we had $\Lambda_{r+1} = \Lambda_{r+2} = \ldots = \Lambda_n = 0$. The second law ($\sigma \geqslant 0$) implies that

$$\Lambda_1 \geqslant 0\,, \Lambda_2 \geqslant 0\,, \ldots, \Lambda_r \geqslant 0\,, \tag{310}$$

since $\psi_1, \psi_2, \ldots, \psi_n$ are independent variables. [It is of interest to consider also the equilibrium conditions which follow from $\sigma = 0$ and the fact that $\psi_1, \psi_2, \ldots, \psi_n$ are independent variables:

$$(\text{grad}\, \psi_i^*)^2 + \Lambda_i (\psi_i^*)^2 = 0\,, \quad (i = 1, 2, \ldots, r)\,, \tag{311}$$

$$(\text{grad}\, \psi_i^*)^2 = 0\,, \quad (i = r+1, r+2, \ldots, n)\,. \tag{312}$$

From these formulae follow the equilibrium conditions

$$\psi_i^* = 0\,, \quad (i = 1, 2, \ldots, r)\,, \tag{313}$$

$$\psi_i^* = \text{constant}\,, \quad (i = r+1, r+2, \ldots, n)\,. \tag{314}$$

The set (313) characterizes chemical equilibrium.]

With (300), (301), (305) and (306) we can write the phenomenological equations (295) and (296) in the form (with a's and b's uniform)

$$\boldsymbol{K}^* = \text{grad}\, \boldsymbol{\psi}^*\,, \quad (\text{or} \quad \boldsymbol{K}_i^* = \text{grad}\, \psi_i^*)\,, \tag{315}$$

$$\boldsymbol{m}^* = \Lambda \cdot \boldsymbol{\psi}^*\,, \quad (\text{or} \quad m_i^* = \Lambda_i \psi_i^*)\,, \tag{316}$$

where the variables ($i = 1, 2, \ldots, n$) are separated.

We shall now especially consider the *stationary state* of a system, enclosed in a reservoir, for which we can neglect the centre of gravity motion ($\boldsymbol{v} = 0$). Then the mass law (II.3) or (II.10) and the energy law (II.36) for this system get the forms

$$\operatorname{div} \boldsymbol{J}_i = \sum_{j=1}^{r} \nu_{ij} J_j \equiv m_i\,, \quad (i = 1, 2, \ldots, n)\,, \tag{317}$$

$$\operatorname{div} \boldsymbol{J}_q = 0\,, \tag{318}$$

where the abbreviation (288) has been used. From (317) it is clear that even in the stationary state the diffusion flow $\boldsymbol{J}_i$ does not necessarily vanish in the bulk of the system if chemical reactions take place, *i.e.* if the right-hand side of (317) differs from zero. Only at the walls we have the boundary condition $\boldsymbol{J}_i = 0$. If we write (317) and (318) in terms of the fluxes $\boldsymbol{K}_i$ of (279), they become, neglecting as usual squares of gradients,

$$\operatorname{div} \boldsymbol{K}_i = m_i\,, \quad (i = 1, 2, \ldots, n)\,, \tag{319}$$

or, with (305), and supposing that the phenomenological coefficients do not sensibly depend on the space coordinates,

$$\operatorname{div} \boldsymbol{K}_i^* = m_i^*\,, \quad (i = 1, 2, \ldots, n)\,. \tag{320}$$

Introducing the phenomenological equations (315) and (316), this equation becomes

$$\triangle \psi_i^* = \Lambda_i \psi_i^*\,, \quad (i = 1, 2, \ldots, n)\,, \quad (\Lambda_{r+1} = \Lambda_{r+2} = \ldots = \Lambda_n = 0)\,, \tag{321}$$

where the variables are again separated. Let us solve this equation for the case of a system in a reservoir with a temperature gradient in the x-direction. The distance between the walls which have different temperatures is called d. We choose the origin $x = 0$ of the coordinate system in the middle of the reservoir. At $x = -\frac{1}{2}d$ there is a wall of the vessel with temperature T and at $x = \frac{1}{2}d$ a wall with temperature $T + \Delta T$. The fluxes $\boldsymbol{J}_i$, $\boldsymbol{J}_q$ and $\boldsymbol{K}_i^*$ have only components in the x-direction and the differential equation (321) becomes

$$\frac{\partial^2 \psi_i^*}{\partial x^2} = \Lambda_i \psi_i^*\,, \quad (i = 1, 2, \ldots, n), \quad (\Lambda_{r+1} = \Lambda_{r+2} = \ldots = \Lambda_n = 0)\,. \tag{322}$$

The boundary condition $\boldsymbol{J}_i = 0$ at the walls becomes with (279)

$$K_i + \frac{c_i}{h} J_q = 0\,, \quad (i = 1, 2, \ldots, n)\,, \quad (x = -\tfrac{1}{2}d, \tfrac{1}{2}d)\,. \tag{323}$$

With the transformed n-dimensional vector

$$\boldsymbol{c}^* = \tilde{Q} \cdot \boldsymbol{c}\,, \tag{324}$$

where the vector $\boldsymbol{c}$ stands for $c_1, c_2, \ldots, c_n$, and with (305) and (315) the boundary condition becomes

$$K_i^* + \frac{c_i^*}{h} J_q = \frac{\partial \psi_i^*}{\partial x} + \frac{c_i^*}{h} J_q = 0\,, \quad (i = 1, 2, \ldots, n)\,, \quad (x = -\tfrac{1}{2}d, \tfrac{1}{2}d)\,, \tag{325}$$

where we take the same value for c_i^*/h at both walls [*cf.* the discussion after formula (329)]. The solution of (322) is a sum of two real exponentials (because $\Lambda_i \geqslant 0$) of which the coefficients are determined by (325). This gives, since J_q is uniform [*cf.* (318)],

$$\psi_i^*(x) = -\frac{c_i^* J_q}{h} \frac{d_i}{2} \frac{\sinh(2x/d_i)}{\cosh(d/d_i)}\,, \quad (i = 1, 2, \ldots, n)\,, \tag{326}$$

where the "penetration depth"

$$d_i = \frac{2}{\sqrt{\Lambda_i}}\,, \quad (i = 1, 2, \ldots, r)\,; \quad d_i = \infty\,, \quad (i = r+1, r+2, \ldots, n) \tag{327}$$

is a characteristic length. The infinite values of d_i for $i = r + 1, r + 2, \ldots, n$ arise from the fact that the corresponding Λ_i vanish. The solution (326) becomes for $d_i = \infty$

$$\psi_i^*(x) = -\frac{c_i^* J_q}{h} x\,, \quad (i = r+1, r+2, \ldots, n)\,. \tag{328}$$

Clearly the results for $i \leqslant r$ and for $i > r$ have an essentially different character. We shall from now on restrict ourselves to the simplest case, *viz. a binary, single reaction system* ($n = 2$, $r = 1$), which has one function $\psi_1^*(x)$ of the type (326) and one function $\psi_2^*(x)$ of the type (328). Formulae for the general n-component, r-reactions type are simply obtained by writing sums over $i = 1, 2, \ldots, r$ and $i = r + 1, r + 2, \ldots, n$ respectively instead of the single terms $i = 1$ and $i = 2$.

Let us first determine the *temperature distribution* in the system. The Gibbs–Duhem relation may be written as

$$- \frac{h}{T^2} \frac{\partial T}{\partial x} + c_1^* \frac{\partial \psi_1^*}{\partial x} + c_2^* \frac{\partial \psi_2^*}{\partial x} = 0 , \tag{329}$$

as follows from (276) with (290), (306) and (324) and uniform a's and b's. If one linearizes this differential equation, the coefficients h/T^2, c_1^* and c_2^* may be considered to be uniform constants. One can then take some average value of these quantities over the vessel for these constants.

One finds now by integration of (329)

$$T(x) = T(0) + \frac{T^2}{h} \{ c_1^* \psi_1^*(x) + c_2^* \psi_2^*(x) \} . \tag{330}$$

With (326) and (328) this becomes

$$T(x) = T(0) - \left(\frac{T}{h} \right)^2 J_q \left\{ (c_1^*)^2 \frac{d_1}{2} \frac{\sinh (2x/d_1)}{\cosh (d/d_1)} + (c_2^*)^2 x \right\} , \tag{331}$$

giving the temperature distribution $T(x)$ for certain values of J_q, d and d_1. It is of interest to consider in particular the two limiting cases of absence of chemical reaction [*i.e.* the chemical drag coefficient l_{11} of (284) is zero] and of chemical equilibrium (*i.e.* the coefficient $l_{11} = \infty$). From (289) and (301) it is seen that these limiting cases correspond to values of $\Lambda_1 = 0$ and $\Lambda_1 = \infty$ respectively and thus, according to (327), to $d_1 = \infty$ and $d_1 = 0$ respectively. Hence we find for these two cases from (331)

$$T_0(x) = T(0) - \left(\frac{T}{h} \right)^2 J_q \{ (c_1^*)^2 + (c_2^*)^2 \} x , \quad (d_1 = \infty) , \tag{332}$$

$$T_\infty(x) = T(0) - \left(\frac{T}{h} \right)^2 J_q (c_2^*)^2 x , \quad (d_1 = 0) , \tag{333}$$

where the index of T indicates the value of l_{11}. So in these two limiting cases T is linear in x, but with a different slope. In the intermediate cases $T(x)$ is not linear as (331) shows.

We can express the temperature difference ΔT in terms of J_q by means of (331), because

$$\Delta T \equiv T(\tfrac{1}{2}d) - T(-\tfrac{1}{2}d) = -d\left(\frac{T}{h}\right)^2 J_q \left\{ (c_1^*)^2 \frac{d_1}{d} \operatorname{tgh} \frac{d}{d_1} + (c_2^*)^2 \right\}. \quad (334)$$

With this formula we can eliminate J_q from (331), which gives

$$T(x) = T(0) + \frac{\Delta T}{d} \frac{(c_1^*)^2 \tfrac{1}{2}d_1 \dfrac{\sinh(2x/d_1)}{\cosh(d/d_1)} + (c_2^*)^2 x}{(c_1^*)^2 \dfrac{\operatorname{tgh}(d/d_1)}{(d/d_1)} + (c_2^*)^2}, \quad (335)$$

where the temperature distribution $T(x)$ is given for fixed values of d, d_1 and ΔT, the most frequently encountered experimental situation. For the two limiting cases quoted above of no chemical reaction and chemical equilibrium one has

$$T_0(x) = T(0) + \Delta T \frac{x}{d}, \quad (d_1 = \infty), \quad (336)$$

$$T_\infty(x) = T(0) + \Delta T \frac{x}{d}, \quad (d_1 = 0). \quad (337)$$

Especially instructive is the temperature gradient distribution for fixed d, d_1 and ΔT, which follows from (335):

$$\frac{\partial T(x)}{\partial x} = \frac{\Delta T}{d} \frac{(c_1^*)^2 \dfrac{\cosh(2x/d_1)}{\cosh(d/d_1)} + (c_2^*)^2}{(c_1^*)^2 \dfrac{\operatorname{tgh}(d/d_1)}{(d/d_1)} + (c_2^*)^2}. \quad (338)$$

Hence for the limiting cases

$$\left(\frac{\partial T(x)}{\partial x}\right)_0 = \frac{\Delta T}{d}, \quad (d_1 = \infty), \quad (339)$$

$$\left(\frac{\partial T(x)}{\partial x}\right)_\infty = \frac{\Delta T}{d} \frac{(c_1^*)^2 \{s(-x - \tfrac{1}{2}d) + s(x - \tfrac{1}{2}d)\} + (c_2^*)^2}{(c_2^*)^2}, \quad (d_1 = 0), \quad (340)$$

where $s(y)$ is the step function (*i.e.* the function $s(y) = 0$ for $y < 0$, and $s(y) = 1$ for $y \geqslant 0$). The variable x can have values in the "open interval" $-\frac{1}{2}d < x < \frac{1}{2}d$. From (340) it follows that in the bulk of the medium the temperature gradient is $\Delta T/d$, whereas at the walls it is greater by a factor $\{ (c_1^*)^2 + (c_2^*)^2 \} / (c_2^*)^2$. There is thus a discontinuity in the temperature gradient when one passes from the wall into the bulk. The intermediate cases can be studied by plotting $\partial T/\partial x$ as a function of x for various values of the parameter d_1 between 0 and ∞.

The heat conductivity $\kappa(x)$ can be found from the equation which defines it

$$J_q = -\kappa(x)\frac{\partial T}{\partial x} \tag{341}$$

and formula (331). This yields

$$\frac{1}{\kappa(x)} = \left(\frac{T}{h}\right)^2 \left\{ (c_1^*)^2 \frac{\cosh(2x/d_1)}{\cosh(d/d_1)} + (c_2^*)^2 \right\}, \tag{342}$$

giving the heat conductivity field for fixed values of d and d_1. The limiting cases are

$$\frac{1}{\kappa_0} = \left(\frac{T}{h}\right)^2 \{ (c_1^*)^2 + (c_2^*)^2 \}, \quad (d_1 = \infty), \tag{343}$$

$$\frac{1}{\kappa_\infty} = \left(\frac{T}{h}\right)^2 [(c_1^*)^2 \{ s(-x-\tfrac{1}{2}d) + s(x-\tfrac{1}{2}d) \} + (c_2^*)^2], \quad (d_1 = 0). \tag{344}$$

The last expression shows that in the bulk $(-\frac{1}{2}d < x < \frac{1}{2}d)$

$$\frac{1}{\kappa_\infty} = \left(\frac{T}{h}\right)^2 (c_2^*)^2, \tag{345}$$

whereas at the walls $(x = \frac{1}{2}d, x = -\frac{1}{2}d)$ we have $\kappa_\infty = \kappa_0$. Both κ_0 and κ_∞ (in the bulk) turn out to be independent of x. Furthermore we find from the comparison of (343) and (345) that

$$\kappa_\infty > \kappa_0, \tag{346}$$

i.e. the chemical reaction enhances the heat conduction. For (342) we can alternatively write, using (343) and (345) (valid in the bulk)

$$\frac{1}{\kappa(x)} = \frac{\cosh(2x/d_1)}{\cosh(d/d_1)}\left(\frac{1}{\kappa_0} - \frac{1}{\kappa_\infty}\right) + \frac{1}{\kappa_\infty}, \tag{347}$$

which gives indeed $\kappa(x) = \kappa_0$ for $d_1 = \infty$, and $\kappa(x) = \kappa_\infty$ for $d_1 = 0$.

An important experimental quantity is also the overall heat conductivity κ defined by

$$J_q = -\kappa \frac{\Delta T}{d}. \tag{348}$$

From this equation and (334) we find

$$\frac{1}{\kappa} = \left(\frac{T}{h}\right)^2 \left\{ (c_1^*)^2 \frac{d_1}{d} \operatorname{tgh} \frac{d}{d_1} + (c_2^*)^2 \right\}. \tag{349}$$

The limiting cases $d_1 = \infty$ and $d_1 = 0$ are again (343) and (345), so that we can write (349) in the form

$$\frac{1}{\kappa} = \frac{\operatorname{tgh}(d/d_1)}{d/d_1}\left(\frac{1}{\kappa_0} - \frac{1}{\kappa_\infty}\right) + \frac{1}{\kappa_\infty}. \tag{350}$$

We have found heat conductivities expressed in terms of c_1^*, c_2^* and d_1, or according to (324) and (327), in terms of c_1 and c_2 and of the elements of Q and of Λ. These, in turn, depend on the elements of a and b, as the formulae (300), (301) and (304) show. For our case of a binary ($n = 2$), single reaction ($r = 1$) system the equation (304) is a quadratic equation of which the two roots are

$$\Lambda_1 = \frac{a_{11}b_{22} + a_{22}b_{11} - 2b_{12}a_{12}}{|a|}, \quad \Lambda_2 = 0, \tag{351}$$

where we have used (285) and (291). With (289) for $n = 2$ and $r = 1$ and $\nu_1 + \nu_2 = 0$ (Chapter II or X) we have

$$b_{11} = b_{22} = -b_{12} = \nu_1^2 l_{11}, \tag{352}$$

where the second index of ν_{ij} has been dropped. Thus (351) gives

$$\Lambda_1 = l_{11}\nu_1^2 \frac{a_{11} + 2a_{12} + a_{22}}{|a|} \tag{353}$$

and with (327) we find

$$d_1 = \frac{2}{\sqrt{\Lambda_1}} = \frac{2}{\nu_1}\sqrt{\frac{|a|}{l_{11}(a_{11} + 2a_{12} + a_{22})}}. \tag{354}$$

From (300) and (301) we find the matrix Q in terms of the elements of the matrix a:

$$Q = \frac{1}{\sqrt{a_{11} + 2a_{12} + a_{22}}} \begin{pmatrix} (a_{12} + a_{22}) \,|\, a \,|^{-\frac{1}{2}} & 1 \\ -(a_{11} + a_{12}) \,|\, a \,|^{-\frac{1}{2}} & 1 \end{pmatrix} \tag{355}$$

The inverse matrix Q^{-1} is

$$Q^{-1} = \frac{1}{\sqrt{a_{11} + 2a_{12} + a_{22}}} \begin{pmatrix} |\, a \,|^{\frac{1}{2}} & -|\, a \,|^{\frac{1}{2}} \\ a_{11} + a_{12} & a_{12} + a_{22} \end{pmatrix} \tag{356}$$

This gives, with (324),

$$c_1^\bullet = Q_{11}c_1 + Q_{21}c_2 = \frac{(a_{12} + a_{22})c_1 - (a_{11} + a_{12})c_2}{\sqrt{|\, a \,|\, (a_{11} + 2a_{12} + a_{22})}}, \tag{357}$$

$$c_2^\bullet = Q_{12}c_1 + Q_{22}c_2 = \frac{1}{\sqrt{a_{11} + 2a_{12} + a_{22}}} \tag{358}$$

With the help of these two expressions we can now write for (343) and (345)

$$\frac{1}{\kappa_0} = \left(\frac{T}{h}\right)^2 \frac{a_{11}c_2^2 - 2a_{12}c_1c_2 + a_{22}c_1^2}{|\, a \,|}, \tag{359}$$

$$\frac{1}{\kappa_\infty} = \left(\frac{T}{h}\right)^2 \frac{1}{a_{11} + 2a_{12} + a_{22}}. \tag{360}$$

We have now found the heat conductivities (347) and (350) in terms of the phenomenological coefficients a_{11}, a_{12}, a_{22} and l_{11}, as (354), (359) and (360) show. We wish, however, to go one step further still, and express the coefficients a_{11}, a_{12} and a_{22} in the ordinary coefficients λ, $D' = D''$ and D which describe the vectorial transport phenomena and which were introduced in § 7 of this chapter. This can be achieved in the following way. In the first place (280) and (282) give for a binary system:

$$\boldsymbol{J}_1 = c_2\boldsymbol{K}_1 - c_1\boldsymbol{K}_2, \tag{361}$$

$$\boldsymbol{J}_q' = -h_1\boldsymbol{K}_1 - h_2\boldsymbol{K}_2, \tag{362}$$

with the phenomenological equations

$$\boldsymbol{K}_1 = -a_{11}\,\mathrm{grad}\,\frac{\mu_1}{T} - a_{12}\,\mathrm{grad}\,\frac{\mu_2}{T}, \tag{363}$$

$$\boldsymbol{K}_2 = -a_{12}\,\mathrm{grad}\,\frac{\mu_1}{T} - a_{22}\,\mathrm{grad}\,\frac{\mu_2}{T}, \tag{364}$$

which follow from (283) and (285). We can express the gradients of the Planck potentials in terms of grad T and grad c_1:

$$\mathrm{grad}\,\frac{\mu_1}{T} = -\frac{h_1}{T^2}\,\mathrm{grad}\,T + \frac{\mu_{11}^c}{T}\,\mathrm{grad}\,c_1\,, \quad \left[\mu_{11}^c \equiv \left(\frac{\partial \mu_1}{\partial c_1}\right)_{p,T}\right], \tag{365}$$

$$\mathrm{grad}\,\frac{\mu_2}{T} = -\frac{h_2}{T^2}\,\mathrm{grad}\,T - \frac{c_1}{c_2}\frac{\mu_{11}^c}{T}\,\mathrm{grad}\,c_1\,, \tag{366}$$

where in the last formula the Gibbs–Duhem relation

$$c_1\,\mathrm{grad}\,\mu_1 + c_2\,\mathrm{grad}\,\mu_2 = 0\,, \quad (p, T \text{ constant}) \tag{367}$$

has been used.

If one inserts (363)–(366) into (361) and (362) one obtains phenomenological equations with the same fluxes and thermodynamic forces as used in (226) and (227) of § 7. Identification of the coefficients gives the desired connexion between D, $D' = D''$ and λ on the one hand and a_{11}, a_{12} and a_{22} on the other hand:

$$D = \frac{(a_{11}c_2^2 - 2a_{12}c_1c_2 + a_{22}c_1^2)\,\mu_{11}^c}{c_2\rho T}, \tag{368}$$

$$D' = \frac{a_{22}c_1h_2 + a_{12}(c_1h_1 - c_2h_2) - a_{11}c_2h_1}{c_1c_2\rho T^2}, \tag{369}$$

$$\lambda = \frac{a_{11}h_1^2 + 2a_{12}h_1h_2 + a_{22}h_2^2}{T^2}. \tag{370}$$

From those formulae follows the relationship

$$D\lambda(\mu_{11}^c)^{-1} - (D')^2c_1^2c_2\rho T = \frac{(a_{11}a_{22} - a_{12}^2)h^2}{c_2\rho T^3} \geqslant 0\,. \tag{371}$$

Now finally one finds with the help of (368)–(371) that the formulae (359), (360) and (354) get the following forms

$$\kappa_0 = \lambda - \frac{\rho c_1^2 c_2 \mu_{11}^c T(D')^2}{D}, \tag{372}$$

$$\kappa_\infty = \lambda + 2\rho c_1 c_2 D' \Delta h + \frac{\rho_2 D(\Delta h)^2}{T\mu_{11}^c}, \quad (\Delta h \equiv h_1 - h_2), \tag{373}$$

$$d_1 = \frac{2}{\sqrt{\Lambda_1}} = \frac{2}{\nu_1}\sqrt{\frac{\rho_2 T\,\{D\lambda(\mu_{11}^c)^{-1} - (D')^2 c_1^2 c_2 \rho T\}}{l_{11}\,\{\lambda + 2\rho c_1 c_2 D'\Delta h + \rho_2 D(\Delta h)^2/T\mu_{11}^c\}}}. \tag{374}$$

The validity of these expressions can be checked by inserting (368)–(371) into them and showing that one gets back (359), (360) and (354).

First of all it must be remarked that with the three preceding formulae we have found the *heat conductivities* $\kappa(x)$, (347), and κ, (350), in terms of the phenomenological constants (the ordinary heat conductivity λ, the thermal diffusion coefficient D', the diffusion coefficient D and the chemical drag coefficient l_{11}) which characterize the irreversible behaviour, and as a function of various thermodynamical variables such as the concentration, the temperature, etc., and the dimension d of the reservoir.

Let us now discuss the set of results (347), (350) and (372)–(374) in somewhat more detail. The two limiting cases, discussed after formula (331), can now be discussed a little more quantitatively.

(i) $d_1 \to \infty$. The chemical reaction is impeded. The precise meaning of this is, according to the expression (374), that the rate constant l_{11} is slow as compared to the rates of the transport phenomena, which are characterized by λ, D' and D. In the limiting case where practically no chemical reaction can take place the characteristic length d_1 tends to infinity. Then according to (347) and (350) we have $\kappa(x) = \kappa = \kappa_0$, of which the value is given by (372). This was the case discussed in § 7 of this chapter. The expression (372) corresponds indeed with the result (249) found there for the stationary state.

(ii) $d_1 \to 0$. Chemical equilibrium. This is the opposite limiting case, where the chemical reaction rate l_{11} is so high that chemical equilibrium is reached before any appreciable transport phenomena have taken place. Then, from (374), we have the limit of d_1 tending to zero. Then (347) and (350) give $\kappa(x) = \kappa = \kappa_\infty$, of which the value is (373). It contains four contributions: one from the ordinary heat conduction λ, two equal contributions arising from thermal diffusion and the Dufour effect, characterized by $D' = D''$, and one caused by the chemical reaction, giving a form involving $\Delta h \equiv h_1 - h_2$, the reaction heat at constant temperature and pressure.

The terms with D' in (372)–(374) are so small that in practice they play almost no role. The reason for this is that in all experimental cases D' is so small, as was discussed in § 7. We shall neglect these terms in the following. So we write instead of (372) and (373)

$$\kappa_0 \simeq \lambda \,, \tag{375}$$

$$\kappa_\infty \simeq \lambda + \frac{\rho_2 D(\Delta h)^2}{T\mu_{11}^c}\,, \quad (\Delta h \equiv h_1 - h_2)\,. \tag{376}$$

Since thermodynamic stability requires $\mu_{11}^c \geqslant 0$, we find again that $\kappa_\infty > \kappa_0$ [*cf.* (346)], *i.e.* the chemical reaction increases the heat conduction. Intuitively this can be understood, if we consider the example of the chemical reaction $A_2 \leftrightarrows 2A$ (a dissociating bi-atomic gas), where the dissociation of the molecule A_2 requires a certain amount of energy. In the system with a non-uniform temperature distribution a sort of cyclic stationary state will be reached with molecules A_2 diffusing to the higher temperature regions where their concentration is lower, dissociating there under absorption of heat, then diffusing back in the form of atoms A to the lower temperature regions where they recombine and give off their dissociation energy. Thus besides the ordinary heat conduction mechanism an additional energy transfer takes place, given by the last, "chemical", term of (376) and enhancing the heat conductivity. Let us now give some orders of magnitude for the relevant quantities. For the case of gases these are: reaction heat $r \equiv M_1 \Delta h = 10^3$ to 10^4 cal/mole, heat conductivity $\lambda \simeq 5.10^{-5}$ cal/cm sec degree and diffusion coefficient $D \simeq 10^{-1}$ cm^2/sec. With the lowest value given for the reaction heat, the second term of (376) becomes already of the same order of magnitude as λ; with the highest, it is therefore appreciably greater than the ordinary heat conduction term λ. For a liquid r and μ_{11}^c are of the same order, and λ is 10^2 times greater than in a gas. The density of a liquid is much higher than that of a gas, but the diffusion coefficient is so much smaller that the net result is that the factor $\rho_2 D$ is somewhat smaller. Therefore, in the case of a liquid the second term of (376) is negligible compared to the first.

To discuss the order of magnitude of the characteristic length d_1, we shall write (374) in the somewhat simplified form

$$\frac{1}{d_1} = \frac{1}{2}\sqrt{k_f\left(\frac{1}{D} + \frac{N_1}{RT^2}\frac{r^2}{\lambda}\right)}, \tag{377}$$

where we have neglected terms with D', supposed the system to be ideal, and left out the factor $\rho^2/N_2NM_2^2$, since it is of the order of unity. Furthermore we have introduced the chemical reaction rate constant k_f instead of the phenomenological constant l_{11} by means of the relation

$$k_f = l_{11}\frac{R}{N_1M_2^2}, \tag{378}$$

which is a special case of (X.49) with $\bar{\nu}_1 = -1$. It is clear from (377) that the depth d_1 is a measure for the relative importance of the chemical reaction rate compared to the transport processes, described by λ and D. A typical reaction rate in gases is $k_f = 10^{-1}$ sec^{-1}. This would lead to a characteristic length d_1 of the order of 1 cm, using reasonable values of the transport coefficients. In liquids one may have a range of values for k_f from, say, 10^{-1} sec^{-1} to 10^{-5} sec^{-1}. This would lead to $d_1 \simeq 10^{-2}$ to 1 cm, roughly.

We note that the theory described here is also valid for the calculation of the influence of molecular vibration (and rotation) relaxation on the heat conductivity. One has simply to consider the various vibrational and rotational quantum states as different chemical species, and describe transitions between these quantum states as chemical reactions. Meixner* proved that one finds thus the results given by Eucken** and by Chapman and Cowling***.

In the preceding pages we studied the temperature field $T(x)$ in a binary, single reaction system. Let us now turn our attention to the

* J. Meixner, Zeitschr. Naturforschung **8**A (1953) 69.

** A. Eucken, Phys. Zeitschr. **14** (1913) 324.

*** S. Chapman and T. G. Cowling, The mathematical theory of non-uniform gases (Cambridge, 1939) p. 237.

concentration distribution $c_1(x)$ in the same system in its stationary state. From (282) and (279) one has with $n = 2$:

$$\boldsymbol{J}_q = \boldsymbol{J}'_q + (\Delta h)\boldsymbol{J}_1 \,, \quad (\Delta h = h_1 - h_2) \,. \tag{379}$$

This gives with (226), (227) and (231)

$$J_q = -\,(\rho c_1 \mu^c_{11} TD' + \rho D \Delta h)\,\frac{\partial c_1(x)}{\partial x} - (\lambda + \rho c_1 c_2 D' \Delta h)\,\frac{\partial T(x)}{\partial x} \,. \tag{380}$$

If this is integrated taking the bracket factors as constants, we obtain, because J_q is independent of x,

$$c_1(x) = c_1(0) - \frac{J_q x + (\lambda + \rho c_1 c_2 D' \Delta h)\,\{\,T(x) - T(0)\,\}}{\rho c_1 \mu^c_{11} TD' + \rho D \Delta h} \,. \tag{381}$$

For the temperature field we have from (331) with (343) and (345)

$$T(x) = T(0) - J_q \left\{ \frac{d_1}{2}\,\frac{\sinh\,(2x/d_1)}{\cosh\,(d/d_1)} \left(\frac{1}{\kappa_0} - \frac{1}{\kappa_\infty} \right) + x\,\frac{1}{\kappa_\infty} \right\} . \tag{382}$$

With this expression, the concentration field (381) becomes

$$c_1(x) = c_1(0) - J_q \frac{x - (\lambda + \rho c_1 c_2 D' \Delta h) \left\{ \dfrac{d_1}{2}\,\dfrac{\sinh\,(2x/d_1)}{\cosh\,(d/d_1)} \left(\dfrac{1}{\kappa_0} - \dfrac{1}{\kappa_\infty} \right) + x\,\dfrac{1}{\kappa_\infty} \right\}}{\rho c_1 \mu^c_{11} TD' + \rho D \Delta h} \,, \tag{383}$$

where $c_1(x)$ is given for a fixed heat flow J_q. We can eliminate J_q with (348), using also (350). Then (383) becomes

$$c_1(x) = c_1(0) + \frac{\Delta T}{d}\,\frac{x - (\lambda + \rho c_1 c_2 D' \Delta h) \left\{ \dfrac{d_1}{2}\,\dfrac{\sinh\,(2x/d_1)}{\cosh\,(d/d_1)} \left(\dfrac{1}{\kappa_0} - \dfrac{1}{\kappa_\infty} \right) + x\,\dfrac{1}{\kappa_\infty} \right\}}{\{\,\rho c_1 \mu^c_{11} TD' + \rho D \Delta h\,\} \left\{ \left(\dfrac{d_1}{d}\,\mathrm{tgh}\,\dfrac{d}{d_1} \right) \left(\dfrac{1}{\kappa_0} - \dfrac{1}{\kappa_\infty} \right) + \dfrac{1}{\kappa_\infty} \right\}} \,, \tag{384}$$

which gives the concentration distribution for fixed ΔT. In the two limiting cases of an impeded chemical reaction ($d_1 = \infty$) and of chemical equilibrium ($d_1 = 0$), this gives

$$c_1(x) = c_1(0) + \frac{\Delta T(\kappa_0 - \lambda - \rho c_1 c_2 D' \Delta h)}{d(\rho c_1 \mu_{11}^c T D' + \rho D \Delta h)} x, \quad (d_1 = \infty), \tag{385}$$

$$c_1(x) = c_1(0) + \frac{\Delta T(\kappa_\infty - \lambda - \rho c_1 c_2 D' \Delta h)}{d(\rho c_1 \mu_{11}^c T D' + \rho D \Delta h)} x, \quad (d_1 = 0). \tag{386}$$

Both these fields are linear in the coordinate x; in the intermediate cases, however, the concentration is not a linear function of x, as (384) shows. With the expressions (372) and (373) for κ_0 and κ_∞ the concentration fields (385) and (386) are given by

$$\frac{\{c_1(x) - c_1(0)\}/x}{\Delta T/d} = -c_1 c_2 \frac{D'}{D}, \quad (d_1 = \infty), \tag{387}$$

$$\frac{\{c_1(x) - c_1(0)\}/x}{\Delta T/d} = \frac{c_2 \Delta h}{\mu_{11}^c T}, \quad (d_1 = 0). \tag{388}$$

An important quantity is also the concentration difference between the two ends of the reservoir

$$\Delta c_1 = c_1(\tfrac{1}{2}d) - c_1(-\tfrac{1}{2}d). \tag{389}$$

One finds for this quantity from (384)

$$\frac{\Delta c_1}{\Delta T} = \frac{\kappa - \lambda - \rho c_1 c_2 D' \Delta h}{\rho c_1 \mu_{11}^c T D' + \rho D \Delta h}, \tag{390}$$

where κ is given by (350) with (372) and (373). For the limiting case of no chemical reactions ($\kappa = \kappa_0$) we find from (390) and (372)

$$\frac{\Delta c_1}{\Delta T} = -c_1 c_2 \frac{D'}{D}, \quad (d_1 = \infty), \tag{391}$$

a result which was also found in § 7. In the limiting case of chemical equilibrium ($\kappa = \kappa_\infty$) we obtain from (390) and (373)

$$\frac{\Delta c_1}{\Delta T} = \frac{c_2 \Delta h}{\mu_{11}^c T}, \quad (d_1 = 0), \tag{392}$$

which shows that the concentration difference is determined by the reaction heat Δh. This result can also directly be obtained* from the condition of chemical equilibrium.

* I. Prigogine and R. Buess, Acad. roy. Belg., Cl. Sc. [5] **38** (1952) 711.

CHAPTER XII

VISCOUS FLOW AND RELAXATION PHENOMENA

§ 1. *Viscous Flow in an Isotropic Fluid*

In the two preceding chapters we studied scalar and vectorial phenomena. In this chapter we wish to consider the phenomena, which are related to the tensorial quantities occurring in the thermodynamics of fluid systems: the pressure tensor and the velocity gradient field. We shall first, in this section, develop the thermodynamics of a single component isotropic fluid system not subjected to any external force. We shall establish the equations of motion for viscous flow in a way similar to the procedure followed in Chapters II, III and IV, but now without omitting rotational phenomena.

The conservation laws of energy and momentum of a fluid system in the absence of external forces can be written as [*cf.* (II.19) and (II.31)]:

$$\rho \frac{\mathrm{d}e}{\mathrm{d}t} = - \operatorname{div} (\boldsymbol{P} \cdot \boldsymbol{v} + \boldsymbol{J}_q) , \tag{1}$$

$$\rho \frac{\mathrm{d}\boldsymbol{v}}{\mathrm{d}t} = - \operatorname{Div} \boldsymbol{P} . \tag{2}$$

In contrast to what was done in Chapter II, we shall not assume that the pressure tensor $\boldsymbol{P}$ is symmetric, but leave open the possibility that $\boldsymbol{P}$ contains an antisymmetric part. Then we need to consider still another conservation law, *viz.* the law of conservation of angular momentum*. We can write the angular momentum per unit mass as an axial vector $\boldsymbol{J}$ with Cartesian components J_1, J_2 and J_3, or alternatively as an antisymmetric tensor J with Cartesian components

* J. Frenkel, Kinetic theory of liquids (Oxford, 1946) Ch. V § 7.
C. F. Curtiss, J. chem. Phys. **24** (1956) 225.
H. Grad, Comm. pure and applied Math., New York, **5** (1952) 455.
L. Waldmann, Encyclopedia of Physics **12** (1958) 301.

$J_{12} = -J_{21} = J_3$ (cycl.). The law of conservation of angular momentum reads

$$\rho \frac{\mathrm{d}J_{\alpha\beta}}{\mathrm{d}t} = -\sum_{\gamma=1}^{3} \frac{\partial}{\partial r_\gamma} (r_\alpha P_{\gamma\beta} - r_\beta P_{\gamma\alpha}), \quad (\alpha, \beta = 1, 2, 3), \tag{3}$$

where the vector $\boldsymbol{r}$ with components r_α is the position vector of a mass element with respect to an arbitrary coordinate system. At the right-hand side a negative divergence of the "angular momentum flow" occurs. This flow is due to the couple (with respect to the origin) exerted on a mass element by the pressure tensor.

The total angular momentum can be considered as the sum of two contributions: an "external angular momentum" and an "internal angular momentum". We can therefore write, using either anti-symmetric tensors or axial vectors:

$$J_{\alpha\beta} = L_{\alpha\beta} + S_{\alpha\beta}, \quad (\alpha, \beta = 1, 2, 3) \quad \text{or} \quad \boldsymbol{J} = \boldsymbol{L} + \boldsymbol{S}. \tag{4}$$

The "external angular momentum" per unit mass $L_{\alpha\beta}$ or $\boldsymbol{L}$ is defined as

$$L_{\alpha\beta} = r_\alpha v_\beta - r_\beta v_\alpha \quad (\alpha, \beta = 1, 2, 3) \quad \text{or} \quad \boldsymbol{L} = \boldsymbol{r} \wedge \boldsymbol{v}. \tag{5}$$

The "internal angular momentum" per unit mass $S_{\alpha\beta}$ or $\boldsymbol{S}$ arises as a consequence of a possible rotational motion of the constituent particles of the system. We may write

$$S_{\alpha\beta} = \Theta \omega_{\alpha\beta}, \quad (\alpha, \beta = 1, 2, 3) \quad \text{or} \quad \boldsymbol{S} = \Theta \boldsymbol{\omega}, \tag{6}$$

where $\boldsymbol{\omega}$ is the mean angular velocity of the constituent particles at each point in the fluid. The quantity Θ is the (average) moment of inertia per unit mass of the constituent particles*.

We find the balance equation for $\boldsymbol{L}$ by vectorial multiplication of (2) with the radius vector $\boldsymbol{r}$:

$$\rho \frac{\mathrm{d}L_{\alpha\beta}}{\mathrm{d}t} = -\sum_{\gamma=1}^{3} \frac{\partial}{\partial r_\gamma} (r_\alpha P_{\gamma\beta} - r_\beta P_{\gamma\alpha}) + P_{\alpha\beta} - P_{\beta\alpha}, \quad (\alpha, \beta = 1, 2, 3). \tag{7}$$

* In equations (1) and (3) we have omitted terms in the energy and angular momentum flow, describing the effect which an "intrinsic couple density" can have. Under normal circumstances [*cf.* H. Grad, Comm. pure and applied Math., New York, **5** (1952) 455] the effect of such terms on these equations and on the entropy production is negligibly small.

The balance equation for $\boldsymbol{S}$ then follows from the difference of (3) and (7):

$$\rho \frac{\mathrm{d}S_{\alpha\beta}}{\mathrm{d}t} = -2P^{\mathrm{a}}_{\alpha\beta}, \quad (\alpha, \beta = 1, 2, 3) \quad \text{or} \quad \rho \frac{\mathrm{d}\boldsymbol{S}}{\mathrm{d}t} = -2\boldsymbol{P}^{\mathrm{a}}, \tag{8}$$

where we have written $P^{\mathrm{a}}_{\alpha\beta}$ for the antisymmetric part $\frac{1}{2}(P_{\alpha\beta}-P_{\beta\alpha})$ of the total pressure tensor $P_{\alpha\beta}$. The axial vector $\boldsymbol{P}^{\mathrm{a}}$ corresponds in the usual way to this antisymmetric tensor $[P^{\mathrm{a}}_1 = P^{\mathrm{a}}_{23}$ (cycl.)].

It is clear from the equations (3), (7) and (8) that if the system possesses no intrinsic internal motion (*i.e.* $\boldsymbol{S} = 0$), then $\boldsymbol{L}$ is conserved and the pressure tensor is symmetric:

$$\mathsf{P} = \tilde{\mathsf{P}} \quad \text{or} \quad \boldsymbol{P}^{\mathrm{a}} = 0\,. \tag{9}$$

On the other hand, if the pressure tensor is symmetric then the external and internal angular momentum $\boldsymbol{L}$ and $\boldsymbol{S}$ are separately conserved.

Let us now split the total energy per unit mass into three parts

$$e = \tfrac{1}{2}\boldsymbol{v}^2 + u_r + u\,, \tag{10}$$

where $\frac{1}{2}\boldsymbol{v}^2$ is the kinetic energy per unit mass, u_r the macroscopic intrinsic rotational energy per unit mass

$$u_r = \tfrac{1}{2}\Theta\omega^2\,, \tag{11}$$

and u the internal energy per unit mass. We can find the balance equation for u_r from (8) with (6)

$$\rho \frac{\mathrm{d}u_r}{\mathrm{d}t} = \rho\boldsymbol{\omega}\cdot\frac{\mathrm{d}\boldsymbol{S}}{\mathrm{d}t} = -2\boldsymbol{\omega}\cdot\boldsymbol{P}^{\mathrm{a}} \quad (= \omega : \mathsf{P}^{\mathrm{a}})\,. \tag{12}$$

The balance equation for $\frac{1}{2}\boldsymbol{v}^2$ follows from scalar multiplication of (2) with $\boldsymbol{v}$

$$\rho \frac{\mathrm{d}\frac{1}{2}\boldsymbol{v}^2}{\mathrm{d}t} = -\boldsymbol{v}\cdot\mathrm{Div}\,\mathsf{P} = -\mathrm{div}\,(\mathsf{P}\cdot\boldsymbol{v}) + \tilde{\mathsf{P}} : \mathrm{Grad}\,\boldsymbol{v}\,. \tag{13}$$

The balance of internal energy which is obtained from (1), if (12) and (13) are subtracted, reads

$$\rho \frac{\mathrm{d}u}{\mathrm{d}t} = -\tilde{\mathsf{P}} : \mathrm{Grad}\,\boldsymbol{v} + 2\boldsymbol{P}^{\mathrm{a}}\cdot\boldsymbol{\omega} - \mathrm{div}\,\boldsymbol{J}_q\,. \tag{14}$$

Let us now split the total pressure tensor into two parts, of which the first is the (scalar) equilibrium pressure p multiplied by the unit tensor U, and the second the viscous pressure tensor $\mathsf{\Pi}$:

$$\mathsf{P} = p\mathsf{U} + \mathsf{\Pi} . \tag{15}$$

Subsequently let us write for the viscous pressure tensor

$$\mathsf{\Pi} = \Pi\mathsf{U} + \overset{\circ}{\mathsf{\Pi}}{}^{s} + \mathsf{\Pi}^{a} , \tag{16}$$

where Π is one third of the trace, $\overset{\circ}{\mathsf{\Pi}}{}^{s}$ the symmetric part with zero trace and $\mathsf{\Pi}^{a}$ the antisymmetric part of $\mathsf{\Pi}$. The latter, which according to (15) and (16) is equal to P^{a} used above, can again be represented by an axial vector $\boldsymbol{\Pi}^{a} = \boldsymbol{P}^{a}$. Similarly let us split the velocity gradient tensor in the following way

$$\text{Grad}\, \boldsymbol{v} = \tfrac{1}{3}(\text{div}\, \boldsymbol{v})\mathsf{U} + (\overset{\circ}{\text{Grad}}\, \boldsymbol{v})^{s} + (\text{Grad}\, \boldsymbol{v})^{a} , \tag{17}$$

where the antisymmetric part corresponds to the axial vector $\frac{1}{2}$ rot $\boldsymbol{v}$. In this way the balance equation for the internal energy becomes

$$\rho \frac{\mathrm{d}u}{\mathrm{d}t} = -(p + \Pi)\, \text{div}\, \boldsymbol{v} - \overset{\circ}{\mathsf{\Pi}}{}^{s} : (\overset{\circ}{\text{Grad}}\, \boldsymbol{v})^{s} - \boldsymbol{\Pi}^{a} \cdot (\text{rot}\, \boldsymbol{v} - 2\boldsymbol{\omega}) - \text{div}\, \boldsymbol{J}_q . \tag{18}$$

With the help of the law of conservation of mass (II.14) it can finally be brought into the form

$$\rho \left(\frac{\mathrm{d}u}{\mathrm{d}t} + p \frac{\mathrm{d}v}{\mathrm{d}t} \right) = -\Pi\, \text{div}\, \boldsymbol{v} - \overset{\circ}{\mathsf{\Pi}}{}^{s} : (\overset{\circ}{\text{Grad}}\, \boldsymbol{v})^{s} - \boldsymbol{\Pi}^{a} \cdot (\text{rot}\, \boldsymbol{v} - 2\boldsymbol{\omega}) - \text{div}\, \boldsymbol{J}_q , \tag{19}$$

where $v = \rho^{-1}$ is the specific volume.

When this expression is inserted into the Gibbs relation

$$T \frac{\mathrm{d}s}{\mathrm{d}t} = \frac{\mathrm{d}u}{\mathrm{d}t} + p \frac{\mathrm{d}v}{\mathrm{d}t} , \tag{20}$$

we obtain the entropy balance equation

$$\rho \frac{\mathrm{d}s}{\mathrm{d}t} = -\text{div}\, \boldsymbol{J}_s + \sigma , \tag{21}$$

with the entropy flow

$$J_s = \frac{J_q}{T} \tag{22}$$

and the entropy production

$$\sigma = - J_q \cdot \frac{\text{grad } T}{T^2} - \frac{\Pi \text{ div } v}{T} - \frac{\mathring{\Pi}^s : (\mathring{\text{Grad}}\, v)^s}{T} - \frac{\Pi^a \cdot (\text{rot } v - 2\omega)}{T} \geqslant 0. \tag{23}$$

At the right-hand side we recognize a sum of scalar products of polar vectors, scalars, symmetric tensors with zero trace and axial vectors respectively. In an isotropic fluid, according to the considerations of Chapter VI, the fluxes and thermodynamic forces of (23) are related by the following simple set of phenomenological equations

$$J_q = - \lambda \text{ grad } T\,, \tag{24}$$

$$\Pi = - \eta_v \text{ div } v\,, \tag{25}$$

$$\mathring{\Pi}_s = - 2\eta\, (\mathring{\text{Grad}}\, v)^s\,, \tag{26}$$

$$\Pi^a = - \eta_r\, (\text{rot } v - 2\omega)\,. \tag{27}$$

The phenomenological coefficients, introduced in these equations, are the heat conductivity λ, the volume viscosity η_v, the ordinary or shear viscosity η, and the "rotational viscosity" η_r. All four quantities are positive, as follows when (24)–(27) are inserted into (23). The so-called second coefficient of viscosity, which is sometimes used, is defined as $\eta_v - \frac{2}{3}\eta$.

We could alternatively have written (27) as

$$\Pi^a = - 2\eta_r \{ (\text{Grad } v)^a - \omega \}\,, \tag{28}$$

using antisymmetric tensors instead of polar vectors.

We may note that if the only motion of the fluid would be an expansion, which can be described by the velocity field,

$$v = ar\,, \quad (a \text{ constant scalar})\,, \tag{29}$$

then $(\mathring{\text{Grad}}\, v)^s$ and rot v vanish, but div v is different from zero, and volume viscosity plays a role. If on the other hand the fluid motion is like the rotation of a rigid body

$$\boldsymbol{v} = \boldsymbol{b} \wedge \boldsymbol{r}, \quad (\boldsymbol{b} \text{ constant vector}), \tag{30}$$

then $(\text{Grad } \boldsymbol{v})^s$ and $\text{div } \boldsymbol{v}$ vanish, but

$$\text{rot } \boldsymbol{v} = 2\boldsymbol{b}, \tag{31}$$

so that only the rotational viscosity could play a role.

With the help of (6) and (27) we can write the balance of internal angular momentum (8) in the following form

$$\frac{\mathrm{d}\boldsymbol{\omega}}{\mathrm{d}t} = -\frac{2\eta_r}{\rho\Theta}(2\boldsymbol{\omega} - \text{rot } \boldsymbol{v}). \tag{32}$$

If we suppose a situation in which $\text{rot } \boldsymbol{v}$ is approximately uniform and constant and $\boldsymbol{\omega}$ vanishing initially, then (32) yields the time behaviour of $\boldsymbol{\omega}$

$$\boldsymbol{\omega} = \tfrac{1}{2} \text{rot } \boldsymbol{v} (1 - \mathrm{e}^{-t/\tau}), \tag{33}$$

involving a relaxation time

$$\tau = \frac{\rho\Theta}{4\eta_r} \tag{34}$$

of the internal angular momentum $\boldsymbol{S}$ (or $\boldsymbol{\omega}$). Thus usually after a short time the quantities $2\boldsymbol{\omega}$ and $\text{rot } \boldsymbol{v}$ are equalized, and the antisymmetric part (27) of the pressure tensor vanishes.

The equation of motion (2), with (15) and (16), becomes by inserting the phenomenological equations (25)–(27)

$$\begin{aligned}\rho \frac{\mathrm{d}\boldsymbol{v}}{\mathrm{d}t} &= -\text{grad}\, p + \text{Div}\{2\eta\, (\text{Grad}\, \boldsymbol{v})^s\} + \text{grad}\{(\eta_v - \tfrac{2}{3}\eta)\, \text{div}\, \boldsymbol{v}\} \\ &\qquad + \text{rot}\{\eta_r (2\boldsymbol{\omega} - \text{rot}\, \boldsymbol{v})\} \\ &= -\text{grad}\, p + \text{Div}[\eta\{2\,(\text{Grad}\, \boldsymbol{v})^s - U \text{div}\, \boldsymbol{v}\}] + \text{grad}\{(\tfrac{1}{3}\eta + \eta_v)\, \text{div}\, \boldsymbol{v}\} \\ &\qquad + \text{rot}\{\eta_r (2\boldsymbol{\omega} - \text{rot}\, \boldsymbol{v})\}.\end{aligned} \tag{35}$$

By performing the differentiations one obtains the general result

$$\rho \frac{\mathrm{d}\boldsymbol{v}}{\mathrm{d}t} = -\operatorname{grad} p + \eta \triangle \boldsymbol{v} + (\tfrac{1}{3}\eta + \eta_v) \operatorname{grad} \operatorname{div} \boldsymbol{v} + \eta_r \operatorname{rot} (2\boldsymbol{\omega} - \operatorname{rot} \boldsymbol{v})$$

$$+ 2(\operatorname{Grad} \boldsymbol{v})^s \cdot \operatorname{grad} \eta + (\operatorname{div} \boldsymbol{v}) \operatorname{grad} (\eta_v - \tfrac{2}{3}\eta) - (2\boldsymbol{\omega} - \operatorname{rot} \boldsymbol{v}) \wedge \operatorname{grad} \eta_r . \qquad (36)$$

An important special case is obtained, when the viscosities do not appreciably depend on the space coordinates:

$$\rho \frac{\mathrm{d}\boldsymbol{v}}{\mathrm{d}t} = -\operatorname{grad} p + \eta \triangle \boldsymbol{v} + (\tfrac{1}{3}\eta + \eta_v) \operatorname{grad} \operatorname{div} \boldsymbol{v} + \eta_r \operatorname{rot} (2\boldsymbol{\omega} - \operatorname{rot} \boldsymbol{v}) . \qquad (37)$$

These are Navier–Stokes equations, involving besides shear and volume viscosity also rotational viscosity. It contains in the first member, besides the local derivatives $\partial \boldsymbol{v}/\partial t$, also the non-linear term

$$\boldsymbol{v} \cdot \operatorname{grad} \boldsymbol{v} = \operatorname{grad} \tfrac{1}{2}\boldsymbol{v}^2 - \boldsymbol{v} \wedge \operatorname{rot} \boldsymbol{v} . \qquad (38)$$

Frequently one eliminates from (37) either grad div or $\triangle$ by means of the operator equality (in Cartesian coordinates)

$$\operatorname{grad} \operatorname{div} = \triangle + \operatorname{rot} \operatorname{rot} . \qquad (39)$$

It is seen that (32) and (37) form a coupled set of differential equations for $\boldsymbol{\omega}$ and $\boldsymbol{v}$. In many cases, however, $\boldsymbol{\omega}$ and $\tfrac{1}{2}$ rot $\boldsymbol{v}$ will be practically equal after a short time of the order of τ, given by (34). Then the term with η_r has disappeared from the differential equation (37), and one is left with the ordinary Navier–Stokes equation

$$\rho \frac{\mathrm{d}\boldsymbol{v}}{\mathrm{d}t} = -\operatorname{grad} p + \eta \triangle \boldsymbol{v} + (\tfrac{1}{3}\eta + \eta_v) \operatorname{grad} \operatorname{div} \boldsymbol{v} . \qquad (40)$$

This equation can be further simplified for the case of dilute gases, where, according to kinetic theory, the volume viscosity η_v vanishes.

In incompressible fluids one has div $\boldsymbol{v} = 0$. Then the last term in (40) vanishes and the volume viscosity thus disappears again from the equations of motion*.

* A formalism can be set up also for mixtures (*cf.* also Ch. II–IV). An important case studied is the so-called "two-fluid theory", where it is assumed that the momentum transfer between two components of a mixture is partially or totally inhibited. This theory has been used for establishing the macroscopic equations, which describe the behaviour of liquid helium II. [I. Prigogine and P. Mazur, Physica **17** (1951) 661; P. Mazur and I. Prigogine, Physica **17** (1951) 680.]

§ 2. *Viscous Flow in a Magnetic Field*

An originally isotropic fluid (*e.g.* an ionized gas), which is submitted to the influence of an external magnetic field, shows a viscous behaviour of a more complicated character than is described by the phenomenological equations of the preceding section. In fact we shall show that, even supposing again that the antisymmetric part of the pressure tensor plays no role, we need to introduce eight viscosity coefficients (instead of just η and η_v) amongst which one Onsager relation exists.

Let us first write down the entropy production which in absence of heat flow and for vanishing $\boldsymbol{\Pi}^a$, reads according to (23)

$$\sigma = -\frac{\Pi \operatorname{div} \boldsymbol{v}}{T} - \frac{\mathring{\mathsf{\Pi}}^s : (\mathrm{Gr\mathring{a}d}\, \boldsymbol{v})^s}{T} = -\frac{\mathsf{\Pi}^s : (\mathrm{Grad}\, \boldsymbol{v})^s}{T}, \tag{41}$$

where

$$\mathsf{\Pi}^s = \Pi \mathsf{U} + \mathring{\mathsf{\Pi}}^s \tag{42}$$

is the total (symmetric) pressure tensor and

$$(\mathrm{Grad}\, \boldsymbol{v})^s = \tfrac{1}{3} (\operatorname{div} \boldsymbol{v})\, \mathsf{U} + (\mathrm{Gr\mathring{a}d}\, \boldsymbol{v})^s \tag{43}$$

the total (symmetric) velocity gradient tensor. With the help of the fluxes and thermodynamic forces of the last member of (41) we can establish the following phenomenological equations

$$\Pi^s_{\alpha\beta} = -\sum_{\gamma,\delta=1}^{3} L_{\alpha\beta\gamma\delta} (\mathrm{Grad}\, \boldsymbol{v})^s_{\delta\gamma}, \quad (\alpha, \beta = 1, 2, 3), \tag{44}$$

which involve 81 phenomenological coefficients $L_{\alpha\beta\gamma\delta}$. Since the fluxes and thermodynamic forces form symmetric tensors, they contain each 6 independent numbers only, and employing these one can obtain a set of phenomenological equations, containing 36 independent phenomenological coefficients. We shall to this end introduce for convenience the notations $\Pi_1, \Pi_2, \ldots, \Pi_6$ for the components $\Pi^s_{xx}, \Pi^s_{yy}, \Pi^s_{zz}, \Pi^s_{yz} = \Pi^s_{zy}$, $\Pi^s_{zx} = \Pi^s_{xz}$ and $\Pi^s_{xy} = \Pi^s_{yx}$ respectively, and similarly $\chi_1, \chi_2, \ldots, \chi_6$ for $(\mathrm{Grad}\, \boldsymbol{v})^s_{xx}$, $(\mathrm{Grad}\, \boldsymbol{v})^s_{yy}$, $(\mathrm{Grad}\, \boldsymbol{v})^s_{zz}$, $2(\mathrm{Grad}\, \boldsymbol{v})^s_{yz} = 2(\mathrm{Grad}\, \boldsymbol{v})^s_{zy}$, etc., respectively. [The factor 2 is written here because the terms with mixed indices occur twice in (41).] Then the entropy production becomes

$$\sigma = -\frac{1}{T} \sum_{i=1}^{6} \Pi_i \chi_i, \tag{45}$$

and thus we can write as phenomenological equations

$$\Pi_i = -\frac{1}{T}\sum_{k=1}^{6} L_{ik}\chi_k\,, \quad (i = 1, 2, \ldots, 6)\,, \tag{46}$$

with 36 independent phenomenological coefficients L_{ik}, in which the $L_{\alpha\beta\gamma\delta}$ can be expressed, by comparing (44) and (46).

Let us choose the direction of the magnetic field $\boldsymbol{B}$ as the x-axis. To find the effect of the rotational symmetry around the x-axis we apply formula (VI.12) to the coefficients $L_{\alpha\beta\gamma\delta}$, using for the matrix A a rotation R around the x-axis. The simplest procedure is to choose an infinitesimal rotation around the x-axis of the form

$$\mathsf{R} = \begin{pmatrix} 1 & 0 & 0 \\ 0 & 1 & -\alpha \\ 0 & \alpha & 1 \end{pmatrix}, \tag{47}$$

where α is an infinitesimal angle. In this way one finds the form of the scheme $L_{\alpha\beta\gamma\delta}$ by straightforward calculation, neglecting squares and higher powers of α, as one should do for an infinitesimal rotation. The result obtained can alternatively be written in terms of the scheme L_{ik}. It becomes

$$\begin{bmatrix} L_{11} & L_{12} & L_{12} & 0 & 0 & 0 \\ L_{21} & L_{22} & L_{23} & L_{24} & 0 & 0 \\ L_{21} & L_{23} & L_{22} & -L_{24} & 0 & 0 \\ 0 & -L_{24} & L_{24} & \frac{1}{2}(L_{22}-L_{23}) & 0 & 0 \\ 0 & 0 & 0 & 0 & L_{55} & L_{56} \\ 0 & 0 & 0 & 0 & -L_{56} & L_{55} \end{bmatrix}, \tag{48}$$

involving eight different numbers.

Since any axis at right angles to the x-direction is a two-fold axis of rotation, the phenomenological equations are invariant for a rotation of the coordinate system by an angle π about the z-axis, ($x \to -x$, $y \to -y$, $z \to z$). In the new coordinate system $\boldsymbol{B}$ points in the opposite direction with respect to the positive x-axis ($\boldsymbol{B} \to -\boldsymbol{B}$). This leads to the result that $L_{\alpha\beta\gamma\delta}$ is an even or odd function of $\boldsymbol{B}$ if the number of times that the index z occurs in indices of $L_{\alpha\beta\gamma\delta}$ is even or odd respectively. Thus we find, in terms of the coefficients L_{ik}, that

$$\begin{aligned} &L_{11}, L_{12}, L_{21}, L_{22}, L_{23} \text{ and } L_{55} \text{ are even functions of } \boldsymbol{B}\,, \\ &L_{24} \text{ an } \; L_{56} \text{ are odd functions of } \boldsymbol{B}\,. \end{aligned} \tag{49}$$

In addition we have for the phenomenological coefficients the Onsager reciprocal relations

$$L_{ik}(\boldsymbol{B}) = L_{ki}(-\boldsymbol{B})\,, \tag{50}$$

which lead in view of (48) and (49) to only one new relation amongst the eight numbers encountered so far:

$$L_{12} = L_{21}\,, \tag{51}$$

leaving seven independent viscosity coefficients.

We can finally write the phenomenological equations with the help of the fluxes and thermodynamic forces occurring in the second member of (41), which were used also in the preceding section. These equations get a simple form if we introduce the following linear combinations of the seven independent coefficients L_{ik}:

$$\begin{aligned} \eta_1 &= \frac{1}{6T}(2L_{11} - 4L_{12} + L_{22} + L_{23})\,, \\ \eta_2 &= \frac{1}{6T}(L_{11} - 2L_{12} + 2L_{22} - L_{23})\,, \\ \eta_3 &= \frac{1}{T}L_{55}\,, \\ \eta_4 &= \frac{1}{T}L_{24}\,, \\ \eta_5 &= \frac{1}{T}L_{56}\,, \\ \zeta &= \frac{1}{9T}(L_{11} + L_{12} - L_{22} - L_{23})\,, \\ \eta_v &= \frac{1}{9T}(L_{11} + 4L_{12} + 2L_{22} + 2L_{23})\,. \end{aligned} \tag{52}$$

With these quantities we obtain with (48) and (51) the following scheme of phenomenological coefficients connecting the sets of fluxes and thermodynamic forces, indicated in the left-hand column and the first line respectively*:

	$(\overset{\circ}{\text{Grad}}\,\boldsymbol{v})^s_{xx}$	$(\overset{\circ}{\text{Grad}}\,\boldsymbol{v})^s_{yy}$	$(\overset{\circ}{\text{Grad}}\,\boldsymbol{v})^s_{zz}$	$(\overset{\circ}{\text{Grad}}\,\boldsymbol{v})^s_{yz}$	$(\overset{\circ}{\text{Grad}}\,\boldsymbol{v})^s_{zx}$	$(\overset{\circ}{\text{Grad}}\,\boldsymbol{v})^s_{xy}$	$\text{div}\,\boldsymbol{v}$
$\overset{\circ}{\Pi}{}^s_{xx}$	$-2\eta_1$	0	0	0	0	0	-2ζ
$\overset{\circ}{\Pi}{}^s_{yy}$	0	$-2\eta_2$	$-2(\eta_1-\eta_2)$	$-2\eta_4$	0	0	ζ
$\overset{\circ}{\Pi}{}^s_{zz}$	0	$-2(\eta_1-\eta_2)$	$-2\eta_2$	$2\eta_4$	0	0	ζ
$\overset{\circ}{\Pi}{}^s_{yz}$	0	η_4	$-\eta_4$	$2\eta_1-4\eta_2$	0	0	0
$\overset{\circ}{\Pi}{}^s_{zx}$	0	0	0	0	$-2\eta_3$	$-2\eta_5$	0
$\overset{\circ}{\Pi}{}^s_{xy}$	0	0	0	0	$2\eta_5$	$-2\eta_3$	0
Π	-2ζ	ζ	ζ	0	0	0	$-\eta_v$

(53)

The coefficients $\eta_1, \eta_2, \ldots, \eta_5$ connect the components of the tensor $\overset{\circ}{\Pi}{}^s$ to the components of $(\overset{\circ}{\text{Grad}}\,\boldsymbol{v})^s$. They can therefore be called coefficients of shear viscosity. The coefficient η_v connects the traces Π and div $\boldsymbol{v}$ and is therefore the volume viscosity. The seventh coefficient, ζ, describes a cross-effect between shear and volume viscosity.

From (49) and (52) it follows that η_1, η_2, η_3, ζ and η_v are even functions of the magnetic field $\boldsymbol{B}$, whereas η_4 and η_5 are odd functions.

The scheme (53) and the symmetry character of the viscosities with respect to reversal of the magnetic field is in agreement with the results of the kinetic theory of gases in a magnetic field**. It may be noted that in the approximation of the kinetic theory ζ and η_v vanish.

It may be checked that in the absence of the magnetic field we have from spatial symmetry

$$\eta_1 = \eta_2 = \eta_3\,, \quad \eta_4 = 0\,, \eta_5 = 0\,, \quad \zeta = 0\,, \tag{54}$$

* G. J. Hooyman, P. Mazur and S. R. de Groot, Physica **21** (1955) 355.

** S. Chapman and T. G. Cowling, The mathematical theory of non-uniform gases (Cambridge, 1952).

which reduces the scheme (53) to a diagonal form, equivalent to the viscous flow equations (25) and (26) for purely isotropic fluids, involving only two viscosity coefficients, the shear viscosity $\eta = \eta_1 = \eta_2 = \eta_3$ and the volume viscosity η_v.

§ 3. *Propagation of Sound*

In this section and the next two we shall study the effect of irreversible phenomena on the propagation of sound*. The irreversible processes, which may accompany the phenomenon of sound propagation are heat conduction, viscous flow, relaxation phenomena and chemical reactions. They will be shown to give rise to sound dispersion and absorption. We shall first, in the present section, discuss a few notions concerning acoustical phenomena in general.

Let us consider an isotropic fluid system in which a single relaxation process (or a single chemical reaction) can take place besides the phenomena of heat conduction and viscous flow. For such a system the Gibbs entropy law reads [*cf.* Chapters III and X, in particular formula (X.18)]

$$T \frac{\mathrm{d}s}{\mathrm{d}t} = \frac{\mathrm{d}u}{\mathrm{d}t} + p \frac{\mathrm{d}v}{\mathrm{d}t} - A \frac{\mathrm{d}\xi}{\mathrm{d}t}, \tag{55}$$

where A is the affinity of the relaxation process and ξ its progress variable. (The generalization to more than one single relaxation phenomenon is straightforward.) The entropy balance law, which follows from the conservation laws and the entropy law (55), has the usual form (21) with (22), but with an entropy source strength given by

$$\sigma = - \boldsymbol{J}_q \cdot \frac{\operatorname{grad} T}{T^2} - \frac{\Pi \operatorname{div} \boldsymbol{v}}{T} - \frac{\mathring{\Pi}^s : (\mathring{\operatorname{Grad}} \boldsymbol{v})^s}{T} - \frac{\rho}{T} \frac{\mathrm{d}\xi}{\mathrm{d}t} A \geqslant 0. \tag{56}$$

The phenomenological equation for the relaxation process will be

$$\frac{\mathrm{d}\xi}{\mathrm{d}t} = - \frac{l}{\rho T} A \equiv - \beta A, \tag{57}$$

if no cross-effects with div $\boldsymbol{v}$ are assumed to exist (*cf.* however Chapter IV, § 2). The coefficients l and β are positive quantities, because σ must be non-negative definite. The other phenomenological equations

* J.Meixner, Ann. Physik [5] **43** (1943) 470; Acoustica **2** (1952) 101.

are (24), (25) (where we omit a possible cross-term with the affinity A) and (26). We shall suppose that the heat conductivity λ, the shear viscosity η and the volume viscosity η_v are uniform quantities. Effects due to rotational viscosity η_r will be neglected. We have thus the conservation law of matter (II.14):

$$\rho \frac{\mathrm{d}v}{\mathrm{d}t} = \operatorname{div} \boldsymbol{v} , \quad (v \equiv \rho^{-1}) , \tag{58}$$

the conservation law of momentum (40):

$$\rho \frac{\mathrm{d}\boldsymbol{v}}{\mathrm{d}t} = - \operatorname{grad} p + \eta \triangle \boldsymbol{v} + (\tfrac{1}{3}\eta + \eta_v) \operatorname{grad} \operatorname{div} \boldsymbol{v} , \tag{59}$$

and the energy law (19) which becomes upon insertion of the phenomenological equations (24)–(26) and omitting the term due to rotational viscosity:

$$\rho \left(\frac{\mathrm{d}u}{\mathrm{d}t} + p \frac{\mathrm{d}v}{\mathrm{d}t} \right) = \eta_v (\operatorname{div} \boldsymbol{v})^2 + 2\eta (\mathring{\mathrm{Grad}}\, \boldsymbol{v})^s : (\mathring{\mathrm{Grad}}\, \boldsymbol{v})^s + \lambda \triangle T . \tag{60}$$

Finally we have at our disposal three equations of state which can, for instance, be written as

$$p = p(v, \xi, s) , \tag{61}$$

$$A = A(v, \xi, s) , \tag{62}$$

$$T = T(v, \xi, s) . \tag{63}$$

We now have eight equations (55) and (57)–(63), of which one has vectorial character, for eight quantities v, $\boldsymbol{v}$, p, u, T, s, A and ξ, of which one is a vector. These equations can in principle be solved if sufficient initial and boundary conditions are given.

In the usual approach to acoustical problems one linearizes the set of equations in such a way that only first order deviations from a reference state, in which the system is at thermodynamic equilibrium and at rest, are taken into account. The set of equations then becomes, if we use the index 0 for the time independent and uniform equilibrium quantities and if we note that the affinity vanishes at equilibrium:

$$\rho_0 \frac{\partial v}{\partial t} = \operatorname{div} \boldsymbol{v} , \tag{64}$$

$$\rho_0 \frac{\partial \boldsymbol{v}}{\partial t} = - \operatorname{grad} p + \eta \triangle \boldsymbol{v} + (\tfrac{1}{3}\eta + \eta_v) \operatorname{grad} \operatorname{div} \boldsymbol{v} , \tag{65}$$

$$\rho_0 \frac{\partial u}{\partial t} = - p_0 \rho_0 \frac{\partial v}{\partial t} + \lambda \triangle T , \tag{66}$$

$$T_0 \frac{\partial s}{\partial t} = \frac{\partial u}{\partial t} + p_0 \frac{\partial v}{\partial t} , \tag{67}$$

$$\frac{\partial \xi}{\partial t} = - \beta A , \tag{68}$$

$$p - p_0 = \left(\frac{\partial p}{\partial v}\right)_{\xi, s} (v - v_0) + \left(\frac{\partial p}{\partial \xi}\right)_{v, s} (\xi - \xi_0) + \left(\frac{\partial p}{\partial s}\right)_{v, \xi} (s - s_0) , \tag{69}$$

$$A = \left(\frac{\partial A}{\partial v}\right)_{\xi, s} (v - v_0) + \left(\frac{\partial A}{\partial \xi}\right)_{v, s} (\xi - \xi_0) + \left(\frac{\partial A}{\partial s}\right)_{v, \xi} (s - s_0) , \tag{70}$$

$$T - T_0 = \left(\frac{\partial T}{\partial v}\right)_{\xi, s} (v - v_0) + \left(\frac{\partial T}{\partial \xi}\right)_{v, s} (\xi - \xi_0) + \left(\frac{\partial T}{\partial s}\right)_{v, \xi} (s - s_0) . \tag{71}$$

The nine derivatives in the last three equations are the relevant equilibrium properties. Of the three diagonal elements the first is connected to the isentropic compressibility $\chi_{\xi,s}$ at constant ξ as $(\partial p/\partial v)_{\xi,s} = -(v_0\chi_{\xi,s})^{-1}$. Thermodynamic stability requires that the compressibility is a positive quantity. The second is the derivative of the affinity A with respect to ξ at constant v and s, or what amounts to the same, at constant v and u, as (55) shows. It occurs, in combination with the rate quantity β, in the "relaxation time" τ which is defined as the characteristic time in which ξ relaxes to a fraction e^{-1} of its initial value at constant v and u (*cf.* Ch. X). It follows indeed from (68) that

$$\frac{\partial \xi}{\partial t} = - \beta \left(\frac{\partial A}{\partial \xi}\right)_{v, u} (\xi - \xi_0) , \tag{72}$$

and hence

$$\tau = \frac{1}{\beta} \left(\frac{\partial \xi}{\partial A}\right)_{v, u} = \frac{1}{\beta} \left(\frac{\partial \xi}{\partial A}\right)_{v, s} \geqslant 0 . \tag{73}$$

The inequality arises from $\beta \geqslant 0$ (because the entropy production is non-negative definite) and $\partial\xi/\partial A \geqslant 0$, which is a thermodynamic stability condition. The third diagonal element $(\partial T/\partial s)_{v,\xi}$ is equal to $T/c_{v,\xi}$, where $c_{v,\xi}$ is the specific heat at constant v and ξ. It is positive again as a result of thermodynamic stability. The six off-diagonal elements are connected in pairs by the Maxwell relations, which follow from (55):

$$\left(\frac{\partial p}{\partial \xi}\right)_{v,s} = -\left(\frac{\partial A}{\partial v}\right)_{\xi,s}, \quad \left(\frac{\partial p}{\partial s}\right)_{v,\xi} = -\left(\frac{\partial T}{\partial v}\right)_{\xi,s},$$

$$\left(\frac{\partial A}{\partial s}\right)_{v,\xi} = \left(\frac{\partial T}{\partial \xi}\right)_{v,s}. \tag{74}$$

The physical meaning of those quantities is apparent from the notations: *e.g.* the first and last give the change of pressure and temperature per unit change of ξ during the relaxation process at constant volume and entropy. The second is connected to the expansion coefficient at constant ξ and s. We note that all equilibrium quantities at constant ξ must be taken at the equilibrium value $\xi = \xi_0$.

Let us now write each of the eight physical quantities $q(\boldsymbol{r}\,;t)$ as a four-fold Fourier integral of the following form

$$q(\boldsymbol{r}\,;t) = \left(\frac{1}{2\pi}\right)^4 \int_{-\infty}^{\infty}\int_{-\infty}^{\infty} \hat{q}(\boldsymbol{k}\,;\omega)\, \mathrm{e}^{-\mathrm{i}\omega t+\mathrm{i}\boldsymbol{k}\cdot\boldsymbol{r}}\, \mathrm{d}\boldsymbol{k}\, \mathrm{d}\omega\,, \tag{75}$$

which is supposed to exist, and which contains the real parameters ω and $\boldsymbol{k}$. The Fourier transform is given by

$$\hat{q}(\boldsymbol{k}\,;\omega) = \int_{-\infty}^{\infty}\int_{-\infty}^{\infty} q(\boldsymbol{r}\,;t)\, \mathrm{e}^{\mathrm{i}\omega t-\mathrm{i}\boldsymbol{k}\cdot\boldsymbol{r}}\, \mathrm{d}\boldsymbol{r}\, \mathrm{d}t\,. \tag{76}$$

We can extend the definition of the Fourier transformation to complex values

$$\boldsymbol{K} = \boldsymbol{k} + \mathrm{i}\gamma\,, \tag{77}$$

and write

$$\hat{q}(\boldsymbol{K}\,;\omega) = \int_{-\infty}^{\infty}\int_{-\infty}^{\infty} q(\boldsymbol{r}\,;t)\, \mathrm{e}^{\mathrm{i}\omega t-\mathrm{i}\boldsymbol{K}\cdot\boldsymbol{r}}\, \mathrm{d}\boldsymbol{r}\, \mathrm{d}t\,, \tag{78}$$

where γ must lie within certain limits to ensure the convergence of (78) if (75), with $\boldsymbol{k}$ replaced by $\boldsymbol{K}$, is convergent for these values of γ*. We can then write for (75)

$$q(\boldsymbol{r};t) = \left(\frac{1}{2\pi}\right)^4 \int_{-\infty}^{\infty} \int_{-\infty}^{\infty} \hat{q}(\boldsymbol{K};\omega)\, \mathrm{e}^{-\mathrm{i}\omega t+\mathrm{i}\boldsymbol{K}\cdot\boldsymbol{r}}\, \mathrm{d}\boldsymbol{k}\, \mathrm{d}\omega\,, \tag{79}$$

or, with (77),

$$q(\boldsymbol{r};t) = \left(\frac{1}{2\pi}\right)^4 \int_{-\infty}^{\infty} \int_{-\infty}^{\infty} \hat{q}(\boldsymbol{K};\omega)\, \mathrm{e}^{-\mathrm{i}\omega t+\mathrm{i}\boldsymbol{k}\cdot\boldsymbol{r}-\boldsymbol{\gamma}\cdot\boldsymbol{r}}\, \mathrm{d}\boldsymbol{k}\, \mathrm{d}\omega\,. \tag{80}$$

From (78) it follows that

$$\mathrm{i}\boldsymbol{K}\hat{q}(\boldsymbol{K};\omega) = \int_{-\infty}^{\infty} \int_{-\infty}^{\infty} \operatorname{grad} q(\boldsymbol{r};t)\, \mathrm{e}^{\mathrm{i}\omega t-\mathrm{i}\boldsymbol{K}\cdot\boldsymbol{r}}\, \mathrm{d}\boldsymbol{r}\, \mathrm{d}t\,, \tag{81}$$

which shows (if we take a scalar for q) that $\boldsymbol{k}$ and $\boldsymbol{\gamma}$ of (77) are parallel vectors. We can therefore write them as

$$\boldsymbol{k} = k\boldsymbol{u}\,, \quad \boldsymbol{\gamma} = \gamma\boldsymbol{u}\,, \tag{82}$$

where $\boldsymbol{u}$ is a unit vector. From (80) and (82) we note that $q(\boldsymbol{r};t)$ can be represented by a superposition of damped partial waves with amplitudes $\hat{q}(\boldsymbol{K};\omega)\,\mathrm{e}^{-\boldsymbol{\gamma}\cdot\boldsymbol{r}}$, directions of propagation $\boldsymbol{u}$ and phase velocities

$$c = \frac{\omega}{k}\,. \tag{83}$$

The amplitude attenuation per unit of length is determined by the factor γ. The amplitude attenuation factor over one wave length $2\pi/k$ is therefore equal to the dimensionless quantity

$$\mu \equiv \frac{2\pi\gamma}{k}\,. \tag{84}$$

By Fourier transformation (78) we now obtain for (64) and (65)

$$\omega\rho_0\hat{v} = -\,\boldsymbol{K}\cdot\hat{\boldsymbol{v}}\,, \tag{85}$$

* B. van der Pol and H. Bremmer, Operational calculus based on the two-sided Laplace integral (Cambridge, 1955) Ch. II.

$$\mathrm{i}\omega\rho_0\hat{\boldsymbol{v}} = \mathrm{i}\boldsymbol{K}\hat{p} - \eta(\boldsymbol{K}\cdot\boldsymbol{K})\hat{\boldsymbol{v}} - (\tfrac{1}{3}\eta + \eta_v)\boldsymbol{K}(\boldsymbol{K}\cdot\hat{\boldsymbol{v}}) , \tag{86}$$

where $\hat{v}$ and $\hat{\boldsymbol{v}}$ are the Fourier transforms of $v - v_0$ and $\boldsymbol{v} - \boldsymbol{v}_0$. Upon elimination of $\hat{\boldsymbol{v}}$ these equations give

$$\hat{p} = -\left\{\frac{\omega^2}{\boldsymbol{K}\cdot\boldsymbol{K}}\,\rho_0^2 - \mathrm{i}\omega(\tfrac{4}{3}\eta + \eta_v)\rho_0\right\}\hat{v} . \tag{87}$$

Fourier transformation of (66)–(71) yields

$$\omega\rho_0\hat{u} = -\,\omega p_0\rho_0\hat{v} + \mathrm{i}\lambda(\boldsymbol{K}\cdot\boldsymbol{K})\hat{T} , \tag{88}$$

$$T_0\hat{s} = \hat{u} + p_0\hat{v} , \tag{89}$$

$$\mathrm{i}\omega = \beta\hat{A} , \tag{90}$$

$$\hat{p} = \left(\frac{\partial p}{\partial v}\right)_{\xi,s}\hat{v} + \left(\frac{\partial p}{\partial \xi}\right)_{v,s}\hat{\xi} + \left(\frac{\partial p}{\partial s}\right)_{v,\xi}\hat{s} , \tag{91}$$

$$\hat{A} = \left(\frac{\partial A}{\partial v}\right)_{\xi,s}\hat{v} + \left(\frac{\partial A}{\partial \xi}\right)_{v,s}\hat{\xi} + \left(\frac{\partial A}{\partial s}\right)_{v,\xi}\hat{s} , \tag{92}$$

$$\hat{T} = \left(\frac{\partial T}{\partial v}\right)_{\xi,s}\hat{v} + \left(\frac{\partial T}{\partial \xi}\right)_{v,s}\hat{\xi} + \left(\frac{\partial T}{\partial s}\right)_{v,\xi}\hat{s} . \tag{93}$$

The set of algebraic equations (88)–(93) in the Fourier transforms $\hat{p}$, $\hat{v}$, etc. of $p - p_0$, $v - v_0$, etc., which are all functions of $\boldsymbol{K}$ and ω, allows only non-vanishing solutions if the coefficient determinant vanishes. This gives a complex relationship between ω and $\boldsymbol{K}$ of the form

$$f(\omega\,;\,\boldsymbol{K}\cdot\boldsymbol{K}) = 0 , \tag{94}$$

which is equivalent to two real relations. Therefore the solutions of our set of equations must be superpositions of partial waves (sound waves) with $\boldsymbol{K}$ and ω related in this way. The velocity c and the amplitude attenuation γ of the sound waves is now uniquely determined by the frequency ω. Indeed for a given direction of propagation $\boldsymbol{u}$, we can solve the two real quantities k and γ from (94) with (77) and (82). In our case of an isotropic system $\boldsymbol{K}$ occurs only as $\boldsymbol{K}\cdot\boldsymbol{K}$ in equation

(94). The latter quantity, which according to (77) and (82) can be written as

$$\boldsymbol{K}\cdot\boldsymbol{K} = k^2 - \gamma^2 + 2\mathrm{i}k\gamma \;, \tag{95}$$

is independent of $\boldsymbol{u}$. This means that k (or the sound velocity $c = \omega/k$) and γ, solved from (94), will only be functions of ω independent of the direction of propagation $\boldsymbol{u}$*.

Let us now turn our attention to the polarization of the sound waves, that is the direction of vibration, given by the vector $\boldsymbol{v}$ for the displacement of the material elements, with respect to the direction of propagation $\boldsymbol{u}$, which latter is the same as the direction of $\boldsymbol{K}$ as we saw above. Now from equation (86) it is seen that the vector $\boldsymbol{v}$ (which according to (78) is parallel to $\hat{\boldsymbol{v}}$) is parallel to $\boldsymbol{K}$. Thus the polarization of the sound waves in the system studied here is longitudinal, *i.e.* the vibration takes place in the same direction as the propagation.

The set of equations (88)–(93) allows to find the quantities $\hat{p}$, $\hat{T}$, $\hat{\xi}$ and $\hat{A}$ (and also $\hat{u}$ and $\hat{s}$) as linear functions of $\hat{v}$:

$$\hat{p} = \hat{\kappa}\hat{v} \,, \tag{96}$$

$$\hat{T} = \hat{\kappa}_T\hat{v} \,, \tag{97}$$

$$\hat{\xi} = \hat{\kappa}_\xi\hat{v} \,, \tag{98}$$

$$\hat{A} = \hat{\kappa}_A\hat{v} \,. \tag{99}$$

The coefficients found in this way will be called the susceptibilities**.

We shall calculate these susceptibilities and the ensuing properties of sound dispersion and absorption in the next two sections: first (in § 4) for the case where a relaxation process exists, but where effects of viscous flow and heat conduction can be neglected, and then (in § 5) for the case where besides relaxation heat conductivity and viscosity play a role.

* In an anisotropic system, however, equation (94) will in general be of the form $f(\omega;\boldsymbol{K}) = 0$. If k and γ are solved from this equation, one will find expressions which depend both on ω and $\boldsymbol{u}$, showing a dependency on the direction of propagation of the sound velocity and attenuation.

** The quantity $\hat{\kappa}$ introduced here would correspond to a *reciprocal* susceptibility in the sense of Chapter VIII. The reason for this is that for convenience we have inverted the role of the α-variable (here v) and the driving force F (here p), because in the systems considered in this chapter we can more easily fix v and find the effect on p than realize the inverse procedure, followed in Chapter VIII. These reciprocal susceptibilities also satisfy causality conditions (*cf.* Ch. VIII).

To finish this section we shall give the theory of the propagation of sound in a fluid in which all irreversibility can be neglected. Then (87)–(93) give

$$\hat{p} = -\frac{\omega^2}{\mathbf{K}\cdot\mathbf{K}}\rho_0^2\hat{v}, \tag{100}$$

$$\hat{s} = 0, \tag{101}$$

$$\hat{p} = \left(\frac{\partial p}{\partial v}\right)_s \hat{v}, \tag{102}$$

$$\hat{T} = \left(\frac{\partial T}{\partial v}\right)_s \hat{v}, \tag{103}$$

showing the isentropic and non-isothermal character of the propagation of sound. The last two expressions mean in view of (96) and (97) that the susceptibilities are

$$\hat{\kappa} = \left(\frac{\partial p}{\partial v}\right)_s \equiv -\frac{1}{v\chi_s}, \tag{104}$$

$$\hat{\kappa}_T = \left(\frac{\partial T}{\partial v}\right)_s \equiv \frac{1}{v\alpha_s}, \tag{105}$$

where χ_s and α_s are the isentropic compressibility and expansion coefficients respectively. Furthermore (94) reads simply

$$\frac{\omega^2}{\mathbf{K}\cdot\mathbf{K}} = -\frac{\hat{\kappa}}{\rho_0^2} = \frac{1}{\rho_0\chi_s}, \tag{106}$$

which means that $\mathbf{K}$ is real. Therefore no attenuation arises ($\gamma = 0$), and the sound velocity becomes

$$c = \frac{\omega}{k} = \frac{1}{\sqrt{\rho_0\chi_s}}, \tag{107}$$

which is independent of ω (*i.e.* shows no dispersion). This is the result of Laplace. For a monatomic perfect gas the last expression becomes

$$c = \sqrt{\frac{5}{3}\frac{RT}{M}} = \sqrt{\frac{5}{3}\frac{kT}{m}}, \tag{108}$$

where $R = Nk$ is the gas constant and $M = Nm$ the molar mass.

§ 4. *Acoustical Relaxation*

If one neglects the effects of heat conductivity and viscosity, but retains the relaxation phenomenon, one obtains from (87)–(93) the following set of equations:

$$\hat{p} = -\frac{\omega^2}{\mathbf{K}\cdot\mathbf{K}}\rho_0^2\hat{v}, \tag{109}$$

$$\hat{s} = 0, \tag{110}$$

$$i\omega\hat{\xi} = \beta\hat{A}, \tag{111}$$

$$\hat{p} = \left(\frac{\partial p}{\partial v}\right)_{\xi,s}\hat{v} + \left(\frac{\partial p}{\partial \xi}\right)_{v,s}\hat{\xi}, \tag{112}$$

$$\hat{A} = \left(\frac{\partial A}{\partial v}\right)_{\xi,s}\hat{v} + \left(\frac{\partial A}{\partial \xi}\right)_{v,s}\hat{\xi}, \tag{113}$$

$$\hat{T} = \left(\frac{\partial T}{\partial v}\right)_{\xi,s}\hat{v} + \left(\frac{\partial T}{\partial \xi}\right)_{v,s}\hat{\xi}. \tag{114}$$

From the equations (111)–(113) with (73) we find a relation (96) with the acoustic susceptibility

$$\begin{aligned}\hat{\kappa}(\omega) &= \left(\frac{\partial p}{\partial v}\right)_{\xi,s} - \left(\frac{\partial p}{\partial \xi}\right)_{v,s}\left(\frac{\partial \xi}{\partial A}\right)_{v,s}\left(\frac{\partial A}{\partial v}\right)_{\xi,s}\frac{1}{1-i\omega\tau} \\ &= \left(\frac{\partial p}{\partial v}\right)_{\xi,s} + \left(\frac{\partial p}{\partial \xi}\right)_{v,s}\left(\frac{\partial \xi}{\partial v}\right)_{A,s}\frac{1}{1-i\omega\tau},\end{aligned} \tag{115}$$

because $(\partial x/\partial y)_z\,(\partial y/\partial z)_x\,(\partial z/\partial x)_y = -1$, if a functional relation exists between x, y and z. The differentiation at constant A is to be taken at $A = 0$, because the reference state with respect to which we have expanded the thermodynamic quantities is the equilibrium state at which the affinity A vanishes.

For the two limiting cases of zero and infinite frequency this formula leads to

$$\hat{\kappa}(0) = \left(\frac{\partial p}{\partial v}\right)_{\xi,s} + \left(\frac{\partial p}{\partial \xi}\right)_{v,s}\left(\frac{\partial \xi}{\partial v}\right)_{A,s} = \left(\frac{\partial p}{\partial v}\right)_{A,s} = -\frac{1}{v\chi_{A,s}} \leqslant 0, \tag{116}$$

$$\hat{\kappa}(\infty) = \left(\frac{\partial p}{\partial v}\right)_{\xi,s} = -\frac{1}{v\chi_{\xi,s}} \leqslant 0, \tag{117}$$

involving the two isentropic compressibilities $\chi_{A,s}$ and $\chi_{\xi,s}$. The first of these has to be taken at constant affinity A and at equilibrium, *i.e.* at $A = 0$. The other isentropic compressibility has to be taken at constant ξ, *i.e.* in a state in which the relaxation process is frozen in at the value ξ_0 of the progress variable. The inequalities of the two last formulae represent thermodynamic stability conditions. From (73), (74), (116) and (117) we can find a few useful expressions for the difference of $\hat{\kappa}(0)$ and $\hat{\kappa}(\infty)$:

$$\hat{\kappa}(0) - \hat{\kappa}(\infty) = -\left(\frac{\partial p}{\partial \xi}\right)_{v,s}\left(\frac{\partial \xi}{\partial A}\right)_{v,s}\left(\frac{\partial A}{\partial v}\right)_{\xi,s} = \left(\frac{\partial \xi}{\partial A}\right)_{v,s}\left(\frac{\partial A}{\partial v}\right)^2_{\xi,s}$$

$$= \beta\tau\left(\frac{\partial A}{\partial v}\right)^2_{\xi,s} \geqslant 0\,. \tag{118}$$

(The inequality follows from $\partial\xi/\partial A \geqslant 0$, which is again a thermodynamic stability condition.)

On the other hand from (73), (116), (117) and a Maxwell relation which follows from the fact that $\mathrm{d}(u - A\xi)$ is a total differential one has also

$$\hat{\kappa}(0) - \hat{\kappa}(\infty) = \left(\frac{\partial p}{\partial v}\right)_{A,s} - \left(\frac{\partial p}{\partial v}\right)_{\xi,s} = -\left(\frac{\partial p}{\partial A}\right)_{v,s}\left(\frac{\partial A}{\partial v}\right)_{\xi,s}$$

$$= -\left(\frac{\partial \xi}{\partial v}\right)_{A,s}\left(\frac{\partial A}{\partial v}\right)_{\xi,s}$$

$$= \left(\frac{\partial A}{\partial \xi}\right)_{v,s}\left(\frac{\partial \xi}{\partial v}\right)^2_{A,s} = \frac{1}{\beta\tau}\left(\frac{\partial \xi}{\partial v}\right)^2_{A,s} \geqslant 0\,. \tag{119}$$

We can then write (115) as

$$\hat{\kappa}(\omega) = \hat{\kappa}(\infty) + \{\hat{\kappa}(0) - \hat{\kappa}(\infty)\}\,\frac{1}{1 - \mathrm{i}\omega\tau}\,, \tag{120}$$

which has the general form of a relaxation spectrum, discussed in Chapters VIII and X. Splitting $\hat{\kappa}(\omega)$ into its real and imaginary parts

$$\hat{\kappa}(\omega) = \hat{\kappa}'(\omega) + \mathrm{i}\hat{\kappa}''(\omega)\,, \tag{121}$$

we obtain from (120)

$$\hat{\kappa}'(\omega) = \hat{\kappa}(\infty) + \{\hat{\kappa}(0) - \hat{\kappa}(\infty)\}\,\frac{1}{1 + \omega^2\tau^2} < 0\,, \tag{122}$$

$$\hat{\kappa}''(\omega) = \{\hat{\kappa}(0) - \hat{\kappa}(\infty)\} \frac{\omega\tau}{1+\omega^2\tau^2} \geqslant 0, \quad (\text{for } \omega > 0). \tag{123}$$

These two functions satisfy the dispersion relations, discussed in Chapter VIII. The inequalities follow from $\tau \geqslant 0$ [see formula (73)], together with the inequalities in (116)–(118).

From (96) and (109) it follows that equation (94) has here the form

$$-\frac{\omega^2 \rho_0^2}{\boldsymbol{K}\cdot\boldsymbol{K}} = \hat{\kappa}(\omega). \tag{124}$$

We can solve k and γ from this equation if (95) and (121) are inserted. This gives the result

$$\left.\begin{matrix} k \\ \gamma \end{matrix}\right\} = \frac{\omega\rho_0}{\sqrt{2}} \left[\mp \frac{\hat{\kappa}'}{(\hat{\kappa}')^2 + (\hat{\kappa}'')^2} + \left\{ \frac{1}{(\hat{\kappa}')^2 + (\hat{\kappa}'')^2} \right\}^{\frac{1}{2}} \right]^{\frac{1}{2}}, \tag{125}$$

where the upper sign is valid for k, the lower for γ, and where the values of $\hat{\kappa}'$ and $\hat{\kappa}''$ are given by (122) and (123). We know thus the velocity of sound $c = \omega/k$ and the attenuation γ. It follows that c increases from $\rho_0^{-1} \mid \hat{\kappa}(0) \mid^{\frac{1}{2}} = (\rho_0 \chi_{A,s})^{-\frac{1}{2}}$ at $\omega = 0$ to $\rho_0^{-1} \mid \hat{\kappa}(\infty) \mid^{\frac{1}{2}} = (\rho_0 \chi_{\xi,s})^{-\frac{1}{2}}$ at $\omega = \infty$, passing through a "dispersion region" in the neighbourhood of $\omega = \tau^{-1}$. Similarly γ increases from zero at $\omega = 0$ to a maximum near $\omega = \tau^{-1}$ and then decreases to zero again at $\omega = \infty$. Measurements of either the dispersion, *i.e.* the function $c(\omega)$, or the absorption γ as a function of ω permit to determine the relaxation time τ with the help of the general formula given above. The most important practical case arises when the relaxation is due to the transfer of energy between vibrational and translational degrees of freedom in poly-atomic molecules. This phenomenon has been extensively studied experimentally. Another important case is when the relaxation process is a chemical reaction, *e.g.* the dissociation of a bi-atomic gas, the case studied originally by Einstein*.

The general formula (125) simplifies in the special case where

$$\hat{\kappa}(0) - \hat{\kappa}(\infty) \ll \mid \hat{\kappa}(\infty) \mid, \tag{126}$$

which is experimentally realized to some approximation in a number of

* A. Einstein, Sitzungsber. Preussische Akad. Wiss., Berlin (1920) 380.

examples. Then for all values of $\omega\tau$ one has

$$\hat{\kappa}''(\omega) \ll |\hat{\kappa}'(\omega)|, \tag{127}$$

and therefore, from (125), using the fact that $\kappa'(\omega)$ is a negative quantity, approximately

$$k \simeq \omega\rho_0 |\hat{\kappa}'(\omega)|^{-\frac{1}{2}}, \quad c^2 \simeq \rho_0^{-2} |\hat{\kappa}'(\omega)|, \tag{128}$$

$$\gamma \simeq \tfrac{1}{2}\omega\rho_0\hat{\kappa}''(\omega) |\hat{\kappa}(0)|^{-\frac{3}{2}}. \tag{129}$$

With the expressions (122) and (123) these formulae become

$$c^2 \simeq \frac{|\hat{\kappa}(\infty)|}{\rho_0^2} + \frac{|\hat{\kappa}(0) - \hat{\kappa}(\infty)|}{\rho_0^2} \frac{1}{1 + \omega^2\tau^2}, \tag{130}$$

$$\gamma \simeq \frac{\rho_0}{2} \frac{\hat{\kappa}(0) - \hat{\kappa}(\infty)}{|\hat{\kappa}(0)|^{\frac{3}{2}}} \frac{\omega^2\tau}{1 + \omega^2\tau^2} \geqslant 0, \tag{131}$$

or with (84)

$$\mu = \frac{2\pi\gamma}{k} \simeq \pi \frac{\hat{\kappa}''(\omega)}{|\hat{\kappa}(0)|} = \pi \frac{\hat{\kappa}(0) - \hat{\kappa}(\infty)}{|\hat{\kappa}(0)|} \frac{\omega\tau}{1 + \omega^2\tau^2} \geqslant 0. \tag{132}$$

(Note that, if $\omega\tau \ll 1$, the coefficient γ is proportional to $\omega^2\tau$ and the coefficient μ is proportional to $\omega\tau$.)

If c^2 and μ are plotted as functions of $\ln \omega$, one finds a dispersion curve with an inflexion point at $\omega = \tau^{-1}$ and an absorption curve with a maximum at that value of ω.

Returning to the general case of this section, we now calculate from (111)–(114) the susceptibilities defined in (97)–(99). We then find, using also (73) and (74), the temperature susceptibility

$$\hat{\kappa}_T(\omega) = \left(\frac{\partial T}{\partial v}\right)_{\xi, s} + \left(\frac{\partial T}{\partial \xi}\right)_{v, s} \left(\frac{\partial \xi}{\partial v}\right)_{A, s} \frac{1}{1 - \mathrm{i}\omega\tau}, \tag{133}$$

and consequently

$$\hat{\kappa}_T(0) = \left(\frac{\partial T}{\partial v}\right)_{A, s} = \frac{\rho_0}{\alpha_{A, s}}, \tag{134}$$

$$\hat{\kappa}_T(\infty) = \left(\frac{\partial T}{\partial v}\right)_{\xi, s} = \frac{\rho_0}{\alpha_{\xi, s}}, \tag{135}$$

involving the adiabatic expansion coefficients $\alpha_{A,s}$ and $\alpha_{\xi,s}$ at constant A and ξ respectively. Furthermore we find the susceptibilities

$$\hat{\kappa}_{\xi}(\omega) = \left(\frac{\partial \xi}{\partial v}\right)_{A,s} \frac{1}{1 - \mathrm{i}\omega\tau}, \tag{136}$$

$$\hat{\kappa}_{A}(\omega) = -\left(\frac{\partial A}{\partial v}\right)_{\xi,s} \frac{\mathrm{i}\omega\tau}{1 - \mathrm{i}\omega\tau} = \left(\frac{\partial p}{\partial \xi}\right)_{v,s} \frac{\mathrm{i}\omega\tau}{1 - \mathrm{i}\omega\tau}, \tag{137}$$

where (74) has been used.

Let us now turn our attention to the dissipation in the sound waves. The energy which is dissipated per unit mass is according to (60), in our case where $\eta_v = 0$, $\eta = 0$ and $\lambda = 0$, equal to

$$\frac{\mathrm{d}u}{\mathrm{d}t} = -p\frac{\mathrm{d}v}{\mathrm{d}t} = -(p - p_0)\frac{\partial v}{\partial t} - (p - p_0)\,\boldsymbol{v}\cdot\operatorname{grad} v - p_0\frac{\mathrm{d}v}{\mathrm{d}t}. \tag{138}$$

The second contribution is a convection term. We shall study the case of a "monochromatic" sound wave of frequency ω_0, given by [*cf.* (VIII.173)]

$$v(\boldsymbol{r}\,;t) - v_0 = \tfrac{1}{2}v(\boldsymbol{r})\,\mathrm{e}^{-\mathrm{i}\omega_0 t} + \tfrac{1}{2}v^{*}(\boldsymbol{r})\,\mathrm{e}^{\mathrm{i}\omega_0 t}, \tag{139}$$

of which the Fourier transform with respect to t is equal to

$$\hat{v}(\boldsymbol{r}\,;\omega) = \pi\,\{\,v(\boldsymbol{r})\,\delta(\omega - \omega_0) + v^{*}(\boldsymbol{r})\,\delta(\omega + \omega_0)\,\}. \tag{140}$$

Writing (138) with a Fourier integral, one has

$$-p\frac{\mathrm{d}v}{\mathrm{d}t} = -\frac{1}{4\pi^2}\int_{-\infty}^{\infty}\int_{-\infty}^{\infty} \mathrm{i}\omega'\hat{p}(\boldsymbol{r}\,;\omega)\,\hat{v}^{*}(\boldsymbol{r}\,;\omega)\,\mathrm{e}^{-\mathrm{i}(\omega-\omega')t}\,\mathrm{d}\omega\,\mathrm{d}\omega' - p_0\frac{\mathrm{d}v}{\mathrm{d}t}$$
$$+ \text{ convection term}. \tag{141}$$

In view of the fact that $\hat{\kappa}$ is independent of $\boldsymbol{K}$, we have from (96)

$$\hat{p}(\boldsymbol{r}\,;\omega) = \hat{\kappa}(\omega)\,\hat{v}(\boldsymbol{r}\,;\omega), \tag{142}$$

where $\hat{p}(\boldsymbol{r};\omega)$ is the Fourier transform of $p(\boldsymbol{r};t) - p_0$ with respect to the time t only. (It is also the inverse Fourier transform of $\hat{p}(\boldsymbol{k};\omega)$

with respect to $\boldsymbol{k}$.) Hence one obtains for (141)

$$-p\frac{\mathrm{d}v}{\mathrm{d}t} = -\frac{1}{4\pi^2}\int_{-\infty}^{\infty}\int_{-\infty}^{\infty} \mathrm{i}\omega'\hat{\kappa}(\omega)\,\hat{v}(\boldsymbol{r};\omega)\,\hat{v}^*(\boldsymbol{r};\omega')\,\mathrm{e}^{-\mathrm{i}(\omega-\omega')t}\mathrm{d}\omega\,\mathrm{d}\omega' - p_0\frac{\mathrm{d}v}{\mathrm{d}t}$$
$$+ \text{ convection term}, \qquad (143)$$

which upon introduction of (140) yields

$$-p\frac{\mathrm{d}v}{\mathrm{d}t} = -\frac{\mathrm{i}\omega_0}{4}\{\hat{\kappa}(\omega_0) - \hat{\kappa}(-\omega_0)\}\,|\,v(\boldsymbol{r})\,|^2 + \text{oscillating terms}, \qquad (144)$$

where we have not explicitly written the time dependent terms (which oscillate with frequencies ω_0 and $3\omega_0$). With $\hat{\kappa}(\omega) = \hat{\kappa}^*(-\omega)$ [see (120) and (VIII.104)] and (121) the last expression becomes

$$-p\frac{\mathrm{d}v}{\mathrm{d}t} = \tfrac{1}{2}\omega_0\hat{\kappa}''(\omega_0)\,|\,v(\boldsymbol{r})\,|^2 + \text{oscillating terms}. \qquad (145)$$

Thus the energy dissipation per unit mass over a period $2\pi/\omega_0$ is equal to

$$-\int_{-\pi/\omega_0}^{\pi/\omega_0} p\frac{\mathrm{d}v}{\mathrm{d}t}\,\mathrm{d}t = \pi\hat{\kappa}''(\omega_0)\,|\,v(\boldsymbol{r})\,|^2 \geqslant 0\,^*. \qquad (146)$$

The inequality follows from (123).

In the special case where the sound wave is a plane wave one has

$$v(\boldsymbol{r}) = v\,\mathrm{e}^{\mathrm{i}\boldsymbol{K}\cdot\boldsymbol{r}} = v\,\mathrm{e}^{\mathrm{i}\boldsymbol{k}\cdot\boldsymbol{r}-\boldsymbol{\gamma}\cdot\boldsymbol{r}}, \qquad (147)$$

and thus

$$|\,v(\boldsymbol{r})\,|^2 = |\,v\,|^2\,\mathrm{e}^{-2\boldsymbol{\gamma}\cdot\boldsymbol{r}}, \qquad (148)$$

showing the damping arising from the attenuation factor γ.

The entropy production formula (56) gives for our case, where heat conduction and viscous flow are neglected, a dissipation per unit mass

$$\frac{T\sigma}{\rho} = -A\frac{\mathrm{d}\xi}{\mathrm{d}t} = -A\frac{\partial\xi}{\partial t} - A\boldsymbol{v}\cdot\operatorname{grad}\xi. \qquad (149)$$

* This general result can be compared with (VIII. 178).

A calculation along the same lines as above gives, with the use of (98) and (99), for the dissipation per unit mass over a period $2\pi/\omega_0$

$$-\int_{-\pi/\omega_0}^{\pi/\omega_0} A \frac{d\xi}{dt} dt = \tfrac{1}{2}\pi i \left\{ \hat{\kappa}_\xi(\omega_0)\hat{\kappa}_A^*(\omega_0) - \hat{\kappa}_\xi^*(\omega_0)\hat{\kappa}_A(\omega_0) \right\} | v(\boldsymbol{r}) |^2 . \tag{150}$$

With the expressions (136) and (137) for the susceptibilities one gets from the last formula

$$-\int_{-\pi/\omega_0}^{\pi/\omega_0} A \frac{d\xi}{dt} dt = \pi \left(\frac{\partial p}{\partial \xi}\right)_{v,s} \left(\frac{\partial \xi}{\partial v}\right)_{A,s} \frac{\omega_0 \tau}{1+\omega_0^2\tau^2} | v(\boldsymbol{r}) |^2 . \tag{151}$$

From (116), (117) and (123) one finds immediately that this expression can be written as

$$-\int_{-\pi/\omega_0}^{\pi/\omega_0} A \frac{d\xi}{dt} dt = \pi\hat{\kappa}''(\omega_0) | v(\boldsymbol{r}) |^2 \geqslant 0 . \tag{152}$$

This result is the same as (146), as it should be.

We shall finally show that in the low frequency region ($\omega\tau \ll 1$) a relaxation phenomenon has formally the same effect as a volume viscosity, and we shall derive an expression for the "effective volume viscosity" $\bar{\eta}_v$ in terms of relaxation quantities*. Let us first remark that in the approximation $\omega\tau \ll 1$ the susceptibility (120) becomes

$$\hat{\kappa}(\omega) = \hat{\kappa}(0) + \{ \hat{\kappa}(0) - \hat{\kappa}(\infty) \} i\omega\tau , \tag{153}$$

or upon insertion of (116) and (119),

$$\hat{\kappa}(\omega) = \left(\frac{\partial p}{\partial v}\right)_{A,s} + \frac{i\omega}{\beta}\left(\frac{\partial \xi}{\partial v}\right)_{A,s}^2 . \tag{154}$$

Similarly for $\omega\tau \ll 1$ the susceptibility (137) gets the form

$$\hat{\kappa}_A(\omega) = -\left(\frac{\partial A}{\partial v}\right)_{\xi,s} i\omega\tau . \tag{155}$$

* M. Leontovich, Zh. eks. teor. fyz. **6** (1936) 561.
L. I. Mandelshtam and M. A. Leontovich, Zh. eks. teor. fyz. **7** (1937) 438.

Using the last two expressions in (96) and (99), and performing the inverse Fourier transformation, one obtains

$$p - p_0 = \left(\frac{\partial p}{\partial v}\right)_{A,s} (v - v_0) - \frac{1}{\beta}\left(\frac{\partial \xi}{\partial v}\right)^2_{A,s} \frac{\partial v}{\partial t}, \tag{156}$$

$$A = \left(\frac{\partial A}{\partial v}\right)_{\xi,s} \tau \frac{\partial v}{\partial t} = \frac{1}{\beta}\left(\frac{\partial A}{\partial v}\right)_{\xi,s}\left(\frac{\partial \xi}{\partial A}\right)_{v,s} \frac{\partial v}{\partial t} = -\frac{1}{\beta}\left(\frac{\partial \xi}{\partial v}\right)_{A,s} \frac{\partial v}{\partial t}, \tag{157}$$

where in the last line (73) has been applied and also the mathematical relation $(\partial A/\partial v)_\xi \ (\partial v/\partial \xi)_A \ (\partial \xi/\partial A)_v = -1$. With (64) the last two formulae can be written alternatively as

$$p - p_0 = \left(\frac{\partial p}{\partial v}\right)_{A,s} (v - v_0) - \frac{1}{\beta \rho_0}\left(\frac{\partial \xi}{\partial v}\right)^2_{A,s} \operatorname{div} \boldsymbol{v}, \tag{158}$$

$$A = -\frac{1}{\beta \rho_0}\left(\frac{\partial \xi}{\partial v}\right)_{A,s} \operatorname{div} \boldsymbol{v}. \tag{159}$$

We can transform the first of these two formulae still further, if we remark that instead of (61) we could have written the equation of state for the pressure as

$$p = p\,(v, A, s)\,, \tag{160}$$

which gives in the linear approximation

$$p - p_0 = \left(\frac{\partial p}{\partial v}\right)_{A,s} (v - v_0) + \left(\frac{\partial p}{\partial A}\right)_{v,s} A + \left(\frac{\partial p}{\partial s}\right)_{A,s} (s - s_0)\,. \tag{161}$$

Here we have used the fact that the affinity vanishes at equilibrium. The two differential quotients at constant affinity A, which appear, are again equilibrium properties and must thus be taken at $A = 0$. The relation (161) reads for a reversible, isentropic transformation ($A = 0$, $s = s_0$)

$$p^{\text{eq}} - p_0 = \left(\frac{\partial p}{\partial v}\right)_{A,s} (v - v_0)\,. \tag{162}$$

We recognize the equilibrium pressure p^{eq}, given by this relation, in equation (158), which can therefore be transformed into

$$p = p^{\text{eq}} - \frac{1}{\beta \rho_0}\left(\frac{\partial \xi}{\partial v}\right)^2_{A,s} \operatorname{div} \boldsymbol{v}\,. \tag{163}$$

Furthermore the entropy production (56) for the case of the present section (*i.e.* without the heat conduction and viscous flow terms), which reads with (57)

$$\sigma = -\frac{\rho_0}{T}\frac{\mathrm{d}\xi}{\mathrm{d}t}A = \frac{\rho_0}{T}\beta A^2\,, \tag{164}$$

gets upon insertion of (159) the form

$$\sigma = \frac{1}{T\beta\rho_0}\left(\frac{\partial\xi}{\partial v}\right)^2_{A,s}(\mathrm{div}\,\boldsymbol{v})^2\,. \tag{165}$$

We can now formally write the relations (163) and (165) as

$$p - p^{\mathrm{eq}} = \Pi = -\bar{\eta}_v\,\mathrm{div}\,\boldsymbol{v}\,, \tag{166}$$

$$\sigma = -\frac{\Pi\,\mathrm{div}\,\boldsymbol{v}}{T} = \frac{\bar{\eta}_v}{T}(\mathrm{div}\,\boldsymbol{v})^2\,, \tag{167}$$

which are the same as (25) and (23) for a fluid with an "effective volume viscosity"

$$\bar{\eta}_v \equiv \frac{1}{\beta\rho_0}\left(\frac{\partial\xi}{\partial v}\right)^2_{A,s} \geqslant 0\,. \tag{168}$$

(This is a non-negative quantity, because the inequality $\sigma \geqslant 0$ for the entropy source strength implied $\beta \geqslant 0$.)

Thus at low frequencies, *i.e.* at frequencies well below the value τ^{-1}, a relaxation phenomenon can be considered as having the same effect as a volume viscosity. In the case of more than one relaxation process occurring inside the system one can always consider those processes which have relaxation times τ much smaller than the experimentally relevant reciprocal frequencies ω^{-1} as contributing to a total effective volume viscosity, while the other relaxation phenomena must be taken into account as such*.

§ 5. *The Influence of Viscosity and Heat Conduction on the Propagation of Sound*

We shall now take into account the influence of the viscosities η and η_v and of the heat conductivity λ, which were so far supposed to have negligible influence on the propagation of sound.

* J. Meixner, Zeitschr. Physik **131** (1951) 456.

To this end one must first solve the complete set of equations (88)–(93). We shall, however, on doing so, neglect higher powers than the first in the parameter λ. Then, using also (74), one obtains the relation $\hat{p} = \hat{\kappa}\hat{v}$ with the acoustic susceptibility (see also problem 17)

$$\hat{\kappa}(\boldsymbol{K};\omega) = \hat{\kappa}^{(0)}(\omega) - \frac{\mathrm{i}\boldsymbol{K}\cdot\boldsymbol{K}}{\omega}\frac{\lambda}{\rho_0 T_0}\{\hat{\kappa}_T^{(0)}(\omega)\}^2, \tag{169}$$

where $\hat{\kappa}^{(0)}(\omega)$ and $\hat{\kappa}_T^{(0)}(\omega)$ are given by formulae (115) and (133), valid for the pure relaxation case. Now with (87) one finds from the preceding formula the equation relating $\boldsymbol{K}$ with ω:

$$\frac{\omega^2}{\boldsymbol{K}\cdot\boldsymbol{K}} = -\frac{\hat{\kappa}^{(0)}(\omega)}{\rho_0^2} + \frac{\mathrm{i}\omega}{\rho_0}\left[\tfrac{4}{3}\eta + \eta_v + \frac{\boldsymbol{K}\cdot\boldsymbol{K}}{\omega^2\rho_0^2}\frac{\lambda}{T_0}\{\hat{\kappa}_T^{(0)}(\omega)\}^2\right]. \tag{170}$$

If we consider linear effects in λ, η and η_v only, we may replace $\boldsymbol{K}\cdot\boldsymbol{K}$ at the right-hand side by the value which follows from (124). Then one obtains

$$\frac{\omega^2}{\boldsymbol{K}\cdot\boldsymbol{K}} = -\frac{\hat{\kappa}^{(0)}(\omega)}{\rho_0^2} + \frac{\mathrm{i}\omega}{\rho_0}\left[\tfrac{4}{3}\eta + \eta_v - \frac{\lambda}{T_0}\frac{\{\hat{\kappa}_T^{(0)}(\omega)\}^2}{\hat{\kappa}^{(0)}(\omega)}\right]. \tag{171}$$

As in the preceding section the case $\omega\tau \ll 1$ is of special interest. The last equation becomes then

$$\frac{\omega^2}{\boldsymbol{K}\cdot\boldsymbol{K}} = -\frac{1}{\rho_0^2}\left(\frac{\partial p}{\partial v}\right)_{A,s} + \frac{\mathrm{i}\omega}{\rho_0}\left\{\tfrac{4}{3}\eta + \eta_v + \frac{1}{\beta\rho_0}\left(\frac{\partial\xi}{\partial v}\right)_{A,s}^2 - \frac{\lambda}{T_0}\left(\frac{\partial v}{\partial p}\right)_{A,s}\left(\frac{\partial T}{\partial v}\right)_{A,s}^2\right\}. \tag{172}$$

With (107) and (168) this equation can be written as (see also problem 18)

$$\frac{\omega^2}{\boldsymbol{K}\cdot\boldsymbol{K}} = c_0^2 + \frac{\mathrm{i}\omega}{\rho_0}\left(\tfrac{4}{3}\eta + \eta_v + \bar{\eta}_v + \lambda\frac{c_{p,A} - c_{v,A}}{c_{p,A}c_{v,A}}\right), \tag{173}$$

where c_0 is the Laplace sound velocity in the absence of irreversible processes, and where the "effective bulk viscosity" $\bar{\eta}_v$ appears. The quantities $c_{p,A}$ and $c_{v,A}$ are the specific heats at constant affinity A.

In the approximation employed throughout here the last equation leads with (95) to a sound velocity

$$c \equiv \frac{\omega}{k} = c_0 , \tag{174}$$

and a sound attenuation

$$\gamma = \frac{\omega^2}{2c_0^3\rho_0}\left(\tfrac{4}{3}\eta + \eta_v + \bar{\eta}_v + \lambda\frac{c_{p,A} - c_{v,A}}{c_{p,A}c_{v,A}}\right), \tag{175}$$

which is proportional to the square of the frequency and shows an additive character of the four irreversible effects of shear viscosity, bulk viscosity, relaxation "viscosity" and heat conductivity. For mono-atomic perfect gases one has $\eta_v = 0$, $\bar{\eta}_v = 0$, $c_{p,A} = 5k/2m$ and $c_{v,A} = 3k/2m$ (with m the atomic mass and k Boltzmann's constant), so that the expression for γ reduces to

$$\gamma = \frac{\omega^2}{c_0^3\rho_0}\left(\tfrac{2}{3}\eta + \tfrac{2}{15}\lambda\frac{m}{k}\right), \tag{176}$$

known as the classical result of Stokes and Kirchhoff.

For higher values of $\omega\tau$ one must employ the complete formula (171)*.

§ 6. *Elastic Relaxation*

In the preceding sections of this chapter isotropic fluids were considered throughout. Let us now study the behaviour of an arbitrary anisotropic elastic medium, in which a relaxation process is possible**. Then the equation of motion, the energy law and the entropy law have the forms

$$\rho\frac{\mathrm{d}\boldsymbol{v}}{\mathrm{d}t} = -\operatorname{Div} \boldsymbol{P} , \tag{177}$$

* There are indications that for real systems at extremely high values of ω, the "collective motion" representing a sound wave breaks down. See G. E. Uhlenbeck and M. Kac, Boulder Lectures.

** J. Meixner, Zeitschr. Naturforsch. **9**A (1954) 654; Proc. roy. Soc., London A **226** (1954) 51.

A. J. Staverman and F. Schwarzl, Proc. roy. Acad. Sc., Amsterdam B **55** (1952) 474, 486.

D. Polder, Philips Res. Rep. **1** (1945) 5.

G. Falk and J. Meixner, Zeitschr. Naturforsch. **11** A (1956) 782.

B. Manz, Thesis, Aachen (1956).

$$\rho \frac{\mathrm{d}u}{\mathrm{d}t} = - P : \mathrm{Grad}\, \boldsymbol{v} - \mathrm{div}\, \boldsymbol{J}_q , \tag{178}$$

$$T \frac{\mathrm{d}s}{\mathrm{d}t} = \frac{\mathrm{d}u}{\mathrm{d}t} + \frac{1}{\rho} p : \frac{\mathrm{d}\epsilon}{\mathrm{d}t} - A \frac{\mathrm{d}\xi}{\mathrm{d}t} . \tag{179}$$

respectively. Here P is the total pressure tensor (supposed to be symmetric again), whereas p is the equilibrium pressure tensor. The viscous pressure tensor Π is then given by

$$P = p + \Pi . \tag{180}$$

Furthermore ϵ is the symmetric elastic deformation tensor, defined as $(\partial/\partial \boldsymbol{a})\boldsymbol{s}$, where $\boldsymbol{s}$ is the displacement vector of a material point labelled by the coordinate vector $\boldsymbol{a}$ ("Lagrange coordinate"). A "substantial" time derivative is defined as a time derivative at constant Lagrange coordinate $\boldsymbol{a}$. Therefore one can write

$$\frac{\mathrm{d}\epsilon}{\mathrm{d}t} = \frac{\partial^2 \boldsymbol{s}}{\partial t \partial \boldsymbol{a}} = \frac{\partial}{\partial \boldsymbol{a}} \frac{\partial \boldsymbol{s}}{\partial t} = \frac{\partial \boldsymbol{v}(\boldsymbol{a}; t)}{\partial \boldsymbol{a}} = \mathrm{Grad}\, \boldsymbol{v}(\boldsymbol{r}; t) . \tag{181}$$

The last equality follows because at each time t the Lagrange coordinate $\boldsymbol{a}$ is taken equal to the (Euler) coordinate $\boldsymbol{r}$.

From (178)–(181) one finds immediately the entropy balance equation (21) with the entropy flux (22) and the entropy source strength

$$\sigma = - \boldsymbol{J}_q \cdot \frac{\mathrm{grad}\, T}{T^2} - \frac{\Pi}{T} : \frac{\mathrm{d}\epsilon}{\mathrm{d}t} - \frac{\rho}{T} \frac{\mathrm{d}\xi}{\mathrm{d}t} A . \tag{182}$$

We shall now, just as in § 4 of this chapter, neglect heat conduction and take $P = p$, *i.e.* $\Pi = 0$. Then the entropy source strength reduces to

$$\sigma = - \frac{\rho}{T} \frac{\mathrm{d}\xi}{\mathrm{d}t} A . \tag{183}$$

The equations of state can be written in the general form

$$p = p(\epsilon, \xi, s) , \tag{184}$$

$$A = A(\epsilon, \xi, s) , \tag{185}$$

$$T = T(\epsilon, \xi, s) . \tag{186}$$

It is convenient to number the components of the symmetric tensors p and ϵ from one to six, and denote these quantities as 6-dimensional vectors to abbreviate the notation. From the preceding formulae one finds, exactly as in § 4, a set of equations for the Fourier transforms of the phvsical quantities; in particular one has now instead of (110)–(114):

$$\hat{s} = 0\,, \tag{187}$$

$$\mathrm{i}\omega\hat{\xi} = \beta\hat{A}\,, \tag{188}$$

$$\hat{\boldsymbol{p}} = \left(\frac{\partial \boldsymbol{p}}{\partial \boldsymbol{\varepsilon}}\right)_{\xi,s} \cdot \hat{\boldsymbol{\varepsilon}} + \left(\frac{\partial \boldsymbol{p}}{\partial \xi}\right)_{\boldsymbol{\varepsilon},s} \hat{\xi}\,, \tag{189}$$

$$\hat{A} = \left(\frac{\partial A}{\partial \boldsymbol{\varepsilon}}\right)_{\xi,s} \cdot \hat{\boldsymbol{\varepsilon}} + \left(\frac{\partial A}{\partial \xi}\right)_{\boldsymbol{\varepsilon},s} \hat{\xi}\,, \tag{190}$$

$$\hat{T} = \left(\frac{\partial T}{\partial \boldsymbol{\varepsilon}}\right)_{\xi,s} \cdot \hat{\boldsymbol{\varepsilon}} + \left(\frac{\partial T}{\partial \xi}\right)_{\boldsymbol{\varepsilon},s} \hat{\xi}\,. \tag{191}$$

Here the dots indicate sums over the six components of $\boldsymbol{\varepsilon}$. The 36 quantities $(\partial \boldsymbol{p}/\partial \boldsymbol{\varepsilon})_{\xi,s}$ are the adiabatic elastic constants at constant ξ. (This set of quantities can show certain symmetry relations, depending on the spatial symmetry of the medium.) We can solve $\hat{\boldsymbol{p}}$ and $\hat{A}$ from the foregoing set, which gives, again in complete analogy with § 4,

$$\hat{\boldsymbol{p}} = \hat{\boldsymbol{\kappa}} \cdot \hat{\boldsymbol{\varepsilon}}\,, \tag{192}$$

$$\hat{A} = \hat{\boldsymbol{\kappa}}_A \cdot \hat{\boldsymbol{\varepsilon}}\,, \tag{193}$$

where the susceptibilities are given by

$$\hat{\kappa}(\omega) = \hat{\kappa}(\infty) + \{\,\hat{\kappa}(0) - \hat{\kappa}(\infty)\,\}\,\frac{1}{1 - \mathrm{i}\omega\tau}\,, \tag{194}$$

with the adiabatic elastic coefficients at constant A and ξ respectively

$$\hat{\kappa}(0) = \left(\frac{\partial \boldsymbol{p}}{\partial \boldsymbol{\varepsilon}}\right)_{A,s}\,, \tag{195}$$

$$\hat{\kappa}(\infty) = \left(\frac{\partial \boldsymbol{p}}{\partial \boldsymbol{\varepsilon}}\right)_{\xi,s}\,, \tag{196}$$

$$\hat{\boldsymbol{\kappa}}_A(\omega) = -\left(\frac{\partial A}{\partial \boldsymbol{\varepsilon}}\right)_{\xi, s} \frac{\mathrm{i}\omega\tau}{1 - \mathrm{i}\omega\tau} = \left(\frac{\partial \boldsymbol{p}}{\partial \xi}\right)_{\boldsymbol{\varepsilon}, s} \frac{\mathrm{i}\omega\tau}{1 - \mathrm{i}\omega\tau} \,. \tag{197}$$

If the formulae (192) and (193) are transformed back by means of an inverse Fourier transformation, they give the time behaviour of $\boldsymbol{p}$ and A. It turns out then (as explained in Chapter VIII) that the physical situation at a certain moment t depends on the whole past history of the phenomena. The equations thus describe elastic "after-effects" and obey the causality principle.

We shall now show that at low frequencies the elastic relaxation phenomenon considered can be described as due to "viscous effects". Indeed in the region where $\omega\tau \ll 1$, we can write for (194) and (197):

$$\hat{\kappa}(\omega) = \hat{\kappa}(0) + \{\, \hat{\kappa}(0) - \hat{\kappa}(\infty) \,\}\, \mathrm{i}\omega\tau \,, \tag{198}$$

$$\hat{\boldsymbol{\kappa}}_A(\omega) = \left(\frac{\partial \boldsymbol{p}}{\partial \xi}\right)_{\boldsymbol{\varepsilon}, s} \mathrm{i}\omega\tau \,. \tag{199}$$

With these susceptibilities equations (192) and (193) yield upon inverse Fourier transformation and with the use of (181), in analogy to (156)–(163)

$$\boldsymbol{p} = \boldsymbol{p}_0 + \hat{\kappa}(0) \cdot (\boldsymbol{\varepsilon} - \boldsymbol{\varepsilon}_0) - \tau \,\{\, \hat{\kappa}(0) - \hat{\kappa}(\infty) \,\} \cdot \frac{\partial \boldsymbol{\varepsilon}}{\partial t}$$

$$= \boldsymbol{p}^{\mathrm{eq}} - \frac{1}{\beta \rho_0} \left(\frac{\partial \xi}{\partial \boldsymbol{\varepsilon}}\right)_{A, s} \left(\frac{\partial \xi}{\partial \boldsymbol{\varepsilon}}\right)_{A, s} \cdot (\mathrm{Grad}\, \boldsymbol{v})^{\mathrm{s}} \,, \tag{200}$$

$$A = \tau \left(\frac{\partial A}{\partial \boldsymbol{\varepsilon}}\right)_{\xi, s} \cdot \frac{\partial \boldsymbol{\varepsilon}}{\partial t} = -\frac{1}{\beta \rho_0} \left(\frac{\partial \xi}{\partial \boldsymbol{\varepsilon}}\right)_{A, s} \cdot (\mathrm{Grad}\, \boldsymbol{v})^{\mathrm{s}} \,, \tag{201}$$

where $(\mathrm{Grad}\, \boldsymbol{v})^{\mathrm{s}}$ now represents a vector with six components.

With the phenomenological equation (57) and the last expression for A we obtain for the entropy source strength (183)

$$\sigma = \frac{\rho}{T} \beta A^2 = \frac{1}{T\beta\rho} \left(\frac{\partial \xi}{\partial \boldsymbol{\varepsilon}}\right)_{A, s} \left(\frac{\partial \xi}{\partial \boldsymbol{\varepsilon}}\right)_{A, s} : (\mathrm{Grad}\, \boldsymbol{v})^{\mathrm{s}} \, (\mathrm{Grad}\, \boldsymbol{v})^{\mathrm{s}} , \tag{202}$$

where double dots now stand for double sums over the six components of $\boldsymbol{\varepsilon}$ and $(\mathrm{Grad}\, \boldsymbol{v})^{\mathrm{s}}$.

We can compare these results with the case where instead of a relaxation phenomenon, one has viscous flow. In that case one has from (182) the entropy source strength

$$\sigma = -\frac{\boldsymbol{\Pi}}{T}\cdot\frac{\mathrm{d}\boldsymbol{\varepsilon}}{\mathrm{d}t} = -\frac{\boldsymbol{\Pi}}{T}\cdot(\mathrm{Grad}\,\boldsymbol{v})^{s}\,, \tag{203}$$

where we have written a single dot, denoting a sum over six components. Correspondingly one has the phenomenological equations

$$\boldsymbol{\Pi} = -\,\eta\cdot(\mathrm{Grad}\,\boldsymbol{v})^{s}\,, \tag{204}$$

where η has in general 36 components. With this equation (203) becomes

$$\sigma = \frac{1}{T}\,\eta : (\mathrm{Grad}\,\boldsymbol{v})^{s}\,(\mathrm{Grad}\,\boldsymbol{v})^{s}\,. \tag{205}$$

Now comparison of $\boldsymbol{p} - \boldsymbol{p}^{eq}$ from (200) with $\boldsymbol{\Pi}$ from (204) and also comparison of (202) with (205) shows that relaxation in the low frequency region can be described with the help of "effective viscosity coefficients", given by

$$\bar{\eta} = \frac{1}{\beta\rho}\left(\frac{\partial\xi}{\partial\boldsymbol{\varepsilon}}\right)_{A,s}\left(\frac{\partial\xi}{\partial\boldsymbol{\varepsilon}}\right)_{A,s}. \tag{206}$$

The scheme of these 36 coefficients has the same symmetry as the scheme of elastic coefficients for the medium in question.

If more than one single relaxation process takes place inside the system, it is always possible to describe those relaxations, for which $\omega\tau_i \ll 1$ in the frequency region of experimental interest, as resulting from "effective viscosities", whereas the other relaxation processes have to be taken into account as such. Only if all relaxation times τ_i are such that in the relevant frequency band the inequalities $\omega\tau_i \ll 1$ are satisfied, one is entitled to speak of elastic relaxation as "visco-elasticity".

CHAPTER XIII

ELECTRICAL CONDUCTION

§ 1. *Introduction*

In sofar as external forces have been considered in the preceding chapters these forces were assumed to be conservative. In this chapter and the next we shall discuss irreversible phenomena due to electromagnetic forces, which are in general not conservative. For this reason we shall first reformulate the basic conservation laws of momentum and energy, taking into account the presence of an electromagnetic field*. In order to set up these laws use must be made of the equations obeyed by the electromagnetic field. These equations, the Maxwell equations, are summarized in the next section. In the remaining part of this chapter we then consider unpolarized systems, whereas the following chapter is devoted to a discussion of polarized systems.

§ 2. *The Maxwell Equations*

The macroscopic electric and magnetic field strengths $\boldsymbol{E}$ and $\boldsymbol{B}$ in matter obey the Maxwell field equations (in the rationalized Gauss system)

$$\operatorname{div} \boldsymbol{D} = \rho z \,, \tag{1}$$

$$\operatorname{div} \boldsymbol{B} = 0 \,, \tag{2}$$

$$\frac{\partial \boldsymbol{D}}{\partial t} - c \operatorname{rot} \boldsymbol{H} = - \boldsymbol{I} \,, \tag{3}$$

$$\frac{\partial \boldsymbol{B}}{\partial t} + c \operatorname{rot} \boldsymbol{E} = 0 \,. \tag{4}$$

Here ρ is the density of mass, z the electrical charge per unit mass and $\boldsymbol{I}$ the electrical current density. The quantities $\boldsymbol{D}$ and $\boldsymbol{H}$ are the electric and magnetic displacement vectors. In a system at rest these quantities are connected with the fields $\boldsymbol{E}$ and $\boldsymbol{B}$ through the relations

* P. Mazur and I. Prigogine, Mem. Acad. roy. Belg. (Cl. Sc.) **23** (1952) 1.

$$\boldsymbol{D} = \boldsymbol{\varepsilon} \cdot \boldsymbol{E} \,, \tag{5}$$

$$\boldsymbol{H} = \boldsymbol{\mu}^{-1} \cdot \boldsymbol{B} \,, \tag{6}$$

where $\boldsymbol{\varepsilon}$ is the dielectric tensor of the system and $\boldsymbol{\mu}$ the magnetic permeability tensor. For isotropic systems one has

$$\boldsymbol{\varepsilon} = \varepsilon \mathsf{U} \,, \tag{7}$$

$$\boldsymbol{\mu} = \mu \mathsf{U} \,, \tag{8}$$

where ε is the dielectric constant and μ the magnetic permeability.

The electric and the magnetic polarization are defined as

$$\boldsymbol{P} = \boldsymbol{D} - \boldsymbol{E} \,, \tag{9}$$

$$\boldsymbol{M} = \boldsymbol{B} - \boldsymbol{H} \,. \tag{10}$$

Combining (9) and (10) with (5) and (6) respectively, we have for a system at rest

$$\boldsymbol{P} = (\boldsymbol{\varepsilon} - \mathsf{U}) \cdot \boldsymbol{E} = \boldsymbol{\kappa} \cdot \boldsymbol{E} \,, \tag{11}$$

$$\boldsymbol{M} = (\boldsymbol{\mu} - \mathsf{U}) \cdot \boldsymbol{H} = \boldsymbol{\chi} \cdot \boldsymbol{H} \,, \tag{12}$$

where $\boldsymbol{\kappa}$ and $\boldsymbol{\chi}$ are the electric and magnetic susceptibility tensors respectively, which in the isotropic case also reduce to scalar multiples of the unit tensor [*cf.* (7) and (8)].

The electric and magnetic fields $\boldsymbol{E}$ and $\boldsymbol{B}$ may be obtained from the scalar potential φ and the vector potential $\boldsymbol{A}$, according to

$$\boldsymbol{E} = -\operatorname{grad} \varphi - \frac{1}{c} \frac{\partial \boldsymbol{A}}{\partial t} \,, \tag{13}$$

$$\boldsymbol{B} = \operatorname{rot} \boldsymbol{A} \,. \tag{14}$$

We have thus introduced the relevant electromagnetic quantities, which will be needed in this chapter and the next.

§ 3. *Conservation Laws and Entropy Balance in Systems without Polarization*

We consider non-reacting mixtures of n (charged or uncharged) components in an electromagnetic field. It will be assumed throughout the rest of this chapter that polarization phenomena may be neglected so that the fields $\boldsymbol{D}$ and $\boldsymbol{E}$ coincide, and also $\boldsymbol{B}$ and $\boldsymbol{H}$.

a. *The law of conservation of mass* for component k is expressed by [*cf.* (II.13)]

$$\rho \frac{\mathrm{d}c_k}{\mathrm{d}t} = - \operatorname{div} \boldsymbol{J}_k, \quad (k = 1, 2, \ldots, n) . \tag{15}$$

The total electric current density $\boldsymbol{I}$ can be written as follows in terms of the velocities $\boldsymbol{v}_k$ of the components:

$$\boldsymbol{I} = \sum_{k=1}^{n} \rho_k z_k \boldsymbol{v}_k = \rho z \boldsymbol{v} + \sum_{k=1}^{n} z_k \boldsymbol{J}_k , \tag{16}$$

where the definition (II.9) of the diffusion flows $\boldsymbol{J}_k$ has been used and where z_k is the charge per unit mass of component k. The total charge per unit mass of the system is

$$z = \rho^{-1} \sum_{k=1}^{n} \rho_k z_k = \sum_{k=1}^{n} c_k z_k . \tag{17}$$

The second term of the last member of (16) is seen to be the electric current due to the relative motion of the various components. Writing

$$\boldsymbol{i} = \sum_{k=1}^{n} z_k \boldsymbol{J}_k , \tag{18}$$

equation (16) becomes

$$\boldsymbol{I} = \rho z \boldsymbol{v} + \boldsymbol{i} , \tag{19}$$

where $\rho z \boldsymbol{v}$ is the electric current due to convection and $\boldsymbol{i}$ the conduction current.

The law of conservation of charge follows from (15) and (17) in the form

$$\rho \frac{\mathrm{d}z}{\mathrm{d}t} = - \operatorname{div} \boldsymbol{i} . \tag{20}$$

b. *The law of conservation of momentum* for a material system (without polarization) in an electromagnetic field may be written as

$$\frac{\partial}{\partial t}\left(\rho \boldsymbol{v} + \frac{1}{c} \boldsymbol{E} \wedge \boldsymbol{H} \right) = - \operatorname{Div} (\rho \boldsymbol{v}\boldsymbol{v} + \mathsf{P} - \mathsf{T}) . \tag{21}$$

Here $\rho \boldsymbol{v}$ is the momentum density of matter and $c^{-1}\boldsymbol{E} \wedge \boldsymbol{H}$ the momentum density of the electromagnetic field. The tensor $\rho \boldsymbol{v}\boldsymbol{v}$ is the convective part of the momentum flow, P is the pressure tensor of the system (assumed to be symmetrical) and T Maxwell's electromagnetic stress tensor. Equation (21) expresses the fact that the total momentum of matter and field is conserved. Actually since, as we shall see in a moment, the tensor T is completely determined by the field equations of § 2, the momentum balance equation *defines* the pressure tensor P (*cf.* Ch. II, § 3).

From (1)–(4), with $\boldsymbol{D} = \boldsymbol{E}$ and $\boldsymbol{H} = \boldsymbol{B}$, it follows that

$$\frac{1}{c}\frac{\partial(\boldsymbol{E} \wedge \boldsymbol{B})}{\partial t} = \operatorname{Div} \mathsf{T} - \rho z \boldsymbol{E} - \frac{1}{c}\boldsymbol{I} \wedge \boldsymbol{B}, \tag{22}$$

where the Maxwell stress tensor is given by

$$\mathsf{T} = \boldsymbol{E}\boldsymbol{E} + \boldsymbol{B}\boldsymbol{B} - \tfrac{1}{2}(\boldsymbol{E}^2 + \boldsymbol{B}^2)\,\mathsf{U}. \tag{23}$$

Subtracting (22) from (21) we obtain the momentum balance equation for matter

$$\frac{\partial \rho \boldsymbol{v}}{\partial t} = -\operatorname{Div}(\rho \boldsymbol{v}\boldsymbol{v} + \mathsf{P}) + \rho z \boldsymbol{E} + \frac{1}{c}\boldsymbol{I} \wedge \boldsymbol{B}. \tag{24}$$

This equation may be written in the form of an equation of motion

$$\rho \frac{\mathrm{d}\boldsymbol{v}}{\mathrm{d}t} = -\operatorname{Div} \mathsf{P} + \sum_{k=1}^{n} \rho_k z_k \left(\boldsymbol{E} + \frac{1}{c}\boldsymbol{v}_k \wedge \boldsymbol{B}\right), \tag{25}$$

where (16) and (17) have been used. Each term of the sum at the right-hand side contains a factor

$$\boldsymbol{F}_k = z_k \left(\boldsymbol{E} + \frac{1}{c}\boldsymbol{v}_k \wedge \boldsymbol{B}\right), \tag{26}$$

which is the Lorentz force acting on component k per unit mass.

Another useful form of (24) follows with (19):

$$\rho \frac{\mathrm{d}\boldsymbol{v}}{\mathrm{d}t} = -\operatorname{Div} \mathsf{P} + \rho z \left(\boldsymbol{E} + \frac{1}{c}\boldsymbol{v} \wedge \boldsymbol{B}\right) + \frac{1}{c}\boldsymbol{i} \wedge \boldsymbol{B}. \tag{27}$$

c. *Conservation of energy.* Multiplying the equation of motion (27) by $\boldsymbol{v}$ one obtains a balance equation for the kinetic energy of the centre of gravity motion

$$\frac{\partial \frac{1}{2}\rho \boldsymbol{v}^2}{\partial t} = -\operatorname{div}\left(\tfrac{1}{2}\rho \boldsymbol{v}^2 \boldsymbol{v} + \mathsf{P}\cdot \boldsymbol{v}\right) + \mathsf{P} : \operatorname{Grad} \boldsymbol{v} + \rho z \boldsymbol{v}\cdot \boldsymbol{E} - \frac{1}{c}\boldsymbol{i}\cdot(\boldsymbol{v} \wedge \boldsymbol{B}) \tag{28}$$

On the other hand one obtains from (3) and (4), with $\boldsymbol{D} = \boldsymbol{E}$ and $\boldsymbol{H} = \boldsymbol{B}$, the Poynting equation

$$\frac{\partial}{\partial t}\tfrac{1}{2}(\boldsymbol{E}^2 + \boldsymbol{B}^2) = -\operatorname{div} c\,(\boldsymbol{E} \wedge \boldsymbol{B}) - \boldsymbol{I}\cdot \boldsymbol{E}\,. \tag{29}$$

Here $\frac{1}{2}(\boldsymbol{E}^2 + \boldsymbol{B}^2)$ is the electromagnetic energy density, $c(\boldsymbol{E} \wedge \boldsymbol{B})$ the Poynting vector and $\boldsymbol{I}\cdot\boldsymbol{E}$ the work performed by the electromagnetic field. Combining (28) and (29) one has

$$\begin{aligned}\frac{\partial}{\partial t}\tfrac{1}{2}(\rho \boldsymbol{v}^2 + \boldsymbol{E}^2 + \boldsymbol{B}^2) = &-\operatorname{div}\left(\tfrac{1}{2}\rho \boldsymbol{v}^2 \boldsymbol{v} + \mathsf{P}\cdot \boldsymbol{v} + c\boldsymbol{E} \wedge \boldsymbol{B}\right) \\ &+ \mathsf{P} : \operatorname{Grad} \boldsymbol{v} - \boldsymbol{i}\cdot\left(\boldsymbol{E} + \frac{1}{c}\boldsymbol{v} \wedge \boldsymbol{B}\right),\end{aligned} \tag{30}$$

where (19) has been used. This equation expresses the fact that the sum of the densities of barycentric kinetic energy and electromagnetic energy is not conserved, but that an amount of energy $\boldsymbol{i}\cdot(\boldsymbol{E} + c^{-1}\,\boldsymbol{v} \wedge \boldsymbol{B}) - \mathsf{P}:\operatorname{Grad} \boldsymbol{v}$ is transformed into some other form of energy.

However, the density of total energy of matter and field e_v must be conserved (*cf.* Ch. II, § 4):

$$\frac{\partial e_v}{\partial t} = -\operatorname{div} \boldsymbol{J}_e\,, \tag{31}$$

where $\boldsymbol{J}_e$ is the total energy current. We therefore now define the density of internal energy ρu by

$$\rho u = e_v - \tfrac{1}{2}(\rho \boldsymbol{v}^2 + \boldsymbol{E}^2 + \boldsymbol{B}^2)\,. \tag{32}$$

Similarly we define a heat flow $\boldsymbol{J}_q$ by

$$\boldsymbol{J}_q = \boldsymbol{J}_e - \left(\tfrac{1}{2}\rho \boldsymbol{v}^2 \boldsymbol{v} + \rho u \boldsymbol{v} + \mathsf{P}\cdot \boldsymbol{v} + c\boldsymbol{E} \wedge \boldsymbol{B}\right). \tag{33}$$

Subtracting (30) from (31) and using (32) and (33) we obtain the balance equation of internal energy

$$\frac{\partial \rho u}{\partial t} = - \operatorname{div} (\rho u \boldsymbol{v} + \boldsymbol{J}_q) - \mathsf{P} : \operatorname{Grad} \boldsymbol{v} + \boldsymbol{i} \cdot \left(\boldsymbol{E} + \frac{1}{c} \boldsymbol{v} \wedge \boldsymbol{B} \right), \tag{34}$$

or alternatively

$$\rho \frac{\mathrm{d}u}{\mathrm{d}t} = - \operatorname{div} \boldsymbol{J}_q - \mathsf{P} : \operatorname{Grad} \boldsymbol{v} + \boldsymbol{i} \cdot \left(\boldsymbol{E} + \frac{1}{c} \boldsymbol{v} \wedge \boldsymbol{B} \right)$$
$$= - \operatorname{div} \boldsymbol{J}_q - \rho p \frac{\mathrm{d}v}{\mathrm{d}t} - \Pi : \operatorname{Grad} \boldsymbol{v} + \boldsymbol{i} \cdot \left(\boldsymbol{E} + \frac{1}{c} \boldsymbol{v} \wedge \boldsymbol{B} \right), \tag{35}$$

where Π is the viscous pressure tensor (*cf.* Ch. II, § 2 and 4); we consider only non-elastic fluids in this chapter and the next. The last form of (35) represents the amount of electromagnetic energy transformed into internal energy (per unit volume and per unit time). The balance equation of internal energy has been derived here for a system acted upon by a non-conservative force (the Lorentz force). The method followed here is therefore slightly different from the treatment of Chapter II, where only conservative forces were taken into account. In particular the total energy could not be considered as being the sum of the kinetic, the internal and a potential energy of the material part of the system alone.

d. *Entropy balance equation.* For an unpolarized system the entropy per unit mass is a function of the internal energy u, the specific volume v and the mass fractions c_k, just as in the case considered in Chapter II, § 2, where only conservative forces were taken into account. Of course the electromagnetic field may modify the spatial distribution of values of the various state parameters. (This is also the case with a conservative force.) In conformity with the principles put forward in Chapter III the entropy balance can now be obtained from the Gibbs equation

$$T \frac{\mathrm{d}s}{\mathrm{d}t} = \frac{\mathrm{d}u}{\mathrm{d}t} + p \frac{\mathrm{d}v}{\mathrm{d}t} - \sum_{k=1}^{n} \mu_k \frac{\mathrm{d}c_k}{\mathrm{d}t}. \tag{36}$$

Introducing (15) and (35) with (18) into this equation one has

$$\rho \frac{\mathrm{d}s}{\mathrm{d}t} = - \operatorname{div} \left(\frac{\boldsymbol{J}_q - \sum_{k=1}^{n} \mu_k \boldsymbol{J}_k}{T} \right) - \frac{1}{T^2} \boldsymbol{J}_q \cdot \operatorname{grad} T$$
$$- \frac{1}{T} \sum_{k=1}^{n} \boldsymbol{J}_k \cdot \left\{ T \operatorname{grad} \frac{\mu_k}{T} - z_k \left(\boldsymbol{E} + \frac{1}{c} \boldsymbol{v} \wedge \boldsymbol{B} \right) \right\} - \frac{1}{T} \Pi : \operatorname{Grad} \boldsymbol{v}. \tag{37}$$

The entropy flow is thus given by the usual expression

$$J_s = \frac{1}{T}\left(J_q - \sum_{k=1}^{n} \mu_k J_k\right), \tag{38}$$

and the entropy source strength by

$$\sigma = -\frac{1}{T^2} J_q \cdot \text{grad}\, T - \frac{1}{T}\sum_{k=1}^{n} J_k \cdot \left\{ T \,\text{grad}\, \frac{\mu_k}{T} - z_k \left(E + \frac{1}{c} v \wedge B \right) \right\} - \frac{1}{T} \Pi : \text{Grad}\, v\,. \tag{39}$$

Expression (39) can be used as a starting point for the discussion of irreversible phenomena connected with electrical conduction. However, an alternative form for the entropy source strength is more adequate for our purpose: eliminating J_q from (39) by means of (38) one obtains

$$T\sigma = -J_s \cdot \text{grad}\, T - \sum_{k=1}^{n} J_k \cdot \left\{ \text{grad}\, \mu_k - z_k \left(E + \frac{1}{c} v \wedge B \right) \right\} - \Pi : \text{Grad}\, v\,, \tag{40}$$

or with (13) and (14)

$$T\sigma = -J_s \cdot \text{grad}\, T - \sum_{k=1}^{n} J_k \cdot \left\{ \text{grad}\, \tilde{\mu}_k + \frac{z_k}{c} \left(\frac{\partial A}{\partial t} - v \wedge \text{rot}\, A \right) \right\} - \Pi : \text{Grad}\, v\,, \tag{41}$$

where

$$\tilde{\mu}_k = \mu_k + z_k \varphi \tag{42}$$

is the electrochemical potential of component k.

Except for the inclusion of terms containing the vector potential A, expression (41) is identical with the form (III.31) for the entropy source strength of a system with conservative forces (omitting chemical reactions).

The expression (41) demonstrates that the thermodynamic force conjugate to the diffusion flow J_k does not solely consist of the gradient of the electrochemical potential, but contains two additional terms of electromagnetic origin*.

* P. Mazur and I. Prigogine, Mém. Acad. roy. Belg. (Cl. Sc.) **23** (1952) 1.
S. R. de Groot, P. Mazur and H. A. Tolhoek, Physica **19** (1953) 549.

At thermodynamic equilibrium ($\sigma = 0$) it follows from (41) and the condition $\Sigma_{k=1}^{n} \boldsymbol{J}_k = 0$ that

$$\operatorname{grad}(\tilde{\mu}_k - \tilde{\mu}_n) = -\frac{z_k - z_n}{c}\left(\frac{\partial \boldsymbol{A}}{\partial t} - \boldsymbol{v} \wedge \operatorname{rot} \boldsymbol{A}\right), \quad (k = 1, 2, \ldots, n-1). \quad (43)$$

In the absence of barycentric motion and if the vector potential is independent of time this condition becomes:

$$\operatorname{grad}(\tilde{\mu}_k - \tilde{\mu}_n) = 0, \quad (k = 1, 2, \ldots, n-1). \quad (44)$$

This is the usual condition of thermodynamic equilibrium for electrochemical systems with conservative electric forces*.

It may be stressed that the electrochemical potential (42) consists of two parts, *viz.* the ordinary chemical potential, which contains contributions due to short range interactions, and the macroscopic scalar electric potential field φ, which arises as a consequence of long range electromagnetic interactions.

In this connexion it is interesting to consider again the Gibbs relation (36). If the chemical potentials μ_k are eliminated by means of (42) this relation takes the form

$$T\frac{\mathrm{d}s}{\mathrm{d}t} = \frac{\mathrm{d}\tilde{u}}{\mathrm{d}t} + p\frac{\mathrm{d}v}{\mathrm{d}t} - \sum_{k=1}^{n} \tilde{\mu}_k \frac{\mathrm{d}c_k}{\mathrm{d}t} - z\frac{\mathrm{d}\varphi}{\mathrm{d}t}, \quad (45)$$

where the energy $\tilde{u}$ is defined by:

$$\tilde{u} = u + z\varphi, \quad (46)$$

and where (17) has been used. The entropy must now be considered to be a function of the energy $\tilde{u}$, the specific volume v, the mass fractions c_k and also of the scalar potential φ. This shows that the electrical potential φ occurs itself explicitly in thermodynamic relations when electrochemical potentials are introduced; the two parts of the electrochemical potentials play a different role in thermodynamics**. A similar remark may be made with regard to the occurence of the scalar potential in the equation of motion: this equation contains three measurable quantities, the acceleration, the pressure gradient and the Lorentz force (which contains the gradient φ itself). Only if the total charge z vanishes does φ occur exclusively in the combination $\tilde{\mu}_k$ and $\tilde{u}$. Note that the energy per unit mass $\tilde{u}$, which contains the "potential energy" $z\varphi$, is not conserved in general.

§ 4. *Entropy Balance (continued)*

In view of the applications to be discussed in the following sections we shall now establish a special form of the entropy source strength (40). Consider an n-component system in which no viscous flow occurs. The

* E. A. Guggenheim, J. phys. Chem. **33** (1929) 842.

** S. R. de Groot and H. A. Tolhoek, Proc. Kon. Ned. Akad. v. Wet. Amsterdam, **B**, **54** (1951) 42.

entropy source strength (40) is then given by

$$T\sigma = -\,\boldsymbol{J}_s\cdot \operatorname{grad} T - \sum_{k=1}^{n} \boldsymbol{J}_k \cdot \left\{ \operatorname{grad} \mu_k - z_k \left(\boldsymbol{E} + \frac{1}{c}\boldsymbol{v} \wedge \boldsymbol{B} \right) \right\}$$

$$= -\,\boldsymbol{J}_s\cdot \operatorname{grad} T - \sum_{k=1}^{n} \boldsymbol{J}_k \cdot \left\{ \operatorname{grad} \mu_k - z_k \left(\boldsymbol{E} + \frac{1}{c}\boldsymbol{v}_k \wedge \boldsymbol{B} \right) \right\}. \quad (47)$$

The last equality follows from the property

$$\boldsymbol{J}_k\cdot(\boldsymbol{J}_k \wedge \boldsymbol{B}) = 0\,. \quad (48)$$

We now assume the system to be in mechanical equilibrium [*cf.* (25) and (V.2)]:

$$\sum_{k=1}^{n} \rho_k \left\{ (\operatorname{grad} \mu_k)_T - z_k \left(\boldsymbol{E} + \frac{1}{c}\boldsymbol{v}_k \wedge \boldsymbol{B} \right) \right\} = 0\,, \quad (49)$$

or

$$\sum_{k=1}^{n} \rho_k \left\{ \operatorname{grad} \mu_k - z_k \left(\boldsymbol{E} + \frac{1}{c}\boldsymbol{v}_k \wedge \boldsymbol{B} \right) \right\} = -\,\rho s \operatorname{grad} T\,. \quad (50)$$

We may then apply Prigogine's theorem (Ch. V, § 2) in a slightly different form. Indeed one finds from (47) with (50)

$$T\sigma = -\,\{\boldsymbol{J}_s + \rho s\,(\boldsymbol{v} - \boldsymbol{v}^{a})\}\cdot \operatorname{grad} T$$

$$- \sum_{k=1}^{n} \boldsymbol{J}_k^{a} \cdot \left\{ \operatorname{grad} \mu_k - z_k \left(\boldsymbol{E} + \frac{1}{c}\boldsymbol{v}_k \wedge \boldsymbol{B} \right) \right\}. \quad (51)$$

Here $\boldsymbol{v}^{a}$ is, as in Chapter V, § 2, an arbitrary reference velocity and $\boldsymbol{J}_k^{a} = \rho_k(\boldsymbol{v}_k - \boldsymbol{v}^{a})$ the diffusion flow with respect to that velocity. Using again (48), but for the vector $\boldsymbol{J}_k^{a}$, formula (51) becomes

$$T\sigma = -\,\{\boldsymbol{J}_s + \rho s\,(\boldsymbol{v} - \boldsymbol{v}^{a})\}\cdot \operatorname{grad} T$$

$$- \sum_{k=1}^{n} \boldsymbol{J}_k^{a} \cdot \left\{ \operatorname{grad} \mu_k - z_k \left(\boldsymbol{E} + \frac{1}{c}\boldsymbol{v}^{a} \wedge \boldsymbol{B} \right) \right\}. \quad (52)$$

In particular we can take as reference velocity $\boldsymbol{v}^{a}$ the velocity of the nth component $\boldsymbol{v}_n$. This can for instance be the positive ion lattice in a metal, or the neutral solvent in an electrolytic solution. For the case of a metal, or a dilute electrolytic solution it is then natural to choose the reference system in such a way that $\boldsymbol{v}^a = \boldsymbol{v}_n = 0$. The fluxes $\boldsymbol{J}_k^a$ are now given by

$$\boldsymbol{J}_k^a \equiv \boldsymbol{J}_k^r = \rho_k \boldsymbol{v}_k\,, \quad (k = 1, 2, \ldots, n-1)\,, \tag{53}$$

$$\boldsymbol{J}_n^a = 0\,. \tag{54}$$

Expression (52) becomes under these circumstances

$$T\sigma = -\boldsymbol{J}_{s,\,\mathrm{tot}} \cdot \operatorname{grad} T - \sum_{k=1}^{n-1} \boldsymbol{J}_k^r \cdot (\operatorname{grad} \mu_k - z_k \boldsymbol{E})\,, \tag{55}$$

where the "total" entropy flux

$$\boldsymbol{J}_{s,\,\mathrm{tot}} = \boldsymbol{J}_s + \rho s \boldsymbol{v} \tag{56}$$

contains a convective part $\rho s \boldsymbol{v}$. This entropy flux is related to the heat flux $\boldsymbol{J}_q'$ by

$$\boldsymbol{J}_{s,\,\mathrm{tot}} = \frac{1}{T} \boldsymbol{J}_q' + \sum_{k=1}^{n-1} s_k \boldsymbol{J}_k^r\,. \tag{57}$$

This follows from (III.26) with (53), (54) and (56).

We now specialize (55) to a metal, which may be considered as a binary system, of which the first component is formed by the electrons and the second by the positive ion lattice. Since fluxes are measured with respect to this lattice, formula (55) gives for this case

$$T\sigma = -\boldsymbol{J}_{s,\,\mathrm{tot}} \cdot \operatorname{grad} T - \boldsymbol{J}_e^r \cdot (\operatorname{grad} \mu_e - z_e \boldsymbol{E})\,. \tag{58}$$

The suffix e refers to quantities associated with the electrons. The total entropy flux (57) is now simply

$$\boldsymbol{J}_{s,\,\mathrm{tot}} = \frac{1}{T} \boldsymbol{J}_q' + s_e \boldsymbol{J}_e^r\,. \tag{59}$$

Since, according to (16), the total electric current $\boldsymbol{I}$ for the case under consideration, is given by

$$\boldsymbol{I} = z_e \boldsymbol{J}_e^r\,, \tag{60}$$

we finally have

$$T\sigma = -\boldsymbol{J}_{s,\,\mathrm{tot}}\cdot\mathrm{grad}\,T - \boldsymbol{I}\cdot\left\{\mathrm{grad}\left(\frac{\mu_e}{z_e}\right) - \boldsymbol{E}\right\}, \tag{61}$$

an expression for the entropy source strength, which we shall use for the discussion of irreversible phenomena connected with electric conduction in metals.

§ 5. *Electric Resistance*

Let us first consider a system in which only electric conduction takes place but no temperature gradient exists. According to (61) the entropy source strength is then given by

$$T\sigma = \boldsymbol{I}\cdot\left\{\boldsymbol{E} - \mathrm{grad}\left(\frac{\mu_e}{z_e}\right)\right\}. \tag{62}$$

The phenomenological equation becomes

$$\boldsymbol{E} - \mathrm{grad}\left(\frac{\mu_e}{z_e}\right) = \mathsf{R}\cdot\boldsymbol{I}. \tag{63}$$

This is Ohm's law, with R the resistance tensor. In an isotropic fluid or a cubic crystal and in the absence of an external magnetic field (*cf.* Ch. IV and VI) this tensor reduces to a scalar multiple of the unit tensor. In an anisotropic crystal the resistivity tensor R obeys the Onsager relations [*cf.* (IV.58)–(IV.60)]

$$\mathsf{R}(\boldsymbol{B}^e) = \tilde{\mathsf{R}}(-\boldsymbol{B}^e). \tag{64}$$

Here $\boldsymbol{B}^e$ refers to the external, applied magnetic field and must be distinguished from the Maxwell field $\boldsymbol{B}$*.

Let us split the resistance tensor into its symmetric and anti-symmetric parts

$$\mathsf{R}^{(s)} = \tfrac{1}{2}(\mathsf{R} + \tilde{\mathsf{R}}), \tag{65}$$

$$\mathsf{R}^{(a)} = \tfrac{1}{2}(\mathsf{R} - \tilde{\mathsf{R}}). \tag{66}$$

* In previous chapters where no confusion could arise between the applied field $\boldsymbol{B}^e$ and the Maxwell field $\boldsymbol{B}$, we have used for the applied magnetic field in the Onsager relations the symbol $\boldsymbol{B}$.

The Onsager relations (64) imply for these quantities

$$\mathsf{R}^{(s)}(\boldsymbol{B}^e) = \mathsf{R}^{(s)}(-\boldsymbol{B}^e), \tag{67}$$

$$\mathsf{R}^{(a)}(\boldsymbol{B}^e) = -\mathsf{R}^{(a)}(-\boldsymbol{B}^e). \tag{68}$$

The antisymmetric tensor can alternatively be represented as an axial vector with components

$$R_1^{(a)} = -R_{23}^{(a)} = R_{32}^{(a)}, \quad \text{(cycl.)}. \tag{69}$$

The phenomenological equation then becomes

$$\boldsymbol{E} - \operatorname{grad}\left(\frac{\mu_e}{z_e}\right) = \mathsf{R}^{(s)} \cdot \boldsymbol{I} + \boldsymbol{R}^{(a)} \wedge \boldsymbol{I}. \tag{70}$$

In the case of a system which would be isotropic in the absence of an external magnetic field, the resistance tensor has the form

$$\mathsf{R} = \begin{pmatrix} R_{xx} & R_{xy} & 0 \\ -R_{xy} & R_{xx} & 0 \\ 0 & 0 & R_{zz} \end{pmatrix}, \tag{71}$$

if a magnetic field is applied parallel to the z-axis. This follows from invariance of (63) for rotations around this axis. Since furthermore any axis at right angles to the z-axis is a two-fold axis of rotation the relations (63) are also invariant for a rotation of the coordinate system by an angle π about the x-axis. This leads to

$$R_{xx}(\boldsymbol{B}^e) = R_{xx}(-\boldsymbol{B}^e), \quad R_{zz}(\boldsymbol{B}^e) = R_{zz}(-\boldsymbol{B}^e), \tag{72}$$

$$R_{xy}(\boldsymbol{B}^e) = -R_{xy}(-\boldsymbol{B}^e), \tag{73}$$

so that the Onsager relations (67) and (68) are trivially satisfied. The form (71) of the resistance tensor shows that a current in the x-direction gives rise to an electric field in the y-direction, and *vice versa*. This phenomenon is called the Hall effect. The coefficient R_{xy} governing this effect is called the Hall coefficient. In general the Hall effect is described by the axial vector $\boldsymbol{R}^{(a)}$, the so-called Hall vector.

For systems with less symmetry than the one considered above the Onsager relations (67) and (68) may yield additional information about the scheme of phenomenological coefficients. Casimir and Gerritsen* have studied the resistance tensor of bismuth experimentally. This substance is appropriate for a check of the validity of (67) and (68), because the symmetry of the crystal alone would not lead to these relations. Casimir and Gerritsen found that the symmetric tensor components were even functions of the magnetic field strength, in agreement with the Onsager relations (67).

§ 6. *Thermo-electric Potential and Peltier Effect***

We shall now restrict ourselves to isotropic systems in the absence of an applied magnetic field and establish the full scheme of phenomenological equations corresponding to the form (61) of the entropy source strength:

$$\boldsymbol{J}_{s,\,\mathrm{tot}} = -L_{11}\,\mathrm{grad}\,T - L_{12}\left\{\mathrm{grad}\left(\frac{\mu_e}{z_e}\right) - \boldsymbol{E}\right\}, \tag{74}$$

$$\boldsymbol{I} = -L_{21}\,\mathrm{grad}\,T - L_{22}\left\{\mathrm{grad}\left(\frac{\mu_e}{z_e}\right) - \boldsymbol{E}\right\}, \tag{75}$$

with the Onsager relation

$$L_{12} = L_{21}\,. \tag{76}$$

In a somewhat different form, the phenomenological equations may be written as

$$\boldsymbol{J}_{s,\,\mathrm{tot}} = -\frac{\lambda}{T}\,\mathrm{grad}\,T + \frac{\pi}{T}\boldsymbol{I}\,, \tag{77}$$

$$\boldsymbol{E} - \mathrm{grad}\left(\frac{\mu_e}{z_e}\right) = -\eta\,\mathrm{grad}\,T + R\boldsymbol{I}\,. \tag{78}$$

The phenomenological coefficients appearing in these equations are directly related to experimental quantities. Indeed from (77) and (78) it follows that λ is the heat conductivity at zero electrical current, whereas R is the isothermal resistivity of the medium. The coefficient η

* H. B. G. Casimir and A. N. Gerritsen, Physica **8** (1941) 1107.

** I. Prigogine, Etude thermodynamique des phénomènes irréversibles (Desoer, Liège, 1947).
H. B. Callen, Phys. Rev. **73** (1948) 1349.

is called the differential thermo-electric power, and will be seen to be related to the thermo-electric power of a thermocouple. As for π/T, the entropy transported at uniform temperature per unit electric current, this coefficient is directly related to the so-called Peltier effect, to be discussed later on in this section.

We first want to observe that as a consequence of the Onsager relation (76) we also have

$$T\eta = -\pi . \tag{79}$$

This can be seen by solving (74) and (75) for $\boldsymbol{J}_{s,\,\mathrm{tot}}$ and $\boldsymbol{E} - \mathrm{grad}\,(\mu_e/z_e)$. Comparison with (77) and (78) yields a set of relations between the coefficients of (74) and (75) and the coefficients λ, π, η and R of (77) and (78). Application of (76) then immediately gives (79)

Let us now consider a thermocouple. A thermocouple consists essentially of two metals A and B with junctions at temperatures T and $T + \Delta T$ in two heat reservoirs. Furthermore a condensor with plates 1 and 2 is inserted in the metal A; the temperature of both condensor plates is the same.

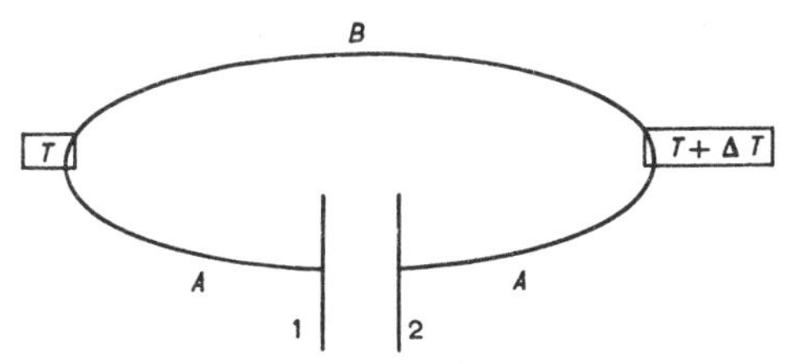

Fig. 1. Thermocouple.

We are interested in the total thermo-electric power or Seebeck effect of such a couple, that is the potential difference $\Delta\varphi$ between the condensor plates at the stationary state of zero electric current ($\boldsymbol{I} = 0$). This potential difference is given by

$$\Delta\varphi = \varphi_1 - \varphi_2 = -\int_1^2 \mathrm{grad}\,\varphi \cdot \mathrm{d}\boldsymbol{l} = \int_1^2 \boldsymbol{E} \cdot \mathrm{d}\boldsymbol{l} , \tag{80}$$

where (13) has been used. (At the stationary state $\partial \boldsymbol{A}/\partial t$ vanishes.) The integration in (80) is performed over any line inside the metallic wires joining the two condensor plates. Inserting relation (78) into (80) with the condition $\boldsymbol{I} = 0$, we get

$$\Delta\varphi = \int_1^2 \operatorname{grad}\left(\frac{\mu_e}{z_e}\right) \cdot \mathrm{d}\boldsymbol{l} - \int_1^2 \eta \operatorname{grad} T \cdot \mathrm{d}\boldsymbol{l}. \tag{81}$$

The first integral on the right-hand side vanishes. This is seen in the following way. The chemical potential μ_e is a function of temperature, pressure and concentration of the electrons only. It has already been mentioned that the temperature is kept uniform across the condensor. Furthermore because of electroneutrality ($z = 0$) the pressure is uniform throughout the system at mechanical equilibrium and when $\boldsymbol{I} = 0$ [*cf.* (25)], and the concentration of electrons on both plates is virtually the same. Thus μ_e must have the same value on both plates and the integral under consideration is indeed zero. The differential thermo-electric power η in each homogeneous wire can be considered in first approximation as a constant independent of temperature, so that according to (81) the final expression for $\Delta\varphi$ is given by

$$\frac{\Delta\varphi}{\Delta T} = \eta_A - \eta_B \,. \tag{82}$$

The following thermo-electric effect which we shall consider here is the Peltier heat, defined as the heat, absorbed at a junction in the state of uniform temperature, per unit electric current across the junctions from metal A to metal B. In order to discuss this effect we shall start from the entropy balance equation applied to the case under consideration. The entropy source has been given by (61). We therefore have

$$\rho \frac{\mathrm{d}s}{\mathrm{d}t} = -\operatorname{div} \boldsymbol{J}_s - \frac{\boldsymbol{J}_{s,\,\mathrm{tot}} \cdot \operatorname{grad} T}{T} - \frac{\boldsymbol{I} \cdot \left\{ \operatorname{grad}\left(\frac{\mu_e}{z_e}\right) - \boldsymbol{E} \right\}}{T}, \tag{83}$$

or alternatively, with (56),

$$\frac{\partial \rho s}{\partial t} = -\operatorname{div} \boldsymbol{J}_{s,\,\mathrm{tot}} - \frac{\boldsymbol{J}_{s,\,\mathrm{tot}} \cdot \operatorname{grad} T}{T} - \frac{\boldsymbol{I} \cdot \left\{ \operatorname{grad}\left(\frac{\mu_e}{z_e}\right) - \boldsymbol{E} \right\}}{T}. \tag{84}$$

If we introduce the phenomenological equations (77) and (78) and use the Onsager relation (79) we obtain

$$\frac{\partial \rho s}{\partial t} = \frac{\mathrm{div}\,(\lambda\,\mathrm{grad}\,T)}{T} - \mathrm{div}\,\frac{\pi \boldsymbol{I}}{T} + \frac{R\boldsymbol{I}^2}{T}. \tag{85}$$

The first term on the right-hand side of (85) represents the change of entropy due to heat conduction; the last term is the Joule heat divided by T; as for the second term, we shall show that it is related to the above-defined Peltier effect.

Consider a junction between the metals A and B, and let us assume first that such a junction can be represented by a small region of volume V between two cross-sections Ω_A and Ω_B of the wire in which the properties and composition of the wire change continuously from those of the pure metal A to those of the pure metal B. At uniform temperature an electric current is flowing from metal A to metal B. The entropy change per unit time $\mathrm{d}S_v/\mathrm{d}t$ in the region between the two cross-sections Ω_A and Ω_B is then given by

$$T\,\frac{\mathrm{d}S_v}{\mathrm{d}t} = -\int_V \mathrm{div}\,\pi \boldsymbol{I}\,\mathrm{d}V + \int_V R\boldsymbol{I}^2\,\mathrm{d}V. \tag{86}$$

Applying Gauss' theorem to the first integral on the right-hand side of (86) we get

$$T\,\frac{\mathrm{d}S_v}{\mathrm{d}t} = \pi_A \int_{\Omega_A} \boldsymbol{I}\cdot\mathrm{d}\boldsymbol{\Omega}_A - \pi_B \int_{\Omega_B} \boldsymbol{I}\cdot\mathrm{d}\boldsymbol{\Omega}_B + \int_V R\boldsymbol{I}^2\,\mathrm{d}V, \tag{87}$$

where π_A and π_B are the values of the coefficient π in the pure metals A and B; $\mathrm{d}\boldsymbol{\Omega}_A$ and $\mathrm{d}\boldsymbol{\Omega}_B$ are both counted positive in the direction of the current $\boldsymbol{I}$.

The corresponding change of entropy for a discontinuous junction is obtained from (87) in the limit $V \to 0$, that is when both cross-sections $\boldsymbol{\Omega}_A$ and $\boldsymbol{\Omega}_B$ coincide. The last integral on the right-hand side of (87) then vanishes since $\boldsymbol{I}$ remains finite, and we find

$$T\,\frac{\mathrm{d}S}{\mathrm{d}t} = (\pi_A - \pi_B) \int_{\Omega} \boldsymbol{I}\cdot\mathrm{d}\boldsymbol{\Omega}, \tag{88}$$

Ω being the cross-section at the junction.

In order to maintain a uniform and constant temperature, this entropy change at a junction will have to be compensated by an absorption of heat from the reservoir, and gives therefore rise to the Peltier effect.

Calling π_{AB} the Peltier heat as defined above [after formula (82)] we have according to (88)

$$\pi_{AB} = \frac{T \dfrac{\mathrm{d}S}{\mathrm{d}t}}{\int \boldsymbol{I} \cdot \mathrm{d}\boldsymbol{\Omega}} = \pi_A - \pi_B \,. \tag{89}$$

Combining (89) and (82), and using again the Onsager relation (79) we get the well-known Thomson relation (second Thomson relation) between the thermo-electric power and the Peltier effect

$$\frac{\Delta\varphi}{\Delta T} = -\frac{\pi_{AB}}{T} \,. \tag{90}$$

A third thermo-electric effect of interest is the Thomson heat. This heat arises simultaneously with the Joule heat, when an electric current flows in a temperature gradient.

This effect is also contained in the entropy balance (85). The second term on the right-hand side of this equation may be transformed into

$$-\operatorname{div}\frac{\pi \boldsymbol{I}}{T} = -\frac{\boldsymbol{I}}{T} \cdot \operatorname{grad}\pi + \frac{\pi \boldsymbol{I}}{T^2} \cdot \operatorname{grad} T \,, \tag{91}$$

because as a consequence of electroneutrality ($z = 0$) the divergence of the electric current vanishes. Writing

$$\operatorname{grad}\pi = (\operatorname{grad}\pi)_T + \frac{\partial \pi}{\partial T} \operatorname{grad} T \,, \tag{92}$$

formula (91) becomes

$$-\operatorname{div}\frac{\pi \boldsymbol{I}}{T} = -\frac{\boldsymbol{I}}{T} \cdot (\operatorname{grad}\pi)_T - \left(\frac{\partial \pi}{\partial T} - \frac{\pi}{T}\right) \frac{\boldsymbol{I}}{T} \cdot \operatorname{grad} T \,. \tag{93}$$

This term thus splits up into two parts: the first gives rise to a heat effect even in the absence of a temperature gradient. In fact only this part was considered above for the discussion of the Peltier effect at a junction. We see that even if no real discontinuity exists, any inhomogeneity of the system gives rise to what may be called a continuous Peltier effect. The second part of (93) corresponds to a heat effect due to the simultaneous presence of an electrical current and a temperature gradient. This is precisely the Thomson heat effect. The coefficient of the Thomson effect, σ_t, is given by

$$\sigma_t = \frac{\pi}{T} - \frac{\partial \pi}{\partial T} = -\eta - \frac{\partial \pi}{\partial T}, \tag{94}$$

where (79) has been used. This relation is known as Thomson's first relation.

Applying (94) to a thermocouple, we find from two such expressions, using also (82), (89) and (90)

$$\sigma_{t_A} - \sigma_{t_B} = T \frac{\partial}{\partial T} \left(\frac{\Delta \varphi}{\Delta T} \right). \tag{95}$$

This is a relation between the Thomson coefficients of the two metals A and B and the temperature coefficient of the thermo-electric power of the thermocouple.

The heat effects in thermocouples have been found here from inspection of the entropy balance. We could, however, have given a similar derivation of these effects starting from a balance equation for the internal energy, a procedure used by different authors*.

§ 7. *Galvanomagnetic and Thermomagnetic Effects***

We shall now consider the system of the previous section, but with an external magnetic field. The phenomenological equations become

$$\boldsymbol{J}_{s,\,\mathrm{tot}} = -\frac{\lambda}{T} \cdot \operatorname{grad} T + \frac{\pi}{T} \cdot \boldsymbol{I}, \tag{96}$$

$$\boldsymbol{E} - \operatorname{grad} \frac{\mu_e}{z_e} = -\eta \cdot \operatorname{grad} T + \mathsf{R} \cdot \boldsymbol{I}, \tag{97}$$

* I. Prigogine, loc. cit. p. 350,
C. A. Domenicali, Rev. mod. Phys. **26** (1954) 237.
** P. Mazur and I. Prigogine, J. Phys. Radium **12** (1951) 616.
H. B. Callen, Phys. Rev. **85** (1952) 16.
R. Fieschi, S. R. de Groot and P. Mazur, Physica **20** (1954) 259.

where $\boldsymbol{\lambda}$, $\boldsymbol{\pi}$, $\boldsymbol{\eta}$ and $\boldsymbol{R}$ are now tensor quantities. For simplicity's sake we take the case that all currents and gradients are parallel to the x–y-plane, whereas the magnetic field is parallel to the z-axis. Since the system is isotropic in the absence of the applied magnetic field the four tensors are of the form [*cf.* Ch. XI, § 1 and formula (71) of this chapter]

$$\boldsymbol{\lambda} = \begin{pmatrix} \lambda_{xx} & \lambda_{xy} \\ -\lambda_{xy} & \lambda_{xx} \end{pmatrix}, \quad \boldsymbol{\pi} = \begin{pmatrix} \pi_{xx} & \pi_{xy} \\ -\pi_{xy} & \pi_{xx} \end{pmatrix}, \tag{98}$$

$$\boldsymbol{\eta} = \begin{pmatrix} \eta_{xx} & \eta_{xy} \\ -\eta_{xy} & \eta_{xx} \end{pmatrix}, \quad \boldsymbol{R} = \begin{pmatrix} R_{xx} & R_{xy} \\ -R_{xy} & R_{xx} \end{pmatrix}. \tag{99}$$

The tensor elements have the parity properties [*cf. e.g.* (72) and (73)]:

$$\begin{aligned} \lambda_{xx}(\boldsymbol{B}^e) &= \lambda_{xx}(-\boldsymbol{B}^e), & \pi_{xx}(\boldsymbol{B}^e) &= \pi_{xx}(-\boldsymbol{B}^e), \\ \lambda_{xy}(\boldsymbol{B}^e) &= -\lambda_{xy}(-\boldsymbol{B}^e), & \pi_{xy}(\boldsymbol{B}^e) &= -\pi_{xy}(-\boldsymbol{B}^e), \end{aligned} \tag{100}$$

$$\begin{aligned} \eta_{xx}(\boldsymbol{B}^e) &= \eta_{xx}(-\boldsymbol{B}^e), & R_{xx}(\boldsymbol{B}^e) &= R_{xx}(-\boldsymbol{B}^e), \\ \eta_{xy}(\boldsymbol{B}^e) &= -\eta_{xy}(-\boldsymbol{B}^e), & R_{xy}(\boldsymbol{B}^e) &= -R_{xy}(-\boldsymbol{B}^e). \end{aligned} \tag{101}$$

Instead of the Onsager relation (79) we have here

$$T\boldsymbol{\eta}(\boldsymbol{B}^e) = -\tilde{\boldsymbol{\pi}}(-\boldsymbol{B}^e), \tag{102}$$

where $\tilde{\boldsymbol{\pi}}$ is the transposed matrix of $\boldsymbol{\pi}$. The form (102) can be found as follows: one first writes down the phenomenological equations (74) and (75) with tensor coefficients which obey the Onsager relation

$$\mathsf{L}_{12}(\boldsymbol{B}^e) = \tilde{\mathsf{L}}_{21}(-\boldsymbol{B}^e). \tag{103}$$

Transforming to the form (96) and (97) one then obtains from (103) the Onsager relation (102). In terms of the tensor components (102) reads

$$T\eta_{xx}(\boldsymbol{B}^e) = -\pi_{xx}(-\boldsymbol{B}^e) = -\pi_{xx}(\boldsymbol{B}^e), \tag{104}$$

$$T\eta_{xy}(\boldsymbol{B}^e) = -\pi_{yx}(-\boldsymbol{B}^e) = \pi_{xy}(-\boldsymbol{B}^e) = -\pi_{xy}(\boldsymbol{B}^e). \tag{105}$$

The last equalities in (104) and (105) are consequences of the parity properties (100). We note that with (104) and (105) relation (102) may also be written as

$$T\eta(\boldsymbol{B}^{e}) = -\pi(\boldsymbol{B}^{e}) . \tag{106}$$

A number of physical effects will now be studied. If these effects are caused by an electric current one speaks of *galvanomagnetic effects*. If they are the result of a heat current they are called *thermomagnetic effects*. The effects may furthermore be divided into *transversal effects*, in which the primary current is perpendicular to the produced effect, and into *longitudinal effects*, in which the primary current and the resulting effect have the same direction. One furthermore speaks of an isothermal effect, when the temperature gradient perpendicular to the primary current vanishes, and of an adiabatic effect, when the heat current perpendicular to the primary current is zero.

We shall give the definitions* of twelve effects and express the relevant coefficients in terms of the coefficients (98) and (99).

I. GALVANOMAGNETIC EFFECTS

a. *Transversal*

1. The isothermal Hall effect $R_{\mathrm{i}}^{\mathrm{t}}$

Definition

$$R_{\mathrm{i}}^{\mathrm{t}} \equiv \frac{\left(E_y - \dfrac{1}{z_{\mathrm{e}}}\dfrac{\partial \mu_{\mathrm{e}}}{\partial y}\right)}{I_x}, \tag{107}$$

with the conditions

$$I_y = 0 , \quad \mathrm{grad}\, T = 0 . \tag{108}$$

According to (97) and (99) one has

$$R_{\mathrm{i}}^{\mathrm{t}} = -R_{xy} . \tag{109}$$

2. The adiabatic Hall effect $R_{\mathrm{a}}^{\mathrm{t}}$

This effect is again defined by the right-hand side of (107), but now with the conditions

$$I_y = 0 , \quad (\boldsymbol{J}_{s,\,\mathrm{tot}})_y = 0 , \quad \frac{\partial T}{\partial x} = 0 . \tag{110}$$

* See for these definitions: W. Meissner, Handbuch der Experimentalphysik, Leipsick, XI, **2** (1935) 311.

According to (96), (97), (98) and (99) one then has

$$R_a^t = -R_{xy} + \frac{\eta_{xx}\pi_{xy}}{\lambda_{xx}}. \tag{111}$$

3. The Ettingshausen effect P^t
This effect is defined as

$$P^t \equiv \frac{\dfrac{\partial T}{\partial y}}{I_x}, \tag{112}$$

with the conditions

$$I_y = 0\,, \quad (\boldsymbol{J}_{s,\,\mathrm{tot}})_y = 0\,, \quad \frac{\partial T}{\partial x} = 0. \tag{113}$$

From (96) and (98) it follows that

$$P^t = -\frac{\pi_{xy}}{\lambda_{xx}}. \tag{114}$$

b. *Longitudinal*

1. The isothermal electric resistivity R_i^l
Definition

$$R_i^l \equiv \frac{\left(E_x - \dfrac{1}{z_e}\dfrac{\partial \mu_e}{\partial x}\right)}{I_x}, \tag{115}$$

when

$$I_y = 0\,, \quad \mathrm{grad}\, T = 0. \tag{116}$$

One has from (97) and (99)

$$R_i^l = R_{xx}. \tag{117}$$

2. The adiabatic electric resistivity R_a^l
This coefficient is again defined by the right-hand side of (115), but with the conditions

$$I_y = 0\,, \quad (\boldsymbol{J}_{s,\,\mathrm{tot}})_y = 0\,, \quad \frac{\partial T}{\partial x} = 0\,. \tag{118}$$

One finds from (96)–(99)

$$R_a^l = R_{xx} + \frac{\eta_{xy}\pi_{xy}}{\lambda_{xx}}. \tag{119}$$

II. THERMOMAGNETIC EFFECTS

a. *Transversal*

1. The Righi–Leduc effect S^{t}
Definition

$$S^{\mathrm{t}} \equiv \frac{\dfrac{\partial T}{\partial y}}{\dfrac{\partial T}{\partial x}}, \tag{120}$$

when

$$\boldsymbol{I} = 0, \quad (\boldsymbol{J}_{s,\,\mathrm{tot}})_y = 0. \tag{121}$$

From (96) and (98) one finds

$$S^{\mathrm{t}} = \frac{\lambda_{xy}}{\lambda_{xx}}. \tag{122}$$

2. The isothermal Nernst effect $Q_{\mathrm{i}}^{\mathrm{t}}$
Definition

$$Q_{\mathrm{i}}^{\mathrm{t}} \equiv \frac{\left(E_y - \dfrac{1}{z_{\mathrm{e}}}\dfrac{\partial \mu_{\mathrm{e}}}{\partial y}\right)}{\dfrac{\partial T}{\partial x}}, \tag{123}$$

with

$$\boldsymbol{I} = 0, \quad \frac{\partial T}{\partial y} = 0. \tag{124}$$

From (97) and (99) it follows that

$$Q_{\mathrm{i}}^{\mathrm{t}} = \eta_{xy}. \tag{125}$$

3. The adiabatic Nernst effect $Q_{\mathrm{a}}^{\mathrm{t}}$
This effect is again defined by the right-hand side of (123) but with

$$\boldsymbol{I} = 0, \quad (\boldsymbol{J}_{s,\,\mathrm{tot}})_y = 0. \tag{126}$$

From (96)–(99) we have

$$Q_{\mathrm{a}}^{\mathrm{t}} = \eta_{xy} - \frac{\eta_{xx}\lambda_{xy}}{\lambda_{xx}}. \tag{127}$$

b. *Longitudinal*

1. The isothermal heat conductivity λ_i

Definition

$$\lambda_i \equiv \frac{-T\,(\boldsymbol{J}_{s,\,\mathrm{tot}})_x}{\dfrac{\partial T}{\partial x}}\,, \tag{128}$$

with

$$\boldsymbol{I} = 0\,, \quad \frac{\partial T}{\partial y} = 0\,. \tag{129}$$

From (96) and (98) one has

$$\lambda_i = \lambda_{xx}\,. \tag{130}$$

2. The adiabatic heat conductivity λ_a

This coefficient is also defined by the right-hand side of (128), but with

$$\boldsymbol{I} = 0\,, \quad (\boldsymbol{J}_{s,\,\mathrm{tot}})_y = 0\,. \tag{131}$$

From (96) and (98) it follows that

$$\lambda_a = \lambda_{xx} + \frac{\lambda_{xy}^2}{\lambda_{xx}}\,. \tag{132}$$

3. The isothermal Ettingshausen–Nernst effect Q_i^l

Definition

$$Q_i^l \equiv \frac{\left(E_x - \dfrac{1}{z_e}\dfrac{\partial \mu_e}{\partial x}\right)}{\dfrac{\partial T}{\partial x}}\,, \tag{133}$$

with

$$\boldsymbol{I} = 0\,, \quad \frac{\partial T}{\partial y} = 0\,. \tag{134}$$

From (97) and (99) one has

$$Q_i^l = -\,\eta_{xx}\,. \tag{135}$$

4. The adiabatic Ettingshausen–Nernst effect Q_a^l

This coefficient is defined by the right-hand side of (133), but with

$$\boldsymbol{I} = 0\,, \quad (\boldsymbol{J}_{s,\,\mathrm{tot}})_y = 0\,. \tag{136}$$

From (96)–(99) one obtains

$$Q_a^l = -\eta_{xx} - \frac{\eta_{xy}\lambda_{xy}}{\lambda_{xx}}. \tag{137}$$

The twelve effects defined above involve seven of the eight coefficients appearing in (98) and (99). (The coefficient π_{xx} does not occur in any of the expressions derived for the galvanomagnetic and thermomagnetic effects considered here; we have seen in § 6 that π_{xx} is related to the Peltier effect.) There must therefore be five relations amongst these effects. These are, as can be verified from the explicit expressions for the coefficients:

$$R_a^t - R_i^t = Q_i^l P^t, \tag{138}$$

$$R_a^l - R_i^l = -Q_i^t P^t, \tag{139}$$

$$Q_a^t - Q_i^t = Q_i^l S^t, \tag{140}$$

$$Q_a^l - Q_i^l = -Q_i^t S^t, \tag{141}$$

$$\lambda_a - \lambda_i = \lambda_i (S^t)^2. \tag{142}$$

Formulae (138) and (140) are known as the Heulinger relations. We note that all these relations connect longitudinal with transversal effects.

Furthermore due to the Onsager relation (105), we must have an additional connexion amongst the effects considered. (The Onsager relation (104), involving the coefficient π_{xx}, does not establish a new connexion between the effects studied in this section; we have shown in § 6 that it leads to the second Thomson relation.) Indeed it follows from (105), (112), (125) and (128) that

$$T\,Q_i^t = \lambda_i\, P^t. \tag{143}$$

This relation between the isothermal Nernst effect and the Ettingshausen effect is known as the Bridgman relation.

All the effects considered above are defined locally. We must still relate these effects to the measurable coefficients. We shall do this for the two effects occurring in the Bridgman relation (143).

Let us consider a system consisting of a rectangular metallic sample A and of the metallic wires B, which connect two points on the opposite faces at right angles to the y-axis to a condensor with plates 1 and 4 (see Fig. 2) Only the metallic sample is subjected to the magnetic

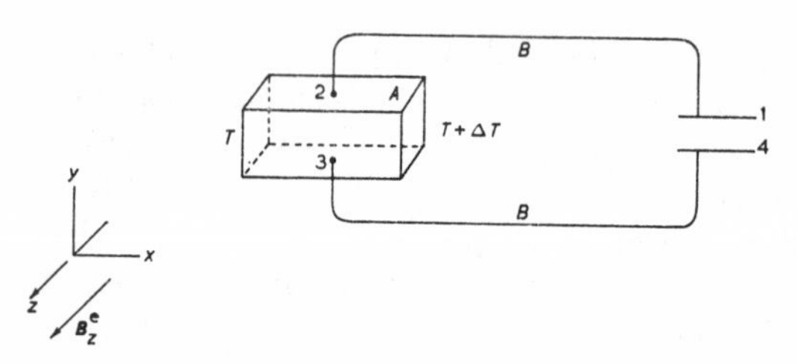

Fig. 2. Sample with connected wires for the measurement of the isothermal Nernst effect.

field $\boldsymbol{B}^e$ in the z-direction. The two faces perpendicular to the x-axis are kept at temperatures T and $T + \Delta T$. The two junctions 2 and 3 are chosen in such a way that without an applied magnetic field no potential difference is measured across the condensor. The measurable isothermal Nernst effect $\overline{Q_i^!}$ is defined as minus the potential difference across the condensor plates divided by ΔT, when the junctions 2 and 3 and also the condensor plates 1 and 4 are kept at the same temperature, while the current $\boldsymbol{I}$ vanishes in all directions. We have

$$\Delta\varphi = \varphi_1 - \varphi_4 = -\int_1^4 \operatorname{grad}\varphi \cdot \mathrm{d}\boldsymbol{l} = \int_1^4 \boldsymbol{E}\cdot\mathrm{d}\boldsymbol{l}$$

$$= \int_1^2 \boldsymbol{E}\cdot\mathrm{d}\boldsymbol{l} + \int_2^3 E_y\,\mathrm{d}y + \int_3^4 \boldsymbol{E}\cdot\mathrm{d}\boldsymbol{l}. \tag{144}$$

Introducing into the first and the third integral of the last member equation (78) with $\boldsymbol{I} = 0$ and grad $T = 0$, and into the second integral (97) with $\boldsymbol{I} = 0$ and $\partial T/\partial y = 0$, we have

$$\Delta\varphi = \int_1^2 \operatorname{grad}\frac{\mu_e}{z_e}\cdot\mathrm{d}\boldsymbol{l} + \int_2^3 \frac{1}{z_e}\frac{\partial\mu_e}{\partial y}\mathrm{d}y + \int_2^3 \eta_{xy}\frac{\partial T}{\partial x}\mathrm{d}y + \int_3^4 \operatorname{grad}\frac{\mu_e}{z_e}\cdot\mathrm{d}\boldsymbol{l}$$

$$= \int_1^4 \operatorname{grad} \frac{\mu_e}{z_e} \cdot d\boldsymbol{l} + \int_2^3 \eta_{xy} \frac{\partial T}{\partial x} dy$$

$$= \frac{1}{z_e} \{ (\mu_e^B)_4 - (\mu_e^B)_1 \} + \eta_{xy} \int_2^3 \frac{\partial T}{\partial x} dy, \tag{145}$$

where η_{xy} has been considered to be a constant independent of temperature in first approximation. The expression between curly brackets vanishes because the chemical potential of the electrons is the same at both condensor plates [see the discussion after formula (81)]. We therefore have, using (125),

$$\overline{Q_i^t} = -\frac{\Delta\varphi}{\Delta T} = \eta_{xy} \frac{1}{\Delta T} \int_3^2 \frac{\partial T}{\partial x} dy$$

$$= Q_i^t \frac{1}{\Delta T} \int_3^2 \frac{\partial T}{\partial x} dy. \tag{146}$$

The measurable Ettingshausen effect $\overline{P^t}$ is defined as the temperature difference $(\Delta T)_y$ between the points 2 and 3, divided by the total current in the x-direction, when the current and the entropy flow (heat flow) vanish in the y-direction, and when the temperature is the same on both faces perpendicular to the x-axis. For the temperature difference $(\Delta T)_y$ we have

$$(\Delta T)_y = T_2 - T_3 = \int_3^2 \frac{\partial T}{\partial y} dy$$

$$= -\frac{\pi_{xy}}{\lambda_{xx}} \int_3^2 I_x \, dy = P^t \int_3^2 I_x \, dy, \tag{147}$$

or for the measurable Ettingshausen coefficient

$$\overline{P^t} = \frac{(\Delta T)_y}{\iint I_x \, dy \, dz} = \frac{P^t}{\Delta z}, \tag{148}$$

where Δz is the length of the sample in the z-direction, and where I_x has been assumed to be independent of z.

Finally the isothermal heat transmission factor $\overline{\lambda_i}$ of the sample is given by

$$\overline{\lambda_i} = -\frac{\iint T(\boldsymbol{J}_{s,\,\mathrm{tot}})_x \,\mathrm{d}y\,\mathrm{d}z}{\Delta T}$$

$$= \lambda_i \frac{\left(\int_3^2 \frac{\partial T}{\partial x}\,\mathrm{d}y\right)\Delta z}{\Delta T}, \tag{149}$$

where $\partial T/\partial x$ has also been assumed to be independent of z.

Using relation (143) we have with (146), (148) and (149)

$$T\overline{Q_i^t} = \overline{\lambda_i}\,\overline{P^t}, \tag{150}$$

i.e. the Bridgman relation for the measurable coefficients.

In a similar way it is possible to relate the other coefficients to measurable effects*.

§ 8. *Sedimentation Potential and Electrophoresis*

In this section we shall apply the thermodynamical theory of irreversible processes to a system which is subjected both to a centrifugal field $\omega^2 \boldsymbol{r}$ and an electric field $\boldsymbol{E}$.** We consider a liquid system of n components of which m carry electrical charges z_k ($k = 1, 2, \ldots, m$) per unit mass, and of which $n - m$ are neutral ($z_{m+1} = z_{m+2} = \ldots = z_n = 0$). The temperature and the concentrations are assumed to be uniform; viscous phenomena are neglected. According to (52) the entropy source strength is then given by

$$T\sigma = -\sum_{k=1}^{n} \boldsymbol{J}_k^{\mathrm{a}} \cdot \left\{ v_k \,\mathrm{grad}\, p - z_k \left(\boldsymbol{E} + \frac{1}{c} \boldsymbol{v}^{\mathrm{a}} \wedge \boldsymbol{B} \right) - \omega^2 \boldsymbol{r} \right\}, \tag{151}$$

where we have used the thermodynamic relation

$$(\mathrm{grad}\, \mu_k)_{T, c_1, c_2, \ldots, c_n} = v_k \,\mathrm{grad}\, p, \tag{152}$$

* R. Fieschi, S. R. de Groot and P. Mazur, loc. cit. p. 355.

** S. R. de Groot, P. Mazur and J. Th. G. Overbeek, J. chem. Phys. **20** (1952) 1825.

and where the conservative force per unit mass $\omega^2 \boldsymbol{r}$ acting on every component k has been added to the thermodynamic force. We now choose as reference velocity the so-called mean volume velocity (*cf.* Ch. XI, § 2), which in the liquid system considered here may be assumed to be negligibly small. We then have

$$\boldsymbol{v}^{a} = \boldsymbol{v}^{0} = \sum_{k=1}^{n} \rho_k v_k \boldsymbol{v}_k = 0 \tag{153}$$

and

$$\boldsymbol{J}_k^{a} = \boldsymbol{J}_k^{0} = \rho_k \boldsymbol{v}_k , \quad (k = 1, 2, \ldots, n) . \tag{154}$$

With these conditions (151) may be written as

$$T\sigma = \boldsymbol{I}\cdot\boldsymbol{E} + \boldsymbol{J}\cdot\omega^2 \boldsymbol{r} , \tag{155}$$

where

$$\boldsymbol{I} = \sum_{k=1}^{m} z_k \boldsymbol{J}_k^{0} = \sum_{k=1}^{m} \rho_k z_k \boldsymbol{v}_k \tag{156}$$

is the total electric current density, and where

$$\boldsymbol{J} = \sum_{k=1}^{n} \boldsymbol{J}_k^{0} = \sum_{k=1}^{n} \rho_k \boldsymbol{v}_k = \rho \boldsymbol{v} \tag{157}$$

is called the total mass flow. Using (153) the total mass flow may also be written as

$$\boldsymbol{J} = \sum_{k=1}^{n-1} \left(1 - \frac{v_k}{v_n}\right) \rho_k \boldsymbol{v}_k . \tag{158}$$

Usually one studies colloidal systems in centrifugal fields. Let us therefore consider an electroneutral mixture of four components ($n = 4$), of which three ($m = 3$) carry electrical charges. We shall assume that the particles of the first charged component (the colloid) are very much larger than those of the other three. The second and third components are ions of opposite charge. The fourth component is a neutral solvent. For such a system, (158) becomes

$$\boldsymbol{J} = \sum_{k=1}^{3} \left(1 - \frac{v_k}{v_4}\right) \rho_k \boldsymbol{v}_k . \tag{159}$$

As a rule the factors $(1 - v_k/v_4)$ are of order 1, furthermore under no circumstances is the velocity $\boldsymbol{v}_1$ of the colloid very much smaller than $\boldsymbol{v}_2$ and $\boldsymbol{v}_3$. Consequently the second and third terms of the right-hand side of (159) may be neglected provided that

$$\rho_1 \gg \rho_2 , \quad \rho_1 \gg \rho_3 , \tag{160}$$

conditions which are frequently fulfilled in colloidal systems. One then has

$$\boldsymbol{J} = \boldsymbol{J}_1^0 \left(1 - \frac{v_1}{v_4}\right), \tag{161}$$

and for the entropy source strength

$$T\sigma = \boldsymbol{I} \cdot \boldsymbol{E} + \boldsymbol{J}_1^0 \cdot \left(1 - \frac{v_1}{v_4}\right) \omega^2 \boldsymbol{r} . \tag{162}$$

The factor $(1 - v_1/v_4)\omega^2 \boldsymbol{r}$ represents as it were the centrifugal field acting on the colloid component modified by the "buoyancy". The phenomenological equations corresponding to (162) are, if one neglects transversal effects

$$\boldsymbol{I} = R^{-1} \boldsymbol{E} + G \left(1 - \frac{v_1}{v_4}\right) \omega^2 \boldsymbol{r} , \tag{163}$$

$$\boldsymbol{J}_1^0 = \rho_1 M_1 U_e \boldsymbol{E} + \rho_1 M_1 U_s \left(1 - \frac{v_1}{v_4}\right) \omega^2 \boldsymbol{r} . \tag{164}$$

Here R is the electric resistance of the solution and G is the sedimentation current per unit modified centrifugal field. The coefficient U_e is called the electrophoretic mobility of the colloid, and U_s is the mobility in the modified centrifugal field. (U_s is related through a conversion factor to the mobility U^a, discussed in Chapter XI, § 6, where the modified centrifugal field was defined differently.)

Amongst these coefficients we have the Onsager relation

$$G = \rho_1 M_1 U_e . \tag{165}$$

An important effect measured in colloidal systems is the so-called sedimentation potential. This effect may be defined as the electric field per unit centrifugal force in the stationary state of zero electrical current. From (163) one has for this effect

$$\left(\frac{\boldsymbol{E}}{\omega^2 \boldsymbol{r}}\right)_{\boldsymbol{I}=0} = -G\left(1-\frac{v_1}{v_4}\right)R. \tag{166}$$

With the Onsager relation (165) we may express this effect in terms of the electrophoretic mobility of the colloid

$$\left(\frac{\boldsymbol{E}}{\omega^2 \boldsymbol{r}}\right)_{\boldsymbol{I}=0} = -\rho_1 M_1 U_e\left(1-\frac{v_1}{v_4}\right)R. \tag{167}$$

This relation thus connects two important electrokinetic phenomena, to wit the sedimentation potential and electrophoresis (the motion of a colloid under the influence of an electric field).

§ 9. *Diffusion and Thermal Diffusion Potentials; Thermopotential of a Thermocell*

To conclude this chapter, we consider an electroneutral multicomponent system of charged or uncharged components (in the absence of an applied magnetic field), of which the entropy source strength is given by (52). We write this expression with (56) in the form

$$T\sigma = -\boldsymbol{J}_{s,\,\mathrm{tot}}\cdot \operatorname{grad} T - \sum_{k=1}^{n} \boldsymbol{J}_k^{\mathrm{abs}}\cdot(\operatorname{grad}\mu_k - z_k\boldsymbol{E}), \tag{168}$$

where the reference velocity $\boldsymbol{v}^{\mathrm{a}}$ has been chosen zero. The fluxes $\boldsymbol{J}_k^{\mathrm{abs}} = \rho_k \boldsymbol{v}_k$ are thus measured with respect to a laboratory coordinate system. An alternative way of writing this expression is

$$T\sigma = -\boldsymbol{J}_{s,\,\mathrm{tot}}\cdot \operatorname{grad} T - \sum_{k=1}^{n} \boldsymbol{J}_k^{\mathrm{abs}}\cdot \operatorname{grad}\mu_k + \boldsymbol{I}\cdot\boldsymbol{E}, \tag{169}$$

where

$$\boldsymbol{I} = \sum_{k=1}^{n} \boldsymbol{I}_k \tag{170}$$

is the total electric current, with

$$\boldsymbol{I}_k = z_k \boldsymbol{J}_k^{\text{abs}}, \quad (k = 1, 2, \ldots, n), \tag{171}$$

the partial electric current due to component k.

We shall now study in particular a system enclosed in a vessel at rest with respect to the laboratory coordinate system, and consider experimental situations such that some weighted mean value of the fluxes $\boldsymbol{J}_k^{\text{abs}}$ vanishes. One may then write

$$\sum_{k=1}^{n} w_k \boldsymbol{J}_k^{\text{abs}} = 0, \tag{172}$$

where w_k is a weight factor, which is practically uniform in space. Eliminating $\boldsymbol{J}_n^{\text{abs}}$ with the help of (172) one obtains

$$T\sigma = - \boldsymbol{J}_{s,\,\text{tot}} \cdot \text{grad}\, T - \sum_{k=1}^{n-1} \boldsymbol{J}_k^{\text{abs}} \cdot \text{grad}\, \mu_k' + \boldsymbol{I} \cdot \boldsymbol{E}, \tag{173}$$

where the abbreviation

$$\mu_k' \equiv \mu_k - \frac{w_k \mu_n}{w_n} \tag{174}$$

has been used. The form (173) contains dependent fluxes and independent thermodynamic forces. It is, however, possible to eliminate a partial current $\boldsymbol{I}_{n-1}$ (for $z_{n-1} \neq 0$) with condition (171) from the entropy production. One then obtains

$$T\sigma = - \boldsymbol{J}_{s,\,\text{tot}} \cdot \text{grad}\, T - \sum_{k=1}^{n-2} \boldsymbol{J}_k^{\text{abs}} \cdot \text{grad}\, \mu_k'' + \boldsymbol{I} \cdot \boldsymbol{E}', \tag{175}$$

with the abbreviations

$$\mu_k'' = \mu_k' - \frac{z_k - \dfrac{w_k z_n}{w_n}}{z_{n-1} - \dfrac{w_{n-1} z_n}{w_n}} \mu_{n-1}', \tag{176}$$

and

$$\boldsymbol{E}' = \boldsymbol{E} - \frac{1}{z_{n-1} - \dfrac{w_{n-1} z_n}{w_n}} \text{grad}\, \mu_{n-1}'. \tag{177}$$

Let us now write down the phenomenological equations for the fluxes and thermodynamic forces of (175), when no external magnetic field is applied, assuming the system to be isotropic

$$\boldsymbol{J}_{s,\,\mathrm{tot}} = - L_{ss} \operatorname{grad} T - \sum_{j=1}^{n-2} L_{sj} \operatorname{grad} \mu_j'' + L_{se} \boldsymbol{E}' , \tag{178}$$

$$\boldsymbol{J}_k^{\mathrm{abs}} = - L_{ks} \operatorname{grad} T - \sum_{j=1}^{n-2} L_{kj} \operatorname{grad} \mu_j'' + L_{ke} \boldsymbol{E}' , \quad (k = 1, 2, \ldots, n-2) , \tag{179}$$

$$\boldsymbol{I} = - L_{es} \operatorname{grad} T - \sum_{j=1}^{n-2} L_{ej} \operatorname{grad} \mu_j'' + L_{ee} \boldsymbol{E}' . \tag{180}$$

The phenomenological coefficients obey the Onsager relations

$$L_{sj} = L_{js} , \quad (j = 1, 2, \ldots, n-2) , \tag{181}$$

$$L_{kj} = L_{jk} , \quad (k, j = 1, 2, \ldots, n-2) , \tag{182}$$

$$L_{se} = L_{es} , \tag{183}$$

$$L_{ke} = L_{ek} , \quad (k = 1, 2, \ldots, n-2) . \tag{184}$$

Solving the phenomenological equations for $\boldsymbol{J}_{s,\,\mathrm{tot}}$, $\boldsymbol{J}_k^{\mathrm{abs}}$ and $\boldsymbol{E}'$ one obtains

$$\boldsymbol{J}_{s,\,\mathrm{tot}} = - \frac{\lambda}{T} \operatorname{grad} T - \sum_{j=1}^{n-2} L'_{sj} \operatorname{grad} \mu_j'' + \frac{\pi}{T} \boldsymbol{I} , \tag{185}$$

$$\boldsymbol{J}_k^{\mathrm{abs}} = - L'_{ks} \operatorname{grad} T - \sum_{j=1}^{n-2} L'_{kj} \operatorname{grad} \mu_j'' + \frac{t_k}{z_k} \boldsymbol{I} , \quad (k = 1, 2, \ldots, n-2) , \tag{186}$$

$$\boldsymbol{E}' = - \eta \operatorname{grad} T - \sum_{j=1}^{n-2} C_j \operatorname{grad} \mu_j'' + R \boldsymbol{I} . \tag{187}$$

The coefficients occurring in these equations are combinations of the coefficients occurring in (178)–(180): λ is the heat conductivity of the system, π its Peltier coefficient, t_k is the so-called transference number of component k, *i.e.* the relative contribution of component k to the transport of electricity (for uncharged components $t_k = 0$, but the ratio t_k/z_k may be finite), η is the differential thermo-electric potential of the system, C_k is the partial diffusion potential of component k, and finally R the electric resistivity of the system. The other coefficients L'_{sj}, L'_{js} and L'_{kj} are related to the Dufour coefficients, Soret coefficients and diffusion coefficients respectively. For all those coefficients one can show that the Onsager relations become

$$L'_{sj} = L'_{js}, \quad (j = 1, 2, \ldots, n-2), \tag{188}$$

$$L'_{kj} = L'_{jk}, \quad (k, j = 1, 2, \ldots, n-2), \tag{189}$$

$$T\eta = -\pi, \tag{190}$$

$$C_k = -\frac{t_k}{z_k}, \quad (k = 1, 2, \ldots, n-2). \tag{191}$$

We will be especially interested in the coefficients obeying the last two relations. Note that the transference numbers of the $(n-1)^{\text{st}}$ and the nth component may be found from

$$\sum_{k=1}^{n} t_k = 1 \tag{192}$$

and

$$\sum_{k=1}^{n} \frac{w_k}{z_k} t_k = 0 \tag{193}$$

in terms of the first $n-2$ transference numbers. These relations follow from (171) and (172).

Note also that in the case of a metal, considered in § 6 and § 7, one has two weight factors $w_1 \equiv w_e$ and $w_2 \equiv w_i$, where the index e refers to the electrons and i to the positive ions. Due to the choice of the reference system one has $w_e = 0$ and $w_i = 1$: equation (187) then reduces with (174) and (177) to the phenomenological equation (78) considered in § 6.

Using the Onsager relations (190) and (191) equation (187) becomes

$$\boldsymbol{E}' = \frac{\pi}{T} \operatorname{grad} T + \sum_{j=1}^{n-2} \frac{t_j}{z_j} \operatorname{grad} \mu''_j + R\boldsymbol{I}, \tag{194}$$

or

$$\boldsymbol{E} = \frac{\pi}{T} \operatorname{grad} T + \sum_{j=1}^{n} \frac{t_j}{z_j} \operatorname{grad} \mu_j + R\boldsymbol{I}. \tag{195}$$

The second form follows with (174), (176), (177), (192) and (193). At zero electric current the electric field is given by

$$\boldsymbol{E} = \frac{\pi}{T}\,\mathrm{grad}\,T + \sum_{k=1}^{n} \frac{t_k}{z_k}\,\mathrm{grad}\,\mu_k\,. \tag{196}$$

If one considers an electrolytic solution of uniform temperature and pressure (mechanical equilibrium) but with a non-uniform concentration of the charged components, the field at zero electric current becomes

$$\boldsymbol{E} = \sum_{k=1}^{n} \frac{t_k}{z_k}\,(\mathrm{grad}\,\mu_k)_{T,\,p} = \sum_{k=1}^{n} \frac{t_k}{z_k} \sum_{j=1}^{n-1} \left(\frac{\partial \mu_k}{\partial c_j}\right)_{p,\,T} \mathrm{grad}\,c_j\,. \tag{197}$$

Thus if two electrolytic solutions of different concentration are brought into contact with each other an electric potential difference arises between the two solutions*. In formula (197) use has already been made of the Onsager reciprocal relations.

In the more general case of non-uniform temperature one has to use the complete expression (196); potential differences which arise as a consequence of temperature and concentration gradients are called thermal diffusion potentials**.

As an application of (196) we shall consider the thermo-electric power of a thermocell***. A thermocell is analogous to a metallic thermocouple: it consists essentially of an electrolyte I (Fig. 3, p. 372) with two electrodes 2 and 3 at temperatures T and $T + \Delta T$ and two metal wires II which connect the electrodes to a condensor with plates 1 and 4, which are kept at the same temperature. The electrolyte may be a solid or a fused salt or an electrolytic solution. We shall treat here the case of a thermocell with identical solid metal electrodes of

* The theory of this potential difference has first been given by M. Planck, Ann. Physik **40** (1890) 561 and P. Henderson, Z. phys. Chem. **59** (1907) 118.

** I. Prigogine, Etude thermodynamique des phénomènes irréversibles, Desoer, Liège (1947).

S. R. de Groot, L'effet Soret, North-Holland Publishing Company, Amsterdam (1955); Thermodynamics of irreversible processes, ibid., (1951).

R. Haase, Erg. exact. Naturw. **26** (1952) 56.

H. Holtan Jr., P. Mazur and S. R. de Groot, Physica **19** (1953) 1109.

K. J. Hansen, Zeitschr. Naturf. **9** A (1954) 323, 919.

*** H. Holtan Jr., Thesis, Utrecht (1953); Proc. Kon. Ned. Akad. v. Wet., Amsterdam, B, 56 (1953) 498, 510.

the same material M as the wires. The electrolyte shall be a solid or fused salt MX. The total thermo-electric power of such a thermocell,

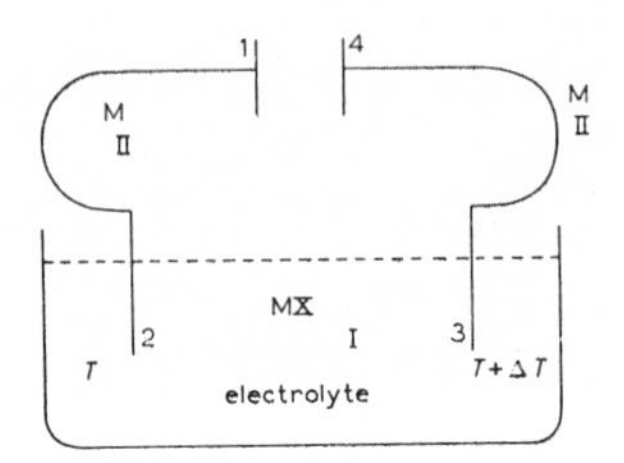

Fig. 3. Thermocell.

that is the potential difference between the condensor plates in the stationary state of vanishing electric current is given by

$$\Delta\varphi = \varphi_1 - \varphi_4 = -\int_1^4 \operatorname{grad}\varphi \cdot \mathrm{d}\boldsymbol{l} = \int_1^4 \boldsymbol{E} \cdot \mathrm{d}\boldsymbol{l}\,, \tag{198}$$

where the integration is performed along any line joining the two condensor plates inside the metallic wires and the electrolyte. The integral in (198) may be split into five parts:

$$\Delta\varphi = \int_1^2 \boldsymbol{E} \cdot \mathrm{d}\boldsymbol{l} + \int_2^3 \boldsymbol{E} \cdot \mathrm{d}\boldsymbol{l} + \int_3^4 \boldsymbol{E} \cdot \mathrm{d}\boldsymbol{l} + \Delta\varphi_2 + \Delta\varphi_3\,, \tag{199}$$

where the integration between points 1 and 2 and also 3 and 4 are performed within the metallic wire and the integration between 2 and 3 within the electrolyte; $\Delta\varphi_2$ and $\Delta\varphi_3$ are the potential jumps at the two electrodes. Into the first and the third integral we introduce equation (196) applied to the metallic wire, and into the second integral (196) applied to the electrolyte. We then obtain

$$\Delta\varphi = \frac{\pi_{\mathrm{I}} - \pi_{\mathrm{II}}}{T}\,\Delta T + \frac{t_{\mathrm{M^+I}}}{z_{\mathrm{M^+}}}\{(\mu_{\mathrm{M^+I}})_3 - (\mu_{\mathrm{M^+I}})_2\}$$

$$+ \frac{t_{\mathrm{X^-I}}}{z_{\mathrm{X^-}}}\{(\mu_{\mathrm{X^-I}})_3 - (\mu_{\mathrm{X^-I}})_2\} - \frac{1}{z_e}\{(\mu_{\mathrm{eII}})_3 - (\mu_{\mathrm{eII}})_2\} + \Delta\varphi_2 + \Delta\varphi_3\,, \tag{200}$$

where we have assumed that η and C can be considered in first approximation as constants. Here the indices I and II refer to quantities of the electrolytic and the metallic phase respectively; M^+ denotes the positive ion, X^- the negative ion, and e the electron. The indices 2 and 3 denote the values of quantities at the electrodes. In obtaining (200) use has been made of the fact that the chemical potential μ_{eII} has the same value on both condensor plates (*cf.* § 6).

Since the electrodes are in thermodynamic equilibrium, the values of the potential jumps follow from the equality of the electrochemical potentials of the ions M^+ in the electrolytic phase I and the metallic phase II (*cf.* end of § 3). Thus

$$z_{M^+} \Delta \varphi_2 = (\mu_{M^+I})_2 - (\mu_{M^+II})_2 , \tag{201}$$

$$z_{M^+} \Delta \varphi_3 = (\mu_{M^+II})_3 - (\mu_{M^+I})_3 . \tag{202}$$

With these relations (200) becomes

$$\begin{aligned} \Delta \varphi = \frac{\pi_I - \pi_{II}}{T} \Delta T + \frac{1}{c_{M^+II} z_{M^+}} \{ (c_{M^+II}\, \mu_{M^+II})_3 + (c_{eII}\, \mu_{eII})_3 \\ - (c_{M^+II}\, \mu_{M^+II})_2 - (c_{eII}\, \mu_{eII})_2 \} \\ - \frac{t_{X^-I}}{c_{M^+I} z_{M^+}} \{ (c_{M^+I} \mu_{M^+I})_3 + (c_{X^-I} \mu_{X^-I})_3 - (c_{M^+I} \mu_{M^+I})_2 - (c_{X^-I} \mu_{X^-I})_2 \} , \end{aligned} \tag{203}$$

where we have eliminated t_{M^+I} by means of

$$t_{M^+I} + t_{X^-I} = 1 , \tag{204}$$

and the specific charges z_{X^-} and z_e by means of the conditions of electroneutrality

$$c_{X^-I}\, z_{X^-} + c_{M^+I}\, z_{M^+} = 0 , \tag{205}$$

$$c_{eII}\, z_e + c_{M^+II}\, z_{M^+} = 0 , \tag{206}$$

where c_{X^-I}, etc., are mass fractions. Introducing the chemical potentials μ_M and μ_{MX} of the metal and of the electrolyte respectively

$$\mu_M = c_{M^+II}\, \mu_{M^+II} + c_{eII}\, \mu_{eII} , \tag{207}$$

$$\mu_{MX} = c_{M^+I}\, \mu_{M^+I} + c_{X^-I}\, \mu_{X^-I} , \tag{208}$$

formula (203) gets the form

$$\Delta\varphi = \frac{\pi_{\mathrm{I}} - \pi_{\mathrm{II}}}{T}\Delta T + \frac{1}{c_{\mathrm{M^+ II}}z_{\mathrm{M^+}}}\{(\mu_{\mathrm{M}})_3 - (\mu_{\mathrm{M}})_2\} - \frac{t_{\mathrm{X^- I}}}{c_{\mathrm{M^+ I}}z_{\mathrm{M^+}}}\{(\mu_{\mathrm{MX}})_3 - (\mu_{\mathrm{MX}})_2\}. \tag{209}$$

Since furthermore the pressure and the concentrations in a binary, electroneutral system in mechanical equilibrium are uniform, one has

$$(\mu_{\mathrm{M}})_3 - (\mu_{\mathrm{M}})_2 = -s_{\mathrm{M}}\Delta T, \tag{210}$$

$$(\mu_{\mathrm{MX}})_3 - (\mu_{\mathrm{MX}})_2 = -s_{\mathrm{MX}}\Delta T, \tag{211}$$

so that (209) finally becomes

$$\frac{\Delta\varphi}{\Delta T} = \frac{\pi_{\mathrm{I}} - \pi_{\mathrm{II}}}{T} - \frac{s_{\mathrm{M}}}{c_{\mathrm{M^+ II}}z_{\mathrm{M^+}}} + \frac{t_{\mathrm{X^- I}}s_{\mathrm{MX}}}{c_{\mathrm{M^+ I}}z_{\mathrm{M^+}}}. \tag{212}$$

It turns out in many practical cases that the terms in (212) containing specific entropies are very much larger in magnitude then the contribution resulting from the difference $\pi_{\mathrm{II}} - \pi_{\mathrm{I}}$.* This empirical fact bears some relation to an experimental rule found by Reinhold, in which comparison is made between the temperature coefficient of the electric potential of an isothermal cell and thermopotentials of two corresponding thermocells**.

Finally it may be noted that the Peltier coefficient π occurring in the phenomenological equation (185) may be related to the so-called "absolute entropies of transfer" (*cf.* Ch. XI § 7). For an n-component system the entropies of transfer are defined by the relation

$$J_{s,\,\mathrm{tot}} = \sum_{k=1}^{n} S^*_{k,\,\mathrm{abs}} J_k^{\mathrm{abs}}, \tag{213}$$

valid at uniform temperature. If both the temperature and the chemical potentials are uniform we have with (186)

* H. Holtan Jr., Proc. Kon. Ned. Acad. v. Wet. Amsterdam, loc. cit. p. 371.
** H. Holtan Jr., Thesis,. loc. cit. p. 371.

$$J_{s,\,tot} = \sum_{k=1}^{n} \frac{t_k S^*_{k,\,abs}}{z_k} I\,. \tag{214}$$

Comparing with (185) at uniform temperature and chemical potentials, one finds the relation

$$\pi = T \sum_{k=1}^{n} \frac{t_k S^*_{k,\,abs}}{z_k}\,. \tag{215}$$

This explicit form of π in terms of the entropies of transfer is sometimes used in treatments of the thermo-electric power of a thermocouple or a thermocell.

CHAPTER XIV

IRREVERSIBLE PROCESSES IN POLARIZED SYSTEMS

§ 1. *Conservation Laws in Polarized Systems*

In this chapter we shall discuss how the treatment of the previous chapter must be modified, if the system considered may be polarized by the electromagnetic field*. Let us first consider the various conservation laws.

a. *The laws of conservation of mass* for the n-component non-reacting mixture are of course still given by (XIII.15).

b. *The law of conservation of momentum* on the other hand must be modified. Let us first consider the balance equation for the momentum density** of the electromagnetic field $c^{-1}\boldsymbol{E} \wedge \boldsymbol{H}$.

We have, with the definitions (XIII.9) and (XIII.10):

$$\frac{1}{c}\frac{\partial(\boldsymbol{E} \wedge \boldsymbol{H})}{\partial t} = \frac{1}{c}\frac{\partial(\boldsymbol{D} \wedge \boldsymbol{B})}{\partial t} - \frac{1}{c}\frac{\partial(\boldsymbol{P} \wedge \boldsymbol{B})}{\partial t} - \frac{1}{c}\frac{\partial(\boldsymbol{E} \wedge \boldsymbol{M})}{\partial t}. \tag{1}$$

On the other hand, using the vector relation

$$(\text{rot}\,\boldsymbol{a}) \wedge \boldsymbol{b} = \text{Div}\,(\boldsymbol{ba} - \boldsymbol{a}\cdot\boldsymbol{b}U) - \boldsymbol{a}\,\text{div}\,\boldsymbol{b} + (\text{Grad}\,\dot{\boldsymbol{b}})\cdot\boldsymbol{a}\,, \tag{2}$$

* P. Mazur and I. Prigogine, Mém. Acad. roy. Belg. (Cl. Sc.) **28** (1953) nr. 1.

** We have chosen here Abraham's definition for the momentum density of the electromagnetic field. This choice enables one to carry through a consistent thermodynamic theory of the behaviour of matter in an electromagnetic field. On the other hand such a consistent scheme has not been set up as yet, using Minkowski's choice $c^{-1}\ \boldsymbol{D} \wedge \boldsymbol{B}$. A relativistic treatment of the behaviour of material systems in an electromagnetic field leads to similar conclusions. For a relativistic treatment of thermodynamics of irreversible processes see: C. Eckart, Phys. Rev. **58** (1940) 919; G. A. Kluitenberg, S. R. de Groot and P. Mazur, Physica **19** (1953) 689, 1079; G. A. Kluitenberg and S. R. de Groot, Physica **20** (1954) 199; **21** (1955) 148, 169; G. M. Rancoita, Suppl. Nuovo Cim. **11** (1959) 183. We wish to stress, however, that in a thermodynamic theory, dealing both with the electromagnetic field and the material system, Minkowski's formalism may well be equivalent to the treatment presented here, while it would correspond to a different definition of the "material" energy-momentum tensor. The situation would then be similar to the one encountered in the definition of the ponderomotive force in a dielectric (*cf.* § 3).

one obtains from the Maxwell equations (XIII.1)–(XIII.4)

$$\frac{1}{c}\frac{\partial(\boldsymbol{D}\wedge\boldsymbol{B})}{\partial t} = \operatorname{Div}\{\boldsymbol{DE}+\boldsymbol{BH}-(\boldsymbol{D}\cdot\boldsymbol{E}+\boldsymbol{B}\cdot\boldsymbol{H})\mathsf{U}\}$$

$$+(\operatorname{Grad}\boldsymbol{D})\cdot\boldsymbol{E}+(\operatorname{Grad}\boldsymbol{B})\cdot\boldsymbol{H}-\rho z\boldsymbol{E}-\frac{1}{c}\boldsymbol{I}\wedge\boldsymbol{B}$$

$$=\operatorname{Div}\{\boldsymbol{DE}+\boldsymbol{BH}-(\tfrac{1}{2}\boldsymbol{E}^2+\tfrac{1}{2}\boldsymbol{B}^2-\boldsymbol{M}\cdot\boldsymbol{B})\mathsf{U}\}$$

$$-(\operatorname{Grad}\boldsymbol{E})\cdot\boldsymbol{P}-(\operatorname{Grad}\boldsymbol{B})\cdot\boldsymbol{M}-\rho z\boldsymbol{E}-\frac{1}{c}\boldsymbol{I}\wedge\boldsymbol{B}. \quad (3)$$

In the last equality we have used the two relations:

$$(\operatorname{Grad}\boldsymbol{D})\cdot\boldsymbol{E}=\operatorname{Div}\{(\boldsymbol{P}\cdot\boldsymbol{E}+\tfrac{1}{2}\boldsymbol{E}^2)\mathsf{U}\}-(\operatorname{Grad}\boldsymbol{E})\cdot\boldsymbol{P}, \quad (4)$$

$$(\operatorname{Grad}\boldsymbol{B})\cdot\boldsymbol{H}=\operatorname{Div}(\tfrac{1}{2}\boldsymbol{B}^2\mathsf{U})-(\operatorname{Grad}\boldsymbol{B})\cdot\boldsymbol{M}. \quad (5)$$

Combining (1) and (3) one has:

$$\frac{1}{c}\frac{\partial(\boldsymbol{E}\wedge\boldsymbol{H})}{\partial t}=\operatorname{Div}\{\boldsymbol{DE}+\boldsymbol{BH}-(\tfrac{1}{2}\boldsymbol{E}^2+\tfrac{1}{2}\boldsymbol{B}^2-\boldsymbol{M}\cdot\boldsymbol{B})\mathsf{U}\}$$

$$-\rho z\boldsymbol{E}-\frac{1}{c}\boldsymbol{I}\wedge\boldsymbol{B}-(\operatorname{Grad}\boldsymbol{E})\cdot\boldsymbol{P}-(\operatorname{Grad}\boldsymbol{B})\cdot\boldsymbol{M}-\frac{1}{c}\frac{\partial(\boldsymbol{P}\wedge\boldsymbol{B})}{\partial t}+\frac{1}{c}\frac{\partial(\boldsymbol{M}\wedge\boldsymbol{E})}{\partial t}. \quad (6)$$

Let us now write down the balance equations for the polarization $\boldsymbol{P}$ and the magnetization $\boldsymbol{M}$:

$$\frac{\partial\boldsymbol{P}}{\partial t}=-\operatorname{Div}(\boldsymbol{vP})+\rho\frac{\mathrm{d}\boldsymbol{p}}{\partial t}, \quad (7)$$

$$\frac{\partial\boldsymbol{M}}{\partial t}=-\operatorname{Div}(\boldsymbol{vM})+\rho\frac{\mathrm{d}\boldsymbol{m}}{\mathrm{d}t}, \quad (8)$$

where

$$\boldsymbol{p}=\rho^{-1}\boldsymbol{P} \quad (9)$$

and

$$\boldsymbol{m}=\rho^{-1}\boldsymbol{M} \quad (10)$$

are the polarization and the magnetization per unit mass. The balance equations (7) and (8) are a consequence of the law of conservation of mass and the definition of the substantial (barycentric) derivative (*cf.* Chapter II).

Furthermore we may write:

$$\frac{\partial \boldsymbol{B}}{\partial t} = \frac{\mathrm{d}\boldsymbol{B}}{\mathrm{d}t} - \boldsymbol{v}\cdot\mathrm{Grad}\,\boldsymbol{B}, \tag{11}$$

$$\frac{\partial \boldsymbol{E}}{\partial t} = \frac{\mathrm{d}\boldsymbol{E}}{\mathrm{d}t} - \boldsymbol{v}\cdot\mathrm{Grad}\,\boldsymbol{E}. \tag{12}$$

From (7) with (11), and (8) with (12) one gets:

$$\frac{\partial(\boldsymbol{P}\wedge\boldsymbol{B})}{\partial t} = -\,\mathrm{Div}\,\{\boldsymbol{v}(\boldsymbol{P}\wedge\boldsymbol{B})\} + \rho\frac{\mathrm{d}(\boldsymbol{p}\wedge\boldsymbol{B})}{\mathrm{d}t}, \tag{13}$$

$$\frac{\partial(\boldsymbol{M}\wedge\boldsymbol{E})}{\partial t} = -\,\mathrm{Div}\,\{\boldsymbol{v}(\boldsymbol{M}\wedge\boldsymbol{E})\} + \rho\frac{\mathrm{d}(\boldsymbol{m}\wedge\boldsymbol{E})}{\mathrm{d}t}. \tag{14}$$

Introducing these relations into (6) we finally obtain:

$$\frac{1}{c}\frac{\partial \boldsymbol{E}\wedge\boldsymbol{H}}{\partial t} = \mathrm{Div}\,T - \boldsymbol{F}, \tag{15}$$

where

$$T = \boldsymbol{DE} + \boldsymbol{BH} + \frac{\boldsymbol{v}}{c}(\boldsymbol{P}\wedge\boldsymbol{B}) - \frac{\boldsymbol{v}}{c}(\boldsymbol{M}\wedge\boldsymbol{E}) - (\tfrac{1}{2}\boldsymbol{E}^2 + \tfrac{1}{2}\boldsymbol{B}^2 - \boldsymbol{M}\cdot\boldsymbol{B})U \tag{16}$$

may be interpreted as the Maxwell stress tensor in a polarized system, and

$$\boldsymbol{F} = \rho z\boldsymbol{E} + \frac{1}{c}\boldsymbol{I}\wedge\boldsymbol{B} + (\mathrm{Grad}\,\boldsymbol{E})\cdot\boldsymbol{P} + (\mathrm{Grad}\,\boldsymbol{B})\cdot\boldsymbol{M} + \frac{\rho}{c}\frac{\mathrm{d}}{\mathrm{d}t}(\boldsymbol{p}\wedge\boldsymbol{B}) - \frac{\rho}{c}\frac{\mathrm{d}}{\mathrm{d}t}(\boldsymbol{m}\wedge\boldsymbol{E}) \tag{17}$$

as the force per unit volume acting on the polarized system.

Let us also introduce the fields $\boldsymbol{E}'$, $\boldsymbol{D}'$, $\boldsymbol{B}'$ and $\boldsymbol{H}'$ and the polarization $\boldsymbol{P}'$ and magnetization $\boldsymbol{M}'$ measured by an observer moving with the velocity $\boldsymbol{v}$ of the medium. In the non-relativistic approximation (neglecting terms of order $\boldsymbol{v}^2/c^2$), one has:

$$\boldsymbol{D} = \boldsymbol{D}' - \frac{1}{c}\boldsymbol{v} \wedge \boldsymbol{H}',$$

$$\boldsymbol{E} = \boldsymbol{E}' - \frac{1}{c}\boldsymbol{v} \wedge \boldsymbol{B}', \tag{18}$$

$$\boldsymbol{P} = \boldsymbol{P}' + \frac{1}{c}\boldsymbol{v} \wedge \boldsymbol{M}',$$

and

$$\boldsymbol{H} = \boldsymbol{H}' + \frac{1}{c}\boldsymbol{v} \wedge \boldsymbol{D}',$$

$$\boldsymbol{B} = \boldsymbol{B}' + \frac{1}{c}\boldsymbol{v} \wedge \boldsymbol{E}', \tag{19}$$

$$\boldsymbol{M} = \boldsymbol{M}' - \frac{1}{c}\boldsymbol{v} \wedge \boldsymbol{P}'.$$

In terms of the dashed quantities the Maxwell stress tensor becomes (neglecting terms of order $\boldsymbol{v}^2/c^2$):

$$\begin{aligned} \mathsf{T} = {} & \boldsymbol{D}'\boldsymbol{E}' + \boldsymbol{B}'\boldsymbol{H}' - \tfrac{1}{2}(\boldsymbol{D}'\cdot\boldsymbol{E}' + \boldsymbol{B}'\cdot\boldsymbol{H}' - \boldsymbol{P}'\cdot\boldsymbol{E}' - \boldsymbol{B}'\cdot\boldsymbol{M}')\mathsf{U} \\ & - c^{-1}\{\boldsymbol{D}'(\boldsymbol{v} \wedge \boldsymbol{B}') + \boldsymbol{v}(\boldsymbol{B}' \wedge \boldsymbol{D}') + \boldsymbol{B}'(\boldsymbol{D}' \wedge \boldsymbol{v}) - \boldsymbol{D}'\cdot(\boldsymbol{v} \wedge \boldsymbol{B}')\mathsf{U}\} \\ & + c^{-1}\{(\boldsymbol{H}' \wedge \boldsymbol{v})\boldsymbol{E}' + (\boldsymbol{v} \wedge \boldsymbol{E}')\boldsymbol{H}' - \boldsymbol{v}(\boldsymbol{E}' \wedge \boldsymbol{H}') - \boldsymbol{H}'\cdot(\boldsymbol{v} \wedge \boldsymbol{E}')\mathsf{U}\}. \end{aligned} \tag{20}$$

Now for any three vectors $\boldsymbol{a}$, $\boldsymbol{b}$ and $\boldsymbol{c}$ the equality

$$\boldsymbol{a}(\boldsymbol{b} \wedge \boldsymbol{c}) + \boldsymbol{b}(\boldsymbol{c} \wedge \boldsymbol{a}) + \boldsymbol{c}(\boldsymbol{a} \wedge \boldsymbol{b}) = \boldsymbol{a}\cdot(\boldsymbol{b} \wedge \boldsymbol{c})\mathsf{U} \tag{21}$$

holds. This follows for instance from inspection of the tensor components of this equality. Applying this property to (20) T reduces to:

$$\begin{aligned} \mathsf{T} = {} & \boldsymbol{D}'\boldsymbol{E}' + \boldsymbol{B}'\boldsymbol{H}' - \tfrac{1}{2}(\boldsymbol{D}'\cdot\boldsymbol{E}' + \boldsymbol{B}'\cdot\boldsymbol{H}' - \boldsymbol{P}'\cdot\boldsymbol{E}' - \boldsymbol{B}'\cdot\boldsymbol{M}')\mathsf{U} \\ & - c^{-1}\{(\boldsymbol{E}' \wedge \boldsymbol{H}')\boldsymbol{v} + \boldsymbol{v}(\boldsymbol{E}' \wedge \boldsymbol{H}')\}. \end{aligned} \tag{22}$$

The force (17) becomes in the non-relativistic approximation

$$\boldsymbol{F} = \rho z \boldsymbol{E}' + \frac{1}{c} \boldsymbol{i} \wedge \boldsymbol{B}' + (\text{Grad}\, \boldsymbol{E}') \cdot \boldsymbol{P}' + (\text{Grad}\, \boldsymbol{B}') \cdot \boldsymbol{M}'$$

$$+ \frac{1}{c} (\text{Grad}\, \boldsymbol{v}) \cdot (\boldsymbol{P}' \wedge \boldsymbol{B}' - \boldsymbol{M}' \wedge \boldsymbol{E}') + \frac{\rho}{c} \frac{\mathrm{d}}{\mathrm{d}t} (\boldsymbol{p}' \wedge \boldsymbol{B}') - \frac{\rho}{c} \frac{\mathrm{d}}{\mathrm{d}t} (\boldsymbol{m}' \wedge \boldsymbol{E}') . \qquad (23)$$

Here we also used the decomposition (XIII.19) of the current $\boldsymbol{I}$.

We note that according to this expression the force is invariant under a Galilei transformation. In a non-relativistic theory this is of course a requirement which a force acting on a material system must satisfy. On the other hand the stress tensor (22) is not invariant under a Galilei transformation. This is a consequence of the relativistic invariance of the Maxwell equations.

Since the total momentum of the system must be conserved, *i.e.* since the law

$$\frac{\partial}{\partial t} \left(\rho \boldsymbol{v} + \frac{1}{c} \boldsymbol{E} \wedge \boldsymbol{H} \right) = - \text{Div}\, (\rho \boldsymbol{v}\boldsymbol{v} + \mathsf{P} - \mathsf{T}) \qquad (24)$$

holds (P is the mechanical pressure tensor of the system), we obtain from (15) and (24) the equation of motion:

$$\rho \frac{\mathrm{d}\boldsymbol{v}}{\mathrm{d}t} = - \text{Div}\, \mathsf{P} + \boldsymbol{F} , \qquad (25)$$

where $\boldsymbol{F}$ is given by (17) or its alternative form (23). Expression (17) for the force contains besides the terms $\rho z \boldsymbol{E}$ and $c^{-1} \boldsymbol{I} \wedge \boldsymbol{B}$, already encountered in the previous chapter, the so-called ponderomotive force terms $(\text{Grad}\ \boldsymbol{E}) \cdot \boldsymbol{P}$, $(\text{Grad}\ \boldsymbol{B}) \cdot \boldsymbol{M}$, $c^{-1} \rho \mathrm{d}(\boldsymbol{p} \wedge \boldsymbol{B})/\mathrm{d}t$ and $-c^{-1} \rho \mathrm{d}(\boldsymbol{m} \wedge \boldsymbol{E})/\mathrm{d}t$, which only arise when the medium is polarized. It must be stressed, however, that the definitions (22) and (23) of T and $\boldsymbol{F}$ are to some extent arbitrary. Indeed through equation (15) only the sum $\text{Div}\ \mathsf{T} - \boldsymbol{F}$ is well defined. On the other hand it follows from the law of conservation of momentum (24) that only the difference $\mathsf{P} - \mathsf{T}$ has a welldefined meaning. Thus if we had defined T in a different way, for instance by subtracting from (22) the tensor $(\boldsymbol{M}' \cdot \boldsymbol{B}')\mathsf{U}$, this would have amounted to a different choice for $\boldsymbol{F}$, which would have contained instead of the term $(\text{Grad}\ \boldsymbol{B}') \cdot \boldsymbol{M}'$ a term $-(\text{Grad}\ \boldsymbol{M}') \cdot \boldsymbol{B}'$. (The force would then still be invariant under a Galilei transformation.) The pressure tensor P would then have contained an additional term of electromagnetic origin $-(\boldsymbol{M}'\cdot\boldsymbol{B}')U$, in such a way that the right-hand side of (25) would have remained unchanged and that P would still be Galilei invariant. In fact even with the choice for $\boldsymbol{F}$ adopted here the pressure tensor P is not independent of the field quantities as we shall see in the following sections. This is due to the intimate interaction between fields and matter in a polarized system. It is precisely this fact which makes the definition of the ponderomotive force to some extent arbitrary. In the previous chapter this difficulty was not encountered, since without polarization the mechanical pressure tensor is not modified by the electromagnetic field. Of course the expression for the ponderomotive force must at least satisfy the requirement that it reduces to $\rho z\boldsymbol{E} + c^{-1}\boldsymbol{I}\wedge\boldsymbol{B}$ when the polarization vanishes. The special form (17) of the force has been adopted here for two reasons. On the one hand it will be shown that this form fits well into the thermodynamic description of polarized systems. On the other hand it reduces essentially to the ponderomotive force proposed by Kelvin in the case of an electrically polarized system. We shall come back to the arbitrariness in the definition of the ponderomotive force and the pressure tensor in polarized systems in § 3, where it will be shown that the more familiar form of the ponderomotive force proposed by Helmholtz is not in disagreement with the point of view adopted here.

c. *The law of conservation of energy.* If we multiply (25) by $\boldsymbol{v}$, we obtain the balance equation for the macroscopic kinetic energy density $\frac{1}{2}\rho\boldsymbol{v}^2$ of the system:

$$\frac{\partial}{\partial t}\tfrac{1}{2}\rho\boldsymbol{v}^2 = -\operatorname{div}(\tfrac{1}{2}\rho\boldsymbol{v}^2\boldsymbol{v} + P\cdot\boldsymbol{v}) + \tilde{P}:\operatorname{Grad}\boldsymbol{v} + \boldsymbol{F}\cdot\boldsymbol{v}\,. \tag{26}$$

Using the explicit expression (23) for $\boldsymbol{F}$ the scalar product $\boldsymbol{F}\cdot\boldsymbol{v}$ becomes:

$$\boldsymbol{F}\cdot\boldsymbol{v} = \rho z\boldsymbol{E}'\cdot\boldsymbol{v} + \frac{\boldsymbol{v}}{c}\cdot(\boldsymbol{i}\wedge\boldsymbol{B}') + \boldsymbol{v}\cdot(\operatorname{Grad}\boldsymbol{E}')\cdot\boldsymbol{P}' + \boldsymbol{v}\cdot(\operatorname{Grad}\boldsymbol{B}')\cdot\boldsymbol{M}'$$

$$+\frac{\boldsymbol{v}}{c}\cdot(\operatorname{Grad}\boldsymbol{v})\cdot(\boldsymbol{P}'\wedge\boldsymbol{B}' - \boldsymbol{M}'\wedge\boldsymbol{E}') + \frac{\rho\boldsymbol{v}}{c}\cdot\frac{\mathrm{d}}{\mathrm{d}t}(\boldsymbol{p}'\wedge\boldsymbol{B}') - \frac{\rho\boldsymbol{v}}{c}\cdot\frac{\mathrm{d}}{\mathrm{d}t}(\boldsymbol{m}'\wedge\boldsymbol{E}'). \tag{27}$$

We also have the relations:

$$\boldsymbol{v}\cdot(\operatorname{Grad}\boldsymbol{E}')\cdot\boldsymbol{P}' = \operatorname{div}(\boldsymbol{P}'\cdot\boldsymbol{E}'\boldsymbol{v}) - \boldsymbol{E}'\cdot\operatorname{Div}\boldsymbol{v}\boldsymbol{P}'$$

$$= \operatorname{div}(\boldsymbol{P}'\cdot\boldsymbol{E}'\boldsymbol{v}) + \boldsymbol{E}'\cdot\frac{\partial\boldsymbol{P}'}{\partial t} - \rho\boldsymbol{E}'\cdot\frac{\mathrm{d}\boldsymbol{p}'}{\mathrm{d}t}, \tag{28}$$

$$\boldsymbol{v}\cdot(\text{Grad}\,\boldsymbol{B}')\cdot\boldsymbol{M}' = \text{div}\,(\boldsymbol{M}'\cdot\boldsymbol{B}'\boldsymbol{v}) - \boldsymbol{B}'\cdot\text{Div}\,\boldsymbol{v}\boldsymbol{M}'$$

$$= \text{div}\,(\boldsymbol{M}'\cdot\boldsymbol{B}'\boldsymbol{v}) + \boldsymbol{B}'\cdot\frac{\partial \boldsymbol{M}'}{\partial t} - \rho\boldsymbol{B}'\cdot\frac{\mathrm{d}\boldsymbol{m}'}{\mathrm{d}t}, \qquad (29)$$

$$\boldsymbol{v}\cdot(\text{Grad}\,\boldsymbol{v})\cdot(\boldsymbol{P}'\wedge\boldsymbol{B}' - \boldsymbol{M}'\wedge\boldsymbol{E}') = \text{div}\,\{(\boldsymbol{P}'\wedge\boldsymbol{B}' - \boldsymbol{M}'\wedge\boldsymbol{E}')\cdot\boldsymbol{v}\boldsymbol{v}\}$$

$$-\,\boldsymbol{v}\cdot\text{Div}\,\boldsymbol{v}(\boldsymbol{P}'\wedge\boldsymbol{B}' - \boldsymbol{M}'\wedge\boldsymbol{E}') = \text{div}\,\{(\boldsymbol{P}'\wedge\boldsymbol{B}' - \boldsymbol{M}'\wedge\boldsymbol{E}')\cdot\boldsymbol{v}\boldsymbol{v}\}$$

$$+\,\boldsymbol{v}\cdot\frac{\partial}{\partial t}(\boldsymbol{P}'\wedge\boldsymbol{B}' - \boldsymbol{M}'\wedge\boldsymbol{E}') - \rho\boldsymbol{v}\cdot\frac{\mathrm{d}}{\mathrm{d}t}(\boldsymbol{p}'\wedge\boldsymbol{B}' - \boldsymbol{m}'\wedge\boldsymbol{E}'), \qquad (30)$$

where the balance equations (7) and (8) and (13) and (14) for $\boldsymbol{P}'$ and $\boldsymbol{M}'$ have been used.

Introducing these relations into (27) one gets (up to order $\boldsymbol{v}/c$)

$$\boldsymbol{F}\cdot\boldsymbol{v} = \text{div}\,\{(\boldsymbol{P}'\cdot\boldsymbol{E} + \boldsymbol{M}'\cdot\boldsymbol{B})\boldsymbol{v}\} + \boldsymbol{E}\cdot\frac{\partial\boldsymbol{P}}{\partial t} - \boldsymbol{M}\cdot\frac{\partial\boldsymbol{B}}{\partial t} + \frac{\partial}{\partial t}\boldsymbol{B}\cdot\boldsymbol{M}'$$

$$-\frac{1}{c}\frac{\partial}{\partial t}\{\boldsymbol{E}'\cdot(\boldsymbol{v}\wedge\boldsymbol{M}')\} + \rho z\boldsymbol{E}'\cdot\boldsymbol{v} - \boldsymbol{i}\cdot\left(\frac{\boldsymbol{v}}{c}\wedge\boldsymbol{B}'\right)$$

$$-\,\rho\boldsymbol{E}'\cdot\frac{\mathrm{d}\boldsymbol{p}'}{\mathrm{d}t} - \rho\boldsymbol{B}'\cdot\frac{\mathrm{d}\boldsymbol{m}'}{\mathrm{d}t}. \qquad (31)$$

Here we have used again the relations (18) and (19) between the dashed and undashed field quantities.

The balance equation for $\frac{1}{2}\rho\boldsymbol{v}^2$ may now be written in the form:

$$\frac{\partial}{\partial t}\tfrac{1}{2}\rho\boldsymbol{v}^2 - \boldsymbol{E}\cdot\frac{\partial\boldsymbol{P}}{\partial t} + \boldsymbol{M}\cdot\frac{\partial\boldsymbol{B}}{\partial t} - \frac{\partial}{\partial t}\boldsymbol{B}\cdot\boldsymbol{M}' + \frac{1}{c}\frac{\partial}{\partial t}\boldsymbol{E}'\cdot(\boldsymbol{v}\wedge\boldsymbol{M}')$$

$$= -\,\text{div}\,\{\tfrac{1}{2}\rho\boldsymbol{v}^2\boldsymbol{v} + \mathsf{P}\cdot\boldsymbol{v} - (\boldsymbol{P}'\cdot\boldsymbol{E} + \boldsymbol{M}'\cdot\boldsymbol{B})\boldsymbol{v}\} + \rho z\boldsymbol{E}'\cdot\boldsymbol{v} - \boldsymbol{i}\cdot\left(\frac{\boldsymbol{v}}{c}\wedge\boldsymbol{B}'\right)$$

$$+\,\tilde{\mathsf{P}}:\text{Grad}\,\boldsymbol{v} - \rho\boldsymbol{E}'\cdot\frac{\mathrm{d}\boldsymbol{p}'}{\mathrm{d}t} - \rho\boldsymbol{B}'\cdot\frac{\mathrm{d}\boldsymbol{m}'}{\mathrm{d}t}. \qquad (32)$$

On the other hand it follows from Maxwell's equations (XIII.3)–(XIII.4) that

$$\boldsymbol{E}\cdot\frac{\partial\boldsymbol{D}}{\partial t} + \boldsymbol{H}\cdot\frac{\partial\boldsymbol{B}}{\partial t} = -\,\text{div}\,c(\boldsymbol{E}\wedge\boldsymbol{H}) - \boldsymbol{I}\cdot\boldsymbol{E}. \qquad (33)$$

This is Poynting's theorem: $c\,\boldsymbol{E} \wedge \boldsymbol{H}$ is the Poynting vector. Adding (32) and (33) one obtains the balance equation:

$$\frac{\partial}{\partial t}\{\tfrac{1}{2}\rho v^2 + \tfrac{1}{2}\boldsymbol{E}^2 + \tfrac{1}{2}\boldsymbol{B}^2 - \boldsymbol{M}'\cdot\boldsymbol{B} + c^{-1}\boldsymbol{E}'\cdot(\boldsymbol{v} \wedge \boldsymbol{M}')\}$$
$$= -\operatorname{div}\{\tfrac{1}{2}\rho v^2\boldsymbol{v} + \mathsf{P}\cdot\boldsymbol{v} - (\boldsymbol{P}'\cdot\boldsymbol{E} + \boldsymbol{M}'\cdot\boldsymbol{B})\boldsymbol{v} + c(\boldsymbol{E} \wedge \boldsymbol{H})\}$$
$$+ \tilde{\mathsf{P}} : \operatorname{Grad}\boldsymbol{v} - \boldsymbol{i}\cdot\boldsymbol{E}' - \rho\boldsymbol{E}'\cdot\frac{\mathrm{d}\boldsymbol{p}'}{\mathrm{d}t} - \rho\boldsymbol{B}'\cdot\frac{\mathrm{d}\boldsymbol{m}'}{\mathrm{d}t}. \tag{34}$$

Equation (34) is the generalization of (XIII.30) to polarized systems. It may be interpreted as follows: the rate of change in time of the sums of the kinetic energy and a density of electromagnetic energy of magnitude $\tfrac{1}{2}\boldsymbol{E}^2 + \tfrac{1}{2}\boldsymbol{B}^2 - \boldsymbol{M}'\cdot\boldsymbol{B} + c^{-1}\boldsymbol{E}'\cdot(\boldsymbol{v} \wedge \boldsymbol{M}')$ is equal to the energy flux plus a source term of magnitude

$$\tilde{\mathsf{P}} : \operatorname{Grad}\boldsymbol{v} - \boldsymbol{i}\cdot\boldsymbol{E}' - \rho\boldsymbol{E}'\cdot\frac{\mathrm{d}\boldsymbol{p}'}{\mathrm{d}t} - \rho\boldsymbol{B}'\cdot\frac{\mathrm{d}\boldsymbol{m}'}{\mathrm{d}t}. \tag{35}$$

This amount of energy is therefore transformed into another form of energy. We note that (35) only contains quantities defined with respect to an observer moving with velocity $\boldsymbol{v}$: as should be expected this energy source term is therefore invariant under a Galilei transformation. In order to obtain the balance of internal energy we can proceed as in Ch. XIII, § 3c: the conservation of total energy is expressed by (XIII.31)

$$\frac{\partial e_v}{\partial t} = -\operatorname{div}\boldsymbol{J}_e. \tag{36}$$

We now define the density of internal energy [*cf.* (XIII.32)]

$$\rho u = e_v - \{\tfrac{1}{2}\rho v^2 + \tfrac{1}{2}\boldsymbol{E}^2 + \tfrac{1}{2}\boldsymbol{B}^2 - \boldsymbol{M}'\cdot\boldsymbol{B} + c^{-1}\boldsymbol{E}'\cdot(\boldsymbol{v} \wedge \boldsymbol{M}')\}$$
$$= e_v - \{\tfrac{1}{2}\rho v^2 + \tfrac{1}{2}\boldsymbol{D}'\cdot\boldsymbol{E}' + \tfrac{1}{2}\boldsymbol{B}'\cdot\boldsymbol{H}' - \tfrac{1}{2}\boldsymbol{P}'\cdot\boldsymbol{E}'$$
$$- \tfrac{1}{2}\boldsymbol{M}'\cdot\boldsymbol{B}' + 2c^{-1}\boldsymbol{v}\cdot(\boldsymbol{E}' \wedge \boldsymbol{H}')\} \tag{37}$$

In the last member we have expressed the density of electromagnetic energy in terms of the dashed field quantities of equations (18) and (19) and neglected terms of order $\boldsymbol{v}^2/c^2$. The reader should note that equation (37) not only defines the internal energy in a system at rest, but also specifies how e_v transforms under a Galilei transformation (the internal energy is by definition invariant under such a transformation). We also define the heat flow [*cf.* (XIII.33)]

$$\boldsymbol{J}_q = \boldsymbol{J}_e - \{\tfrac{1}{2}\rho v^2 \boldsymbol{v} + \rho u \boldsymbol{v} + \mathsf{P}\cdot\boldsymbol{v} - (\boldsymbol{P}'\cdot\boldsymbol{E} + \boldsymbol{M}'\cdot\boldsymbol{B})\boldsymbol{v} + c(\boldsymbol{E} \wedge \boldsymbol{H})\}. \quad (38)$$

Subtracting (34) from (36) and using (37) and (38) one obtains the balance equation of internal energy

$$\rho \frac{\mathrm{d}u}{\mathrm{d}t} = -\operatorname{div} \boldsymbol{J}_q - \tilde{\mathsf{P}} : \operatorname{Grad} \boldsymbol{v} + \boldsymbol{i}\cdot\boldsymbol{E}' + \rho \boldsymbol{E}' \cdot \frac{\mathrm{d}\boldsymbol{p}'}{\mathrm{d}t} + \rho \boldsymbol{B}' \cdot \frac{\mathrm{d}\boldsymbol{m}'}{\mathrm{d}t}. \quad (39)$$

This equation generalizes (XIII.35) to systems with polarization.

§ 2. *The Entropy Balance Equation in Polarized Systems*

In order to set up the entropy balance equation for polarized systems, we shall first have to generalize the Gibbs equation (XIII.45) to such systems.

For this purpose we consider first a one-component system. Equation (39) then reads:

$$\rho \frac{\mathrm{d}u}{\mathrm{d}t} = -\operatorname{div} \boldsymbol{J}_q - \tilde{\mathsf{P}} : \operatorname{Grad} \boldsymbol{v} + \rho \boldsymbol{E}' \cdot \frac{\mathrm{d}\boldsymbol{p}'}{\mathrm{d}t} + \rho \boldsymbol{B}' \cdot \frac{\mathrm{d}\boldsymbol{m}'}{\mathrm{d}t}, \quad (40)$$

or

$$\rho \frac{\mathrm{d}u}{\mathrm{d}t} = \rho \frac{\mathrm{d}q}{\mathrm{d}t} - \tilde{\mathsf{P}} : \operatorname{Grad} \boldsymbol{v} + \rho \boldsymbol{E}' \cdot \frac{\mathrm{d}\boldsymbol{p}'}{\mathrm{d}t} + \rho \boldsymbol{B}' \cdot \frac{\mathrm{d}\boldsymbol{m}'}{\mathrm{d}t}, \quad (41)$$

where $\mathrm{d}q$ is the heat added to a mass element of unit mass in time $\mathrm{d}t$. For a reversible transformation of the system we have

$$\frac{1}{T}\frac{\mathrm{d}q}{\mathrm{d}t} = \frac{\mathrm{d}s}{\mathrm{d}t}, \quad (42)$$

with s the entropy per unit mass and T the temperature. Equation (41) may then be written as

$$\rho \frac{\mathrm{d}u}{\mathrm{d}t} = \rho T \frac{\mathrm{d}s}{\mathrm{d}t} - \tilde{\mathsf{P}}_{\mathrm{eq}} : \operatorname{Grad} \boldsymbol{v} + \rho \boldsymbol{E}'_{\mathrm{eq}} \cdot \frac{\mathrm{d}\boldsymbol{p}'}{\mathrm{d}t} + \rho \boldsymbol{B}'_{\mathrm{eq}} \cdot \frac{\mathrm{d}\boldsymbol{m}'}{\mathrm{d}t}. \quad (43)$$

Here P_{eq} represents the pressure (stress) tensor in a reversible (or equilibrium) transformation. Similarly $\boldsymbol{E}'_{eq}$ and $\boldsymbol{B}'_{eq}$ are the electric and magnetic fields in such a transformation.

Now at equilibrium the electric and magnetic displacement vectors $\boldsymbol{D}'$ and $\boldsymbol{H}'$ are related to the fields $\boldsymbol{E}'$ and $\boldsymbol{B}'$ through the relations

$$\boldsymbol{D}' = \varepsilon \cdot \boldsymbol{E}' , \tag{44}$$

$$\boldsymbol{H}' = \mu^{-1} \cdot \boldsymbol{B}' , \tag{45}$$

where ε is the dielectric tensor of the system and μ the magnetic permeability tensor. For the electric and magnetic polarization one therefore has:

$$\boldsymbol{P}' = (\varepsilon - U) \cdot \boldsymbol{E}' = \kappa \cdot \boldsymbol{E}' , \tag{46}$$

$$\boldsymbol{M}' = (\mu - U) \cdot \boldsymbol{H}' = \chi \cdot \boldsymbol{H}' = \chi \cdot (\chi + U)^{-1} \cdot \boldsymbol{B}' , \tag{47}$$

where κ and χ are the electric and magnetic susceptibility tensors respectively. Therefore in (43) the fields $\boldsymbol{E}'_{eq}$ and $\boldsymbol{B}'_{eq}$ are related to the instantaneous values of $\boldsymbol{p}'$ and $\boldsymbol{m}'$ through

$$\boldsymbol{E}'_{eq} = \rho \kappa^{-1} \cdot \boldsymbol{p}' , \tag{48}$$

$$\boldsymbol{B}'_{eq} = \rho (\chi + U) \cdot \chi^{-1} \cdot \boldsymbol{m}' , \tag{49}$$

For isotropic systems:

$$\kappa = \kappa U , \tag{50}$$

$$\chi = \chi U , \tag{51}$$

with κ and χ the electric and magnetic susceptibilities. Relations (48) and (49) then read:

$$\boldsymbol{E}'_{eq} = \frac{\rho}{\kappa} \boldsymbol{p}' , \tag{52}$$

$$\boldsymbol{B}'_{eq} = \frac{\rho(\chi + 1)}{\chi} \boldsymbol{m}' . \tag{53}$$

In these relations κ and χ depend only on the thermodynamic variables characterizing the local (equilibrium) state of the system.

Finally we note that according to equation (XII.181) we have:

$$\text{Grad}\, \boldsymbol{v} = \frac{\mathrm{d}\epsilon}{\mathrm{d}t} , \quad \rho \frac{\mathrm{d}v}{\mathrm{d}t} = U : \frac{\mathrm{d}\epsilon}{\mathrm{d}t} , \tag{54}$$

with ϵ the (instantaneous) deformation tensor of the system. Thus (43) becomes:

$$\rho \frac{\mathrm{d}u}{\mathrm{d}t} = T\rho \frac{\mathrm{d}s}{\mathrm{d}t} - \tilde{P}_{\mathrm{eq}} : \frac{\mathrm{d}\epsilon}{\mathrm{d}t} + \rho \boldsymbol{E}'_{\mathrm{eq}} \cdot \frac{\mathrm{d}\boldsymbol{p}'}{\mathrm{d}t} + \rho \boldsymbol{B}'_{\mathrm{eq}} \cdot \frac{\mathrm{d}\boldsymbol{m}'}{\mathrm{d}t} . \tag{55}$$

We shall consider in what follows fluid systems, *i.e.* systems which are isotropic in the absence of polarization and magnetization, and whose scalar electric and magnetic susceptibilities are functions of temperature and densities (or entropies and densities) alone.

Let us then write for u:

$$u = u^0 + \Delta u , \tag{56}$$

where u^0 is the energy of the system at zero polarization and magnetization but with the same values of s and ρ. From (55), (52) and (53) we then have:

$$\Delta u = \frac{\rho}{\kappa} \int_0^{\boldsymbol{p}'} \boldsymbol{x} \cdot \mathrm{d}\boldsymbol{x} + \frac{\rho(\chi + 1)}{\chi} \int_0^{\boldsymbol{m}'} \boldsymbol{y} \cdot \mathrm{d}\boldsymbol{y} = \frac{\rho}{2\kappa} \boldsymbol{p}'^2 + \frac{\rho(\chi + 1)}{2\chi} \boldsymbol{m}'^2 . \tag{57}$$

Therefore Δu and also u are functions of the vector variables $\boldsymbol{p}'$ and $\boldsymbol{m}'$ and through κ and χ of s and ρ only. It follows that u does not depend on the off-diagonal elements of ϵ, whatever the choice of Cartesian coordinate axes, and from (55) that P_{eq} reduces to a scalar hydrostatic pressure p as should be expected for a fluid system even when polarized.

The Gibbs law for the system under consideration then has the form:

$$\frac{\mathrm{d}s}{\mathrm{d}t} = \frac{1}{T} \frac{\mathrm{d}u}{\mathrm{d}t} + \frac{p}{T} \frac{\mathrm{d}v}{\mathrm{d}t} - \frac{\boldsymbol{E}'_{\mathrm{eq}}}{T} \cdot \frac{\mathrm{d}\boldsymbol{p}'}{\mathrm{d}t} - \frac{\boldsymbol{B}'_{\mathrm{eq}}}{T} \cdot \frac{\mathrm{d}\boldsymbol{m}'}{\mathrm{d}t} . \tag{58}$$

For multi-component systems this relation becomes by straightforward generalizations [*cf.* also (XIII.36)]

$$\frac{\mathrm{d}s}{\mathrm{d}t} = \frac{1}{T} \frac{\mathrm{d}u}{\mathrm{d}t} + \frac{p}{T} \frac{\mathrm{d}v}{\mathrm{d}t} - \frac{\boldsymbol{E}'_{\mathrm{eq}}}{T} \cdot \frac{\mathrm{d}\boldsymbol{p}'}{\mathrm{d}t} - \frac{\boldsymbol{B}'_{\mathrm{eq}}}{T} \cdot \frac{\mathrm{d}\boldsymbol{m}'}{\mathrm{d}t} - \sum_{i=1}^{n} \frac{\mu_i}{T} \frac{\mathrm{d}c_i}{\mathrm{d}t} , \tag{59}$$

where μ_i is by definition the chemical potential of component i in the polarized system.

Introducing (XIII.15) and (39) with (XIII.18) into (59) one obtains the entropy balance equation

$$\rho \frac{\mathrm{d}s}{\mathrm{d}t} = -\operatorname{div} \frac{(\boldsymbol{J}_q - \sum_k \mu_k \boldsymbol{J}_k)}{T} - \frac{1}{T^2} \boldsymbol{J}_q \cdot \operatorname{grad} T$$

$$- \frac{1}{T} \sum_{k=1}^{n} \boldsymbol{J}_k \cdot \left\{ T \operatorname{grad} \frac{\mu_k}{T} - z_k \boldsymbol{E}' \right\} - \frac{1}{T} \tilde{\Pi} : \operatorname{Grad} \boldsymbol{v}$$

$$- \frac{\rho}{T} \frac{\mathrm{d}\boldsymbol{p}'}{\mathrm{d}t} \cdot (\boldsymbol{E}'_{\mathrm{eq}} - \boldsymbol{E}') - \frac{\rho}{T} \frac{\mathrm{d}\boldsymbol{m}'}{\mathrm{d}t} \cdot (\boldsymbol{B}'_{\mathrm{eq}} - \boldsymbol{B}') . \quad (60)$$

Here $\Pi = P - pU$ is the viscous pressure tensor.

We see that the entropy flow is still given formally by (XIII.38):

$$\boldsymbol{J}_s = \frac{1}{T} \left(\boldsymbol{J}_q - \sum_{k=1}^{n} \mu_k \boldsymbol{J}_k \right) , \quad (61)$$

and that the entropy source strength is

$$\sigma = - \frac{1}{T^2} \boldsymbol{J}_q \cdot \operatorname{grad} T - \frac{1}{T} \sum_{k=1}^{n} \boldsymbol{J}_k \cdot \left\{ T \operatorname{grad} \frac{\mu_k}{T} - z_k \boldsymbol{E}' \right\}$$

$$- \frac{\tilde{\Pi}}{T} : \operatorname{Grad} \boldsymbol{v} - \frac{\rho}{T} \frac{\mathrm{d}\boldsymbol{p}'}{\mathrm{d}t} \cdot (\boldsymbol{E}'_{\mathrm{eq}} - \boldsymbol{E}') - \frac{\rho}{T} \frac{\mathrm{d}\boldsymbol{m}'}{\mathrm{d}t} \cdot (\boldsymbol{B}'_{\mathrm{eq}} - \boldsymbol{B}') . \quad (62)$$

The difference between these last expressions and the corresponding ones of Chapter XIII [(XIII.38) and (XIII.39)] is the following. The chemical potential μ_k appearing in (61) and (62) refers to a component in a polarized medium. As we shall see (*cf.* § 4) this quantity differs from the chemical potential of the unpolarizable medium of Chapter XIII which is not modified by the electromagnetic field. Furthermore two additional terms occur in the entropy source strength (62). They are related to the relaxations of the electric and magnetic polarization. As should be expected these terms vanish when the fields $\boldsymbol{E}'$ and $\boldsymbol{B}'$ coincide with $\boldsymbol{E}'_{\mathrm{eq}}$ and $\boldsymbol{B}'_{\mathrm{eq}}$ respectively, *i.e.* when the polarization and magnetization have their equilibrium values (46) and (47) corresponding to the instantaneous values of $\boldsymbol{E}'$ and $\boldsymbol{B}'$.

Expression (62) must be used to set up the phenomenological

equations in a polarized system. The heat conduction and diffusion equations for this system are those of Chapter XIII and most of the results obtained there for various physical applications are also valid in a polarized medium. We shall now first consider in detail the concept of pressure in a polarized material and then discuss in a later section the relaxation effects connected with electric and magnetic polarization.

§ 3. *Pressure and Ponderomotive Force*

In section 1 we have already mentioned that pressure and ponderomotive force are not defined unambiguously in a polarized medium. We shall now discuss this point in greater detail. Let us first show that the form (17) for the force reduces in the case of a neutral one-component system for which $\boldsymbol{M}' = 0$, essentially to the ponderomotive force proposed by Kelvin. Indeed in the case mentioned, (17) may also be written (neglecting relativistic terms)

$$\boldsymbol{F} = (\boldsymbol{P}' \cdot \mathrm{Grad})\, \boldsymbol{E} + \frac{\rho}{c} \frac{\mathrm{d}\boldsymbol{p}'}{\mathrm{d}t} \wedge \boldsymbol{B} + \frac{1}{c} \boldsymbol{v} \wedge (\boldsymbol{P}' \cdot \mathrm{Grad})\, \boldsymbol{B}\,. \tag{63}$$

Use has been made here of the relations $\boldsymbol{P} = \boldsymbol{P}'$, $\boldsymbol{M} = c^{-1}\, \boldsymbol{P}' \wedge \boldsymbol{v}$, which follow from (18) and (19) with $\boldsymbol{M}' = 0$, of relation (11), the vector relation (21), and of the vector equation

$$\begin{aligned} (\boldsymbol{P}' \cdot \mathrm{Grad})\, \boldsymbol{E} &= (\mathrm{Grad}\, \boldsymbol{E}) \cdot \boldsymbol{P}' - \boldsymbol{P}' \wedge \mathrm{rot}\, \boldsymbol{E} \\ &= (\mathrm{Grad}\, \boldsymbol{E}) \cdot \boldsymbol{P}' + \frac{1}{c} \boldsymbol{P}' \wedge \frac{\partial \boldsymbol{B}}{\partial t}\,. \end{aligned} \tag{64}$$

The second member of this last equation follows with the Maxwell equation (XIII.4).

The first term $(\boldsymbol{P}' \cdot \mathrm{Grad})\, \boldsymbol{E}$ on the right-hand side of (63) is equal to the force which an electric field $\boldsymbol{E}$ would exert on a dipole with a moment of magnitude $\boldsymbol{P}'$. This term was proposed by Kelvin as ponderomotive force in a dielectric by analogy with the electric force acting on a dipole in vacuo. The second term $c^{-1}\rho(\mathrm{d}\boldsymbol{p}'/\mathrm{d}t) \wedge \boldsymbol{B}$ of (63) represents the magnetic Lorentz force on a variable dipole. The third term is connected with the motion of the fictitious dipole of moment $\boldsymbol{P}'$. We thus see that apart from the (in general small) last two terms, the ponderomotive force obtained here from general conservation laws is in fact identical with the Kelvin force.

By rearrangement of terms with the use of (19), equation (63) also reads (again up to order $\boldsymbol{v}/c$):

$$\boldsymbol{F} = (\boldsymbol{P}'\cdot\text{Grad})\,\boldsymbol{E}' + \frac{\rho}{c}\frac{\mathrm{d}\boldsymbol{p}'}{\mathrm{d}t} \wedge \boldsymbol{B}' + \frac{1}{c}\,\boldsymbol{B}' \wedge (\boldsymbol{P}'\cdot\text{Grad})\,\boldsymbol{v}\,. \qquad (65)$$

Now if $\boldsymbol{v}$ is uniform, $\boldsymbol{F}$ contains only two terms: the Kelvin force due to the field $\boldsymbol{E}'$ and the magnetic Lorentz force due to the field $\boldsymbol{B}'$ acting on the current $\rho\mathrm{d}\boldsymbol{p}'/\mathrm{d}t$.

For stationary fields, in the absence of relaxation (in other words when $\mathrm{d}\boldsymbol{p}'/\mathrm{d}t = 0$, *i.e.* in "polarization equilibrium") and if $\boldsymbol{v}$ is uniform, $\boldsymbol{F}$ is simply given by:

$$\boldsymbol{F} = (\text{Grad}\,\boldsymbol{E}')\cdot\boldsymbol{P}'\,. \qquad (66)$$

This result follows also directly from (23).

Now due to the ambiguity in the definition of the ponderomotive force, one is free to add to (63) or (66), for instance a term $\frac{1}{2}\alpha\,\text{grad}\,(\boldsymbol{P}')^2$ where α is some constant numerical factor, if one adds at the same time a term $\frac{1}{2}\alpha(\boldsymbol{P}')^2\mathsf{U}$ to the pressure tensor P (*cf.* § 1). This leads to the new quantities

$$\boldsymbol{F}^* = \boldsymbol{F} + \tfrac{1}{2}\alpha\,\text{grad}\,(\boldsymbol{P}')^2 = \{\text{Grad}\,(\boldsymbol{E}' + \alpha\boldsymbol{P}')\}\cdot\boldsymbol{P}'\,, \qquad (67)$$

$$\mathsf{P}^* = \mathsf{P} + \tfrac{1}{2}\alpha\,(\boldsymbol{P}')^2\,\mathsf{U}\,. \qquad (68)$$

In this way the ponderomotive force is the Kelvin force in some effective electric field $\boldsymbol{E}' + \alpha\boldsymbol{P}'$; for instance with $\alpha = 1$, this effective field would be the displacement vector $\boldsymbol{D}'$, with $\alpha = \frac{1}{3}$, the Lorentz spherical cavity field $\boldsymbol{E}' + \frac{1}{3}\boldsymbol{P}'$. All these forms for the force are equally satisfactory if they are used in connexion with a proper, in each case different, definition of the pressure tensor. A discussion of the concept of ponderomotive force alone, or of the relative merits of the various possible effective fields in the Kelvin force is therefore fruitless*.

* The conclusion arrived at here (*cf.* P. Mazur and I. Prigogine, loc. cit. p. 376) has also been reached by M. Jouguet (Traité d'électricité théorique, volume I, Ch. V and VI, Gauthier-Villars, Paris 1952) and is corroborated by the statistical mechanical theory of body forces in a dielectric (P. Mazur and S. R. de Groot, Physica **22** (1956) 657). For the controversy on defining the ponderomotive force in a dielectric see for instance W. B. Smith-White, Phil. Mag. **40** (1949) 466; Journal and Proceedings of the roy. Soc. of N. S. Wales **85** (1951) 82; R. Cade, Proc. phys. Soc. A **64** (1951) 665; A **65** (1952) 287.

We have already seen that in an isotropic fluid P reduces in equilibrium to a scalar hydrostatic pressure p. This must then also be true for the various pressure tensors P^* of (68). Thus the possible choices for $\boldsymbol{F}$ correspond to the different definitions

$$p^* = p + \tfrac{1}{2}\alpha\,(\boldsymbol{P}')^2 \tag{69}$$

of the hydrostatic pressure at equilibrium. The question then arises: what is the meaning of pressure in a dielectric, or rather what does one measure in a pressure experiment of a dielectric fluid?

In order to find an answer to this question let us first relate the pressure p, which occurs in the Gibbs equation (58) (and which corresponds to the form (23) of the ponderomotive force) to the pressure that the fluid would have at the same values of density $\rho = v^{-1}$, temperature T and concentrations c_k, but at zero electric and magnetic field strengths. (We shall now consider again a multicomponent isotropic system which can be polarized electrically as well as magnetically.) We introduce the free energy per unit mass of the system: $f = u - Ts$. Its total derivative can be found from (59):

$$\mathrm{d}f \doteq -s\,\mathrm{d}T - p\,\mathrm{d}v + \boldsymbol{E}'_{\mathrm{eq}}\cdot\mathrm{d}\boldsymbol{p}' + \boldsymbol{B}'_{\mathrm{eq}}\cdot\mathrm{d}\boldsymbol{m}' + \sum_{i=1}^{n}\mu_i\,\mathrm{d}c_i \tag{70}$$

Integrating at constant T, ρ and c_i, one has with (52) and (53)

$$f = f^0 + \tfrac{1}{2}(\boldsymbol{p}'\cdot\boldsymbol{E}'_{\mathrm{eq}} + \boldsymbol{m}'\cdot\boldsymbol{B}'_{\mathrm{eq}})\,, \tag{71}$$

where f^0 is the free energy at zero fields $\boldsymbol{E}'$ and $\boldsymbol{B}'$ and at the same temperature, density and concentrations. For the pressure we then have

$$p = -\left(\frac{\partial f}{\partial v}\right)_{T,\,c_i,\,\boldsymbol{p}',\,\boldsymbol{m}'}$$
$$= p^0 + \tfrac{1}{2}\left(\boldsymbol{P}'\cdot\boldsymbol{E}'_{\mathrm{eq}} + \boldsymbol{M}'\cdot\boldsymbol{B}'_{\mathrm{eq}} + v\boldsymbol{E}'^2_{\mathrm{eq}}\frac{\partial\kappa}{\partial v} + v\boldsymbol{H}'^2_{\mathrm{eq}}\frac{\partial\chi}{\partial v}\right). \tag{72}$$

Here p^0 is the pressure at zero fields, and the same values of the other thermodynamic variables. If the equation of state of the fluid at zero fields is known and if furthermore κ and χ are known functions of T, ρ and the c's, then from measurements of T, ρ, c_i $(i = 1, \ldots, n)$ and of

$\boldsymbol{E}'$ and $\boldsymbol{B}'$ one may calculate with (71) the pressure p or any pressure p^*, related to p according to (69). But relation (72) is not only useful for this indirect determination of p. It also permits us to define the ponderomotive force in still another way. Let us rewrite the equation of motion (25) in the form

$$\rho \frac{\mathrm{d}\boldsymbol{v}}{\mathrm{d}t} = \boldsymbol{F} - \operatorname{grad} p - \operatorname{Div} \Pi \,, \tag{73}$$

with $\Pi \equiv P - pU$ the viscous pressure tensor.
Introducing (72) into this equation we have

$$\rho \frac{\mathrm{d}\boldsymbol{v}}{\mathrm{d}t} = \boldsymbol{F} - \tfrac{1}{2} \operatorname{grad} \left(\boldsymbol{P}' \cdot \boldsymbol{E}'_{\mathrm{eq}} + \boldsymbol{M}' \cdot \boldsymbol{B}'_{\mathrm{eq}} + v \boldsymbol{E}'^2_{\mathrm{eq}} \frac{\partial \kappa}{\partial v} + v \boldsymbol{H}'^2_{\mathrm{eq}} \frac{\partial \chi}{\partial v} \right) - \operatorname{grad} p^0 - \operatorname{Div} \Pi \,. \tag{74}$$

Consequently with the convention that p^0 represents the "pressure" in a dielectric, we arrive at the new expression for the force

$$\boldsymbol{F}^{**} = \boldsymbol{F} - \tfrac{1}{2} \operatorname{grad} \left(\boldsymbol{P}' \cdot \boldsymbol{E}'_{\mathrm{eq}} + \boldsymbol{M}' \cdot \boldsymbol{B}'_{\mathrm{eq}} + v \boldsymbol{E}'^2_{\mathrm{eq}} \frac{\partial \kappa}{\partial v} + v \boldsymbol{H}'^2_{\mathrm{eq}} \frac{\partial \chi}{\partial v} \right) . \tag{75}$$

If no viscous flow occurs (Grad $\boldsymbol{v} = 0$) and if the system is in "polarization equilibrium" (stationary fields and polarization) we have from (75) with (23), and (45), (46) and (47) (and omitting the subscripts)

$$\boldsymbol{F}^{**} = \rho z \boldsymbol{E}' + \frac{1}{c} \boldsymbol{i} \wedge \boldsymbol{B}' - \tfrac{1}{2} \boldsymbol{E}'^2 \operatorname{grad} \kappa - \operatorname{grad} \left(\tfrac{1}{2} v \boldsymbol{E}'^2 \frac{\partial \kappa}{\partial v} \right) - \tfrac{1}{2} \boldsymbol{H}'^2 \operatorname{gra} \ \chi - \operatorname{grad} \left(\tfrac{1}{2} v \boldsymbol{H}'^2 \frac{\partial \chi}{\partial v} \right) . \tag{76}$$

The ponderomotive force is then

$$\boldsymbol{F}^{(H)} = - \tfrac{1}{2} \boldsymbol{E}'^2 \operatorname{grad} \kappa - \operatorname{grad} \left(\tfrac{1}{2} v \boldsymbol{E}'^2 \frac{\partial \kappa}{\partial v} \right) - \tfrac{1}{2} \boldsymbol{H}'^2 \operatorname{grad} \chi - \operatorname{grad} \left(\tfrac{1}{2} v \boldsymbol{H}'^2 \frac{\partial \chi}{\partial v} \right) . \tag{77}$$

For $\boldsymbol{H}' = 0$ this is precisely the ponderomotive force in a dielectric proposed by Helmholtz. It is usually obtained from (free) energy considerations applied to reversible transformations of polarizable systems. It should, however, only be considered to represent the "force" in connexion with the pressure p^0. We note that according to the present derivation, the Helmholtz expression may be used in connexion with p^0, even if some irreversible phenomena occur in the fluid provided that

1. no electric or magnetic relaxation phenomena take place, while the fields are stationary or almost stationary,
2. no viscous flow occurs.

Due to the relative order of magnitude of the various characteristic times involved, these two conditions will be satisfied approximately in many physical situations long before complete equilibrium is reached.

In complete mechanical and thermodynamical equilibrium of a one-component uncharged system ($\mathrm{d}\boldsymbol{v}/\mathrm{d}t = 0$, $\Pi = 0$, $\boldsymbol{i} = 0$, $z = 0$), equation (74) reduces with (75) and (76) to

$$\operatorname{grad} p^0 = -\tfrac{1}{2}\boldsymbol{E}'^2 \operatorname{grad} \kappa - \operatorname{grad}\left(\tfrac{1}{2}\boldsymbol{E}'^2 v \frac{\partial \kappa}{\partial v}\right) - \tfrac{1}{2}\boldsymbol{H}'^2 \operatorname{grad} \chi - \operatorname{grad}\left(\tfrac{1}{2}\boldsymbol{H}'^2 v \frac{\partial \chi}{\partial v}\right) = \rho \operatorname{grad}\left\{\tfrac{1}{2}\boldsymbol{E}'^2 \frac{\partial \kappa}{\partial \rho} + \tfrac{1}{2}\boldsymbol{H}'^2 \frac{\partial \chi}{\partial \rho}\right\}, \tag{78}$$

since in equilibrium, *i.e.* at uniform temperature,

$$\operatorname{grad} \kappa = \frac{\partial \kappa}{\partial \rho} \operatorname{grad} \rho, \tag{79}$$

$$\operatorname{grad} \chi = \frac{\partial \chi}{\partial \rho} \operatorname{grad} \rho. \tag{80}$$

Equation (78) demonstrates that if for instance $\boldsymbol{H}' = 0$, the pressure p^0 of the fluid, and therefore also its density, will be highest, where the quantity $\tfrac{1}{2}\boldsymbol{E}'^2 \partial\kappa/\partial\rho$ is largest. In general $\boldsymbol{E}'^2(\partial\kappa/\partial\rho)$ will be largest where $\boldsymbol{E}'^2$ is largest. This is the phenomenon of electrostriction. The corresponding phenomenon in the magnetic case is called magnetostriction. Let us integrate equation (78) when $\boldsymbol{H}' = 0$. This gives

$$\tfrac{1}{2}\boldsymbol{E}'^2 \frac{\partial \kappa}{\partial \rho} = \int \frac{\mathrm{d}p^0}{\rho} + \text{constant}$$

$$= \int \frac{\partial p^0}{\partial \rho}\, \mathrm{d} \ln \rho + \text{constant} . \tag{81}$$

If both p^0 and κ are known as functions of ρ (and Π) we thus get a relation between the density and the field strength. It is instructive to consider the case of a non-polar ideal gas. One then has the equation of state

$$p^0 = \frac{kT\rho}{m}, \tag{82}$$

(with k Boltzmann's constant and m the mass of a constituent particle), while the dependence of κ on ρ may be assumed to be given by the Clausius–Mossotti expression

$$\frac{\kappa}{\kappa + 3} = \frac{\rho\alpha}{3m}. \tag{83}$$

Here α is the polarizability of a constituent particle. From (83) one finds

$$\frac{\partial \kappa}{\partial \rho} = \frac{\alpha}{m}\left(\frac{\kappa + 3}{3}\right)^2 = \frac{\kappa(\kappa + 3)}{3\rho}, \tag{84}$$

and therefore

$$\tfrac{1}{2}\boldsymbol{E}'^2 \frac{\partial \kappa}{\partial \rho} = \tfrac{1}{2}\frac{\alpha}{m}\boldsymbol{E}_{\mathrm{L}}'^2 = \tfrac{1}{2}\boldsymbol{p}' \cdot \boldsymbol{E}_{\mathrm{L}}', \tag{85}$$

where

$$\boldsymbol{E}_{\mathrm{L}}' = \frac{\kappa + 3}{3}\boldsymbol{E}', \tag{86}$$

the so-called Lorentz field, is the electric field that would be measured in a spherical cavity in the fluid. It is also within the domain of validity of (83), the effective electric field acting on a specific polarizable particle in the fluid. The quantity $-\tfrac{1}{2}\alpha \boldsymbol{E}_{\mathrm{L}}'^2$ represents the electrostatic energy U of a particle in this field.

Using (82) and (85) we obtain from (81) for the ratio of the densities ρ_1 and ρ_2 at two positions 1 and 2 in the fluid:

$$\frac{\rho_1}{\rho_2} = \exp - \frac{U_1 - U_2}{kT}, \tag{87}$$

where $U_1 = -\frac{1}{2}\alpha(\boldsymbol{E}_L^2)_1$ and $U_2 = -\frac{1}{2}\alpha(\boldsymbol{E}_L^2)_2$ are the electrostatic energies per particle at 1 and 2 respectively.

The meaning of the pressure p of the medium used in connexion with the Kelvin force can be further elucidated by considering the equilibrium conditions between two isotropic systems I and II, at rest, with different susceptibilities κ_{I}, χ_{I} and κ_{II}, χ_{II}. From (73) with $\Pi = 0$ and $\mathrm{d}\boldsymbol{v}/\mathrm{d}t = 0$ (thermodynamic and mechanical equilibrium) it then follows that at any point of the interface between I and II

$$\mathrm{grad}\, p = \boldsymbol{F}, \tag{88}$$

or with (15) (since the fields are stationary in equilibrium)

$$\mathrm{Div}\,(\mathsf{T} - p\mathsf{U}) = 0, \tag{89}$$

where T is given by (22) (with $\boldsymbol{v} = 0$). Application of Gauss' theorem leads to

$$\boldsymbol{n}\cdot(\mathsf{T}_{\mathrm{I}} - p_{\mathrm{I}}\mathsf{U}) = \boldsymbol{n}\cdot(\mathsf{T}_{\mathrm{II}} - p_{\mathrm{II}}\mathsf{U}), \tag{90}$$

where $\boldsymbol{n}$ is a unit vector perpendicular to the interface and where T_{I}, p_{I} and T_{II} and p_{II} are the values of T and p in medium I and II respectively, at the same point of the interface. With (22) we then obtain, since the dashed field quantities reduce to the fields measured by an observer at rest,

$$\begin{aligned} p_{\mathrm{II}} - p_{\mathrm{I}} &= \boldsymbol{n}\cdot(\mathsf{T}_{\mathrm{II}} - \mathsf{T}_{\mathrm{I}})\cdot\boldsymbol{n} \\ &= [D_{\perp\,\mathrm{II}} E_{\perp\,\mathrm{II}} + B_{\perp\,\mathrm{II}} H_{\perp\,\mathrm{II}} - \tfrac{1}{2}(E^2_{\perp\,\mathrm{II}} + E^2_{\parallel\,\mathrm{II}} + B^2_{\perp\,\mathrm{II}} + B^2_{\parallel\,\mathrm{II}} \\ &\quad - 2M_{\perp\,\mathrm{II}} B_{\perp\,\mathrm{II}} - 2M_{\parallel\,\mathrm{II}} B_{\parallel\,\mathrm{II}})] - [D_{\perp\,\mathrm{I}} E_{\perp\,\mathrm{I}} + B_{\perp\,\mathrm{I}} H_{\perp\,\mathrm{I}} \\ &\quad - \tfrac{1}{2}(E^2_{\perp\,\mathrm{I}} + E^2_{\parallel\,\mathrm{I}} + B^2_{\perp\,\mathrm{I}} + B^2_{\parallel\,\mathrm{I}} - 2M_{\perp\,\mathrm{I}} B_{\perp\,\mathrm{I}} - 2M_{\parallel\,\mathrm{I}} B_{\parallel\,\mathrm{I}})], \end{aligned} \tag{91}$$

where the subscripts $\perp$ and $\parallel$ denote the perpendicular and parallel components respectively of the field quantities at the interface.

With the well-known conditions (which follow from Maxwell's equations) at an interface between two electroneutral media

$$E_{\parallel\,\mathrm{I}} = E_{\parallel\,\mathrm{II}}, \quad H_{\parallel\,\mathrm{I}} = H_{\parallel\,\mathrm{II}}, \tag{92}$$

$$D_{\perp\,\mathrm{I}} = D_{\perp\,\mathrm{II}}, \quad B_{\perp\,\mathrm{I}} = B_{\perp\,\mathrm{II}}, \tag{93}$$

(91) becomes:

$$p_{\mathrm{II}} - p_{\mathrm{I}} = \tfrac{1}{2}(P^2_{\perp\,\mathrm{I}} + M^2_{\parallel\,\mathrm{I}} - P^2_{\perp\,\mathrm{II}} - M^2_{\parallel\,\mathrm{II}})$$

$$= \tfrac{1}{2}\left(\frac{\kappa_{\mathrm{I}}^2}{\varepsilon_{\mathrm{I}}^2} - \frac{\kappa_{\mathrm{II}}^2}{\varepsilon_{\mathrm{II}}^2}\right) D_\perp^2 + \tfrac{1}{2}(\chi_{\mathrm{I}}^2 - \chi_{\mathrm{II}}^2) H_\parallel^2 . \tag{94}$$

In the last member subscripts I and II have been omitted for the components $D_\perp$ and $H_\parallel$ which are continuous across the interface. If system I is not polarizable ($\kappa_1 = 0$, $\chi_1 = 0$) we have

$$p_{\mathrm{II}} = p_{\mathrm{I}} - \tfrac{1}{2}\frac{\kappa_{\mathrm{II}}^2}{\varepsilon_{\mathrm{II}}^2} D_\perp^2 - \tfrac{1}{2}\chi_{\mathrm{II}}^2 H_\parallel^2 . \tag{95}$$

Thus for instance if $\boldsymbol{H} = 0$, one measures with an electrically and magnetically inert manometer, in a direction perpendicular to the fields $\boldsymbol{E}$ and $\boldsymbol{D}$ ($\boldsymbol{D}\cdot\boldsymbol{n} = D_\perp = 0$), a pressure equal to the pressure p of the polarizable medium. (In the direction parallel to the fields, *i.e.* when $\boldsymbol{D}\cdot\boldsymbol{n} = |\boldsymbol{D}|$, the manometer will register the pressure p increased by a surface force due to the Maxwell stresses.) The pressure p defined in connexion with the Kelvin force is therefore at least in principle, a measurable quantity.

From a microscopic point of view the possibility of defining the pressure of a polarizable medium in various ways can be understood as follows. The pressure p^0 in the absence of an electromagnetic field arises as a consequence of the kinetic energy and the short range interactions of the constituent particles of the medium. Now if an electromagnetic field is applied, keeping the density and the temperature of the system constant, additional interactions of electromagnetic origin are set up in the system. These new interactions give rise on the one hand to long range forces (*i.e.* they contribute to the ponderomotive force), on the other hand they modify the (average) short range interactions between the particles and thus modify what one may call the "pressure" [*cf.* eq. (72)]. However, there remains a certain arbitrariness in the precise definition of short range interactions in such a medium, which makes it possible to define the concept of pressure (and with it the concept of ponderomotive force) in various ways* [*cf.* eq. (67) and (68)].

§ 4. *The Chemical Potential in a Polarized Medium***

The chemical potential μ_i in equation (59) is defined as [*cf.* Appendix II, formula (3)]

$$\mu_i = \left(\frac{\partial G^*}{\partial M_i}\right)_{p,\,T,\,\boldsymbol{E}',\,\boldsymbol{B}'}, \tag{96}$$

where M_i is the total mass of component i and were the quantity G^* given by

$$G^* = U - TS + pV - V(\boldsymbol{P}'\cdot\boldsymbol{E}' + \boldsymbol{M}'\cdot\boldsymbol{B}'), \tag{97}$$

is the total Gibbs function of a uniform system corresponding to an internal energy $U^* \equiv U - V(\boldsymbol{E}'\cdot\boldsymbol{P}' + \boldsymbol{M}'\cdot\boldsymbol{B}')$. Its total differential is

$$\mathrm{d}G^* = \mathrm{d}U - S\mathrm{d}T + V\mathrm{d}p - V(\boldsymbol{P}'\cdot\mathrm{d}\boldsymbol{E}' + \boldsymbol{M}'\cdot\mathrm{d}\boldsymbol{B}') + \sum_{k=1}^{n} \mu_k\,\mathrm{d}M_k\,. \tag{98}$$

The quantities U and S are the total energy and entropy respectively of a uniform system of total volume V.

We have in a polarized medium the following Euler relation [*cf.* Appendix II, formula (4)]

$$\sum_k c_k\mu_k = u - Ts + pv - \boldsymbol{p}'\cdot\boldsymbol{E}' - \boldsymbol{m}'\cdot\boldsymbol{B}' \equiv g^* \tag{99}$$

Now since the specific Gibbs function satisfies the equation

$$\mathrm{d}g^* = -\,s\mathrm{d}T + v\mathrm{d}p - \boldsymbol{p}'\cdot\mathrm{d}\boldsymbol{E}' - \boldsymbol{m}'\cdot\mathrm{d}\boldsymbol{B}' + \sum_k \mu_k\,\mathrm{d}c_k\,, \tag{100}$$

* P. Mazur and S. R. de Groot, loc. cit. p. 389.
** I. Prigogine, P. Mazur and R. Defay, J. Chim. phys. **50** (1953) 146.
R. Defay and P. Mazur, Bull. Soc. Chim. Belg. **63** (1954) 562.

the chemical potentials μ_k obey the Gibbs–Duhem relation

$$\sum_k \rho_k \operatorname{grad} \mu_k = - \rho s \operatorname{grad} T + \operatorname{grad} p - (\operatorname{Grad} \boldsymbol{E}') \cdot \boldsymbol{P}' - (\operatorname{Grad} \boldsymbol{B}') \cdot \boldsymbol{M}', \quad (101)$$

or

$$\sum_k \rho_k (\operatorname{grad} \mu_k)_T = \operatorname{grad} p - (\operatorname{Grad} \boldsymbol{E}') \cdot \boldsymbol{P}' - (\operatorname{Grad} \boldsymbol{B}') \cdot \boldsymbol{M}'. \quad (102)$$

But according to (23) and (25) the condition of mechanical equilibrium for a polarized system with no viscous flow ($P = pU$, Grad $\boldsymbol{v} = 0$), in which the fields are stationary ($\mathrm{d}\boldsymbol{E}/\mathrm{d}t = 0$, $\mathrm{d}\boldsymbol{B}/\mathrm{d}t = 0$) while no relaxation phenomena take place ($\mathrm{d}\boldsymbol{p}/\mathrm{d}t = 0$, $\mathrm{d}\boldsymbol{m}/\mathrm{d}t = 0$), is expressed by

$$\operatorname{grad} p - (\operatorname{Grad} \boldsymbol{E}') \cdot \boldsymbol{P}' - (\operatorname{Grad} \boldsymbol{B}') \cdot \boldsymbol{M}' + \rho z \boldsymbol{E}' + \frac{1}{c} \boldsymbol{i} \wedge \boldsymbol{B}' = 0. \quad (103)$$

This gives in combination with (101) or (102) and using (18) and (19), (up to terms of order $\boldsymbol{v}/c$)

$$\sum_k \rho_k \{ \operatorname{grad} \mu_k - z_k (\boldsymbol{E} + \frac{1}{c} \boldsymbol{v}_k \wedge \boldsymbol{B}) \} = - \rho s \operatorname{grad} T, \quad (104)$$

or

$$\sum_k \rho_k \{ (\operatorname{grad} \mu_k)_T - z_k (\boldsymbol{E} + \frac{1}{c} \boldsymbol{v}_k \wedge \boldsymbol{B}) \} = 0, \quad (105)$$

where also (XIII.16)–(XIII.18) have been applied. We have here again the condition of mechanical equilibrium of the previous chapter [*cf.* (XIII.49) and (XIII.50)]: therefore although the chemical potentials μ_k are modified, the various expressions for the entropy source strength derived in Chapter XIII, § 4 remain valid in the present case, under the conditions specified above. The conditions are clearly satisfied in the usual heat conduction and diffusion experiments.

The gradient of the chemical potential is related to the gradients of the intensive thermodynamic variables in the following way (*cf.* Appendix II):

$$\operatorname{grad}\mu_k = -s_k \operatorname{grad} T + v_k \operatorname{grad} p - (\operatorname{Grad} \boldsymbol{E}')\cdot\boldsymbol{p}'_k$$

$$- (\operatorname{Grad} \boldsymbol{B}')\cdot\boldsymbol{m}'_k + \sum_{j=1}^{n-1} \frac{\partial\mu_k}{\partial c_j} \operatorname{grad} c_j . \tag{106}$$

Here $\boldsymbol{p}'_k$ and $\boldsymbol{m}'_k$ are the partial specific electric and magnetic polarizations of component k:

$$\boldsymbol{p}'_k = \left(\frac{\partial V\boldsymbol{P}'}{\partial M_k}\right)_{T,\,p,\,\boldsymbol{E}',\,\boldsymbol{B}'} , \tag{107}$$

$$\boldsymbol{m}'_k = \left(\frac{\partial V\boldsymbol{M}'}{\partial M_k}\right)_{T,\,p,\,\boldsymbol{E}',\,\boldsymbol{B}'} \tag{108}$$

If the fields $\boldsymbol{E}'$ and $\boldsymbol{B}'$ are uniform, one thus has the same thermodynamic force for diffusion as in the case without polarization. However, we note that, according to (106), non-uniform fields may also induce diffusion in a polarizable medium. We shall discuss this effect in an alternative way. According to (70) and (71) we have

$$\mu_k - \mu_n = \left(\frac{\partial f}{\partial c_k}\right)_{T,\,v,\,\boldsymbol{p}',\,\boldsymbol{m}'} = \mu_k^0 - \mu_n^0 - \tfrac{1}{2}v\boldsymbol{E}'^2 \left(\frac{\partial \kappa}{\partial c_k}\right)_{T,\,v}$$

$$- \tfrac{1}{2}v\boldsymbol{H}'^2 \left(\frac{\partial \chi}{\partial c_k}\right)_{T,\,v} , \quad (k = 1, 2, \ldots, n-1) , \tag{109}$$

where μ_k^0 is the chemical potential of component k at zero fields $\boldsymbol{E}'$ and $\boldsymbol{B}'$ and at the same temperature, density and concentrations (the differentiations with respect to c_k are carried out at constant mass fractions c_i with $i \neq k$ and $i \neq n$). Let us in particular consider a system of two uncharged electrically polarizable ($\boldsymbol{P}' \neq 0$, $\boldsymbol{M}' = 0$) components. Equation (109) then becomes

$$\mu_1 - \mu_2 = \mu_1^0 - \mu_2^0 - \tfrac{1}{2}v\boldsymbol{E}'^2 \left(\frac{\partial \kappa}{\partial c_1}\right)_{T,\,v} \tag{110}$$

At thermodynamic equilibrium when $\operatorname{grad}\mu_1 = \operatorname{grad}\mu_2 = 0$, we thus have

$$\frac{\partial(\mu_1^0 - \mu_2^0)}{\partial c_1} \operatorname{grad} c_1 + (v_1 - v_2) \operatorname{grad} p^0 = \operatorname{grad}\left\{\tfrac{1}{2}v\boldsymbol{E}'^2 \left(\frac{\partial \kappa}{\partial c_1}\right)_{T,\,v}\right\}. \tag{111}$$

We could now proceed to eliminate from this relation the gradient of the pressure p^0 by means of the condition of mechanical equilibrium [*cf.* equation (78)]. Equation (111), however, demonstrates already how the relative mass distribution of the two components will be influenced through a non-homogeneous field, for instance if $v_1 \simeq v_2$, the concentration of the component with the highest dielectric constant will be enhanced in regions of high field strength. This situation may arise in the electric double layers around colloidal particles, where strongly non-homogeneous fields are set up. The above formulae together with the equations derived in the previous section for the phenomenon of electro- (or magneto-) striction completely determine the distribution of matter of a multi-component polarizable system in a non-homogeneous field. To conclude this section we mention that in a system consisting of charged components (111) becomes

$$\frac{\partial(\mu_1^0 - \mu_2^0)}{\partial c_1} \operatorname{grad} c_1 + (v_1 - v_2) \operatorname{grad} p^0 = (z_1 - z_2)\, \boldsymbol{E}' + \operatorname{grad}\left\{ \tfrac{1}{2} v \boldsymbol{E}'^2 \left(\frac{\partial \kappa}{\partial c_1}\right)_{T,v} \right\}, \tag{112}$$

since the condition of thermodynamic equilibrium then reads $\operatorname{grad}(\mu_1 - \mu_2) = (z_1 - z_2)\, \boldsymbol{E}'$.

§ 5. *Dielectric and Magnetic Relaxation*

We have seen in § 2 that the only new irreversible phenomena, occurring in a polarized system are the relaxations of the electric and magnetic polarizations. Let us consider then separately the phenomenon of dielectric relaxation (the magnetic case is analogous). In a system in which no viscous flows, thermal conduction or diffusion occurs, and in which magnetization effects may be neglected, the entropy source strength is according to (62) given by

$$\sigma_P = -\frac{\rho}{T} \frac{\mathrm{d}\boldsymbol{p}'}{\mathrm{d}t} \cdot (\boldsymbol{E}'_{\mathrm{eq}} - \boldsymbol{E}') \geqslant 0\,. \tag{113}$$

If the system has furthermore uniform density and is at rest ($\boldsymbol{v} = 0$), this expression becomes

$$\sigma_P = -\frac{1}{T} \frac{\partial \boldsymbol{P}}{\partial t} \cdot (\boldsymbol{E}_{\mathrm{eq}} - \boldsymbol{E}), \tag{114}$$

since the dashed field quantities then reduce to the fields measured by an observer at rest.

The linear phenomenological law corresponding to (114) is for an isotropic system

$$\frac{\partial \boldsymbol{P}}{\partial t} = -\frac{L}{T}(\boldsymbol{E}_{\mathrm{eq}} - \boldsymbol{E}) = -\frac{L}{T\kappa}(\boldsymbol{P} - \kappa\boldsymbol{E}) . \tag{115}$$

where L is a (positive) phenomenological coefficient characterizing the relaxation phenomenon. This equation is the well-known Debije equation for dielectric relaxation*. If an electric field $\boldsymbol{E}$ of constant magnitude and direction is applied at time $t = 0$ and $\boldsymbol{P}$ is initially zero, then integration of (115) leads to:

$$\boldsymbol{P}(t) = \kappa \boldsymbol{E}(1 - \mathrm{e}^{-t/\tau}) , \tag{116}$$

where the relaxation time τ is given by**

$$\tau = \frac{\kappa T}{L} . \tag{117}$$

If a time dependent field $\boldsymbol{E}(t)$ acts on the system, the most convenient way to solve (115) is to expand first $\boldsymbol{P}(t)$ and $\boldsymbol{E}(t)$ into Fourier integrals

$$\boldsymbol{P}(t) = \frac{1}{2\pi} \int_{-\infty}^{+\infty} \hat{\boldsymbol{P}}(\omega)\, \mathrm{e}^{-\mathrm{i}\omega t}\, \mathrm{d}\omega , \tag{118}$$

* Equation (115) could also have been obtained from the rotational diffusion equation discussed in Chapter X, § 6 by calculating the rate of change of the average dipole moment and linearizing the resulting equation with respect to the external field. Thermodynamically therefore the present discussion is one in terms of the average polarization only, whereas in the treatment of Chapter X the complete distribution over all possible orientations is taken into account. For distributions which are not too far from the equilibrium distribution, both descriptions lead to the same values for thermodynamic functions like the entropy (*cf.* Ch. VII, § 8 on entropy and Gaussian Markoff processes, where this is also the case).

** In fact in the corresponding experiment one does not *apply* a constant Maxwell field $\boldsymbol{E}$, but a constant external field $\boldsymbol{E}^0$. Since the Maxwell field $\boldsymbol{E}$ contains apart from the field $\boldsymbol{E}^0$ contributions from the polarization throughout the system, equation (115) becomes in general, in terms of $\boldsymbol{E}^0$, an integro-differential equation for $\boldsymbol{P}$. For various shapes of the system this integro-differential equation again reduces to an ordinary differential equation, containing, however, a different relaxation time (which is then shape dependent).

$$\boldsymbol{E}(t) = \frac{1}{2\pi} \int_{-\infty}^{+\infty} \hat{\boldsymbol{E}}(\omega)\, \mathrm{e}^{-\mathrm{i}\omega t}\, \mathrm{d}\omega . \tag{119}$$

In terms of the Fourier transforms $\hat{\boldsymbol{P}}(\omega)$ and $\hat{\boldsymbol{E}}(\omega)$

$$\hat{\boldsymbol{P}}(\omega) = \int_{-\infty}^{+\infty} \boldsymbol{P}(t)\, \mathrm{e}^{\mathrm{i}\omega t}\, \mathrm{d}t , \tag{120}$$

$$\hat{\boldsymbol{E}}(\omega) = \int_{-\infty}^{+\infty} \boldsymbol{E}(t)\, \mathrm{e}^{\mathrm{i}\omega t}\, \mathrm{d}t , \tag{121}$$

the phenomenological equation (115) becomes

$$- \mathrm{i}\omega \hat{\boldsymbol{P}} = - \frac{1}{\tau} \hat{\boldsymbol{P}} + \frac{\kappa}{\tau} \hat{\boldsymbol{E}} , \tag{122}$$

or

$$\hat{\boldsymbol{P}}(\omega) = \hat{\kappa}(\omega)\, \hat{\boldsymbol{E}}(\omega) , \tag{123}$$

where the electric susceptibility is given by

$$\hat{\kappa}(\omega) = \frac{\kappa}{1 - \mathrm{i}\omega\tau} . \tag{124}$$

Of course for $\omega = 0$, the susceptibility $\hat{\kappa}(\omega)$ reduces to κ, the static (equilibrium) electric susceptibility, as it should be. The real and imaginary parts of $\hat{\kappa}(\omega)$,

$$\hat{\kappa}'(\omega) = \frac{\kappa}{1 + \omega^2\tau^2} , \tag{125}$$

$$\hat{\kappa}''(\omega) = \frac{\omega\tau\kappa}{1 + \omega^2\tau^2} , \tag{126}$$

satisfy the Kramers–Kronig relations (VIII.102) and (VIII.103). This is not surprising, since we expect $\boldsymbol{P}$ and $\boldsymbol{E}$ to obey a causality condition (*cf.* Ch. VIII, § 3; it has also been mentioned at the end of Chapter VIII that the dissipative character of a process implies a causal relation between response and driving force).

It must be noted that, according to (125), the real part of $\hat{\kappa}(\omega)$ vanishes as ω tends to infinity. In the Debije theory of dielectric relaxation on the other hand $\hat{\kappa}(\omega)$ tends to some constant value $\hat{\kappa}(\infty)$. One assumes in this theory that the polarization $\boldsymbol{P}$ may be written as a sum of two parts

$$\boldsymbol{P} = \boldsymbol{P}_{\text{dip}} + \boldsymbol{P}_{\text{def}}, \tag{127}$$

where $\boldsymbol{P}_{\text{dip}}$ is the polarization due to the orientation of the permanent dipoles of the constituent particles and $\boldsymbol{P}_{\text{def}}$ the polarization due to the deformation of the molecules. It is thought that at all relevant frequencies the deformation polarization instantaneously achieves its equilibrium value in the field $\boldsymbol{E}$:

$$\boldsymbol{P}_{\text{def}} = \kappa_{\text{def}}\, \boldsymbol{E}, \tag{128}$$

whereas $\boldsymbol{P}_{\text{dip}}$ obeys an equation of the type (115). One writes indeed for the variation in time of $\boldsymbol{P}_{\text{dip}}$:

$$\frac{\partial \boldsymbol{P}_{\text{dip}}}{\partial t} = -\frac{1}{\tau}(\boldsymbol{P} - \kappa \boldsymbol{E}). \tag{129}$$

One puts therefore this rate proportional to the deviation of the *total* polarization from its equilibrium value. In view of (128) we also have

$$\frac{\partial \boldsymbol{P}_{\text{dip}}}{\partial t} = -\frac{1}{\tau}\{\boldsymbol{P}_{\text{dip}} - (\kappa - \kappa_{\text{def}})\, \boldsymbol{E}\}, \tag{130}$$

which for a constant field $\boldsymbol{E}$ gives upon integration, with the initial condition $\boldsymbol{P}_{\text{dip}} = 0$ for $t = 0$:

$$\boldsymbol{P}_{\text{dip}} = (\kappa - \kappa_{\text{def}})\, \boldsymbol{E}\, (1 - \mathrm{e}^{-t/\tau}). \tag{131}$$

For time dependent fields $\boldsymbol{E}(t)$ one now finds for the relation between the Fourier transforms $\hat{\boldsymbol{P}}(\omega)$ and $\hat{\boldsymbol{E}}(\omega)$ using (127), (128) and (129),

$$\hat{\boldsymbol{P}}(\omega) = \hat{\kappa}(\omega)\, \hat{\boldsymbol{E}}(\omega), \tag{132}$$

with the generalized susceptibility

$$\hat{\kappa}(\omega) = \frac{\kappa - \mathrm{i}\omega\tau\kappa_{\text{def}}}{1 - \mathrm{i}\omega\tau}. \tag{133}$$

The real and imaginary parts of $\hat{\kappa}(\omega)$ are now given by

$$\hat{\kappa}'(\omega) = \frac{\kappa - \kappa_{\text{def}}}{1 + \omega^2\tau^2} + \kappa_{\text{def}} , \tag{134}$$

$$\hat{\kappa}''(\omega) = \frac{\omega\tau(\kappa - \kappa_{\text{def}})}{1 + \omega^2\tau^2} , \tag{135}$$

so that one has $\hat{\kappa}'(0) \neq \kappa$ and $\hat{\kappa}'(\infty) = \kappa_{\text{def}}$. Here also $\hat{\kappa}'(\omega)$ and $\hat{\kappa}''(\omega)$ satisfy the Kramers–Kronig relations.

Thus the results we have so far derived from thermodynamics of irreversible processes coincide with the Debije theory only if κ_{def} is equal to zero, *i.e.* if the system consists of particles which have permanent dipoles but are not deformable (such a system is called purely polar). The question then arises, how in general the Debije theory may be fitted into the scheme of non-equilibrium thermodynamics. The answer can be given by inspection of the phenomenological Debije theory itself: obviously one must consider as thermodynamic variable describing the state of the system not only the total polarization, but in addition one or more internal variables related to the deformation polarization. Consequently the irreversible behaviour of the system is characterized by a whole set of relaxation times. If then the situation is such that one relaxation time is much larger than all the others, one finds up to relatively large frequencies a spectrum of the type (133), *i.e.* a Debije spectrum. Strictly speaking this implies that $\hat{\kappa}(\infty)$ is not really equal to the constant κ_{def}. However, deviations from the form (133) can only be measured at those very high frequencies for which (133) does not hold any more. The α- or β-type character of the additional internal variables required could be inferred from an analysis of the spectrum at these high frequencies. (Frequently even at low frequencies one does find experimentally a Debije spectrum due to structural relaxation.) In this way the theory of dielectric relaxation becomes an example of the general relaxation theory considered in Chapter X.

Note finally, that from (118), (119) and (123) one finds the desired relation between $\boldsymbol{P}(t)$ and $\boldsymbol{E}(t)$:

$$\boldsymbol{P}(t) = \int_{-\infty}^{+\infty} \kappa(t')\, \boldsymbol{E}(t - t')\, \mathrm{d}t' , \tag{136}$$

with

$$\kappa(t) = \frac{1}{2\pi} \int_{-\infty}^{+\infty} \hat{\kappa}(\omega)\, e^{-i\omega t}\, d\omega\,. \tag{137}$$

According to (124) the electric susceptibility $\kappa(t)$ for the purely polar system is given by

$$\kappa(t) = \left\{ \begin{matrix} \frac{K}{\tau} e^{-t/\tau} & \text{if} \quad t > 0 \\ 0 & \text{if} \quad t < 0 \end{matrix} \right\}. \tag{138}$$

Thus as expected $\kappa(t)$ satisfies the causality principle.

The theory of paramagnetic relaxation is completely analogous: it is contained in the above formulae, if $\boldsymbol{P}(t)$ is replaced by the magnetization $\boldsymbol{M}(t)$ and $\boldsymbol{E}(t)$ by the magnetic field $\boldsymbol{B}(t)$ (κ then represents the magnetic susceptibility χ).

CHAPTER XV

DISCONTINUOUS SYSTEMS

§ 1. *Introduction*

In the preceding chapters systems were studied, of which the state variables are arbitrary continuous functions of space coordinates and of time. Such systems are sometimes referred to as *continuous systems*.

In a number of cases, however, one prefers to consider irreversible processes that take place in a particular experimental set-up, which consists essentially of two large reservoirs connected by a small capillary, hole, porous wall or membrane. Let us suppose that the whole apparatus is filled with an isotropic fluid n-component ($k =$

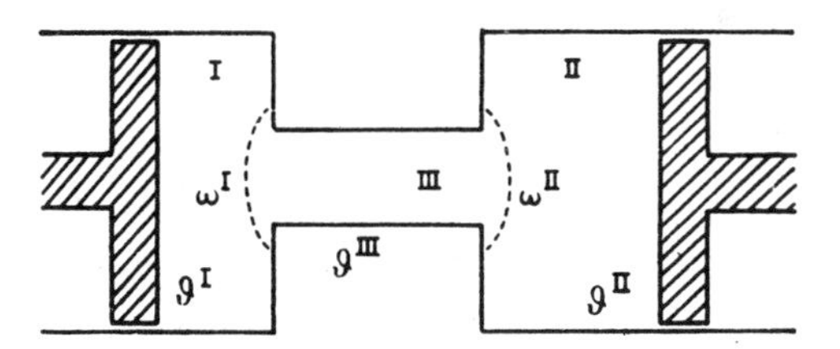

Fig. 4. The discontinuous system.

$1, 2, \ldots, n$) mixture, in which a number of r chemical reactions ($j = 1, 2, \ldots, r$) can take place. Furthermore the system can be subject to external forces, and the volumes of the two large vessels can be changed by means of pistons. The total system is closed, *i.e.* no matter is exchanged with the surroundings. The various parts in which we shall divide the total system are, however, open systems, since they can exchange matter. Heat can pass through the pistons, but the other outer walls are adiabatically insulated from the surroundings. Work can be performed on the system by means of the pistons. It is supposed that accelerations in any part of the system can be neglected. At some initial moment the system is uniform in the volumes V^{I} and V^{II}, forming the bulk of the large reservoirs, but non-uniform in the volume

V^{III} which comprises the capillary and small adjacent parts of the fluid present in the reservoirs. We shall assume that such a situation is maintained in the course of time. (It is known, for instance, that for the flow of a fluid in a tube of varying diameter the pressure gradient is inversely proportional to the cross-section. Since the capillary is very much narrower than the big reservoirs the pressure drop will occur almost exclusively in the capillary. One can argue that the other intensive properties behave in a similar way.)

If one considers the sub-system III as so small that it can be disregarded altogether, then in passing from I to II the state variables suffer discontinuous jumps. For this reason a fluid system of the type described above is usually referred to as a "*discontinuous system*". One can, however, conceive more complicated systems, consisting of several large parts of which the properties vary discontinuously from one sub-system to another if one disregards small connecting capillaries, membranes or holes. All such systems are called "discontinuous".

The laws which describe the phenomena taking place in the discontinuous system, introduced here, are expressed in terms of the variables characterizing the state in volumes I and II. In general an extensive (or total) quantity is obtained by integrating an intensive (or local) quantity (density) over a volume:

$$F^\alpha(t) = \int^{V^\alpha(t)} f(\boldsymbol{r}, t)\, \mathrm{d}V, \quad (\alpha = \mathrm{I, II, III})\,. \tag{1}$$

In accordance with what was said above, f is independent of $\boldsymbol{r}$ in I and II, but not in III. We shall consider so-called "quasi-stationary" states, *i.e.* states in which one can neglect the time dependence of F^{III} as compared to the changes of F^{I} and F^{II}. The situation described here is thus altogether characterized by

$$f \neq f(\boldsymbol{r})\,, \quad F^\alpha = F^\alpha(t)\,, \quad (\alpha = \mathrm{I, II})\,, \tag{2}$$

$$f = f(\boldsymbol{r})\,, \quad F^{III} = \text{constant}\;; \quad V^{III} \ll V^{I}\,, V^{II}\,. \tag{3}$$

Two classes of discontinuous systems, which are different with respect to the nature of the part III, can be distinguished.

a) The portion of the system in III, which may be a capillary or porous wall or the like, has a dimension large compared to the mean free path of the molecules. We can then treat the part of the system in III as a macroscopic system, just as the parts in I and II.

b) The portion of the system in III is of the order or smaller than the mean free path of the molecules; it may for instance be a small hole or a very narrow capillary.

Let us write down formally the boundary conditions for the discontinuous system. The total surface Ω^α of the sub-system α (= I, II, and in case *a*, also III) can be split up as follows

$$\Omega^\alpha = \vartheta^\alpha + \omega^\alpha , \quad [\alpha = \text{I, II, (III)}] , \tag{4}$$

where ϑ^α is the surface at the surroundings and at the piston, whereas ω^α is the internal surface (see Fig. 4 on p. 405). Let us write $\boldsymbol{v}_\Omega$ for the velocity of the surface. We then have for the sub-systems I and II:

$$\boldsymbol{v} = \boldsymbol{v}_k = \boldsymbol{v}_\Omega \quad (k = 1, 2, \ldots, n) \quad \text{at} \quad \vartheta^\alpha ; \quad \boldsymbol{v}_\Omega = 0 \quad \text{at} \quad \omega^\alpha , \quad (\alpha = \text{I, II}) , \tag{5}$$

where $\boldsymbol{v}$ is the barycentric velocity and $\boldsymbol{v}_k$ the velocity of component k. For the case *a* we have also boundary conditions for sub-system III:

$$\boldsymbol{v} = \boldsymbol{v}_k = \boldsymbol{v}_\Omega = 0 , \quad \boldsymbol{J}_q = 0 \quad \text{at} \quad \vartheta^{\text{III}} ; \quad \boldsymbol{v}_\Omega = 0 \quad \text{at} \quad \omega^{\text{III}} , \tag{6}$$

where $\boldsymbol{J}_q$ is the heat flow normal to the surface. The surface ω^{III} is the sum of ω^{I} and ω^{II}.

In this chapter we wish in the first place to derive the conservation laws and the entropy balance for the discontinuous system from the corresponding local laws applied to the sub-systems I and II, and in case *a* also to III, in order to obtain equations between the state variables of I and II only*. This is possible both for case *a* and for case *b*. Furthermore in case *a* we can find the phenomenological equations from the local phenomenological equations. The phenomenological coefficients are functions of the local phenomenological coefficients. Reciprocal relations which are based on the Onsager relations between the local phenomenological coefficients can be proved. In case *b* phenomenological equations are directly set up between the fluxes

* These equations were first set up without using the local differential equations, by I. Prigogine, Etude thermodynamique des phénomènes irréversibles (Desoer, Liège, 1947); see also S. R. de Groot, Thermodynamics of irreversible processes (North-Holland Publishing Company, Amsterdam, 1951).

and thermodynamic forces found for the discontinuous system. Onsager relations follow then immediately from the theory of chapter VII.

§ 2. *Conservation Laws*

For the derivation of the conservation laws for discontinuous systems from the local laws, one needs to know the change in time of an extensive quantity (1):

$$\frac{\mathrm{d}F(t)}{\mathrm{d}t} = \frac{\mathrm{d}}{\mathrm{d}t}\int^{V(t)} f(\boldsymbol{r}, t)\,\mathrm{d}V\,, \tag{7}$$

where the volume $V(t)$ is contained in the closed moving surface $\Omega(t)$. The right-hand side of (7) can be written as

$$\begin{aligned}
&\lim_{\Delta t\to 0}\frac{1}{\Delta t}\left\{\int^{V(t+\Delta t)} f(\boldsymbol{r}, t+\Delta t)\,\mathrm{d}V - \int^{V(t)} f(\boldsymbol{r}, t)\,\mathrm{d}V\right\} \\
&\qquad = \lim_{\Delta t\to 0}\left[\frac{1}{\Delta t}\int^{V(t+\Delta t)}\{f(\boldsymbol{r}, t+\Delta t) - f(\boldsymbol{r}, t)\}\,\mathrm{d}V\right. \\
&\qquad\qquad \left. + \frac{1}{\Delta t}\left\{\int^{V(t+\Delta t)} f(\boldsymbol{r}, t)\,\mathrm{d}V - \int^{V(t)} f(\boldsymbol{r}, t)\,\mathrm{d}V\right\}\right].
\end{aligned} \tag{8}$$

With the last expression formula (7) becomes

$$\frac{\mathrm{d}F(t)}{\mathrm{d}t} = \int^{V(t)} \frac{\partial f(\boldsymbol{r}, t)}{\partial t}\,\mathrm{d}V + \int^{\Omega(t)} f(\boldsymbol{r}, t)\boldsymbol{v}_{\Omega}(\boldsymbol{r}, t)\cdot \mathrm{d}\boldsymbol{\Omega}\,, \tag{9}$$

where $\mathrm{d}\boldsymbol{\Omega}$ is a vector with magnitude $\mathrm{d}\Omega$, normal to the surface and counted positive from the inside to the outside.

Conservation of mass. For the mass of component k contained in the sub-system I, II or III (the latter only in case *a*):

$$M_k^{\alpha} = \int^{V^{\alpha}} \rho_k\,\mathrm{d}V\,, \quad (k = 1, 2, \ldots, n;\, \alpha = \mathrm{I, II, III})\,, \tag{10}$$

we find from (9) and (II.2)

$$\frac{\mathrm{d}M_k^\alpha}{\mathrm{d}t} = -\int^{\Omega^\alpha} \rho_k(\boldsymbol{v}_k - \boldsymbol{v}_\Omega)\cdot \mathrm{d}\boldsymbol{\Omega}^\alpha + \int^{V^\alpha} \sum_{j=1}^{r} \nu_{kj} J_j \,\mathrm{d}V\,, \quad (\alpha = \mathrm{I, II, III})\,. \tag{11}$$

With (2)–(6) this gives

$$\frac{\mathrm{d}M_k^\alpha}{\mathrm{d}t} = -\int^{\omega^\alpha} \rho_k \boldsymbol{v}_k \cdot \mathrm{d}\boldsymbol{\Omega}^\alpha + \sum_{j=1}^{r} \nu_{kj} J_j^\alpha V^\alpha\,, \quad (\alpha = \mathrm{I, II})\,, \tag{12}$$

$$0 = \frac{\mathrm{d}M_k^{\mathrm{III}}}{\mathrm{d}t} = -\int^{\omega^{\mathrm{III}}} \rho_k \boldsymbol{v}_k \cdot \mathrm{d}\boldsymbol{\Omega}^{\mathrm{III}} = \sum_{\alpha=\mathrm{I}}^{\mathrm{II}} \int^{\omega^\alpha} \rho_k \boldsymbol{v}_k \cdot \mathrm{d}\boldsymbol{\Omega}^\alpha\,. \tag{13}$$

In formula (12) we recognize at the right-hand side an "internal transfer" and a "chemical reaction" change of M_k^α ($\alpha = \mathrm{I, II}$), which we shall denote as

$$\frac{\mathrm{d_i}M_k^\alpha}{\mathrm{d}t} \equiv -\int^{\omega^\alpha} \rho_k \boldsymbol{v}_k \cdot \mathrm{d}\boldsymbol{\Omega}^\alpha\,, \quad (\alpha = \mathrm{I, II})\,, \tag{14}$$

$$\frac{\mathrm{d_c}M_k^\alpha}{\mathrm{d}t} = \sum_{j=1}^{r} \frac{\mathrm{d_c}M_{kj}^\alpha}{\mathrm{d}t} \quad \text{with} \quad \frac{\mathrm{d_c}M_{kj}^\alpha}{\mathrm{d}t} \equiv \nu_{kj} J_j^\alpha V^\alpha\,, \quad (\alpha = \mathrm{I, II})\,. \tag{15}$$

From (13) and (14) we have the mass conservation law for discontinuous systems

$$\frac{\mathrm{d_i}M_k^{\mathrm{I}}}{\mathrm{d}t} + \frac{\mathrm{d_i}M_k^{\mathrm{II}}}{\mathrm{d}t} = 0\,. \tag{16}$$

From $\sum_k \nu_{kj} = 0$ [see (II.4)] and (15) we have

$$\sum_{k=1}^{n} \frac{\mathrm{d_c}M_{kj}^\alpha}{\mathrm{d}t} = 0\,, \quad (j = 1, 2, \ldots, r;\, \alpha = \mathrm{I, II})\,, \tag{17}$$

which expresses that mass is conserved in each separate chemical reaction.

Conservation of energy. For the kinetic energy

$$L^\alpha = \int^{V^\alpha} \tfrac{1}{2}\rho \boldsymbol{v}^2 \,\mathrm{d}V\,, \quad (\alpha = \mathrm{I, II, III})\,, \tag{18}$$

we find from (II.25) and (9), with (II.19) and the condition of negligible acceleration, applying also Gauss' theorem

$$\frac{\mathrm{d}L^\alpha}{\mathrm{d}t} = -\int^{\Omega^\alpha} \tfrac{1}{2}\rho v^2(\boldsymbol{v} - \boldsymbol{v}_\Omega)\cdot \mathrm{d}\boldsymbol{\Omega}^\alpha\,, \quad (\alpha = \mathrm{I, II, III})\,. \tag{19}$$

This becomes with (3), (5) and (6)

$$\frac{\mathrm{d}L^\alpha}{\mathrm{d}t} = -\int^{\omega^\alpha} \tfrac{1}{2}\rho v^2 \boldsymbol{v}\cdot \mathrm{d}\boldsymbol{\Omega}^\alpha\,, \quad (\alpha = \mathrm{I, II})\,, \tag{20}$$

$$0 = \frac{\mathrm{d}L^{\mathrm{III}}}{\mathrm{d}t} = \sum_{\alpha=1}^{\mathrm{II}} \int^{\omega^\alpha} \tfrac{1}{2}\rho v^2 \boldsymbol{v}\cdot \mathrm{d}\boldsymbol{\Omega}^\alpha\,. \tag{21}$$

From these two equations we find

$$\frac{\mathrm{d}L^{\mathrm{I}}}{\mathrm{d}t} + \frac{\mathrm{d}L^{\mathrm{II}}}{\mathrm{d}t} = 0\,. \tag{22}$$

For the potential energy

$$\Psi^\alpha = \int^{V^\alpha} \rho\psi\, \mathrm{d}V\,, \quad (\alpha = \mathrm{I, II, III})\,, \tag{23}$$

where ψ is given by (II.20) and $\rho\psi = \sum_k \rho_k\psi_k$, the formula (9) yields with (II.9), (II.26), (II.27) and Gauss' theorem

$$\frac{\mathrm{d}\Psi^\alpha}{\mathrm{d}t} = -\int^{\Omega^\alpha}\left[\left(\sum_{k=1}^{n} \psi_k\rho_k\boldsymbol{v}_k\right) - \psi\rho\boldsymbol{v}_\Omega\right]\cdot \mathrm{d}\boldsymbol{\Omega}^\alpha - \int^{V^\alpha} \sum_{k=1}^{n} \rho_k \boldsymbol{F}_k\cdot\boldsymbol{v}_k\, \mathrm{d}V\,,$$
$$(\alpha = \mathrm{I, II, III})\,. \tag{24}$$

With (II.20), (2)–(6) and (14) this gives

$$\frac{\mathrm{d}\Psi^\alpha}{\mathrm{d}t} = -\int^{\omega^\alpha} \sum_k \psi_k\rho_k\boldsymbol{v}_k\cdot \mathrm{d}\boldsymbol{\Omega}^\alpha = \sum_k \psi_k^\alpha \frac{\mathrm{d_i}M_k^\alpha}{\mathrm{d}t}\,, \quad (\alpha = \mathrm{I, II})\,, \tag{25}$$

$$0 = \frac{\mathrm{d}\Psi^{\mathrm{III}}}{\mathrm{d}t} = \sum_{\alpha=1}^{\mathrm{II}} \int^{\omega^\alpha} \sum_k \psi_k\rho_k\boldsymbol{v}_k\cdot \mathrm{d}\boldsymbol{\Omega}^\alpha - \int^{V^{\mathrm{III}}} \sum_k \rho_k\boldsymbol{F}_k\cdot\boldsymbol{v}_k\, \mathrm{d}V\,. \tag{26}$$

The total energy

$$E^{\alpha} = \int^{V^{\alpha}} e\rho \, \mathrm{d}V \,, \quad (\alpha = \mathrm{I, II, III}) \,, \tag{27}$$

has, according to (9), (II.31) and Gauss' theorem, a time derivative

$$\frac{\mathrm{d}E^{\alpha}}{\mathrm{d}t} = - \int^{\Omega^{\alpha}} (\boldsymbol{J}_e - e\rho \boldsymbol{v}_{\Omega}) \cdot \mathrm{d}\boldsymbol{\Omega}^{\alpha} \,, \quad (\alpha = \mathrm{I, II, III}) \,, \tag{28}$$

or, introducing (II.32) and (II.33),

$$\frac{\mathrm{d}E^{\alpha}}{\mathrm{d}t} = - \int^{\Omega^{\alpha}} \{ \boldsymbol{J}_q + \boldsymbol{P} \cdot \boldsymbol{v} + \sum_{k=1}^{n} \psi_k \boldsymbol{J}_k + \rho(u + \tfrac{1}{2}v^2 + \psi)(\boldsymbol{v} - \boldsymbol{v}_{\Omega}) \} \cdot \mathrm{d}\boldsymbol{\Omega}^{\alpha} \,, \quad (\alpha = \mathrm{I, II, III}) \,. \tag{29}$$

With (2)–(6), (20), (21), (25), (26), (II.9) and (II.35) (where we can neglect Π in I and II because practically no viscous flow arises there) we obtain

$$\frac{\mathrm{d}E^{\alpha}}{\mathrm{d}t} = \frac{\mathrm{d_e}Q^{\alpha}}{\mathrm{d}t} - p^{\alpha}\frac{\mathrm{d}V^{\alpha}}{\mathrm{d}t} + \frac{\mathrm{d_i}Q^{\alpha}}{\mathrm{d}t} + h^{\alpha}\frac{\mathrm{d}M^{\alpha}}{\mathrm{d}t} + \frac{\mathrm{d}L^{\alpha}}{\mathrm{d}t} + \frac{\mathrm{d}\Psi^{\alpha}}{\mathrm{d}t} \,, \quad (\alpha = \mathrm{I, II}) \,, \tag{30}$$

$$0 = \frac{\mathrm{d}E^{\mathrm{III}}}{\mathrm{d}t} = - \sum_{\alpha=\mathrm{I}}^{\mathrm{II}} \left(\frac{\mathrm{d_i}Q^{\alpha}}{\mathrm{d}t} + h^{\alpha}\frac{\mathrm{d}M^{\alpha}}{\mathrm{d}t} + \frac{\mathrm{d}L^{\alpha}}{\mathrm{d}t} + \frac{\mathrm{d}\Psi^{\alpha}}{\mathrm{d}t} \right) , \tag{31}$$

where

$$h = u + p\rho^{-1} \tag{32}$$

is the specific enthalpy. The following notations were introduced for the external and internal heat supply

$$\frac{\mathrm{d_e}Q^{\alpha}}{\mathrm{d}t} = - \int^{\vartheta^{\alpha}} \boldsymbol{J}_q \cdot \mathrm{d}\boldsymbol{\Omega}^{\alpha} \,, \quad (\alpha = \mathrm{I, II}) \,, \tag{33}$$

$$\frac{\mathrm{d_i}Q^{\alpha}}{\mathrm{d}t} = - \int^{\omega^{\alpha}} \boldsymbol{J}_q \cdot \mathrm{d}\boldsymbol{\Omega}^{\alpha} \,, \quad (\alpha = \mathrm{I, II}) \,. \tag{34}$$

The change of volume appearing in (30) and (31) can be deduced with the help of (5) and (9):

$$\frac{\mathrm{d}V^\alpha}{\mathrm{d}t} = \int^{\Omega^\alpha} \boldsymbol{v}_\Omega \cdot \mathrm{d}\boldsymbol{\Omega}^\alpha = \int^{\vartheta^\alpha} \boldsymbol{v} \cdot \mathrm{d}\boldsymbol{\Omega}^\alpha, \quad (\alpha = \mathrm{I}, \mathrm{II}) . \tag{35}$$

The mass change is

$$\frac{\mathrm{d}M^\alpha}{\mathrm{d}t} \equiv \sum_k \frac{\mathrm{d}M_k^\alpha}{\mathrm{d}t} = -\int^{\omega^\alpha} \rho \boldsymbol{v} \cdot \mathrm{d}\boldsymbol{\Omega}^\alpha, \quad (\alpha = \mathrm{I}, \mathrm{II}) . \tag{36}$$

This follows from summation over $k = 1, 2, \ldots, n$ of formula (12) and from (II.4) and (II.7).

One can also introduce the internal energy

$$U^\alpha = \int^{V^\alpha} \rho u \, \mathrm{d}V, \quad (\alpha = \mathrm{I}, \mathrm{II}, \mathrm{III}) . \tag{37}$$

One then has from (II.32)

$$\frac{\mathrm{d}E^\alpha}{\mathrm{d}t} = \frac{\mathrm{d}U^\alpha}{\mathrm{d}t} + \frac{\mathrm{d}L^\alpha}{\mathrm{d}t} + \frac{\mathrm{d}\Psi^\alpha}{\mathrm{d}t}, \quad (\alpha = \mathrm{I}, \mathrm{II}, \mathrm{III}) . \tag{38}$$

From (30) and (38) we conclude that

$$\frac{\mathrm{d}U^\alpha}{\mathrm{d}t} = \frac{\mathrm{d_e}U^\alpha}{\mathrm{d}t} + \frac{\mathrm{d_i}U^\alpha}{\mathrm{d}t}, \quad (\alpha = \mathrm{I}, \mathrm{II}) , \tag{39}$$

where

$$\frac{\mathrm{d_e}U^\alpha}{\mathrm{d}t} = \frac{\mathrm{d_e}Q^\alpha}{\mathrm{d}t} - p^\alpha \frac{\mathrm{d}V^\alpha}{\mathrm{d}t}, \quad (\alpha = \mathrm{I}, \mathrm{II}) , \tag{40}$$

$$\frac{\mathrm{d_i}U^\alpha}{\mathrm{d}t} = \frac{\mathrm{d_i}Q^\alpha}{\mathrm{d}t} + h^\alpha \frac{\mathrm{d}M^\alpha}{\mathrm{d}t}, \quad (\alpha = \mathrm{I}, \mathrm{II}) , \tag{41}$$

give the external and the internal change of the internal energy. One finds from (22), (31) and (41)

$$\frac{\mathrm{d_i}U^{\mathrm{I}}}{\mathrm{d}t} + \frac{\mathrm{d}\Psi^{\mathrm{I}}}{\mathrm{d}t} + \frac{\mathrm{d_i}U^{\mathrm{II}}}{\mathrm{d}t} + \frac{\mathrm{d}\Psi^{\mathrm{II}}}{\mathrm{d}t} = 0 , \tag{42}$$

which is the form of the energy law for discontinuous systems.

In the foregoing we considered case *a*, where III is a macroscopic system. In the case *b* the system III has a negligible extension, so that total quantities and their derivatives do not exist for III. This means that we still have the same conservation laws as above, in particular (16) and (42).

§ 3. *Entropy Law and Entropy Balance*

The change in time of the entropy

$$S^\alpha = \int^{V^\alpha} s\rho \, \mathrm{d}V \,, \quad (\alpha = \mathrm{I, II, III}) \,, \tag{43}$$

is found from (9) with (III.10), (III.13) and (III.20)

$$\frac{\mathrm{d}S^\alpha}{\mathrm{d}t} = -\int^{\Omega^\alpha} \left\{ \frac{\boldsymbol{J}_q - \sum_{k=1}^{n} \mu_k \boldsymbol{J}_k}{T} + s\rho(\boldsymbol{v} - \boldsymbol{v}_\Omega) \right\} \cdot \mathrm{d}\boldsymbol{\Omega}^\alpha + \int^{V^\alpha} \sigma \, \mathrm{d}V \,, \quad (\alpha = \mathrm{I, II, III}) \,. \tag{44}$$

With (2)–(6), $\boldsymbol{J}_k = \rho_k(\boldsymbol{v}_k - \boldsymbol{v})$ and the Euler relation

$$\sum_{k=1}^{n} \mu_k \rho_k = (h - Ts)\rho \,, \tag{45}$$

we find from the equation (44)

$$\frac{\mathrm{d}S^\alpha}{\mathrm{d}t} = \frac{1}{T^\alpha}\left(\frac{\mathrm{d_e}Q^\alpha}{\mathrm{d}t} + \frac{\mathrm{d_i}Q^\alpha}{\mathrm{d}t} + h^\alpha \frac{\mathrm{d}M^\alpha}{\mathrm{d}t} - \sum_{k=1}^{n} \mu_k^\alpha \frac{\mathrm{d_i}M_k^\alpha}{\mathrm{d}t} \right) + \int^{V^\alpha} \sigma \, \mathrm{d}V \,, \quad (\alpha = \mathrm{I, II}) \,, \tag{46}$$

$$0 = \frac{\mathrm{d}S^{\mathrm{III}}}{\mathrm{d}t} = - \sum_{\alpha=\mathrm{I}}^{\mathrm{II}} \frac{1}{T^\alpha}\left(\frac{\mathrm{d_i}Q^\alpha}{\mathrm{d}t} + h^\alpha \frac{\mathrm{d}M^\alpha}{\mathrm{d}t} - \sum_{k=1}^{n} \mu_k^\alpha \frac{\mathrm{d_i}M_k^\alpha}{\mathrm{d}t} \right) + \int^{V^{\mathrm{III}}} \sigma \, \mathrm{d}V \,, \tag{47}$$

where we have employed again the notations (14), (33), (34) and (36). We note also that

$$\int^{V^\alpha} \sigma \, \mathrm{d}V = - \frac{1}{T^\alpha} \sum_{j=1}^{r} J_j^\alpha A_j^\alpha V^\alpha \,, \quad (\alpha = \mathrm{I, II}) \tag{48}$$

follows from (2), (II.20) and (III.21), because gradients vanish in the uniform sub-systems I and II. We recognize in (46) an external and internal part:

$$\frac{\mathrm{d_e}S^\alpha}{\mathrm{d}t} = \frac{1}{T^\alpha}\frac{\mathrm{d_e}Q^\alpha}{\mathrm{d}t}, \quad (\alpha = \mathrm{I, II}), \tag{49}$$

$$\frac{\mathrm{d_i}S^\alpha}{\mathrm{d}t} = \frac{1}{T^\alpha}\left(\frac{\mathrm{d_i}Q^\alpha}{\mathrm{d}t} + h^\alpha \frac{\mathrm{d}M^\alpha}{\mathrm{d}t} - \sum_{k=1}^{n} \mu_k^\alpha \frac{\mathrm{d_i}M_k^\alpha}{\mathrm{d}t}\right) + \int^{V^\alpha} \sigma \, \mathrm{d}V, \quad (\alpha = \mathrm{I, II}). \tag{50}$$

With (40) and (41) one can write these relations alternatively as

$$\frac{\mathrm{d_e}S^\alpha}{\mathrm{d}t} = \frac{1}{T^\alpha}\frac{\mathrm{d_e}U^\alpha}{\mathrm{d}t} + \frac{p^\alpha}{T^\alpha}\frac{\mathrm{d}V^\alpha}{\mathrm{d}t}, \quad (\alpha = \mathrm{I, II}), \tag{51}$$

$$\frac{\mathrm{d_i}S^\alpha}{\mathrm{d}t} = \frac{1}{T^\alpha}\frac{\mathrm{d_i}U^\alpha}{\mathrm{d}t} - \sum_{k=1}^{n} \frac{\mu_k^\alpha}{T^\alpha}\frac{\mathrm{d_i}M_k^\alpha}{\mathrm{d}t} + \int^{V^\alpha} \sigma \, \mathrm{d}V, \quad (\alpha = \mathrm{I, II}). \tag{52}$$

We can now find the total entropy production in the system

$$\sigma_{\mathrm{tot}} \equiv \sum_{\alpha=\mathrm{I}}^{\mathrm{III}} \int^{V^\alpha} \sigma \, \mathrm{d}V \tag{53}$$

from (47) and (50). This gives

$$\sigma_{\mathrm{tot}} = \frac{\mathrm{d_i}S^{\mathrm{I}}}{\mathrm{d}t} + \frac{\mathrm{d_i}S^{\mathrm{II}}}{\mathrm{d}t}, \tag{54}$$

which yields with (52) and (48)

$$\sigma_{\mathrm{tot}} = \sum_{\alpha=\mathrm{I}}^{\mathrm{II}} \left(\frac{1}{T^\alpha}\frac{\mathrm{d_i}U^\alpha}{\mathrm{d}t} - \sum_{k=1}^{n} \frac{\mu_k^\alpha}{T^\alpha}\frac{\mathrm{d_i}M_k^\alpha}{\mathrm{d}t} - \frac{1}{T^\alpha}\sum_{j=1}^{r} J_j^\alpha A_j^\alpha V^\alpha\right). \tag{55}$$

Adding and subtracting the same quantity, one obtains with (25)

$$\sigma_{\mathrm{tot}} = \sum_{\alpha=\mathrm{I}}^{\mathrm{II}} \frac{1}{T^\alpha}\left(\frac{\mathrm{d_i}U^\alpha}{\mathrm{d}t} + \frac{\mathrm{d}\Psi^\alpha}{\mathrm{d}t}\right) - \sum_{k=1}^{n} \frac{\mu_k^\alpha + \psi_k^\alpha}{T^\alpha}\frac{\mathrm{d_i}M_k^\alpha}{\mathrm{d}t} - \frac{1}{T^\alpha}\sum_{j=1}^{r} J_j^\alpha A_j^\alpha V^\alpha. \tag{56}$$

Let us now eliminate $\mathrm{d_i}U^{\mathrm{II}} + \mathrm{d}\Psi^{\mathrm{II}}$ and $\mathrm{d_i}M_k{}^{\mathrm{II}}$ with the help of the conservation laws (16) and (42). Then (56) gets the form of sum of a product of fluxes and thermodynamic forces:

$$\sigma_{\text{tot}} = \tilde{j}_u x_u + \sum_{k=1}^{n} j_k x_k + \sum_{\alpha=\text{I}}^{\text{II}} \sum_{j=1}^{r} j_j^\alpha x_j^\alpha , \tag{57}$$

where

$$j_k = -\frac{\mathrm{d_i} M_k^{\text{I}}}{\mathrm{d}t}, \quad (k = 1, 2, \ldots, n) , \tag{58}$$

$$\tilde{j}_u = -\left(\frac{\mathrm{d_i} U^{\text{I}}}{\mathrm{d}t} + \frac{\mathrm{d}\Psi^{\text{I}}}{\mathrm{d}t}\right) \equiv j_u + \sum_{k=1}^{n} \psi_k j_k , \tag{59}$$

$$j_j^\alpha = J_j^\alpha V^\alpha , \quad (j = 1, 2, \ldots, r; \alpha = \text{I, II}) , \tag{60}$$

$$x_k = -\Delta\left(\frac{\mu_k + \psi_k}{T}\right) = -\Delta\left(\frac{\tilde{\mu}_k}{T}\right), \quad (k = 1, 2, \ldots, n) , \tag{61}$$

$$x_u = \Delta\left(\frac{1}{T}\right), \tag{62}$$

$$x_j^\alpha = -\frac{A_j^\alpha}{T^\alpha}, \quad (j = 1, 2, \ldots, r; \alpha = \text{I, II}) . \tag{63}$$

In the last member of (59) j_u is by definition equal to the first term in the second member. The last term of (59) follows from (25) and (58). The index I of ψ_k has been deleted since it is immaterial whether we put I or II here; it would change σ_{tot} only in third order terms. The symbol Δ in (61) and (62) stands for a difference of a quantity in sub-system II and the corresponding in sub-system I.

The thermodynamic forces are not all independent, because one has the connexion

$$x_j^{\text{II}} - x_j^{\text{I}} = -\Delta\left(\frac{A_j}{T}\right) = -\sum_{k=1}^{n} \nu_{kj} \Delta\left(\frac{\tilde{\mu}_k}{T}\right), \tag{64}$$

where (II.27) and (III.18) have been employed.

We note also that the thermodynamic force (61) can be written explicitly as

$$\Delta\left(\frac{\tilde{\mu}_k}{T}\right) = (h_k + \psi_k)\,\Delta\left(\frac{1}{T}\right) + \frac{v_k \Delta p}{T} + \frac{(\Delta\mu_k)_{T,p}}{T} + \frac{\Delta\psi_k}{T}, \tag{65}$$

where h_k and v_k are the partial specific enthalpy and the partial specific volume of component k. The quantities $(\Delta\mu_k)_{T,p}$, the differences of the chemical potential at constant temperature and pressure, depend only on the concentration differences:

$$(\Delta\mu_k)_{T,p} = \sum_{m=1}^{n-1} \mu^c_{km}\,\Delta c_m\,, \quad \left[\mu^c_{km} \equiv \left(\frac{\partial \mu_k}{\partial c_m}\right)_{T,p,c_l}\right], \quad (k = 1, 2, \ldots, n)\,. \tag{66}$$

They are connected through the Gibbs–Duhem relation

$$\sum_{k=1}^{n} c_k(\Delta\mu_k)_{T,p} = 0\,. \tag{67}$$

In the foregoing we have tacitly supposed that we were treating case *a*, in which it makes sense to attribute the macroscopic concept of entropy to sub-system III. In case *b*, where the sub-system III has negligible dimensions, we still have formulae (43)–(46) and (48)–(52) for I and II. With formula (54) for the total entropy production σ_{tot}, all the preceding results (55)–(67) are obtained again. Mathematically case *b* can also be obtained as a limit of case *a* as V^{III} tends to zero.

In the following sections we shall be concerned mainly with two particular cases. We shall give alternative forms of the entropy production (57) for these special cases.

1. An n-component system, not subject to external forces, in which j chemical reactions take place. The entropy production (57) becomes for this case, with (59), (61)–(63) and (65), putting $\psi_k = 0$,

$$\sigma_{\text{tot}} = -\left(j_u - \sum_{k=1}^{n} h_k j_k\right)\frac{\Delta T}{T^2} - \sum_{k=1}^{n} j_k \frac{(\Delta\mu_k)_{T,p}}{T} - \sum_{k=1}^{n} v_k j_k \frac{\Delta p}{T} - \sum_{\alpha=\text{I}}^{\text{II}} \sum_{j=1}^{r} j^\alpha_j \frac{A^\alpha_j}{T}\,. \tag{68}$$

If we now eliminate $(\Delta\mu_n)_{T,p}$ with the help of (67), we obtain the alternative form for the entropy production

$$\sigma_{\text{tot}} = -j'_q \frac{\Delta T}{T^2} - \sum_{k=1}^{n-1} j'_k \frac{(\Delta\mu_k)_{T,p}}{T} - j_v \frac{\Delta p}{T} - \sum_{\alpha=\text{I}}^{\text{II}} \sum_{j=1}^{r} j^\alpha_j \frac{A^\alpha_j}{T}\,, \tag{69}$$

where we have introduced the "reduced heat flow" j'_q, the diffusion flow j'_k and the volume flow j_v, defined as

$$j_q' = j_u - \sum_{k=1}^{n} h_k j_k , \tag{70}$$

$$j_k' = j_k - \frac{c_k}{c_n} j_n , \quad (k = 1, 2, \ldots, n-1) , \tag{71}$$

$$j_v = \sum_{k=1}^{n} v_k j_k . \tag{72}$$

Since, according to (63) and (64), the thermodynamic forces A_j^{II} can be expressed in terms of the other forces, we have as a set of independent forces: ΔT, Δc_k ($k = 1, 2, \ldots, n-1$), Δp and A_j^{I} ($j = 1, 2, \ldots, r$).

2. An n-component, non-reacting system at uniform temperature and concentrations, under the influence of electrostatic forces, *i.e.* with

$$\psi_k = z_k \varphi , \tag{73}$$

where z_k is the electric charge per unit mass of component k and φ is the electrostatic potential. Now the entropy production (57) becomes with (61), (62), (65) and (66), putting $\Delta c_k = 0$ and $\Delta T = 0$ and delating chemical terms,

$$\sigma_{\mathrm{tot}} = -j_v \frac{\Delta p}{T} - i \frac{\Delta \varphi}{T} , \tag{74}$$

where j_v is given by (72) and where

$$i = \sum_{k=1}^{n} z_k j_k \tag{75}$$

is the total electric current.

§ 4. *Phenomenological Equations and Onsager Reciprocal Relations*

As the next step in the thermodynamic theory of discontinuous systems we have to establish the phenomenological equations and Onsager relations between the phenomenological coefficients. We shall derive these for the case *a*, where the sub-system III is macroscopic, from the corresponding local equations and relations. As illustrative examples we choose the two systems described at the end of the preceding section. For the case *b*, where the sub-system III is macroscopically negligible we shall also establish the phenomenological equations and give the Onsager relations.

First example. Case a (sub-system III *macroscopic).*

Let us consider an n-component, reacting system, of which the entropy production is given by (69). We wish to derive the phenomenological equations between the fluxes and thermodynamic forces occurring in this entropy production from the local phenomenological equations. We shall start by expressing the fluxes (70)–(72) in terms of the local fluxes. This is most conveniently done by applying first the formulae (11) and (29) to the portion of sub-system III between the surface ω^{I} and an arbitrary normal cross-section ω of the capillary. This gives along the same lines of reasoning which led from (11) to (13) using (14),

$$j_k \equiv -\frac{\mathrm{d_i} M_k^{\mathrm{I}}}{\mathrm{d}t} = \int^{\omega} (\boldsymbol{J}_k + \rho_k \boldsymbol{v}) \cdot \mathrm{d}\boldsymbol{\Omega}\,, \quad (k = 1, 2, \ldots, n)\,, \tag{76}$$

and, just as (31) and (42) were obtained from (29) with the help of (21) and (41),

$$j_u \equiv -\frac{\mathrm{d_i} U^{\mathrm{I}}}{\mathrm{d}t} = \int^{\omega} (\boldsymbol{J}_q + h\rho\boldsymbol{v} + \boldsymbol{\Pi}\cdot\boldsymbol{v}) \cdot \mathrm{d}\boldsymbol{\Omega}\,. \tag{77}$$

The term $\boldsymbol{\Pi}\cdot\boldsymbol{v}$ arises from the fact that we may not neglect viscous flow inside the capillary. The vector $\mathrm{d}\boldsymbol{\Omega}$ is counted positive from the inside to the outside of the portion of sub-system III considered here. With the last two relations we find for (70)–(72), using also $\sum_k h_k \rho_k = h\rho$ and $\sum_k \rho_k v_k = 1$,

$$j_q' = \int^{\omega} \left(\boldsymbol{J}_q - \sum_{k=1}^{n} h_k \boldsymbol{J}_k + \boldsymbol{\Pi}\cdot\boldsymbol{v}\right) \cdot \mathrm{d}\boldsymbol{\Omega}\,, \tag{78}$$

$$j_k' = \int^{\omega} \left(\boldsymbol{J}_k - \frac{c_k}{c_n} \boldsymbol{J}_n\right) \cdot \mathrm{d}\boldsymbol{\Omega}\,, \quad (k = 1, 2, \ldots, n-1)\,, \tag{79}$$

$$j_v = \int^{\omega} \left(\sum_{k=1}^{n} v_k \boldsymbol{J}_k + \boldsymbol{v}\right) \cdot \mathrm{d}\boldsymbol{\Omega}\,. \tag{80}$$

If we introduce the reduced heat flow $\boldsymbol{J}_q' = \boldsymbol{J}_q - \sum_k h_k \boldsymbol{J}_k$ into (78), and eliminate $\boldsymbol{J}_n$ from (79) and (80) with $\sum_k \boldsymbol{J}_k = 0$, we get

$$j'_q = \int^{\omega} (\boldsymbol{J}'_q + \boldsymbol{\Pi}\cdot\boldsymbol{v})\cdot\mathrm{d}\boldsymbol{\Omega}\,, \tag{81}$$

$$j'_k = \int^{\omega} \sum_{m=1}^{n-1} \tilde{A}_{km}\boldsymbol{J}_m\cdot\mathrm{d}\boldsymbol{\Omega}\,, \quad (k = 1, 2, \ldots, n-1)\,, \tag{82}$$

$$j_v = \int^{\omega} \left\{ \sum_{k=1}^{n-1} (v_k - v_n)\boldsymbol{J}_k + \boldsymbol{v} \right\}\cdot\mathrm{d}\boldsymbol{\Omega}\,, \tag{83}$$

where

$$\tilde{A}_{km} \equiv A_{mk} \equiv \delta_{mk} + \frac{c_k}{c_n}\,, \quad (k, m = 1, 2, \ldots, n-1)\,. \tag{84}$$

We notice that the fluxes (81)–(83) and (60) are now given in terms of the local fluxes, for which local phenomenological equations are written down in (IV.14)–(IV.18) and of the velocity $\boldsymbol{v}$. In the local equations we shall write explicitly

$$\{\operatorname{grad}(\mu_k - \mu_n)\}_T = \{\operatorname{grad}(\mu_k - \mu_n)\}_{T,p} + (v_k - v_n)\operatorname{grad} p$$

$$= \sum_{m=1}^{n-1} A_{km}(\operatorname{grad}\mu_m)_{T,p} + (v_k - v_n)\operatorname{grad} p\,, \tag{85}$$

where μ_n has been eliminated with the help of the Gibbs–Duhem relation

$$\sum_{k=1}^{n} c_k(\operatorname{grad}\mu_k)_{T,p} = 0\,. \tag{86}$$

The phenomenological equations (IV.14), (IV.15), (IV.16) and (IV.18) become then for our case, if we consider the flow to be incompressible (div $\boldsymbol{v} = 0$),

$$\boldsymbol{J}'_q = -L_{qq}\frac{\operatorname{grad} T}{T^2} - \sum_{k=1}^{n-1} L_{qk}\cdot\frac{\sum_{m=1}^{n-1} A_{km}(\operatorname{grad}\mu_m)_{T,p} + (v_k - v_n)\operatorname{grad} p}{T}\,, \tag{87}$$

$$\boldsymbol{J}_i = -L_{iq}\frac{\operatorname{grad} T}{T^2} - \sum_{k=1}^{n-1} L_{ik}\frac{\sum_{m=1}^{n-1} A_{km}(\operatorname{grad}\mu_m)_{T,p} + (v_k - v_n)\operatorname{grad} p}{T}\,,$$

$$(i = 1, 2, \ldots, n-1)\,, \tag{88}$$

$$\Pi = -2\eta\,(\text{Grad}\,\boldsymbol{v})^s, \quad (\eta = L/2T), \tag{89}$$

$$J_j = -\sum_{j'=1} l_{jj'} \frac{A_{j'}}{T}. \tag{90}$$

We shall linearize the problem by supposing that in good approximation the gradients of all quantities may be written as differences between the values which these quantities have in the sub-systems II and I, divided by the length of the capillary

$$\text{grad}\, T = \frac{\Delta T}{l}, \quad \text{grad}\, \mu_k = \frac{\Delta \mu_k}{l}, \quad \text{grad}\, p = \frac{\Delta p}{l}. \tag{91}$$

The equation of motion (II.19) becomes for our system, where accelerations are neglected, with (89) and (91) and div $\boldsymbol{v} = 0$

$$\eta \nabla^2 \boldsymbol{v} = \frac{\Delta p}{l}, \tag{92}$$

where ∇^2 is the Laplace operator. For a circular capillary of radius a the solution of this equation is Poiseuille's law

$$\boldsymbol{v} = -\frac{1}{4\eta}(a^2 - r^2)\frac{\Delta p}{l}, \tag{93}$$

where r is the distance from the central axis. The total flow through a section of the capillary is thus

$$\int^{\omega} \boldsymbol{v} \cdot \mathrm{d}\boldsymbol{\Omega} = -\frac{\alpha}{\eta}\frac{\Delta p}{l}, \tag{94}$$

where $\alpha = \frac{1}{8}\pi a^4$ for the circular tube. For other tubes with the same section everywhere along their length formula (94) is still valid with a different value for the geometrical factor α. The foregoing analysis shows also that Π is proportional to Δp and thus negligible in (77) and (78), since it would only give rise to second order terms in the phenomenological equations to be derived.

If we insert the expressions (87), (88), (90) and (94) into (81)–(83)

and (60), we obtain, using also (91), the desired phenomenological equations for the discontinuous system

$$j'_q = -\Lambda_{qq}\frac{\Delta T}{T^2} - \sum_{m=1}^{n-1}\Lambda_{qm}\frac{(\Delta\mu_m)_{T,p}}{T} - \Lambda_{qv}\frac{\Delta p}{T}, \qquad (95)$$

$$j'_k = -\Lambda_{kq}\frac{\Delta T}{T^2} - \sum_{m=1}^{n-1}\Lambda_{km}\frac{(\Delta\mu_m)_{T,p}}{T} - \Lambda_{kv}\frac{\Delta p}{T}, \quad (k = 1, 2, \ldots, n-1), \qquad (96)$$

$$j_v = -\Lambda_{vq}\frac{\Delta T}{T^2} - \sum_{m=1}^{n-1}\Lambda_{vm}\frac{(\Delta\mu_m)_{T,p}}{T} - \Lambda_{vv}\frac{\Delta p}{T}, \qquad (97)$$

$$j_j^\alpha = -\sum_{j'=1}^{r}\Lambda_{jj'}M^\alpha\frac{A_{j'}^\alpha}{T}, \quad (j = 1, 2, \ldots, r; \alpha = \mathrm{I}, \mathrm{II}), \qquad (98)$$

with the phenomenological coefficients

$$\Lambda_{qq} = \overline{L_{qq}}\frac{\omega}{l}, \qquad (99)$$

$$\Lambda_{qm} = \sum_{k=1}^{n-1}\overline{L_{qk}}A_{km}\frac{\omega}{l}, \; \Lambda_{mq} = \sum_{k=1}^{n-1}\tilde{A}_{mk}\overline{L_{kq}}\frac{\omega}{l}, \quad (m = 1, 2, \ldots, n-1), \qquad (100)$$

$$\Lambda_{km} = \sum_{i,j=1}^{n-1}\tilde{A}_{ki}\overline{L_{ij}}A_{jm}\frac{\omega}{l}, \quad (k, m = 1, 2, \ldots, n-1), \qquad (101)$$

$$\Lambda_{qv} = \sum_{k=1}^{n-1}\overline{L_{qk}}(v_k - v_n)\frac{\omega}{l}, \; \Lambda_{vq} = \sum_{k=1}^{n-1}(v_k - v_n)\overline{L_{kq}}\frac{\omega}{l}, \qquad (102)$$

$$\Lambda_{kv} = \sum_{i,j=1}^{n-1}\tilde{A}_{ki}\overline{L_{ij}}(v_j - v_n)\frac{\omega}{l}, \; \Lambda_{vk} = \sum_{i,j=1}^{n-1}(v_j - v_n)\overline{L_{ji}}A_{ik}\frac{\omega}{l},$$

$$(k = 1, 2, \ldots, n-1), \qquad (103)$$

$$\Lambda_{vv} = \sum_{i,j=1}^{n-1}(v_i - v_n)\overline{L_{ij}}(v_j - v_n)\frac{\omega}{l} + \frac{\alpha T}{\eta l}, \qquad (104)$$

$$\Lambda_{jj'} = l_{jj'}\rho^\alpha, \quad (j, j' = 1, 2, \ldots, r; \alpha = \mathrm{I}, \mathrm{II}). \qquad (105)$$

The bars over the local phenomenological coefficients indicate averaging over the cross-section of the capillary. All coefficients (99)–(103)

contain the geometrical factor ω/l, and (104) contains besides also the factor α/η arising from (94). We have introduced a factor M^α in (98), because j_j^α is proportional to the dimension of the sub-system α (α = I, II). The coefficient (105) is then still a local quantity.

The $(n+1)^2$ coefficients Λ, which occur in (95)–(97), are expressed in terms of n^2 local coefficients L by means of the equalities (99)–(104). Thus there exist $2n+1$ relations between the Λ's. One can write these as

$$\Lambda_{qv} = \sum_{m,i=1}^{n-1} \Lambda_{qm}\Lambda_{mi}^{-1}(v_i - v_n)\,, \quad \Lambda_{vq} = \sum_{i,m=1}^{n-1} (v_i - v_n)\tilde{\Lambda}_{im}^{-1}\Lambda_{mq}\,, \tag{106}$$

$$\Lambda_{kv} = \sum_{m,i=1}^{n-1} \Lambda_{km}\Lambda_{mi}^{-1}(v_i - v_n)\,, \quad \Lambda_{vk} = \sum_{i,m=1}^{n-1} (v_i - v_n)\tilde{\Lambda}_{im}^{-1}\Lambda_{mk}\,,$$

$$(k = 1, 2, \ldots, n-1)\,, \tag{107}$$

$$\Lambda_{vv} - \frac{\alpha T}{\eta l} = \sum_{i,j,k,m=1}^{n-1} (v_i - v_n)\tilde{\Lambda}_{ij}^{-1}\Lambda_{jk}\Lambda_{km}^{-1}(v_m - v_n)\,, \tag{108}$$

where the index -1 denotes the reciprocal matrix. Simple corollaries of these identities are

$$\Lambda_{qv} = \sum_{i,k=1}^{n-1} \Lambda_{qi}\Lambda_{ik}^{-1}\Lambda_{kv}\,, \quad \Lambda_{vq} = \sum_{i,k=1}^{n-1} \Lambda_{vi}\Lambda_{ik}^{-1}\Lambda_{kq}\,, \tag{109}$$

$$\Lambda_{vv} - \sum_{i,k=1}^{n-1} \Lambda_{vi}\Lambda_{ik}^{-1}\Lambda_{kv} = \frac{\alpha T}{\eta l}\,. \tag{110}$$

The Onsager relations (IV.54), (IV.55) and (IV.57) have as a result that the following reciprocal relations hold for the coefficients (100)–(103) and (105)

$$\Lambda_{qk} = \Lambda_{kq}\,, \quad (k = 1, 2, \ldots, n-1)\,, \tag{111}$$

$$\Lambda_{km} = \Lambda_{mk}\,, \quad (k, m = 1, 2, \ldots, n-1)\,, \tag{112}$$

$$\Lambda_{qv} = \Lambda_{vq}\,, \tag{113}$$

$$\Lambda_{kv} = \Lambda_{vk}\,, \quad (k = 1, 2, \ldots, n-1)\,, \tag{114}$$

$$\Lambda_{jj'} = \Lambda_{j'j}\,, \quad (j, j' = 1, 2, \ldots, r\,; \alpha = \text{I, II})\,. \tag{115}$$

First example. Case b (sub-system III *not macroscopic).*

If the capillary has dimensions of the order of or smaller than the mean free path, the above derivation cannot be given, since the sub-system III is then not a macroscopic system. We can, however, directly establish phenomenological equations (95)–(98) as linear relationships between the fluxes and thermodynamic forces, that occur in the entropy production (68), taking into account only the Curie principle. The phenomenological coefficients are now all independent, *i.e.* no relations as (106)–(108) exist between them. The Onsager relations (111)–(115) are still valid, since they can be considered as directly following from the proof given in Chapter VII.

Second example. Case a (sub-system III *macroscopic).*

Let us now study the second example mentioned at the end of the preceding section, in which external electric forces play a role. The fluxes, occurring in the entropy production (74), defined by (72) and (75), can be written as

$$j_v = \int^{\omega} \left\{ \sum_{k=1}^{n-1} (v_k - v_n)\boldsymbol{J}_k + \boldsymbol{v} \right\} \cdot \mathrm{d}\boldsymbol{\Omega} \,, \tag{116}$$

$$i = \int^{\omega} \left\{ \sum_{k=1}^{n-1} (z_k - z_n)\boldsymbol{J}_k + z\rho\boldsymbol{v} \right\} \cdot \mathrm{d}\boldsymbol{\Omega} \,, \tag{117}$$

where (76) has been used, and where $\boldsymbol{J}_n$ has been eliminated with the aid of $\sum_k \boldsymbol{J}_k = 0$. Use has been made of the fact that the sum $\sum_k \rho_k v_k$ is equal to unity. The factor $z\rho = \sum_k z_k \rho_k$ is the total electric charge density in the capillary. We need to find the local flows in the integrands of (116) and (117). For the fluxes $\boldsymbol{J}_k$ we have from (IV.15), (III.29) and (73), taking into account the fact that the temperature and the concentrations are supposed to be uniform throughout the system, and linearizing according to (91),

$$\boldsymbol{J}_k = -\sum_{m=1}^{n-1} L_{km} \frac{(v_m - v_n)\Delta p + (z_m - z_n)\Delta\varphi}{lT} \,. \tag{118}$$

The equation of motion (II.19) becomes with (II.20), (II.35), (73) and (IV.16), neglecting acceleration, considering the flow as incompressible, and applying (91)

$$\eta \nabla^2 \boldsymbol{v} = \frac{\Delta p}{l} + z\rho \frac{\Delta\varphi}{l} \,. \tag{119}$$

Before solving this equation we must enter into somewhat more detail concerning the charge distribution inside the capillary*. In general in the fluid, at the interface with the capillary wall, an electric double layer will be present, characterized by the charge density $z\rho$. We have Poisson's equation for the relation between the charge density and the electric potential φ

$$z\rho = -\varepsilon\nabla^2\varphi\,, \tag{120}$$

where ε is the dielectric constant. The equation which results from insertion of (120) into (119),

$$\eta\nabla^2\boldsymbol{v} = \frac{\Delta p}{l} - \varepsilon\frac{\Delta\varphi}{l}\nabla^2\varphi\,, \tag{121}$$

can be readily solved. It may be noticed that Δp and $\Delta\varphi$ are constant quantities. One finds for a capillary of circular cross-section with radius a

$$\boldsymbol{v} = -\frac{1}{4\eta}(a^2 - r^2)\frac{\Delta p}{l} - \frac{\varepsilon}{\eta}(\varphi - \zeta)\frac{\Delta\varphi}{l}\,, \tag{122}$$

where ζ is the potential in the double layer on the surface of shear, where the velocity vanishes. The total transport of fluid is given by

$$\int^{\omega}\boldsymbol{v}\cdot\mathrm{d}\boldsymbol{\Omega} = -\frac{\alpha}{\eta}\frac{\Delta p}{l} - \frac{\beta}{\eta}\frac{\Delta\varphi}{l}\,, \tag{123}$$

where $\alpha = \frac{1}{8}\pi a^4$ for the case of the circular diameter, and where

$$\beta = \varepsilon\zeta\int^{\omega}\left(\frac{\varphi}{\zeta} - 1\right)\mathrm{d}\Omega\,, \tag{124}$$

in which the integral depends on the dimensions of the capillary and – usually only slightly – on the structure of the double layer. The transport (123) is one of the terms of (116); it is seen to consist of two parts, *viz.* a Poiseuille flow caused by a pressure difference and an electro-osmotic transport of fluid caused by the potential difference $\Delta\varphi$

* P. Mazur and J. Th. G. Overbeek, Rec. Trav. chim. Pays-Bas **70** (1951) 83.

acting on the charge density $z\rho$ and which drags the fluid in which the charge is embedded along.

We must still calculate a transport of electricity, which appears in formula (117). With Poisson's equation (120) we can write it as

$$\int^{\omega} z\rho \boldsymbol{v} \cdot \mathrm{d}\boldsymbol{\Omega} = -\varepsilon \int^{\omega} (\nabla^2 \varphi) \boldsymbol{v} \cdot \mathrm{d}\boldsymbol{\Omega}. \tag{125}$$

After two integrations by parts this becomes

$$\int^{\omega} z\rho \boldsymbol{v} \cdot \mathrm{d}\boldsymbol{\Omega} = -\varepsilon \int^{\omega} (\varphi - \zeta) \nabla^2 \boldsymbol{v} \cdot \mathrm{d}\boldsymbol{\Omega}, \tag{126}$$

which with the equation of motion (119) gets the form

$$\int^{\omega} z\rho \boldsymbol{v} \cdot \mathrm{d}\boldsymbol{\Omega} = -\frac{\beta}{\eta} \frac{\Delta p}{l} - \frac{\gamma}{\eta} \frac{\Delta \varphi}{l}, \tag{127}$$

where the coefficient β, given by (124), appears again, and where γ is

$$\gamma = \varepsilon \zeta \int \left(\frac{\varphi}{\zeta} - 1 \right) z\rho \, \mathrm{d}\Omega = -\varepsilon^2 \zeta \int \left(\frac{\varphi}{\zeta} - 1 \right) \nabla^2 \varphi \, \mathrm{d}\Omega . \tag{128}$$

The last member follows with Poisson's equation (120). By inserting the results (118), (123) and (127) into (116) and (117) we obtain finally the phenomenological equations for the discontinuous system

$$j_v = -\Lambda_{vv} \frac{\Delta p}{T} - \Lambda_{ve} \frac{\Delta \varphi}{T}, \tag{129}$$

$$i = -\Lambda_{ev} \frac{\Delta p}{T} - \Lambda_{ee} \frac{\Delta \varphi}{T}, \tag{130}$$

where the phenomenological coefficients are given by

$$\Lambda_{vv} = \sum_{k,m=1}^{n-1} (v_k - v_n) \overline{L_{km}} (v_m - v_n) \frac{\omega}{l} + \frac{\alpha T}{\eta l}, \tag{131}$$

$$\Lambda_{ve} = \sum_{k,m=1}^{n-1} (v_k - v_n)\overline{L_{km}}(z_m - z_n)\frac{\omega}{l} + \frac{\beta T}{\eta l}, \tag{132}$$

$$\Lambda_{ev} = \sum_{k,m=1}^{n-1} (z_k - z_n)\overline{L_{km}}(v_m - v_n)\frac{\omega}{l} + \frac{\beta T}{\eta l}, \tag{133}$$

$$\Lambda_{ee} = \sum_{k,m=1}^{n-1} (z_k - z_n)\overline{L_{km}}(z_m - z_n)\frac{\omega}{l} + \frac{\gamma T}{\eta l}. \tag{134}$$

The Onsager relations (IV.55) and the fact that the same coefficient β appears in both (132) and (133) ensure the validity of the reciprocal relation

$$\Lambda_{ve} = \Lambda_{ev}. \tag{135}$$

Second example. Case b.

In practice the sub-system III is quite often not just simply a macroscopic system enclosed in a single tube with constant cross-section. The sub-system III may, for instance be enclosed in a diaphragm, which consists of a complicated network of capillaries. Even if these capillaries are wide enough so that the system enclosed in them can be considered as macroscopic the mathematical analysis of the physical situation becomes very complicated*. When the diameter of the capillaries becomes of the dimension of the double layer, and *a fortiori*, when the diameter becomes of the order of the mean free path of the molecules, no derivation along the lines, outlined above, can be given. But in the framework of thermodynamics of irreversible processes we are still capable of establishing the linear phenomenological relations (129) and (130) between the fluxes and thermodynamic forces occurring in the entropy production (74). Furthermore the reciprocal relation (135) is still valid, because it is an example of the Onsager relations, as derived in Chapter VII.

§ 5. *Thermomolecular Pressure Effect, Thermal Effusion and Mechanocaloric Effect***

If we study the behaviour of a discontinuous multi-component, non-reacting system, several physical effects are found in different experimental situations.

* P. Mazur and J. Th. G. Overbeek, loc. cit. p. 424.
** I. Prigogine, loc. cit. p. 407; S. R. de Groot, loc. cit. p. 407.

A first, particularly frequently studied, experimental situation is the stationary state which arises if we fix the temperature difference $\Delta T = T^{\mathrm{II}} - T^{\mathrm{I}}$ between the two reservoirs. Since the system as a whole is materially closed we shall end up with a state in which the fluxes j'_k $(k = 1, 2, \ldots, n-1)$ and j_v [*cf.* formulae (71)–(72)] disappear. Then the equations (96) and (97) read

$$0 = \Lambda_{kq} \frac{\Delta T}{T^2} + \sum_{l=1}^{n-1} \Lambda_{kl} \frac{(\Delta \mu_l)_{p,T}}{T} + \Lambda_{kv} \frac{\Delta p}{T}, \quad (k = 1, 2, \ldots, n-1), \tag{136}$$

$$0 = \Lambda_{vq} \frac{\Delta T}{T^2} + \sum_{m=1}^{n-1} \Lambda_{vm} \frac{(\Delta \mu_m)_{p,T}}{T} + \Lambda_{vv} \frac{\Delta p}{T}. \tag{137}$$

We first solve Δp from these linear equations. To this end we multiply (136) with $\Lambda_{vm}\Lambda^{-1}_{mk}$, where Λ^{-1}_{mk} is an element of the reciprocal matrix of the matrix of the Λ_{mk} $(m, k = 1, 2, \ldots, n-1)$, and sum over m and k. If the result is subtracted from (137) we obtain

$$\frac{\Delta p}{\Delta T} = - \frac{\Lambda_{vq} - \sum_{m,k=1}^{n-1} \Lambda_{vm}\Lambda^{-1}_{mk}\Lambda_{kq}}{\Lambda_{vv} - \sum_{m,k=1}^{n-1} \Lambda_{vm}\Lambda^{-1}_{mk}\Lambda_{kv}} \frac{1}{T}, \quad (j'_k = 0, j_v = 0), \tag{138}$$

which is called the "*thermomolecular pressure*" *effect*. In case *a*, when the connecting capillary is supposed to be a macroscopic system, we notice that according to the second identity of (109) the numerator of the right-hand side of (138) vanishes, whereas according to (110) the denominator does not. Thus we find

$$\Delta p = 0, \tag{139}$$

i.e. no thermomolecular pressure difference arises. This result is not surprising since in a continuum theory one finds from the equation of motion that in the absence of flow and of external forces the pressure gradient vanishes. Formula (139), valid for case *a*, simply reflects this fact. In case *b*, however, we find a non-vanishing result. Let us write down two special cases of (138), *viz.* for the one-component $(n = 1)$ and the binary system $(n = 2)$:

$$\frac{\Delta p}{\Delta T} = -\frac{\Lambda_{vq}}{\Lambda_{vv}}\frac{1}{T}, \quad (n = 1), \tag{140}$$

$$\frac{\Delta p}{\Delta T} = -\frac{\Lambda_{11}\Lambda_{vq} + \Lambda_{v1}\Lambda_{1q}}{\Lambda_{11}\Lambda_{vv} - \Lambda_{v1}\Lambda_{1v}}\frac{1}{T}, \quad (n = 2). \tag{141}$$

We can also solve the $(\Delta\mu_m)_{p,T}$ $(m = 1, 2, \ldots, n-1)$ from (136) and (137). These quantities are related to the differences of mass fractions $\Delta c_m = c_m^{\mathrm{II}} - c_m^{\mathrm{I}}$ according to (66). For a binary system $(n = 2)$ we find in case *a*

$$\frac{\Delta c_1}{\Delta T} = -\frac{\Lambda_{1q}}{\Lambda_{11}}\frac{1}{\mu_{11}^c T} = -\frac{\Lambda_{vq}}{\Lambda_{v1}}\frac{1}{\mu_{11}^c T}, \tag{142}$$

and in case *b*

$$\frac{\Delta c_1}{\Delta T} = -\frac{\Lambda_{vv}\Lambda_{1q} - \Lambda_{1v}\Lambda_{vq}}{\Lambda_{vv}\Lambda_{11} - \Lambda_{1v}\Lambda_{v1}}\frac{1}{\mu_{11}^c T}. \tag{143}$$

This separation effect is called "*thermal effusion*".

A second experimental situation of practical importance is the stationary state reached when one keeps the temperature uniform $(\Delta T = 0)$ and fixes the pressure difference at a certain value Δp. The final state is then characterized by the vanishing of the fluxes j'_k $(k = 1, 2, \ldots, n-1)$. The equation (96) becomes

$$(\Delta\mu_i)_{p,T} = -\sum_{k=1}^{n-1} \Lambda_{ik}^{-1}\Lambda_{kv}\Delta p. \tag{144}$$

With this relation, and bearing in mind that $\Delta T = 0$, we can deduce from (95) and (97) that

$$\frac{j'_q}{j_v} = \frac{\Lambda_{qv} - \sum_{m,k=1}^{n-1} \Lambda_{qm}\Lambda_{mk}^{-1}\Lambda_{kv}}{\Lambda_{vv} - \sum_{m,k=1}^{n-1} \Lambda_{vm}\Lambda_{mk}^{-1}\Lambda_{kv}}. \tag{145}$$

We shall make use of this result in the following, but we wish first to define the so-called "*mechanocaloric effect*" as

$$q^* = \frac{j_{qe}}{j_v}, \quad (\text{at} \quad \Delta T = 0, \quad j'_k = 0), \tag{146}$$

where
$$j_{qe} \equiv \frac{\mathrm{d_e}Q^{\mathrm{I}}}{\mathrm{d}t} \tag{147}$$

is the heat flowing from the surroundings to vessel I (or from vessel II to the surroundings) and where

$$j \equiv \sum_{k=1}^{n} j_k = \sum_{k=1}^{n} \int^{\omega} (\boldsymbol{J}_k + \rho_k \boldsymbol{v}) \cdot \mathrm{d}\boldsymbol{\Omega} = \int^{\omega} \rho \boldsymbol{v} \cdot \mathrm{d}\boldsymbol{\Omega} \tag{148}$$

is the total mass flow from I to II. [In (148) use has been made of expression (76).]

We can give an alternative expression for q^*, making use of the fact that from the vanishing of the fluxes j'_k and the definitions (71) of these fluxes we can infer that

$$\frac{j_1}{c_1} = \frac{j_2}{c_2} = \ldots = \frac{j_n}{c_n} = j\,, \tag{149}$$

which means also that in the stationary state considered the mass fractions c_k $(k = 1, 2, \ldots, n)$ are constant in time everywhere. Now several conclusions can be drawn from this fact. In the first place we can give a simple form to the following law, which is obtained from the sum of (40) and (41) both for $\alpha = \mathrm{I}$,

$$\frac{\mathrm{d}U^{\mathrm{I}}}{\mathrm{d}t} = \frac{\mathrm{d}Q^{\mathrm{I}}}{\mathrm{d}t} - p^{\mathrm{I}} \frac{\mathrm{d}V^{\mathrm{I}}}{\mathrm{d}t} + h^{\mathrm{I}} \frac{\mathrm{d}M^{\mathrm{I}}}{\mathrm{d}t}\,. \tag{150}$$

As a matter of fact, since we had already uniform temperature and pressure, and now also uniform concentrations, one can write

$$\mathrm{d}U^{\mathrm{I}} = u^{\mathrm{I}}\,\mathrm{d}M^{\mathrm{I}} \quad , \quad \mathrm{d}V^{\mathrm{I}} = v^{\mathrm{I}}\,\mathrm{d}M^{\mathrm{I}}\,. \tag{151}$$

With these relations the law (150) becomes simply

$$(u^{\mathrm{I}} + p^{\mathrm{I}}v^{\mathrm{I}} - h^{\mathrm{I}}) \frac{\mathrm{d}M^{\mathrm{I}}}{\mathrm{d}t} = \frac{\mathrm{d}Q^{\mathrm{I}}}{\mathrm{d}t}\,. \tag{152}$$

Since the bracket expression in the first member vanishes, we have also

$$\frac{\mathrm{d}Q^{\mathrm{I}}}{\mathrm{d}t} \equiv \frac{\mathrm{d_e}Q^{\mathrm{I}}}{\mathrm{d}t} + \frac{\mathrm{d_i}Q^{\mathrm{I}}}{\mathrm{d}t} = 0\,, \tag{153}$$

and thus with the notations (147) and

$$j_q \equiv -\frac{\mathrm{d_i}Q^{\mathrm{I}}}{\mathrm{d}t}, \tag{154}$$

[*cf.* formula (34)] the relationship

$$j_{qe} = j_q \,. \tag{155}$$

A second corollary of (149) follows by using it in relation (70):

$$j'_q = j_u - \sum_{k=1}^{n} h_k j_k = j_u - \sum_{k=1}^{n} c_k h_k j = j_u - hj = j_q \,. \tag{156}$$

The last equality is essentially (41) and follows with the definitions (58), (59), (148) and (154). Finally we have from (72) and (149)

$$j_v = \sum_{k=1}^{n} v_k j_k = \sum_{k=1}^{n} c_k v_k j = vj \,. \tag{157}$$

With the help of (155), (156) and (157) we can write for the mechanocaloric effect

$$q^* = \frac{j'_q}{j_v} v \,, \quad (\text{at} \quad \Delta T = 0, \;\; j'_k = 0) \,. \tag{158}$$

This is the alternative expression for q^*, which we wanted to derive. Now with (145) this expression becomes

$$q^* = \frac{\Lambda_{qv} - \sum\limits_{m,k=1}^{n-1} \Lambda_{qm} \Lambda_{mk}^{-1} \Lambda_{kv}}{\Lambda_{vv} - \sum\limits_{m,k=1}^{n-1} \Lambda_{vm} \Lambda_{mk}^{-1} \Lambda_{kv}} v \,, \tag{159}$$

giving the mechanocaloric effect in terms of the phenomenological coefficients.

With the Onsager relations (111)–(114) we find from (138) and (159) the relation

$$\frac{\Delta p}{\Delta T} = -\frac{q^*}{vT} \tag{160}$$

between the thermomolecular pressure effect $\Delta p/\Delta T$ and the mechano-caloric effect q^* †.

In case *a* (macroscopic capillary) we have from (159) and the first identity of (109)

$$q^* = 0, \tag{161}$$

so that (160) is trivially satisfied, as (139) shows.

In case *b* (narrow capillary) the relation (160) has experimentally been verified for several physical systems. A first example is the Knudsen gas. The system consists then of a perfect gas at such low pressure that the mean free path is large compared to the dimensions of the capillary. A simple kinetic calculation (*cf.* problem 16 of this chapter) yields for the quantity q^* the result

$$q^* = -\frac{1}{2}\frac{RT}{M}, \tag{162}$$

where R is the gas constant and M the molecular mass. Then (160) gives

$$\frac{\Delta p}{\Delta T} = \frac{1}{2}\frac{R}{vM}. \tag{163}$$

This leads with the perfect gas law ($pv = RT/M$) to the well-known equation

$$\frac{p^{\mathrm{I}}}{\sqrt{T^{\mathrm{I}}}} = \frac{p^{\mathrm{II}}}{\sqrt{T^{\mathrm{II}}}} \tag{164}$$

for the pressures and temperatures of a Knudsen gas in the two communicating reservoirs.

A second system for which both the thermomolecular pressure and the mechanocaloric effect have been observed is liquid helium II. The relation (160) has been empirically verified by Kapitza* and by Meyer and Mellink**.

A third case where (160) was verified is the "thermo-osmosis" of gases through a rubber membrane***. It turns out that q^* can be

† S. R. de Groot, Physica **13** (1947) 188; Comptes rendus Acad. Sc., Paris **225** (1947) 173.

* P. L. Kapitza, Journ. Phys., Moscow **5** (1941) 59.

** L. Meyer and J. M. Mellink, Physica **13** (1947) 197.

*** K. G. Denbigh, Nature **163** (1949) 60.

K. G. Denbigh and G. Raumann, Nature **165** (1950) 199; Proc. roy. Soc. A **210** (1951) 377, 518.

R. Haase and C. Steinert, Zeitschr. phys. Chem. **21** (1959) 270.

positive and negative depending on the nature of the gas and the membrane.

Let us finish this section with a few general remarks on the characteristics of the "stationary states" of which we encountered examples in the above treatment. The entropy production of the discontinuous system has the general form

$$\sigma = \sum_{k=1}^{N} j_k x_k \,, \tag{165}$$

where x_k $(k = 1, 2, \ldots, N)$ are the thermodynamic forces, and j_k the fluxes. The state of the system was described by the variables A_k^{I} and A_k^{II} $(k = 1, 2, \ldots, N)$ of which the time derivatives were split into an "internal" and an "external" change, *e.g.*

$$\frac{\mathrm{d}A_k^{\mathrm{I}}}{\mathrm{d}t} = \frac{\mathrm{d_i}A_k^{\mathrm{I}}}{\mathrm{d}t} + \frac{\mathrm{d_e}A_k^{\mathrm{I}}}{\mathrm{d}t}, \quad (k = 1, 2, \ldots, N) \,. \tag{166}$$

The fluxes j_k were always defined as internal changes of the state variables:

$$j_k \equiv -\frac{\mathrm{d_i}A_k^{\mathrm{I}}}{\mathrm{d}t} = \frac{\mathrm{d_i}A_k^{\mathrm{II}}}{\mathrm{d}t} = \frac{1}{2}\frac{\mathrm{d_i}(A_k^{\mathrm{II}} - A_k^{\mathrm{I}})}{\mathrm{d}t}, \quad (k = 1, 2, \ldots, N) \,. \tag{167}$$

Let us now consider the stationary state in which the system will finally arrive, when we fix the thermodynamic forces x_k with $k = 1, 2, \ldots, p$:

$$\frac{\mathrm{d}x_k}{\mathrm{d}t} = 0 \,, \quad (k = 1, 2, \ldots, p) \,. \tag{168}$$

Furthermore we keep the system closed for the quantities A_k^{I} with $k = p + 1, p + 2, \ldots, N$, *i.e.* the external changes of these quantities vanish:

$$\frac{\mathrm{d_e}A_k^{\mathrm{I}}}{\mathrm{d}t} = 0 \,, \quad (k = p + 1, p + 2, \ldots, N) \,. \tag{169}$$

In the final state, which we shall call a *stationary state of order p* (p is one of the numbers 0, 1, 2, ..., N) the quantities A_k^{I} are all constant in time:

$$\frac{\mathrm{d}A_k^{\mathrm{I}}}{\mathrm{d}t} = 0 \,, \quad (k = 1, 2, \ldots, N). \tag{170}$$

With (166), (167) and (169) we can conclude from the last formula:

$$j_k \equiv -\frac{\mathrm{d_i}A_k^{\mathrm{I}}}{\mathrm{d}t} = 0\,, \quad (k = p+1, p+2, \ldots, N)\,. \tag{171}$$

When the phenomenological equations

$$j_k = \sum_{m=1}^{N} \Lambda_{km} x_m\,, \quad (k = 1, 2, \ldots, N)\,, \tag{172}$$

are inserted into (171), one obtains a set of $N - p$ linear equations from which the values of x_k with $k = p+1,\ p+2, \ldots, N$ can be solved as functions of the fixed x_k ($k = 1, 2, \ldots, p$) and of the phenomenological coefficients Λ_{km} ($k = p+1,\ p+2, \ldots, N;\ m = 1, 2, \ldots, N$).

We can now prove the following theorem (which is analogous to a theorem for continuous systems, given in Chapter V): *the stationary states are states of minimum entropy production which are stable with respect to internal changes, if we assume that the phenomenological coefficients* Λ_{km} *are constants.* The proof of this theorem is furthermore based on the linear phenomenological equations (172) and on the Onsager relations

$$\Lambda_{km} = \Lambda_{mk}\,, \quad (m, k = 1, 2, \ldots, N)\,. \tag{173}$$

Let us first show that the entropy production has a minimum value in the stationary state. We have in fact from (165), (171), (172) and (173)

$$\frac{\partial \sigma}{\partial x_i} = \sum_{k=1}^{N} (\Lambda_{ik} + \Lambda_{ki}) x_k = 2 \sum_{k=1}^{N} \Lambda_{ik} x_k = 2 j_i = 0\,, \quad (i = p+1, p+2, \ldots, N)\,. \tag{174}$$

The stability of the stationary state can be demonstrated by writing first an expression for the time derivative of the entropy production, which follows from (165), (168), (172) and (173)

$$\frac{\mathrm{d}\sigma}{\mathrm{d}t} = \sum_{i,k=1}^{N} x_i \Lambda_{ik} \frac{\mathrm{d}x_k}{\mathrm{d}t} + \frac{\mathrm{d}x_k}{\mathrm{d}t} \Lambda_{ki} x_i = 2 \sum_{i,k=1}^{N} \frac{\mathrm{d}x_k}{\mathrm{d}t} \Lambda_{ki} x_i$$

$$= 2 \sum_{k=1}^{N} \frac{\mathrm{d}x_k}{\mathrm{d}t} j_k = 2 \sum_{k=p+1}^{N} \frac{\mathrm{d}x_k}{\mathrm{d}t} j_k\,, \tag{175}$$

where in the last member only a reduced sum is left. Now for the parameters A_k^{I} and A_k^{II} with $k = p+1, p+2, \ldots, N$, for which the system in closed, we can apply the formalism of Chapter IV and VII according to which the so-called α-variables (*i.e.* the deviations of the state variables from their equilibrium values) are linear functions of the so-called X-variables. In the present case we can thus write

$$\alpha_k^{\mathrm{I}} \equiv A_k^{\mathrm{I}} - A_k^{\mathrm{eq}} = -\sum_{i=1}^{N} g_{ki}^{-1} X_i^{\mathrm{I}}, \quad (k = p+1, p+2, \ldots, N), \tag{176}$$

(where the coefficients g_{ik}^{-1} with $i, k = 1, 2, \ldots, N$ are the elements of a symmetric, non-negative definite matrix) and a similar equation for reservoir II. From the difference of these two equations we find

$$A_k^{\mathrm{II}} - A_k^{\mathrm{I}} = \alpha_k^{\mathrm{II}} - \alpha_k^{\mathrm{I}} = -\sum_{i=1}^{N} g_{ki}^{-1}(X_i^{\mathrm{II}} - X_i^{\mathrm{I}}) \equiv -\sum_{i=1}^{N} g_{ki}^{-1} x_i,$$

$$(k = p+1, p+2, \ldots, N), \tag{177}$$

where we have employed the definition of the thermodynamic force x_i as a difference of a quantity X_i^{II} referring to vessel II and the corresponding quantity X_i^{I} in vessel I.

From (177) we find with the help of (166)–(169) for $k = p+1$, $p+2, \ldots, N$:

$$j_k = \frac{1}{2}\frac{\mathrm{d}(A_k^{\mathrm{II}} - A_k^{\mathrm{I}})}{\mathrm{d}t} = \frac{1}{2}\frac{\mathrm{d}(\alpha_k^{\mathrm{II}} - \alpha_k^{\mathrm{I}})}{\mathrm{d}t}$$

$$= -\frac{1}{2}\sum_{i=1}^{N} g_{ki}^{-1}\frac{\mathrm{d}x_i}{\mathrm{d}t} = -\frac{1}{2}\sum_{i=p+1}^{N} g_{ki}^{-1}\frac{\mathrm{d}x_i}{\mathrm{d}t}, \quad (k = p+1, p+2, \ldots, N). \tag{178}$$

Finally one can introduce (178) into (175), which gives

$$\frac{\mathrm{d}\sigma}{\mathrm{d}t} = -\sum_{i,k=p+1}^{N} g_{ki}^{-1}\frac{\mathrm{d}x_k}{\mathrm{d}t}\frac{\mathrm{d}x_i}{\mathrm{d}t} \leqslant 0. \tag{179}$$

This is a non-positive definite expression, because the minor, formed by the elements g_{ki}^{-1} with $k, i = p+1, p+2, \ldots, N$ of the complete non-negative definite matrix g^{-1} with N rows and columns, is also non-negative definite. This result completes the proof of the theorem, in a fashion analogous to the proof of the corresponding theorem for continuous systems, given in Chapter V.

The theorem proved here can be considered as an extension* of Le Chatelier's principle, which was derived originally for equilibrium states only, to stationary states. That this is true follows from the inequality (179) which states that, if a system in a stationary state is disturbed, it will tend to come back to its state of minimum entropy production. This is alternatively expressed by saying that the evolution in time "moderates" the perturbation. We may mention that Le Chatelier's principle was proved for the equilibrium state or, in the terminology of this section, for the stationary state of order zero, *i.e.* the state in which the system ultimately arrives when no external constraints are exerted on the system. In this final state all fluxes and thermodynamic forces, and thus also the (minimal) entropy production, vanish.

§ 6. *Osmotic Pressure and Permeability of Membranes*

In a binary, non-reacting system ($n = 2$) of which the components carry no electrical charges and which has uniform temperature the entropy production (69) reduces to

$$\sigma_{\text{tot}} = -j_1' \frac{(\Delta\mu_1)_{T,p}}{T} - j_v \frac{\Delta p}{T}. \tag{180}$$

The fluxes (71) and (72) are here simply

$$j_1' = j_1 - \frac{c_1}{c_2} j_2 , \tag{181}$$

$$j_v = v_1 j_1 + v_2 j_2 , \tag{182}$$

and the thermodynamic force (66) contains only one term

$$(\Delta\mu_1)_{T,p} = \mu_{11}^c \Delta c_1 . \tag{183}$$

The phenomenological equations are

$$j_1' = -\Lambda_{11} \frac{(\Delta\mu_1)_{T,p}}{T} - \Lambda_{1v} \frac{\Delta p}{T}, \tag{184}$$

$$j_v = -\Lambda_{v1} \frac{(\Delta\mu_1)_{T,p}}{T} - \Lambda_{vv} \frac{\Delta p}{T}, \tag{185}$$

* I. Prigogine, Bull. Acad. roy. Belg., Cl. Sc. [5] **31** (1945) 600.

and we have the Onsager relation

$$\Lambda_{v1} = \Lambda_{1v} . \tag{186}$$

An important experimental situation is the "stationary state" (of the first order) in which the volume flow j_v vanishes. One then obtains from (185), with (183),

$$\frac{\Delta p}{(\Delta \mu_1)_{T,p}} = \frac{\Delta p}{\mu_{11}^c \Delta c_1} = -\frac{\Lambda_{v1}}{\Lambda_{vv}}, \quad (j_v = 0) , \tag{187}$$

the *osmotic pressure* corresponding to a mass fraction difference Δc_1 between the two reservoirs connected by a membrane. It may be noted that such a stationary state is never exactly realized in actual experiments since Δc_1 cannot be maintained strictly constant. However, a pressure difference Δp establishes itself at a much quicker rate than the slow change of Δc_1 in the course of time so that at least a quasi-stationary state of the type described above in always reached.

Another experimental situation is the stationary state (of the second order) with fixed $\Delta c_1 = 0$ [or $(\Delta \mu_1)_{T,p} = 0$] and fixed Δp. Then (184) with (185) yields

$$\frac{j_1'}{j_v} = \frac{\Lambda_{1v}}{\Lambda_{vv}}, \quad (\Delta c_1 = 0) . \tag{188}$$

Experimentally such a stationary state is only approximately realized because one cannot maintain Δc_1 equal to zero exactly.

We can without loss of generality always choose for component 1 the substance which passes easier than component 2 (or in a limiting case just as easy as component 2) through the membrane. In practice this means that component 1 will be the solvent and component 2 the solute, which consists in general of heavier molecules than the solvent. Now we can imagine two limiting cases: first the membrane may have the same permeability for both components:

$$\frac{j_1}{c_1} = \frac{j_2}{c_2} \quad \text{or} \quad j_1' = 0 , \tag{189}$$

where the second formula follows from the first with the help of the definition (181). The other limiting case is when component 2 cannot pass through the membrane. Then j_2 vanishes and we have from (181) and (182)

$$\frac{j_1'}{j_v} = \frac{j_1}{v_1 j_1} = \frac{1}{v_1}. \tag{190}$$

In the general case, comprising the intermediate case between (189) and (190) we may write

$$\frac{j_1'}{j_v} = \frac{a}{v_1}, \quad (0 \leqslant a \leqslant 1), \tag{191}$$

where we can call a (or rather $1 - a$) the relative permeability of the membrane to the solute molecules. The limiting cases (189) and (190) correspond clearly to $a = 0$ and $a = 1$ respectively.

The Onsager relation (186) has a consequence that the effects (187) and (188) are connected as

$$\frac{\Delta p}{(\Delta \mu_1)_{T,p}} = -\frac{j_1'}{j_v}. \tag{192}$$

With (183) and (191) we can write for the osmotic pressure

$$\frac{\Delta p}{\Delta c_1} = -a \frac{\mu_{11}^c}{v_1}, \quad (0 \leqslant a \leqslant 1). \tag{193}$$

In the limiting case when a vanishes (the membrane has the same permeability to solvent and solute) this gvies

$$\Delta p = 0, \tag{194}$$

i.e. no osmotic pressure arises. In the other limiting case $a = 1$ (membrane is impenetrable to the solute) we find from (193)

$$\frac{\Delta p}{\Delta c_1} = -\frac{\mu_{11}^c}{v_1}. \tag{195}$$

This is the well-known expression for the osmotic pressure which is also found from equilibrium thermodynamics. One may check that indeed the entropy production (180) vanishes if we put $j_2 \equiv 0$ and use formula (195).

The intermediate case, *i.e.* formula (193) with $0 < a < 1$ is important for measurements with non-ideal membranes, *i.e.* membranes through which also the solute may leak. The point is that if one measures first the permeation coefficient a, then experiments on Δp still permit, just as in the ideal equilibrium case (195), to determine the molecular mass of the solute molecules (which is usually the purpose of osmotic pressure measurements) from formula (193)*. This idea was first put forward by Staverman, who also treated the case of multi-component mixtures.

§ 7. *Electrokinetic Effects*

At the end of section 3 of this chapter we gave the entropy production (74) in an n-component non-reacting mixture at uniform temperature and concentrations under the influence of electrostatic forces. We found in section 4 the phenomenological equations (129) and (130) which describe the behaviour of this system, and the Onsager relation (135). The following treatment is valid both for case a (macroscopic sub-system III) and case b (more complicated structure of III).

This formalism allows us to study the "electrokinetic effects" which can occur in the system.

In a situation with fixed pressure difference Δp and vanishing electric current i, we have from (130)

$$\left(\frac{\Delta\varphi}{\Delta p}\right)_{i=0} = -\frac{\Lambda_{ev}}{\Lambda_{ee}}, \tag{196}$$

which effect is called the *streaming potential* (or mechano-electric effect).

In the stationary state (of the second order) with fixed $\Delta p = 0$ and fixed $\Delta\varphi$ we find from (129) and (130)

$$\left(\frac{j_v}{i}\right)_{\Delta p=0} = \frac{\Lambda_{ve}}{\Lambda_{ee}}. \tag{197}$$

This effect is called *electro-osmosis*. The two effects introduced here

* A. J. Staverman, Rec. Trav. chim. Pays-Bas **70** (1951) 344; Trans. Faraday Soc. **48** (1952) 176.

See also P. Mazur, Thermodynamics symposium in Utrecht (1952), [published in: Chem. Weekblad **50** (1954) 324].

are related according to the Onsager relation (135) as

$$\left(\frac{\Delta\varphi}{\Delta p}\right)_{i=0} = -\left(\frac{j_v}{i}\right)_{\Delta p=0}. \tag{198}$$

This relation, which has experimentally been found and which is known under the name of Saxén's relation, is here derived independent of special assumptions on the structure of the membrane*.

Two other experimental situations can be realized, in which the roles of the two fluxes and also of the two thermodynamic forces are inverted as compared to the discussion given above. So in the state with $\Delta\varphi$ fixed and vanishing j_v we have from (129)

$$\left(\frac{\Delta p}{\Delta\varphi}\right)_{j_v=0} = -\frac{\Lambda_{ve}}{\Lambda_{vv}}, \tag{199}$$

the *electro-osmotic pressure*. In the state with fixed $\Delta\varphi = 0$ and Δp, one gets from (129) and (130)

$$\left(\frac{i}{j_v}\right)_{\Delta\varphi=0} = \frac{\Lambda_{ev}}{\Lambda_{vv}}, \tag{200}$$

the *streaming current*. The Onsager relation (135) connects also these two effects:

$$\left(\frac{\Delta p}{\Delta\varphi}\right)_{j_v=0} = -\left(\frac{i}{j_v}\right)_{\Delta\varphi=0}. \tag{201}$$

Both in Saxén's relation (198) and (201) the Onsager relation connects a streaming effect and an osmotic effect.

§ 8. *Thermomolecular Pressure Effect, Thermal Effusion and Mechanocaloric Effect in Reacting Mixtures*

The phenomena studied in § 5 of this chapter for non-reacting mixtures can also be considered in multi-component systems in which chemical reactions do occur**. We then have the set of phenomenological equations (95)–(98).

* P. Mazur and J. Th. G. Overbeek, Rec. Trav. chim. Pays-Bas **70** (1951) **83**. P. Mazur, J. Chim. phys. **49** (1952) C 130.

** S. R. de Groot, L. Jansen and P. Mazur, Physica **16** (1950) 691. E. P. Rastogi and R. C. Srivastava, Physica **25** (1959) 391.

Before starting the discussion of the various effects it is useful to write the entropy production (69) as a function of independent thermodynamic forces. This can be achieved by expressing A_j^{II} $(j=1,2,\ldots,r)$ in terms of the other thermodynamic forces. We can write, keeping in mind that the symbol Δ stands for the difference of the value of a quantity in reservoir II and the same quantity in reservoir I,

$$\frac{A_j^{\mathrm{II}}}{T^{\mathrm{II}}} = \frac{\sum_{k=1}^{n} \mu_k^{\mathrm{II}} \nu_{kj}}{T^{\mathrm{II}}} = \frac{\sum_{k=1}^{n} \mu_k^{\mathrm{I}} \nu_{kj}}{T^{\mathrm{I}}} + \sum_{k=1}^{n} \Delta \left(\frac{\mu_k}{T} \right) \nu_{kj}$$

$$= \frac{A_j^{\mathrm{I}}}{T^{\mathrm{I}}} + \sum_{k=1}^{n} \frac{(\Delta\mu_k)_{T,p}}{T} \nu_{kj} - \sum_{k=1}^{n} h_k \nu_{kj} \frac{\Delta T}{T^2} + \sum_{k=1}^{n} v_k \nu_{kj} \frac{\Delta p}{T} . \tag{202}$$

By applying the Gibbs–Duhem relation (67) we find the alternative form

$$\frac{A_j^{\mathrm{II}}}{T^{\mathrm{II}}} = \frac{A_j^{\mathrm{I}}}{T^{\mathrm{I}}} + \sum_{k=1}^{n-1} \left(\nu_{kj} - \frac{c_k}{c_n} \nu_{nj} \right) \frac{(\Delta\mu_k)_{T,p}}{T} - \sum_{k=1}^{n} h_k \nu_{kj} \frac{\Delta T}{T^2} + \sum_{k=1}^{n} v_k \nu_{kj} \frac{\Delta p}{T} . \tag{203}$$

In this way the affinities A_j^{II} $(j = 1, 2, \ldots, r)$ are expressed in terms of the other thermodynamic forces. If we introduce (203) into (69) we obtain the entropy production as a function of a set of independent thermodynamic forces

$$\sigma_{\mathrm{tot}} = - \left(j_q' - \sum_{j=1}^{r} j_j^{\mathrm{II}} \sum_{k=1}^{n} h_k \nu_{kj} \right) \frac{\Delta T}{T^2}$$

$$- \sum_{k=1}^{n-1} \left\{ j_k' + \sum_{j=1}^{r} j_j^{\mathrm{II}} \left(\nu_{kj} - \frac{c_k}{c_n} \nu_{nj} \right) \right\} \frac{(\Delta\mu_k)_{T,p}}{T}$$

$$- \left(j_v + \sum_{j=1}^{r} j_j^{\mathrm{II}} \sum_{k=1}^{n} v_k \nu_{kj} \right) \frac{\Delta p}{T} - \sum_{j=1}^{r} \left(j_j^{\mathrm{I}} + j_j^{\mathrm{II}} \right) \frac{A_j^{I}}{T} . \tag{204}$$

Let us now first consider the stationary state of first order which arises when ΔT is fixed. In accordance with discussions in § 5, the stationary state will be characterized by the vanishing of the fluxes corresponding to the forces different from ΔT. Formula (204) shows that we have thus

$$j'_k + \sum_{j=1}^{r} j_j^{\text{II}} \left(\nu_{kj} - \frac{c_k}{c_n} \nu_{nj} \right) = 0 \, , \quad (k = 1, 2, \ldots, n-1) \, , \tag{205}$$

$$j_v + \sum_{j=1}^{r} j_j^{\text{II}} \sum_{k=1}^{n} v_k \nu_{kj} = 0 \, , \tag{206}$$

$$j_j^{\text{I}} + j_j^{\text{II}} = 0 \, , \qquad (j = 1, 2, \ldots, r) \, . \tag{207}$$

In order to find what these equations mean, we shall reintroduce the mass fluxes j_k by using the definitions (71) and (72). Then (205) and (206) become

$$\frac{\left(j_k + \sum_{j=1}^{r} \nu_{kj} j_j^{\text{II}} \right)}{c_k} = \text{independent of } k \, , \quad (k = 1, 2, \ldots, n) \, , \tag{208}$$

$$\sum_{k=1}^{n} v_k \left(j_k + \sum_{j=1}^{r} \nu_{kj} j_j^{\text{II}} \right) = 0 \, . \tag{209}$$

This set of equations is equivalent with the set

$$j_k + \sum_{j=1}^{r} \nu_{kj} j_j^{\text{II}} = 0 \, , \quad (k = 1, 2, \ldots, n) \, . \tag{210}$$

With the help of (207) the last equations become

$$j_k - \sum_{j=1}^{r} \nu_{kj} j_j^{\text{I}} = 0 \, , \quad (k = 1, 2, \ldots, n) \, , \tag{211}$$

or, equivalently according to (58), (60) and (15),

$$\frac{\mathrm{d_i} M_k^{\text{I}}}{\mathrm{d}t} + \frac{\mathrm{d_e} M_k^{\text{I}}}{\mathrm{d}t} = \frac{\mathrm{d} M_k^{\text{I}}}{\mathrm{d}t} = 0 \, , \quad (k = 1, 2, \ldots, n) \, . \tag{212}$$

The stationary state with fixed ΔT is thus characterized by the fact that transport phenomena and chemical reactions take place in such a way that their effects just cancel in so far that the masses of all components have constant values in both reservoirs (*cf.* Chapter X, § 7).

If the phenomenological equations (96)–(98) are inserted into equations (205)–(207) we obtain with the help of (203) the set

$$- \Lambda_{kq} \frac{\Delta T}{T^2} - \sum_{m=1}^{n-1} \Lambda_{km} \frac{(\Delta\mu_m)_{T,p}}{T} - \Lambda_{kv} \frac{\Delta p}{T}$$

$$+ \sum_{j,j'=1}^{r} \left(\nu_{kj} - \frac{c_k}{c_n} \nu_{nj} \right) \Lambda_{jj'} M^{\text{I}} \frac{A_{j'}^{\text{I}}}{T} = 0 \,, \quad (k = 1, 2, \ldots, n-1) \,, \tag{213}$$

$$- \Lambda_{vq} \frac{\Delta T}{T^2} - \sum_{m=1}^{n-1} \Lambda_{vm} \frac{(\Delta\mu_m)_{T,p}}{T} - \Lambda_{vv} \frac{\Delta p}{T} + \sum_{j,j'=1}^{r} \sum_{k=1}^{n} v_k \nu_{kj} \Lambda_{jj'} M^{\text{I}} \frac{A_{j'}^{\text{I}}}{T} = 0 \,, \tag{214}$$

$$\left(1 + \frac{M^{\text{I}}}{M^{\text{II}}} \right) \frac{A_j^{\text{I}}}{T} + \left\{ \sum_{k=1}^{n-1} \left(\nu_{kj} - \frac{c_k}{c_n} \nu_{nj} \right) \frac{(\Delta\mu_k)_{T,p}}{T} - \sum_{k=1}^{n} h_k \nu_{kj} \frac{\Delta T}{T^2} \right.$$

$$\left. + \sum_{k=1}^{n} v_k \nu_{kj} \frac{\Delta p}{T} \right\} = 0 \,, \qquad (j = 1, 2, \ldots, r) \,. \tag{215}$$

From these linear equations one can solve $(\Delta\mu_k)_{T,p}/\Delta T$ $(k = 1, 2, \ldots, n-1)$, $\Delta p/\Delta T$ and $A_j^{\text{I}}/\Delta T$ $(j = 1, 2, \ldots, r)$. We shall, for the explicit evaluation, confine ourselves to the simplest case: the binary, single reaction mixture $(n = 2, r = 1)$. Then (213)–(215) reduce to

$$- \Lambda_{1q} \frac{\Delta T}{T^2} - \Lambda_{11} \frac{(\Delta\mu_1)_{T,p}}{T} - \Lambda_{1v} \frac{\Delta p}{T} + c' \Lambda \left(1 + \frac{M^{\text{I}}}{M^{\text{II}}} \right) \frac{A^{\text{I}}}{T} = 0 \,, \tag{216}$$

$$- \Lambda_{vq} \frac{\Delta T}{T^2} - \Lambda_{v1} \frac{(\Delta\mu_1)_{T,p}}{T} - \Lambda_{vv} \frac{\Delta p}{T} + v' \Lambda \left(1 + \frac{M^{\text{I}}}{M^{\text{II}}} \right) \frac{A^{\text{I}}}{T} = 0 \,, \tag{217}$$

$$\left(1 + \frac{M^{\text{I}}}{M^{\text{II}}} \right) \frac{A^{\text{I}}}{T} = - c' \frac{(\Delta\mu_1)_{T,p}}{T} + h' \frac{\Delta T}{T^2} - v' \frac{\Delta p}{T} \,, \tag{218}$$

where we have used the following abbreviations (eliminating also ν_1 with the help of $\nu_1 + \nu_2 = 0$)

$$c' = - \frac{\nu_2}{c_2} \,, \quad h' = \nu_2 (h_2 - h_1) \,, \quad v' = \nu_2 (v_2 - v_1) \,,$$

$$\Lambda = \Lambda_{11}^{\text{c}} \frac{M^{\text{I}} M^{\text{II}}}{M^{\text{I}} + M^{\text{II}}} \,, \qquad A^{\text{I}} = A_1^{\text{I}} \,. \tag{219}$$

In the last form Λ_{11}^{c} is the chemical coefficient $\Lambda_{jj'}$ with $j = j' = 1$. It should not be confused with the diffusion coefficient Λ_{11} occurring in (216).

We find in the first place from (216)–(218) the *thermomolecular pressure effect*

$$\frac{\Delta p}{\Delta T} = - \frac{\Lambda_{11}\Lambda_{vq} - \Lambda_{v1}\Lambda_{1q} + (c'^2\Lambda_{vq} - v'h'\Lambda_{11} + c'h'\Lambda_{v1} - c'v'\Lambda_{1q})\Lambda}{\Lambda_{11}\Lambda_{vv} - \Lambda_{v1}\Lambda_{1v} + (v'^2\Lambda_{11} + c'^2\Lambda_{vv} - c'v'\Lambda_{v1} - c'v'\Lambda_{1v})\Lambda} \frac{1}{T} \tag{220}$$

In case *a* (capillary macroscopic system) we have the identities (106)–(110) which for a binary mixture reduce to

$$\Lambda_{qv} = c_2(v_1 - v_2)\Lambda_{q1}\,, \quad \Lambda_{vq} = c_2(v_1 - v_2)\Lambda_{1q}\,, \tag{221}$$

$$\Lambda_{1v} = c_2(v_1 - v_2)\Lambda_{11}\,, \quad \Lambda_{v1} = c_2(v_1 - v_2)\Lambda_{11}\,, \tag{222}$$

$$\Lambda_{vv} = c_2^2(v_1 - v_2)^2\Lambda_{11} + \frac{\alpha T}{\eta l}\,, \tag{223}$$

$$\Lambda_{qv} = \frac{\Lambda_{q1}\Lambda_{1v}}{\Lambda_{11}}\,, \quad \Lambda_{vq} = \frac{\Lambda_{v1}\Lambda_{1q}}{\Lambda_{11}}\,, \tag{224}$$

$$\Lambda_{vv} = \frac{\Lambda_{v1}\Lambda_{1v}}{\Lambda_{11}} + \frac{\alpha T}{\eta l}\,. \tag{225}$$

With (221), (222) and (224) introduced into (220) we obtain the result

$$\Delta p = 0\,, \tag{226}$$

as could be expected (*cf.* Chapter X, § 7).

In case *b* (narrow capillary) we have no identities and the full formula (220) is to be used.

Let us now turn to *thermal effusion*. In case *a* we can solve $(\Delta\mu_1)_{T,p}$ from (216)–(218), taking into account the fact that Δp vanishes. This gives

$$\frac{\Delta c_1}{\Delta T} = \frac{1}{\mu_{11}^c} \frac{(\Delta\mu_1)_{T,p}}{\Delta T} = - \frac{\Lambda_{1q} - c'h'\Lambda}{\Lambda_{11} + c'^2\Lambda} \frac{1}{\mu_{11}^c T} = - \frac{\Lambda_{vq} - v'h'\Lambda}{\Lambda_{v1} + c'v'\Lambda} \frac{1}{\mu_{11}^c T}\,. \tag{227}$$

In the more general case *b* we find from (216)–(218)

$$\frac{\Delta c_1}{\Delta T} = \frac{1}{\mu_{11}^c} \frac{(\Delta \mu_1)_{T,p}}{\Delta T}$$

$$= - \frac{A_{vv}A_{1q} - A_{1v}A_{vq} + (v'^2 A_{1q} - c'h' A_{vv} + v'h' A_{1v} - c'v' A_{vq})A}{A_{11}A_{vv} - A_{1v}A_{v1} + (v'^2 A_{11} + c'^2 A_{vv} - c'v' A_{1v} - c'v' A_{v1})A} \frac{1}{\mu_{11}^c T}. \tag{228}$$

One could also have found (227) from (228) using the identities (221), (222) and (224).

It is clear that there is still a third physical effect, namely the "*chemical effect*" $A^{\text{I}}/\Delta T$. In case *a* ($\Delta p = 0$) the equations (216)–(218) yield

$$\left(1 + \frac{M^{\text{I}}}{M^{\text{II}}}\right) \frac{A^{\text{I}}}{\Delta T} = \frac{c' A_{1q} + h' A_{11}}{A_{11} + c'^2 A} \frac{1}{T} = \frac{c' A_{vq} + h' A_{v1}}{A_{v1} + c'v' A} \frac{1}{T}. \tag{229}$$

In case *b* ($\Delta p \neq 0$) the same equations give

$$\left(1 + \frac{M^{\text{I}}}{M^{\text{II}}}\right) \frac{A^{\text{I}}}{\Delta T}$$

$$= \frac{c'(A_{vv}A_{1q} - A_{1v}A_{vq}) + v'(A_{11}A_{vq} - A_{v1}A_{1q}) + h'(A_{11}A_{vv} - A_{1v}A_{v1})}{(A_{11}A_{vv} - A_{1v}A_{v1}) + (v'^2 A_{11} + c'^2 A_{vv} - c'v' A_{1v} - c'v' A_{v1})A} \frac{1}{T}. \tag{230}$$

Again (229) can also be derived from (230) using the identities valid amongst the phenomenological constants in case *a*.

It is interesting to consider explicitly the two limiting cases $A \to 0$ (no chemical reactions occur) and $A \to \infty$ (chemical equilibrium). In the first case we find back the formulae of § 5 for Δp and Δc_1, as it should. In the second case one obtains from (226), (220), (227), (228), (229) and (230) respectively (we give first the result for case *a*, and then for *b*):

$$\Delta p = 0, \quad \frac{\Delta p}{\Delta T} = - \frac{c'^2 A_{vq} - v'h' A_{11} + c'h' A_{v1} - c'v' A_{1q}}{v'^2 A_{11} - c'^2 A_{vv} - c'v' A_{v1} - c'v' A_{1v}} \frac{1}{T}, \tag{231}$$

$$\frac{\Delta c_1}{\Delta T} = \frac{h'}{c'} \frac{1}{\mu_{11}^c T}, \quad \frac{\Delta c_1}{\Delta T} = - \frac{v'^2 A_{1q} - c'h' A_{vv} + v'h' A_{1v} - c'v' A_{vq}}{v'^2 A_{11} + c'^2 A_{vv} - c'v' A_{1v} - c'v' A_{v1}} \frac{1}{\mu_{11}^c T}, \tag{232}$$

$$A^{\mathrm{I}} = 0\,, \quad \text{(both for case } a \text{ and } b)\,. \tag{233}$$

The last relation is, as it should be, at chemical equilibrium.

Let us now turn to the stationary state of second order with fixed $\Delta T = 0$ and fixed Δp. Then in the final state the fluxes which, in the entropy production (204), are multiplied by the thermodynamic forces other than ΔT and Δp, will vanish:

$$\dot{j}'_k + \sum_{j=1}^{r} \dot{j}_j^{\mathrm{II}} \left(\nu_{kj} - \frac{c_k}{c_n} \nu_{nj} \right) = 0\,, \quad (k = 1, 2, \ldots, n-1)\,, \tag{234}$$

$$\dot{j}_j^{\mathrm{I}} + \dot{j}_j^{\mathrm{II}} = 0\,, \quad (j = 1, 2, \ldots, r)\,. \tag{235}$$

If we use the definitions (71) and (148) we obtain from these relations

$$\dot{j}_k = c_k \dot{j} + \sum_{j=1}^{r} \nu_{kj} \dot{j}_j^{\mathrm{I}}\,, \quad (k = 1, 2, \ldots, n)\,, \tag{236}$$

or, equivalently, according to (58), (60) and (15)

$$\frac{1}{c_k} \frac{\mathrm{d}M_k^{\mathrm{I}}}{\mathrm{d}t} = \text{independent of } k\,, \tag{237}$$

or, in other words, the mass fractions c_k are constant in time in the stationary state considered.

If we introduce the phenomenological equations (96) and (98) into (234) and (235), we get for the binary ($n = 2$), single reaction ($r = 1$) mixture in the state with fixed Δp and $\Delta T = 0$, using also (203) and (219):

$$-\Lambda_{11}(\Delta\mu_1)_{T,p} - \Lambda_{1v}\Delta p + c'\Lambda \left(1 + \frac{M^{\mathrm{I}}}{M^{\mathrm{II}}}\right) \frac{A^{\mathrm{I}}}{T} = 0\,, \tag{238}$$

$$\left(1 + \frac{M^{\mathrm{I}}}{M^{\mathrm{II}}}\right) A^{\mathrm{I}} = -c'(\Delta\mu_1)_{T,p} - v'\Delta p\,. \tag{239}$$

As in § 5 of this chapter we define the *mechanocaloric effect* as

$$q^{*} = \frac{\dot{j}_{qe}}{\dot{j}}\,, \tag{240}$$

in the stationary state of second order with $\Delta T = 0$ and Δp fixed. The proof of the equality

$$j_{qe} = j_q, \tag{241}$$

given in § 5 remains valid also in the presence of chemical reactions. The following relations, based on (236), are, however, different from what was found in § 5 (we omit again the superfluous indices referring to the chemical reaction, since we have chosen $r = 1$):

$$j'_q \equiv j_u - \sum_k h_k j_k = j_u - \sum_k h_k(c_k j + \nu_k j^{\mathrm{I}}) = j_u - hj - h'j^{\mathrm{I}} = j_q - h'j^{\mathrm{I}}, \tag{242}$$

$$j_v = \sum_k v_k j_k = \sum_k v_k(c_k j + \nu_k j^{\mathrm{I}}) = vj + v'j^{\mathrm{I}}. \tag{243}$$

With the help of (241)–(243) we obtain instead of (240)

$$q^* = \frac{j'_q + h'j^{\mathrm{I}}}{j_v - v'j^{\mathrm{I}}} v, \tag{244}$$

which on introducing the phenomenological equations (95), (97) and (98), and keeping in mind that ΔT vanishes, becomes with (219)

$$q^* = \frac{\Lambda_{q1}(\Delta\mu_1)_{T,p} + \Lambda_{qv}\Delta p + h'\Lambda(1 + M^{\mathrm{I}}/M^{\mathrm{II}})A^{\mathrm{I}}}{\Lambda_{v1}(\Delta\mu_1)_{T,p} + \Lambda_{vv}\Delta p - v'\Lambda(1 + M^{\mathrm{I}}/M^{\mathrm{II}})A^{\mathrm{I}}} v, \tag{245}$$

or, finally, with the help of (238) and (239):

$$q^* = \frac{\Lambda_{11}\Lambda_{qv} - \Lambda_{q1}\Lambda_{1v} + (c'^2\Lambda_{qv} - v'h'\Lambda_{11} + c'h'\Lambda_{1v} - c'v'\Lambda_{q1})\Lambda}{\Lambda_{11}\Lambda_{vv} - \Lambda_{v1}\Lambda_{1v} + (v'^2\Lambda_{11} + c'^2\Lambda_{vv} - c'v'\Lambda_{v1} - c'v'\Lambda_{1v})\Lambda} v. \tag{246}$$

From the Onsager relations (111), (113) and (114) it follows that the thermomolecular pressure effect (220) and the mechanocaloric effect (246) are related as

$$\frac{\Delta p}{\Delta T} = -\frac{q^*}{vT}, \tag{247}$$

i.e. in exactly the same way as in non-reacting mixtures. The proof was given here for the binary, single reaction mixture, but the generalization to the n-component mixture with r chemical reactions is trivial.

It is of some interest to study a special case of the binary, single reaction system, namely a system of which one of the components, say number 1, cannot pass through the capillary.

$$j_1 \equiv 0\,, \tag{248}$$

which gives with (71) and (72)

$$j_1' = -\left(\frac{c_1}{c_2}\right) j_2\,, \quad j_v = v_2 j_2\,. \tag{249}$$

This shows that amongst the phenomenological coefficients of the equations (96) and (97) there exist the following identities

$$\frac{\Lambda_{1q}}{\Lambda_{vq}} = \frac{\Lambda_{11}}{\Lambda_{v1}} = \frac{\Lambda_{1v}}{\Lambda_{vv}} = -\frac{c_1}{c_2 v_2}\,. \tag{250}$$

These equalities and the Onsager relations $\Lambda_{q1} = \Lambda_{1q}$, $\Lambda_{qv} = \Lambda_{vq}$ and $\Lambda_{v1} = \Lambda_{1v}$ permit to express all phenomenological coefficients of (95)–(97) in terms of Λ_{qq}, Λ_{vq} and Λ_{vv} alone. In such a way the expression (246) for the mechanocaloric effect reduces simply to

$$q^* = \frac{\Lambda_{vq}}{\Lambda_{vv}} v_2 + c_1(h_2 - h_1)\,, \tag{251}$$

if also (219) is used. The value of the thermomolecular pressure effect is then also given, since we have always the relation (247).

The interest of the system studied here resides in the fact that it bears some relation to a phenomenological model for liquid helium II.* In this model component 1 is called the "normal fluid", which is not capable to pass through a narrow capillary, whereas component 2, the "superfluid" can flow through it. The chemical reaction is simply the transformation of superfluid into normal fluid and vice versa. In order to obtain a description of the properties of liquid helium II, one must furthermore assume that in this system no reduced heat flow

* C. J. Gorter and J. H. Mellink, Physica **15** (1949) 285.
C. J. Gorter, Physica **15** (1949) 523.
C. J. Gorter, P. W. Kasteleijn and J. H. Mellink, Physica **16** (1950) 113.

j'_q takes place in the isothermal state with $\Delta T = 0$. This means, in view of (95), that we suppose

$$\Lambda_{q1} \equiv 0\,, \quad \Lambda_{qv} \equiv 0\,. \tag{252}$$

(We note that the total heat flow j_q, which appears in (242), does not vanish in the isothermal state.) With these new conditions and the Onsager relation (113), equation (251) reduces further to

$$q^* = c_1(h_2 - h_1) = h_2 - h\,, \tag{253}$$

and with (247) one finds

$$\frac{\Delta p}{\Delta T} = -\frac{c_1(h_2 - h_1)}{vT} = -\frac{h_2 - h}{vT}\,. \tag{254}$$

The last two expressions contain only equilibrium quantities.

Still a further-reaching assumption, which has also been made in the framework of the two-fluid model for liquid helium II, is to suppose that the chemical equilibrium is practically immediately established. Then the chemical affinity vanishes:

$$A \equiv \mu_1 \nu_1 + \mu_2 \nu_2 \equiv \nu_2(\mu_2 - \mu_1) = 0\,. \tag{255}$$

From this we have

$$\left(\frac{\partial g}{\partial c_1}\right)_{T,p} = h_1 - h_2 - T\left(\frac{\partial s}{\partial c_1}\right)_{T,p} = h_1 - h_2 - T\,(s_1 - s_2) = 0\,, \tag{256}$$

where g is the specific Gibbs function. With the help of this last relation one finds from (253) and (254)

$$q^* = -\,Tc_1\left(\frac{\partial s}{\partial c_1}\right)_{T,p}, \tag{257}$$

$$\frac{\Delta p}{\Delta T} = \frac{c_1}{v}\left(\frac{\partial s}{\partial c_1}\right)_{T,p}, \tag{258}$$

as expressions for the mechanocaloric effect and the thermomolecular pressure in liquid helium II. Both these expressions have been extensively tested experimentally. If one assumes furthermore that the partial specific entropy of the superfluid vanishes ($s_2 = 0$), then (257) and (258) reduce further to

$$q^* = - Tc_1 s_1 = - Ts \,, \tag{259}$$

$$\frac{\Delta p}{\Delta T} = \frac{1}{v} c_1 s_1 = \frac{s}{v} \,, \tag{260}$$

which are H. London's equations.

It should still be remarked that the assumption (252) was not derived here from the theory of continuous media, as were the other relevant equations. In fact the theory for a discontinuous system containing two "fluids" should be derived from a continuum two-fluid theory in a fashion analogous to the derivation of the ordinary theory for discontinuous systems from the ordinary continuum theory, as presented in the first sections of this chapter. A two-fluid continuum theory has indeed been worked out*. It is based on the definition of "fluids" as components of a liquid or gas between which the transfer of momentum is partially or wholly inhibited. The property (252) is then a result of such a theory.

§ 9. *Electrochemistry*

The phenomena of electrochemistry can also be treated within the framework of non-equilibrium thermodynamics of discontinuous systems. We shall adopt the following model which is characteristic for electrochemical processes. A single species $k = 1$ of a set of n components, carrying electrical charges z_k per unit mass, is transferred from sub-system I to sub-system II:

$$j_1 \neq 0, \quad j_k = 0 \,, \quad (k = 2, 3, \ldots, n) \,. \tag{261}$$

Inside the sub-systems I and II a chemical reaction can take place between a number of the components (which includes component 1 in both cases), say, between $k = 1, 2, \ldots, m$ in I, and between $k = 1, m+1, m+2, \ldots, n$ in II. In a typical example component 1 could

* I. Prigogine and P. Mazur, Physica **17** (1951) 661.
P. Mazur and I. Prigogine, Physica **17** (1951) 680.

be a silver ion, which is transferred from I to II:

$$Ag^{+I} \rightarrow Ag^{+II}, \tag{262}$$

whereas the chemical reactions in I and II could be

$$Ag^{I} \rightarrow Ag^{+I} + e^{-I}, \tag{263}$$

$$Ag^{+II} + Cl^{-II} \rightarrow AgCl^{II}, \tag{264}$$

in both of which the silver ion participates.

The rate of these chemical reactions in I and II is so quick compared to the transport phenomenon of component 1, that one can assume that chemical equilibrium is immediately established. This means that the chemical affinities vanish both in I and II:

$$A^{I} \equiv \sum_{k=1}^{m} \mu_k^{I} \nu_k^{I} = 0, \tag{265}$$

$$A^{II} \equiv \sum_{k=1,m+1}^{n} \mu_k^{II} \nu_k^{II} = 0. \tag{266}$$

The entropy production (57) for such a system, which will moreover be supposed to be isothermal, becomes in view of (73), (261), (265) and (266):

$$\sigma_{\text{tot}} = -\dot{\jmath}_1 \frac{\Delta\tilde{\mu}_1}{T} = -\dot{\jmath}_1 \frac{\Delta(\mu_1 + z_1\varphi)}{T}, \tag{267}$$

where φ is the electro-static potential. If we eliminate μ_1^{I} and μ_1^{II} with the help of (265) and (266), we obtain for the entropy production

$$\sigma_{\text{tot}} = -i\left(\frac{A}{z_1 T} + \frac{\Delta\varphi}{T}\right), \tag{268}$$

where we have introduced the electric current

$$i = z_1 \dot{\jmath}_1, \tag{269}$$

and the "total chemical affinity" of the electrochemical reaction

$$A \equiv \frac{\sum_{k=2}^{m} \mu_k^{\mathrm{I}} \nu_k^{\mathrm{I}}}{\nu_1^{\mathrm{I}}} - \frac{\sum_{k=m+1}^{n} \mu_k^{\mathrm{II}} \nu_k^{\mathrm{II}}}{\nu_1^{\mathrm{II}}} . \tag{270}$$

The latter quantity can be written in standard form as

$$A \equiv \sum_{k=2}^{n} \mu_k \nu_k , \tag{271}$$

if we define the new stoichiometric coefficients

$$\nu_k = \frac{\nu_k^{\mathrm{I}}}{\nu_1^{\mathrm{I}}} , \qquad (k = 2, 3, \ldots, m) , \tag{272}$$

$$\nu_k = -\frac{\nu_k^{\mathrm{II}}}{\nu_1^{\mathrm{II}}} , \quad (k = m+1, m+2, \ldots, n) , \tag{273}$$

and omit the indices I and II of the chemical potentials, since these are strictly spoken superfluous if one keeps in mind that components $2, 3, \ldots, m$ are contained in sub-system I, and components $m + 1$, $m + 2, \ldots, n$ in sub-system II. The quantity A can be considered as the affinity of the total electrochemical reaction, which results from the sum of (262), (263) and (264), *viz.*

$$\mathrm{Ag}^{\mathrm{I}} + \mathrm{Cl}^{-\mathrm{II}} \rightarrow \mathrm{AgCl}^{\mathrm{II}} + \mathrm{e}^{-\mathrm{I}} , \tag{274}$$

from which component 1 (the silver ion) has dropped.

One can alternatively express the entropy production in terms of the electrochemical potentials

$$\tilde{\mu}_k^{\mathrm{I}} \equiv \mu_k^{\mathrm{I}} + z_k \varphi^{\mathrm{I}} , \quad (k = 1, 2, \ldots, m) , \tag{275}$$

$$\tilde{\mu}_k^{\mathrm{II}} \equiv \mu_k^{\mathrm{II}} + z_k \varphi^{\mathrm{II}} , \quad (k = 1, m+1, m+2, \ldots, n) . \tag{276}$$

Indeed, since electric charge is conserved in the chemical reactions

$$\sum_{k=1}^{m} z_k \nu_k^{\mathrm{I}} = 0 , \tag{277}$$

$$\sum_{k=1, m+1}^{n} z_k \nu_k^{\mathrm{II}} = 0 , \tag{278}$$

one can write the chemical affinities as follows

$$A^{\mathrm{I}} = \sum_{k=1}^{m} \tilde{\mu}_k^{\mathrm{I}} \nu_k^{\mathrm{I}} \,, \tag{279}$$

$$A^{\mathrm{II}} = \sum_{k=1,m+1}^{n} \tilde{\mu}_k^{\mathrm{II}} \nu_k^{\mathrm{II}} \,. \tag{280}$$

With the help of these relations, and using (265) and (266), we can eliminate $\tilde{\mu}_1^{\mathrm{I}}$ and $\tilde{\mu}_1^{\mathrm{II}}$ from (267). This gives

$$\sigma_{\mathrm{tot}} = - i \frac{\tilde{A}}{z_1 T} \,, \tag{281}$$

with the "electrochemical affinity"

$$\tilde{A} = \frac{\sum_{k=2}^{m} \tilde{\mu}_k^{\mathrm{I}} \nu_k^{\mathrm{I}}}{\nu_1^{\mathrm{I}}} - \frac{\sum_{k=m+1}^{n} \tilde{\mu}_k^{\mathrm{II}} \nu_k^{\mathrm{II}}}{\nu_1^{\mathrm{II}}} = \sum_{k=2}^{n} \tilde{\mu}_k \nu_k \,, \tag{282}$$

of which the relation with the total chemical affinity A is

$$\tilde{A} = A + z_1 \Delta\varphi \,, \tag{283}$$

as can easily be checked.

The phenomenological equation for the flux and the thermodynamic force in (268) or (281) is

$$i = - \Lambda \frac{\tilde{A}}{z_1 T} = - \Lambda \left(\frac{A}{z_1 T} + \frac{\Delta\varphi}{T} \right) , \tag{284}$$

describing a single irreversible process. The equilibrium state is described by the well-known equation

$$\tilde{A} = 0 \,, \tag{285}$$

or

$$\Delta\varphi = - \frac{A}{z_1} \,, \tag{286}$$

discussed in equilibrium treatments of electrochemical phenomena.

The generalization to more complicated cases, where more than one component is transferred and several chemical reactions occur, is straightforward.

APPENDIX I

ON MATRIX AND TENSOR NOTATION

Throughout this book we use a system of tensor notation, in which different letter-types denote tensors of successive order*: italics for scalars (or tensors of zero order), Clarendon for vectors (or tensors of the first order) and sanserif for tensors (of the second order). Tensors of still higher order will be either represented by their components, or, if no confusion can arise, also by a sanserif letter-type. Thus we write for a vector, of which the components are v_i $(i = 1, 2, \ldots, n)$, the symbol $\boldsymbol{v}$:

$$\boldsymbol{v} \rightarrow v_i \,, \quad (i = 1, 2, \ldots, n) \,.$$

For a tensor (of the second order) we write

$$\mathsf{T} \rightarrow T_{ik} \,, \quad (i, k = 1, 2, \ldots, n) \,.$$

The above tensors are defined in Euclidian space of n dimensions (frequently in ordinary three-dimensional space).

1. *Products of Tensors*

The exterior or ordered product of two tensors of order m and n gives a tensor of order $m + n$. Thus:

$$\boldsymbol{v}\boldsymbol{w} \rightarrow (\boldsymbol{v}\boldsymbol{w})_{ik} = v_i w_k \,,$$

$$\boldsymbol{v}\mathsf{T} \rightarrow (\boldsymbol{v}\mathsf{T})_{ikl} = v_i T_{kl} \,,$$

$$\mathsf{T}\boldsymbol{v} \rightarrow (\mathsf{T}\boldsymbol{v})_{ikl} = T_{ik} v_l \,.$$

The exterior product $\boldsymbol{v}\boldsymbol{w}$ of two vectors is called a dyad.

The interior or contracted product of two tensors is obtained from the exterior product by putting equal two neighbouring indices

* Some authors use the term *rank* or *degree* of a tensor. We follow here the same nomenclature as A. Lichnerowicz, Algèbre et analyse linéaires, Paris (1956).

belonging to each of the tensors respectively, and by summing over the resulting "dummy" index. Each interior product is indicated by inserting a dot between the symbols of the tensors:

$$\boldsymbol{v}\cdot\boldsymbol{w} = \sum_i v_i w_i ,$$

$$\boldsymbol{v}\cdot\mathsf{T} \rightarrow (\boldsymbol{v}\cdot\mathsf{T})_i = \sum_k v_k T_{ki} ,$$

$$\mathsf{T}\cdot\boldsymbol{v} \rightarrow (\mathsf{T}\cdot\boldsymbol{v})_i = \sum_k T_{ik} v_k ,$$

$$\mathsf{S}\cdot\mathsf{T} \rightarrow (\mathsf{S}\cdot\mathsf{T})_{ik} = \sum_l S_{il} T_{lk} .$$

In an analogous way one has for the scalar product of two tensors:

$$\mathsf{S} : \mathsf{T} = \sum_{i,k} S_{ik} T_{ki} .$$

With the unit tensor U, which has components δ_{ik} ($\delta_{ik} = 1$ if $i = k$, $\delta_{ik} = 0$ if $i \neq k$), one can form the interior product

$$\mathsf{T} : \mathsf{U} = \sum_i T_{ii} ,$$

which is called the trace of the tensor T.

2. *Symmetric and Antisymmetric Tensors*

By transposing the indices of a tensor T, one obtains the transposed tensor $\tilde{\mathsf{T}}$

$$\tilde{T}_{ik} = T_{ki} .$$

For a dyad one has

$$\widetilde{\boldsymbol{v}\boldsymbol{w}} = \boldsymbol{w}\boldsymbol{v} ,$$

and for the interior product of two tensors

$$\widetilde{\mathsf{S}\cdot\mathsf{T}} = \tilde{\mathsf{T}}\cdot\tilde{\mathsf{S}} .$$

Tensors are called symmetric if

$$\mathsf{T} = \tilde{\mathsf{T}} ,$$

and antisymmetric if
$$T = -\tilde{T}.$$

Every tensor may be split up into a symmetric and an antisymmetric part, according to
$$T = T^s + T^a,$$
with
$$T^s = \tfrac{1}{2}(T + \tilde{T}),$$
$$T^a = \tfrac{1}{2}(T - \tilde{T}).$$

In particular we have for a dyad $D \equiv \boldsymbol{vw}$
$$D^s = \tfrac{1}{2}(\boldsymbol{vw} + \boldsymbol{wv}),$$
$$D^a = \tfrac{1}{2}(\boldsymbol{vw} - \boldsymbol{wv}).$$

In the case of ordinary three-dimensional space D^a is connected with the vector product of the two vectors, which we denote by
$$\boldsymbol{v} \wedge \boldsymbol{w} \rightarrow (\boldsymbol{v} \wedge \boldsymbol{w})_x = v_y w_z - v_z w_y, \quad \text{(cycl.)}.$$

3. *Spatial Derivations*

The operator of spatial derivation may be treated as a vector
$$\frac{\partial}{\partial \boldsymbol{x}} \rightarrow \frac{\partial}{\partial x_i}, \quad (i = 1, 2, \ldots, n).$$

In three-dimensional space one often denotes the operator $\partial/\partial \boldsymbol{x}$ by $\boldsymbol{\nabla}$ (the so-called nabla-operator). The following operations of the vector $\partial/\partial \boldsymbol{x}$ on tensors of various order occur, which define gradients and divergences:
$$\frac{\partial}{\partial \boldsymbol{x}} a \equiv \text{grad}\, a \rightarrow \left(\frac{\partial}{\partial \boldsymbol{x}} a\right)_i = \frac{\partial a}{\partial x_i},$$
$$\frac{\partial}{\partial \boldsymbol{x}} \boldsymbol{v} \equiv \text{Grad}\, \boldsymbol{v} \rightarrow \left(\frac{\partial}{\partial \boldsymbol{x}} \boldsymbol{v}\right)_{ik} = \frac{\partial v_k}{\partial x_i},$$
$$\frac{\partial}{\partial \boldsymbol{x}} \cdot \boldsymbol{v} \equiv \text{div}\, \boldsymbol{v} = \sum_i \frac{\partial v_i}{\partial x_i},$$

$$\frac{\partial}{\partial \boldsymbol{x}} \cdot \mathsf{T} \equiv \text{Div}\, \mathsf{T} \rightarrow \left(\frac{\partial}{\partial \boldsymbol{x}} \cdot \mathsf{T}\right)_i = \sum_k \frac{\partial T_{ki}}{\partial x_k},$$

$$\frac{\partial}{\partial \boldsymbol{x}} \cdot \frac{\partial}{\partial \boldsymbol{x}} = \text{div grad} = \sum_i \frac{\partial^2}{\partial x_i^2}.$$

In three-dimensional space the last operator is called the Laplacian and denoted by $\triangle$.

The gradient of a vector may be split up into its symmetric and antisymmetric part

$$\text{Grad}\, \boldsymbol{v} = (\text{Grad}\, \boldsymbol{v})^{\text{s}} + (\text{Grad}\, \boldsymbol{v})^{\text{a}},$$

with

$$(\text{Grad}\, \boldsymbol{v})^{\text{s}}_{ik} = \frac{1}{2}\left(\frac{\partial v_k}{\partial x_i} + \frac{\partial v_i}{\partial x_k}\right),$$

$$(\text{Grad}\, \boldsymbol{v})^{\text{a}}_{ik} = \frac{1}{2}\left(\frac{\partial v_k}{\partial x_i} - \frac{\partial v_i}{\partial x_k}\right).$$

In three-dimensional space $(\text{Grad}\, \boldsymbol{v})^{\text{a}}$ is related to the rotation of the vector $\boldsymbol{v}$:

$$\frac{\partial}{\partial \boldsymbol{x}} \wedge \boldsymbol{v} = \text{rot}\, \boldsymbol{v} \rightarrow \left(\frac{\partial}{\partial \boldsymbol{x}} \wedge \boldsymbol{v}\right)_x = \frac{\partial v_z}{\partial y} - \frac{\partial v_y}{\partial z}, \quad \text{(cycl.)}.$$

The system of tensor notation used here is essentially that of Milne and Chapman as outlined in L. Rosenfeld's monograph on the theory of electrons, Amsterdam 1951.

APPENDIX II

ON THERMODYNAMIC RELATIONS

1. *Partial Specific Thermodynamic Quantities and Euler Relations*

The Gibbs function G

$$G = U - TS + pV, \tag{1}$$

where U and S are the total internal energy and the total entropy of a system of volume V, obeys the relation

$$dG = - S\, dT + V\, dp + \sum_{k=1}^{n} \mu_k\, dM_k, \tag{2}$$

with M_k the total mass of component k.

At constant pressure p and temperature T the function G is of the first degree and homogeneous in the masses $M_1, M_2, \ldots, M_n$. Therefore according to Euler's theorem

$$G = \sum_{k=1}^{n} \left(\frac{\partial G}{\partial M_k}\right)_{p,T} M_k = \sum_{k=1}^{n} \mu_k M_k. \tag{3}$$

From (1) and (3) it follows that

$$\sum_{k=1}^{n} \mu_k c_k = u - Ts + pv, \tag{4}$$

where the specific quantities c_k, u, s and v are related to the quantities M_k, U, S and V by

$$M_k = c_k M, \tag{5}$$

$$U = uM, \tag{6}$$

$$S = sM, \tag{7}$$

$$V = vM, \tag{8}$$

with $M = \sum_k M_k$ the total mass of all components.

The Gibbs relation for extensive quantities

$$T\,\mathrm{d}S = \mathrm{d}U + p\,\mathrm{d}V - \sum_{k=1}^{n} \mu_k\,\mathrm{d}M_k \tag{9}$$

may now be rewritten for the specific quantities, applying (4)–(8), as

$$T\,\mathrm{d}s = \mathrm{d}u + p\,\mathrm{d}v - \sum_{k=1}^{n} \mu_k\,\mathrm{d}c_k\,. \tag{10}$$

It is this form, which has been used in Chapter III.

For an arbitrary extensive function of $p, T, M_1, M_2, \ldots M_n$ Euler's theorem gives

$$A = \sum_{k=1}^{n} \left(\frac{\partial A}{\partial M_k}\right)_{p,T} M_k = \sum_{k=1}^{n} a_k M_k\,, \tag{11}$$

where a_k is the partial specific quantity corresponding to A. Since one also has

$$(\mathrm{d}A)_{p,T} = \sum_{k=1}^{n} a_k\,\mathrm{d}M_k\,, \tag{12}$$

we have for the specific quantity $a = A/M$:

$$(\mathrm{d}a)_{p,T} = \sum_{k=1}^{n} a_k\,\mathrm{d}c_k\,, \tag{13}$$

or, since $\sum_{k=1}^{n} c_k = 1$,

$$(\mathrm{d}a)_{p,T} = \sum_{k=1}^{n-1} (a_k - a_n)\,\mathrm{d}c_k\,. \tag{14}$$

The partial specific quantities a_k are related to the specific quantity a by means of

$$a_k - a_n = \left(\frac{\partial a}{\partial c_k}\right)_{p,T}, \quad (k = 1, 2, \ldots, n-1)\,. \tag{15}$$

Equalities of this type are used in Chapters V and XI.

Note also that with $A = V$ equation (13) becomes

$$(\mathrm{d}v)_{p,T} = \sum_{k=1}^{n} v_k\,\mathrm{d}c_k\,. \tag{16}$$

Since $c_k = \rho_k/\rho = \rho_k v$, we also have

$$(\mathrm{d}v)_{p,T} = \mathrm{d}v + v \sum_{k=1}^{n} v_k \, \mathrm{d}\rho_k \,, \tag{17}$$

where we have used the relation $\sum_{k=1}^{n} \rho_k v_k = 1$, which follows from (11) with $A = V$. From equation (17) it follows that

$$\sum_{k=1}^{n} v_k \, (\mathrm{d}\rho_k)_{p,T} = 0 \,, \tag{18}$$

a relation which has been used in Chapter XI.

2. *Thermodynamic Stability with respect to Diffusion**

Consider two n-component systems I and II in contact with each other, which may exchange matter, but are otherwise isolated from their surroundings. We assume that the pressure p and the temperature T of the two systems are kept equal and constant. As a consequence of the second law of thermodynamics the Gibbs function G of the two systems together must have a minimum value at equilibrium. Let us choose as independent variables for the sub-systems, besides p and T, their total masses M^{I} and M^{II}, and their compositions $c_1^{\mathrm{I}}, c_2^{\mathrm{I}}, \ldots, c_{n-1}^{\mathrm{I}}$ and $c_1^{\mathrm{II}}, c_2^{\mathrm{II}}, \ldots, c_{n-1}^{\mathrm{II}}$. The Gibbs functions of the sub-systems are of the form

$$G^{\mathrm{I}} = M^{\mathrm{I}} g^{\mathrm{I}} (p, T, c_1^{\mathrm{I}}, c_2^{\mathrm{I}}, \ldots, c_{n-1}^{\mathrm{I}}) \,, \tag{19}$$

$$G^{\mathrm{II}} = M^{\mathrm{II}} g^{\mathrm{II}} (p, T, c_1^{\mathrm{II}}, c_2^{\mathrm{II}}, \ldots, c_{n-1}^{\mathrm{II}}) \,. \tag{20}$$

Here g^{I} and g^{II} are the specific Gibbs functions of I and II. The deviation of the Gibbs function $G = G^{\mathrm{I}} + G^{\mathrm{II}}$ from its equilibrium value at constant p and T may be expanded into a Taylor series

$$\begin{aligned} \Delta G = \Delta G^{\mathrm{I}} + \Delta G^{\mathrm{II}} &= g^{\mathrm{I}} \Delta M^{\mathrm{I}} + g^{\mathrm{II}} \Delta M^{\mathrm{II}} \\ &+ \sum_{k=1}^{n-1} M^{\mathrm{I}} \left(\frac{\partial g^{\mathrm{I}}}{\partial c_k^{\mathrm{I}}} \right)_{p,T} \Delta c_k^{\mathrm{I}} + \sum_{k=1}^{n-1} M^{\mathrm{II}} \left(\frac{\partial g^{\mathrm{II}}}{\partial c_k^{\mathrm{II}}} \right)_{p,T} \Delta c_k^{\mathrm{II}} \\ &+ \sum_{k=1}^{n-1} \left(\frac{\partial g^{\mathrm{I}}}{\partial c_k^{\mathrm{I}}} \right)_{p,T} \Delta M^{\mathrm{I}} \Delta c_k^{\mathrm{I}} + \sum_{k=1}^{n-1} \left(\frac{\partial g^{\mathrm{II}}}{\partial c_k^{\mathrm{II}}} \right)_{p,T} \Delta M^{\mathrm{II}} \Delta c_k^{\mathrm{II}} \end{aligned}$$

* I. Prigogine and R. Defay, Chemical thermodynamics, Ch. XV, London (1954).

$$+\frac{1}{2}\sum_{k,l=1}^{n-1} M^{\mathrm{I}}\left(\frac{\partial^2 g^{\mathrm{I}}}{\partial c_k^{\mathrm{I}}\partial c_l^{\mathrm{I}}}\right)_{p,T} \Delta c_k^{\mathrm{I}}\Delta c_l^{\mathrm{I}}$$

$$+\frac{1}{2}\sum_{k,l=1}^{n-1} M^{\mathrm{II}}\left(\frac{\partial^2 g^{\mathrm{II}}}{\partial c_k^{\mathrm{II}}\partial c_l^{\mathrm{II}}}\right)_{p,T} \Delta c_k^{\mathrm{II}}\Delta c_l^{\mathrm{II}} + \dots, \qquad (21)$$

where Δ denotes the deviation from an equilibrium value. Because of conservation of mass, one has

$$\Delta M_k^{\mathrm{I}} + \Delta M_k^{\mathrm{II}} = 0\,, \quad (k = 1, 2, \dots, n)\,, \qquad (22)$$

or, alternatively,

$$\Delta M^{\mathrm{I}} + \Delta M^{\mathrm{II}} = 0\,, \qquad (23)$$

$$M^{\mathrm{I}}\Delta c_k^{\mathrm{I}} + M^{\mathrm{II}}\Delta c_k^{\mathrm{II}} + (c_k^{\mathrm{I}} - c_k^{\mathrm{II}})\,\Delta M^{\mathrm{I}} = 0\,, \quad (k = 1, 2, \dots, n-1)\,. \qquad (24)$$

Since at equilibrium G has a minimum, terms linear in the independent quantities ΔM^{I} and Δc_k^{I} $(k = 1, 2, \dots, n-1)$ must vanish in (21). With (23) and (24) we then find the following equilibrium conditions

$$\left(\frac{\partial g^{\mathrm{I}}}{\partial c_k^{\mathrm{I}}}\right)_{p,T} = \left(\frac{\partial g^{\mathrm{II}}}{\partial c_k^{\mathrm{II}}}\right)_{p,T}, \quad (k = 1, 2, \dots, n-1)\,, \qquad (25)$$

$$g^{\mathrm{I}} - \sum_{k=1}^{n-1} c_k^{\mathrm{I}}\left(\frac{\partial g^{\mathrm{I}}}{\partial c_k^{\mathrm{I}}}\right)_{p,T} = g^{\mathrm{II}} - \sum_{k=1}^{n-1} c_k^{\mathrm{II}}\left(\frac{\partial g^{\mathrm{II}}}{\partial c_k^{\mathrm{II}}}\right)_{p,T} \qquad (26)$$

With the relations [*cf.* (3) and (15)]

$$g = \sum_{k=1}^{n} c_k\mu_k\,, \qquad (27)$$

$$\left(\frac{\partial g}{\partial c_k}\right)_{p,T} = \mu_k - \mu_n\,, \quad (k = 1, 2, \dots, n-1)\,, \qquad (28)$$

the conditions (25) and (26) may alternatively be written as

$$\mu_k^{\mathrm{I}} = \mu_k^{\mathrm{II}}\,, \quad (k = 1, 2, \dots, n)\,. \qquad (29)$$

If the chemical potentials μ_k^{I} and μ_k^{II} $(k = 1, 2, \dots, n)$ of both subsystems are the same functions of the concentrations c_j^{I} and c_j^{II}

($j = 1, 2, \ldots, n-1$) and of p and T respectively, these conditions imply:

$$c_k^{\mathrm{I}} = c_k^{\mathrm{II}}, \tag{30}$$

$$\Delta G = \left(1 + \frac{M^{\mathrm{I}}}{M^{\mathrm{II}}}\right) \sum_{k=1}^{n-1} \left(\frac{\partial g}{\partial c_k}\right)_{p,T} \Delta M^{\mathrm{I}} \Delta c_k^{\mathrm{I}}$$

$$+ \tfrac{1}{2} M^{\mathrm{I}} \left(1 + \frac{M^{\mathrm{I}}}{M^{\mathrm{II}}}\right) \sum_{k,l=1}^{n-1} \left(\frac{\partial^2 g}{\partial c_k \partial c_l}\right)_{p,T} \Delta c_k^{\mathrm{I}} \Delta c_l^{\mathrm{I}}. \tag{31}$$

Since ΔG is positive or zero, we can conclude that

$$\sum_{k,l=1}^{n-1} \left(\frac{\partial^2 g}{\partial c_k \partial c_l}\right)_{p,T} \Delta c_k^{\mathrm{I}} \Delta c_l^{\mathrm{I}} \geqslant 0, \tag{32}$$

or, using (25),

$$\sum_{k,l=1}^{n-1} \left\{\frac{\partial(\mu_k - \mu_n)}{\partial c_l}\right\}_{p,T} \Delta c_k \Delta c_l \geqslant 0, \tag{33}$$

which is the thermodynamic condition of stability with respect to diffusion, used in Chapter V, § 3 [see formula (V.51)].

APPENDIX III

THE GAUSSIAN DISTRIBUTION FOR MACROSCOPIC VARIABLES

1. We wish to derive in a simple case the Gaussian distribution for a macroscopic state variable α. For this purpose we consider a large adiabatically insulated one-component system of energy $E = A$, volume V and particle number N. The system is supposed to consist of two sub-systems having energies E_1 and $E_2 = E - E_1$, volumes V_1 and $V_2 = V - V_1$, and particle numbers N_1 and $N_2 = N - N_1$ respectively, with $N_1/N_2 = V_1/V_2$. We shall assume that the sub-systems have constant volume and particle number, but may exchange energy. It will furthermore be assumed that both N_1 and N_2 are very large numbers and that $N_1 \ll N_2$. The total system may be described by a micro-canonical ensemble for which the normalized probability density $\rho(\boldsymbol{r}^N, \boldsymbol{p}^N)$ in $6N$-dimensional phase space is a constant ρ_0 between the energy surfaces A and $A + \Delta A$, and zero elsewhere. We ask for the probability $f(A_1)\,\mathrm{d}A_1$ that the energy E_1 of the small sub-system, which is a function of the coordinates and momenta of its N_1 particles, lies between the values A_1 and $A_1 + \mathrm{d}A_1$. This probability is given by

$$f(A_1)\,\mathrm{d}A_1 = \rho_0 \int\limits_{\substack{A_1 < E_1 < A_1 + \mathrm{d}A_1 \\ A < E < A + \Delta A}} \mathrm{d}\boldsymbol{r}^N\,\mathrm{d}\boldsymbol{p}^N\,, \tag{1}$$

with

$$\rho_0^{-1} = \int\limits_{A < E < A + \Delta A} \mathrm{d}\boldsymbol{r}^N\,\mathrm{d}\boldsymbol{p}^N = \Omega_N(A)\,\Delta A\,. \tag{2}$$

Here $\Omega_N(A)\,\Delta A$ is the volume of the energy shell $(A, A + \Delta A)$ in $6N$-dimensional phase space. We may rewrite the integral (1) in the form

$$f(A_1)\,\mathrm{d}A_1 = \frac{1}{\Omega_N(A)\,\Delta A} \int\limits_{A_1 < E_1 < A_1 + \mathrm{d}A_1} \mathrm{d}\boldsymbol{r}^{N_1}\,\mathrm{d}\boldsymbol{p}^{N_1} \int\limits_{A - E_1 < E_2 < A - E_1 + \Delta A} \mathrm{d}\boldsymbol{r}^{N_2}\,\mathrm{d}\boldsymbol{p}^{N_2}\,. \tag{3}$$

Since the integral

$$\int_{A-E_1<E_2<A-E_1+\Delta A} \mathrm{d}\boldsymbol{r}^{N_2}\,\mathrm{d}\boldsymbol{p}^{N_2} = \Omega_{N_2}(A - E_1)\,\Delta A \tag{4}$$

represents the volume of the energy shell $(A - E_1, A - E_1 + \Delta A)$ in the $6N_2$-dimensional phase space of the large sub-system, we also have

$$\begin{aligned} f(A_1)\,\mathrm{d}A_1 &= \int_{A_1<E_1<A_1+\mathrm{d}A_1} \mathrm{d}\boldsymbol{r}^{N_1}\,\mathrm{d}\boldsymbol{p}^{N_1}\,\frac{\Omega_{N_2}(A - E_1)}{\Omega_N(A)} \\ &= \frac{\Omega_{N_2}(A - A_1)}{\Omega_N(A)} \int_{A_1<E_1<A_1+\mathrm{d}A_1} \mathrm{d}\boldsymbol{r}^{N_1}\,\mathrm{d}\boldsymbol{p}^{N_1} \\ &= \frac{\Omega_{N_2}(A - A_1)\,\Omega_{N_1}(A_1)}{\Omega_N(A)}\,\mathrm{d}A_1\,, \end{aligned} \tag{5}$$

where

$$\Omega_{N_1}(A_1)\,\mathrm{d}A_1 = \int_{A_1<E_1<A_1+\mathrm{d}A_1} \mathrm{d}\boldsymbol{r}^{N_1}\,\mathrm{d}\boldsymbol{p}^{N_1} \tag{6}$$

is the volume of the energy shell $(A_1, A_1 + \mathrm{d}A_1)$ in the $6N_1$-dimensional phase space of the small sub-system.

We shall now assume that the system is ideal, *i.e.* that

$$E = \sum_{i=1}^{N_1} \frac{\boldsymbol{p}_i^2}{2m} + \sum_{i=N_1+1}^{N} \frac{\boldsymbol{p}_i^2}{2m} = E_1 + E_2\,, \tag{7}$$

where m is the mass of a particle. The volume $\Omega_N(A)\Delta A$ can then easily be evaluated. The integration over coordinates gives $V_1^{N_1}(V - V_1)^{N-N_1}$, since the first N_1 particles are restricted to the volume V_1, whereas the remaining N_2 particles are enclosed in the volume V_2. The integration over momenta yields the volume of a shell of thickness ΔA of a $3N$-dimensional hypersphere of radius $\sqrt{2mA}$. In this way one obtains

$$\begin{aligned} \Omega_N(A) &= V_1^{N_1}(V - V_1)^{N-N_1}(2\pi m)^{\frac{3}{2}N}\,\frac{A^{\frac{3}{2}N-1}}{(\frac{3}{2}N - 1)!} \\ &= \frac{1}{\sqrt{2\pi}}\,V_1^{N_1}(V - V_1)^{N-N_1}(2\pi m)^{\frac{3}{2}N}A^{\frac{3}{2}N-1}\,\frac{\mathrm{e}^{\frac{3}{2}N-1}}{(\frac{3}{2}N - 1)^{\frac{3}{2}N-\frac{1}{2}}}\,. \end{aligned} \tag{8}$$

Here Stirling's approximation $N! \simeq N^N e^{-N}\sqrt{2\pi N}$ has been used. Similarly one has

$$\Omega_{N_2}(A - A_1) = \frac{1}{\sqrt{2\pi}}(V - V_1)^{N-N_1}(2\pi m)^{\frac{3}{2}(N-N_1)}(A - A_1)^{\frac{3}{2}(N-N_1)-1} \frac{e^{\frac{3}{2}(N-N_1)-1}}{\{\frac{3}{2}(N - N_1) - 1\}^{\frac{3}{2}(N-N_1)-\frac{1}{2}}}, \tag{9}$$

$$\Omega_{N_1}(A_1) = \frac{1}{\sqrt{2\pi}} V_1^{N_1}(2\pi m)^{\frac{3}{2}N_1} A_1^{\frac{3}{2}N_1-1} \frac{e^{\frac{3}{2}N_1-1}}{(\frac{3}{2}N_1 - 1)^{\frac{3}{2}N_1-\frac{1}{2}}}. \tag{10}$$

Consider first the ratio

$$\frac{\Omega_{N_2}(A - A_1)}{\Omega_N(A)} = \frac{\{(\frac{3}{2}N - 1)/A\}^{\frac{3}{2}N_1}(1 - A_1/A)^{\frac{3}{2}(N-N_1)-1} e^{-\frac{3}{2}N_1}}{V_1^{N_1}(2\pi m)^{\frac{3}{2}N_1}\{1 - \frac{3}{2}N_1/(\frac{3}{2}N - 1)\}^{\frac{3}{2}(N-N_1)-\frac{1}{2}}}. \tag{11}$$

Introducing the parameter

$$\beta = \frac{3}{2}\frac{N}{A}, \tag{12}$$

formula (11) becomes

$$\frac{\Omega_{N_2}(A - A_1)}{\Omega_N(A)} = \frac{1}{V_1^{N_1}}\left\{\frac{\beta(1 - 2/3N)}{2\pi m}\right\}^{\frac{3}{2}N_1} \frac{(1 - \beta A_1/\frac{3}{2}N)^{\frac{3}{2}(N-N_1)-1} e^{-\frac{3}{2}N_1}}{\{1 - \frac{3}{2}N_1/(\frac{3}{2}N - 1)\}^{\frac{3}{2}(N-N_1)-\frac{1}{2}}}. \tag{13}$$

Let us take the limit of this ratio for $N \to \infty$, $A \to \infty$, $N/A = \frac{2}{3}\beta$ finite and N_1 finite. This means that the large sub-system becomes a "heat bath" for the small sub-system. We then obtain

$$\lim_{N\to\infty} \frac{\Omega_{N_2}(A - A_1)}{\Omega_N(A)} = \frac{1}{V_1^{N_1}}\left(\frac{\beta}{2\pi m}\right)^{\frac{3}{2}N_1} e^{-\beta A_1}, \tag{14}$$

where use has been made of the identity

$$\lim_{N\to\infty}\left(1 - \frac{x}{N}\right)^N = e^{-x}. \tag{15}$$

According to (5) the ratio (11), as a function of E_1 instead of A_1, represents the probability density or distribution function in the $6N_1$-dimensional phase space of the small sub-system. The result (14) shows that this distribution is canonical for a system in contact with a heat bath.

Note that according to statistical thermodynamics β is equal to $(kT)^{-1}$, where T is the temperature of the heat bath (or the total adiabatically isolated system).

From (5), (10) and (14) we find for the probability density $f(A_1)$

$$f(A_1) = \frac{1}{\sqrt{2\pi}} \beta^{\frac{3}{2}N_1} A_1^{\frac{3}{2}N_1 - 1} e^{-\beta A_1} \frac{e^{\frac{3}{2}N_1 - 1}}{(\frac{3}{2}N_1 - 1)^{\frac{3}{2}N_1 - \frac{1}{2}}} . \tag{16}$$

The average value of A_1 is given by

$$\langle A_1 \rangle = \int A_1 f(A_1) \, dA_1 = \frac{3}{2} \frac{N_1}{\beta} , \tag{17}$$

the most probable value of A_1 is given by

$$A_1^{\max} = \frac{\frac{3}{2}N_1 - 1}{\beta} . \tag{18}$$

Since $N_1 \gg 1$ we have

$$\langle A_1 \rangle \simeq A_1^{\max} . \tag{19}$$

Furthermore

$$\langle (A_1 - \langle A_1 \rangle)^2 \rangle = \int (A_1 - \langle A_1 \rangle)^2 f(A_1) \, dA_1 = \frac{3}{2} \frac{N_1}{\beta^2} . \tag{20}$$

It follows from (18)–(20) that the distribution $f(A_1)$ is sharply peaked and thus that this probability density is only important for values of $A_1 - \langle A_1 \rangle$ of the order $\sqrt{N_1}/\beta$ or smaller.

Let us now take the logarithm of both sides of (16)

$$\ln f(A_1) = (\tfrac{3}{2}N_1 - 1) \ln A_1 - \beta A_1 + \tfrac{3}{2}N_1 \ln \beta + \tfrac{3}{2}N_1 - 1$$
$$- (\tfrac{3}{2}N_1 - \tfrac{1}{2}) \ln (\tfrac{3}{2}N_1 - 1) - \tfrac{1}{2} \ln 2\pi . \tag{21}$$

Introducing the deviation $\alpha_1 = A_1 - A_1^{\max} \simeq A_1 - \langle A_1 \rangle$ of A_1, we have

$$\ln f(\alpha_1) = (\tfrac{3}{2}N_1 - 1)\ln\left(1 + \frac{\beta\alpha_1}{\tfrac{3}{2}N_1 - 1}\right) - \beta\alpha_1 + \ln\beta$$
$$- \tfrac{1}{2}\ln(\tfrac{3}{2}N_1 - 1) - \tfrac{1}{2}\ln 2\pi\,. \qquad (22)$$

If we only consider such values of α_1, which are of the order $\sqrt{N_1}/\beta$ or smaller, we can expand the logarithmic term in (22) as a power series in α_1:

$$\ln f(\alpha_1) = -\frac{\beta^2\alpha_1^2}{3N_1} + \ln\frac{\beta}{(3\pi N_1)^{\frac{1}{2}}} + O\left(\frac{1}{N_1^{\frac{1}{2}}}\right). \qquad (23)$$

Thus

$$f(\alpha_1) = \left(\frac{\beta^2}{3\pi N_1}\right)^{\frac{1}{2}} e^{-\beta^2\alpha_1^2/3N_1} + O\left(\frac{1}{N_1}\right). \qquad (24)$$

If we neglect terms of the order N_1^{-1} we have obtained the Gaussian distribution for the energy fluctuations in the small sub-system*.

2. In Chapter VII, § 2, we have used Boltzmann's entropy postulate in order to connect the entropy of a state with its probability. According to this postulate the entropy of the *total* system, considered above, in a state in which the energy of the small sub-system has the value A_1, is proportional to the logarithm of the volume occupied by this state in $6N$-dimensional phase space. Therefore

$$S^B(A_1\,;A) = k\ln\Omega_{N_1}(A_1)\Omega_{N_2}(A - A_1)\,dA_1\Delta A$$
$$= k\ln\Omega_{N_1}(A_1)\,dA_1 + k\ln\Omega_{N_2}(A - A_1)\Delta A\,, \qquad (25)$$

as can be seen from (1), (2) and (5). Therefore, using again (5),

$$\Delta S^B(\alpha_1) \equiv S^B(\alpha_1\,;A) - S^B(0\,;A) = k\ln\left\{\frac{f(\alpha_1)}{f(0)}\right\}, \qquad (26)$$

where the independent variable α_1 has been introduced. This expression is equivalent to formula (47) of Chapter VII.

* In case one does not consider a perfect gas one still obtains the canonical distribution (14) (with a different normalization constant) and the Gaussian distribution (24) (with a different variance and normalization constant) by applying the central limit theorem of probability theory (*cf.* A. I. Khinchin, Mathematical foundations of statistical mechanics, Ch. V, New York (1949); J. van der Linden and P. Mazur, Physica **27** (1961) 609).

In the Gaussian approximation (24) for $f(\alpha_1)$ expression (26) becomes

$$\Delta S^{\mathrm{B}}(\alpha_1) = \frac{-k\beta^2\alpha_1^2}{3N_1}. \tag{27}$$

We have furthermore defined in Chapter VII, § 2, intensive state variables conjugate to extensive state variables α by the relation

$$X_1 = \frac{\partial \Delta S^{\mathrm{B}}(\alpha_1)}{\partial \alpha_1}. \tag{28}$$

For the present case we have from (27)

$$X_1 = \frac{-2k\beta^2\alpha_1}{3N_1}. \tag{29}$$

Conceptually one should define the values of the entropy and the intensive thermodynamic variables of the sub-systems by determining these quantities when one suddenly isolates these sub-systems for given values of the extensive parameters α and permits equilibrium to be reached. In the case of our total system of energy $E = A$, we therefore isolate the sub-systems in a state in which the energy E_1 of the small sub-system has the value A_1. The sub-systems may then be described by two micro-canonical ensembles in $6N_1$- and $6N_2$-dimensional phase spaces, corresponding to energy ranges $(A_1, A_1 + \Delta A_1)$ and $(A - A_1, A - A_1 + \Delta A)$ respectively. According to the Gibbs entropy postulate* and in agreement with thermodynamics we have for the entropies of the sub-systems, when they are isolated in a state A_1,

$$S_1(A_1) = k \ln \Omega_{N_1}(A_1)\Delta A_1\,, \tag{30}$$

* The Gibbs entropy of a system is defined as:

$$S = -k \int \rho \ln \rho \, \mathrm{d}\tau\,,$$

where ρ is the probability density in phase space and where the integration is extended over the whole of phase space. For a micro-canonical ensemble corresponding to an adiabatically insulated system, this definition leads to formulae (30) and (31).

$$S_2(A - A_1) = k \ln \Omega_{N_2}(A - A_1)\Delta A\,. \tag{31}$$

Comparing with (25), we see that

$$S^{\mathrm{B}}(A_1\,;A) = S_1(A_1) + S_2(A - A_1) + k \ln \frac{\mathrm{d}A_1}{\Delta A_1}\,. \tag{32}$$

This proves that the Boltzmann entropy of a (non-equilibrium) state, in which the energy of the small sub-system is A_1, is apart from an additive constant equal to the sum of the "local" equilibrium values of the entropy of both sub-systems.

We also note that according to (5)

$$\begin{aligned} k \ln f(A_1)\,\mathrm{d}A_1 &= S_1(A_1) + S_2(A - A_1) - S(A) + k \ln \frac{\mathrm{d}A_1}{\Delta A_1} \\ &= S^{\mathrm{B}}(A_1\,;A) - S(A)\,, \end{aligned} \tag{33}$$

where

$$S(A) = k \ln \Omega(A)\Delta A\,, \tag{34}$$

is the (Gibbs) entropy of the total system.

For the most probable state we have

$$k \ln f(A_1^{\max})\,\mathrm{d}A_1 = S^{\mathrm{B}}(A_1^{\max}\,;A) - S(A)\,, \tag{35}$$

or, with the variable α_1 instead of A_1,

$$S^{\mathrm{B}}(\alpha_1 = 0\,;A) - S(A) = -k \ln \frac{\sqrt{3\pi N_1/\beta^2}}{\mathrm{d}A_1}\,, \tag{36}$$

where (24) has been used. The quotient in (36) is roughly equal to the ratio of the width of the Gaussian distribution (24) and the width $\mathrm{d}A_1$, which is the smallest energy interval considered. Assume $\mathrm{d}A_1$, which must be much smaller than the width of the Gaussian, to be of the order of β^{-1}. Then the difference $S(A) - S^{\mathrm{B}}(\alpha_1 = 0; A)$ is of the order of $\frac{1}{2}k \ln N_1$. This quantity is negligible compared to the magnitude of $S(A)$, which is of the order kN [*cf.* formula (34) with (8)]. This illustrates the well-known fact that for systems with many degrees of freedom the Gibbs entropy and the Boltzmann entropy of the most probable state are almost equal.

Let us now turn to the definition of the intensive thermodynamic variables. According to thermodynamics the temperatures of the isolated sub-systems are given by

$$\frac{1}{T_1(A_1)} = \frac{\partial S_1(A_1)}{\partial A_1}, \tag{37}$$

$$\frac{1}{T_2(A - A_1)} = -\frac{\partial S_2(A - A_1)}{\partial A_1}. \tag{38}$$

Comparing with (28) and (32) we have

$$X_1 = \frac{\partial \Delta S^{\mathrm{B}}(\alpha_1)}{\partial \alpha_1} = \frac{\partial S_1(A_1)}{\partial A_1} + \frac{\partial S_2(A - A_1)}{\partial A_1} = \frac{1}{T_1(A_1)} - \frac{1}{T_2(A - A_1)}. \tag{39}$$

This shows that the intensive variable X_1, conjugate to the energy fluctuations α_1 according to (28), is the difference between the reciprocal temperatures of the two sub-systems, when insulated at fixed α_1. If we assume that the second sub-system is infinitely large compared to the first one (in other words that it becomes a heat bath) it may easily be verified, using the explicit expression for S_2 that the temperature T_2 becomes equal to the equilibrium temperature $T = (k\beta)^{-1}$ of the total system. Then X_1 is the fluctuation of the reciprocal temperature of the small sub-system, temperatures being always defined in insulated systems.

PROBLEMS

CHAPTER VII

1. Prove formula (VII.13). *Hint*: Transform g to diagonal form.
2. Establish the form (VII.63) if a magnetic field is present, and derive then formula (VII.93).
3. Prove formulae (VII.111)–(VII.113).
4. Prove that $| \exp (Mt) | = \exp \{(M\colon U)t\}$. *Hint*: Transform M to diagonal form and note that the determinant value and the trace of a matrix are invariant under a similarity transformation.
5. Derive (VII.232) from (VII.231), using the explicit forms of the distribution functions. *Hint*: Use (VII.231) for $t \to \infty$.
6. Prove that the general Gaussian form

$$P(\boldsymbol{\alpha} \mid \boldsymbol{\alpha}'; t) = c\, \mathrm{e}^{-\frac{1}{2}(A\colon \boldsymbol{\alpha}\boldsymbol{\alpha} + 2\, B\colon \boldsymbol{\alpha}\boldsymbol{\alpha}' + C\colon \boldsymbol{\alpha}'\boldsymbol{\alpha}')} \tag{1}$$

where A and C are symmetric matrices, reduces to

$$P(\boldsymbol{\alpha} \mid \boldsymbol{\alpha}'; t) = c\, \mathrm{e}^{-\frac{1}{2}C\colon (\boldsymbol{\alpha}' - D\cdot\boldsymbol{\alpha})(\boldsymbol{\alpha}' - D\cdot\boldsymbol{\alpha})} \tag{2}$$

with

$$D = -\tilde{B}^{-1}\cdot A \quad \text{and} \quad C = B\cdot A^{-1}\cdot\tilde{B}\,, \tag{3}$$

if the condition (VII.81) is imposed.

7. Show that in formula (2) of the preceding problem

$$C^{-1} = k\,(g^{-1} - D\cdot g^{-1}\cdot\tilde{D})\,, \tag{4}$$

when the condition (VII.82) is used.

8. Show that $D(t)$ in the Gaussian distribution (2) must be of the form

$$D(t) = \mathrm{e}^{-Mt}\,, \tag{5}$$

if the condition (VII.179), expressing the Markoff character, is employed. *Hint*: Calculation of $\bar{\boldsymbol{\alpha}}^{\boldsymbol{\alpha}_0}$, first with the left-hand side of (VII.179), and then with the right-hand side, leads to

$$D(t + t') = D(t')\cdot D(t)\,, \tag{6}$$

of which (5) is the solution.

9. Derive formula (VII.245) from (VII.244) and the conditions (VII.246).

CHAPTER VIII

1. Find the spectral density function (VIII.64) of a process described by a single α-variable, obeying the linear regression law (VII.94), by establishing the differential equation for the correlation function $\rho(\tau)$ and by performing a Fourier transformation of this equation.

Hints: Prove first that the differential equation for the correlation function is

$$\frac{\partial \rho(\tau)}{\partial \tau} = -M\rho(\tau)u(\tau), \quad (\text{all } \tau),$$

where $u(\tau) = -1$ for $\tau < 0$ and 1 for $\tau > 0$. One obtains a second order differential equation by differentiating this equation with respect to τ, in which the delta function $\delta(\tau)$ appears, because $\partial u/\partial \tau = 2\delta(\tau)$. If one eliminates $\partial \rho/\partial \tau$ from this equation with the help of the original differential equation given above, one obtains an equation, which, if submitted to a Fourier transformation, yields an algebraic equation for the spectral density function.

2. Find the spectral density (VIII.80) for a process described by a single α-variable and a single β-variable $\beta = \dot{\alpha}$, obeying the regression laws (VIII.65) and (VIII.66), by establishing the differential equation for the correlation function $\rho_{\alpha\alpha}(\tau)$ and by performing a Fourier transformation of this equation.

Hints: Prove first that $\rho_{\alpha\alpha}(\tau)$ obeys

$$\left(\frac{\partial^2}{\partial \tau^2} + \frac{g}{h}\right) \rho_{\alpha\alpha}(\tau) = -M \frac{\partial \rho_{\alpha\alpha}}{\partial \tau} u(\tau), \quad (\text{all } \tau),$$

where $u(\tau) = -1$ for $\tau < 0$ and 1 for $\tau > 0$. One obtains a fourth order differential equation for $\rho_{\alpha\alpha}$ if one lets the operator $(\partial^2/\partial^2\tau + gh^{-1})^2$ work on this second order differential equation. Using again the original second order equation, written above, at the right-hand side of the fourth order equation, one arrives at an equation of the form

$$\left(\frac{\partial^2}{\partial\tau^2}+\frac{g}{h}\right)^2\rho_{\alpha\alpha}(\tau)$$

$$= M^2\frac{\partial^2\rho_{\alpha\alpha}}{\partial\tau^2}+2M^2\frac{\partial\rho_{\alpha\alpha}}{\partial\tau}u(\tau)\delta(\tau)-2M\frac{\partial^2\rho_{\alpha\alpha}}{\partial\tau^2}\delta(\tau)-2M\frac{\partial}{\partial\tau}\left\{\frac{\partial\rho_{\alpha\alpha}}{\partial\tau}\delta(\tau)\right\}.$$

If one keeps in mind that $(\partial^2\rho_{\alpha\alpha}/\partial\tau^2)_{\tau=0}=\langle\beta^2\rangle=-kh^{-1}$ and $(\partial\rho_{\alpha\alpha}/\partial\tau)_{\tau=0}=\langle\alpha\beta\rangle=0$, one obtains a differential equation, which upon Fourier transformation becomes a simple algebraic equation for the spectral density.

3. Prove that for a system, described by a single α-variable, the following connections exist between the real and imaginary parts of the generalized susceptibility $\hat{\kappa}(\omega)$ and the correlation function $\rho(t)$:

$$kT\hat{\kappa}'(\omega)=-\int_0^\infty \cos\omega t\,\frac{\partial\rho}{\partial t}\,\mathrm{d}t\,,$$

$$kT\hat{\kappa}''(\omega)=-\int_0^\infty \sin\omega t\,\frac{\partial\rho}{\partial t}\,\mathrm{d}t\,.$$

Hints: The correlation function (VIII.140) can be written as

$$\rho(t)=\frac{kT}{\pi}\mathscr{P}\int_{-\infty}^{\infty}\mathrm{e}^{-\mathrm{i}\omega t}\,\frac{\hat{\kappa}''(\omega)}{\omega}\,\mathrm{d}\omega$$

for systems described by a single variable. Now take the time derivative of this equation and perform a Fourier inversion. This leads directly to the second relation to be proved, if one notes that $\partial\rho/\partial t$ is an odd function of time. The first relation then follows from the second with the help of the Kramers–Kronig relation (VIII.102) and the lemma

$$\cos ut=\frac{1}{\pi}\mathscr{P}\int_{-\infty}^{\infty}\frac{\sin\omega t}{\omega-u}\,\mathrm{d}\omega,\quad (t>0)\,.$$

4. Find the relations, generalizing (VIII.145)–(VIII.149), which express microscopic reversibility for systems, described by both **α**- and **β**-type variables, in a magnetic field.

CHAPTER IX

1. The Fokker–Planck equation (IX.117) for the Brownian motion is satisfied by the following distribution function, which is Gaussian both in the coordinates $\boldsymbol{r}$ and the velocities $\boldsymbol{u}$:

$$f(\boldsymbol{r}, \boldsymbol{u}\,; t) = \frac{N}{A^{\frac{3}{2}}} \left(\frac{\beta m}{2\pi k T^{\mathrm{eq}}} \right)^3 \exp\left(- \frac{m}{k} \frac{F\boldsymbol{S}^2 - 2H\boldsymbol{S}\cdot\boldsymbol{R} + G\boldsymbol{R}^2}{2AT^{\mathrm{eq}}} \right).$$

[This formula can be derived from a result obtained by S. Chandrasekhar, Rev. mod. Phys. **15** (1943) 1.] In this expression we have used the abbreviations

$$\boldsymbol{S} \equiv \boldsymbol{u} - \boldsymbol{u}_0\, \mathrm{e}^{-\tau}, \quad (\tau = \beta t),$$

$$\boldsymbol{R} \equiv \beta(\boldsymbol{r} - \boldsymbol{r}_0) - \boldsymbol{u}_0\, (1 - \mathrm{e}^{-\tau}),$$

where $\boldsymbol{u}_0$ and $\boldsymbol{r}_0$ are the mean initial velocities and positions. The dimensionless functions F, G, H and $A = FG - H^2$ depend on τ only, (*cf.* problems 3 and 4 on p. 474)

Prove with this solution f that (IX.118), (IX.121) and (IX.126) become

$$\rho(\boldsymbol{r}\,; t) = N\beta^3 \left(\frac{m}{2\pi F k T^{\mathrm{eq}}} \right)^{\frac{3}{2}} \exp\left(- \frac{m\boldsymbol{R}^2}{2FkT^{\mathrm{eq}}} \right),$$

$$\boldsymbol{v}(\boldsymbol{r}\,; t) = \frac{H}{F} \boldsymbol{R} + \boldsymbol{u}_0\, \mathrm{e}^{-\tau},$$

$$T(t) = \frac{A}{F} T^{\mathrm{eq}}.$$

2. Prove with the results of the preceding problem and with (IX.141) and (IX.144), that

$$\frac{(T - T^{\mathrm{eq}})^2}{TT^{\mathrm{eq}}} = \frac{(F - A)^2}{FA},$$

$$\operatorname{grad} \mu = - \beta \frac{A}{F^2} \boldsymbol{R},$$

$$\frac{\mathrm{d}\boldsymbol{v}}{\mathrm{d}t} = - \beta \left(\frac{FH - A}{F^2} \boldsymbol{R} + \boldsymbol{u}_0\, \mathrm{e}^{-\tau} \right).$$

3. Prove that the fact, that the distribution function $f(\boldsymbol{r}, \boldsymbol{u}; t)$ of problem 1 satisfies the Fokker–Planck equation, implies:

$$\frac{\mathrm{d}F}{\mathrm{d}\tau} = 2H\,, \quad \frac{\mathrm{d}G}{\mathrm{d}\tau} = 2(1 - G)\,, \quad \frac{\mathrm{d}H}{\mathrm{d}\tau} = G - H\,.$$

4. Prove that if these relations are integrated with the help of the initial conditions

$$F(0) = a \geqslant 0\,, \quad G(0) = b \geqslant 0\,, \quad H(0) = 0\,,$$

one obtains:

$$F(\tau) = a + b\,(1 - \mathrm{e}^{-\tau})^2 - 3 + 2\tau + 4\,\mathrm{e}^{-\tau} - \mathrm{e}^{-2\tau} \geqslant 0\,,$$

$$G(\tau) = b\,\mathrm{e}^{-2\tau} + 1 - \mathrm{e}^{-2\tau} \geqslant 0\,,$$

$$H(\tau) = b\,\mathrm{e}^{-\tau}\,(1 - \mathrm{e}^{-\tau}) + (1 - \mathrm{e}^{-\tau})^2 \geqslant 0\,.$$

5. Prove that the initial conditions of the preceding problem imply for the results of problem 1 at the initial time $t = 0$:

$$\rho(\boldsymbol{r}\,;0) = N\beta^3 \left(\frac{m}{2\pi akT^{\mathrm{eq}}}\right)^{\frac{3}{2}} \exp\left\{-\frac{\beta^2 m(\boldsymbol{r} - \boldsymbol{r}_0)^2}{2akT^{\mathrm{eq}}}\right\},$$

$$\boldsymbol{v}(\boldsymbol{r}\,;0) = \boldsymbol{u}_0\,,$$

$$T(0) = bT^{\mathrm{eq}}\,.$$

Note that according to the first of these three results the quantity a is related to the half-width $\Delta\boldsymbol{r}$ of $\rho(\boldsymbol{r}; 0)$ in the following way

$$a = \frac{\beta^2 m(\Delta\boldsymbol{r})^2}{2kT^{\mathrm{eq}}}\,.$$

Note also that the case $b = 0$ corresponds to a delta-function type initial distribution in the velocities $\delta(\boldsymbol{u} - \boldsymbol{u}_0)$, and that the case $b = 1$ corresponds to $T = T^{\mathrm{eq}}$ at $t = 0$.

6. Prove that the functions of problem 4 have the following forms for the limiting cases of large and small times:

$\tau \gg 1$:

$$F(\tau) \simeq a + b - 3 + 2\tau ,$$
$$G(\tau) \simeq 1 ,$$
$$H(\tau) \simeq 1 ,$$

$\tau \ll 1$:

$$F(\tau) \simeq a ,$$
$$G(\tau) \simeq b - 2b\tau + 2\tau ,$$
$$H(\tau) \simeq b\tau .$$

7. Show that from formula (IX.146) for the entropy source strength one has for the proportion of the two contributions σ_1 and σ_2:

$$\frac{\sigma_1}{\sigma_2} = \frac{m\boldsymbol{v}^2 T}{3k(T - T^{\text{eq}})^2} .$$

Show using the results of problems 1 and 2 that

$$\frac{\sigma_1}{\sigma_2} = \frac{m}{3kT^{\text{eq}}} \left(\frac{H^2}{F^2} \mathbf{R}^2 + 2 \frac{H}{F} \mathbf{R} \cdot \boldsymbol{u}_0 \, \mathrm{e}^{-\tau} + \boldsymbol{u}_0^2 \, \mathrm{e}^{-2\tau} \right) \frac{FA}{(F - A)^2} .$$

Prove that for $\tau \gg 1$ one has, with the results of the preceding problem,

$$\frac{\sigma_1}{\sigma_2} = \frac{m\beta^2 (\boldsymbol{r} - \boldsymbol{r}_0 - \beta^{-1} \boldsymbol{u}_0)^2}{3kT^{\text{eq}}} = \frac{1}{3} \frac{(\boldsymbol{r} - \boldsymbol{r}_0 - \beta^{-1} \boldsymbol{u}_0)^2}{D\beta^{-1}} ,$$

where $D = kT^{\text{eq}}/m\beta$.

Let us now suppose that the friction constant is 10^9 sec^{-1}, which is roughly the value for Brownian particles of diameter 10^{-5} cm and mass 10^{-15} gramme immersed in water. Check that for room temperature one has then $kT^{\text{eq}}/m \simeq 40 \text{ cm}^2 \text{ sec}^{-2}$. Show that for $|\boldsymbol{r} - \boldsymbol{r}_0 - \beta^{-1} \boldsymbol{u}_0| > 10^{-7}$cm one has $\sigma_1/\sigma_2 > 10^2$, or

$$\sigma_1 \gg \sigma_2 ,$$

a result mentioned in the text of Chapter IX, § 8, which holds therefore practically everywhere in the fluid, except in a region of negligible dimensions (smaller than the size of one Brownian particle).

8. Prove that the general formula of problem 7, together with the result of problem 6 for $\tau \ll 1$ leads to

$$\frac{\sigma_1}{\sigma_2} = \frac{m}{3kT^{\mathrm{eq}}} \frac{1}{(b-1)^2} \left\{ \frac{2b^2\tau}{a} \beta(\boldsymbol{r} - \boldsymbol{r}_0) \cdot \boldsymbol{u}_0 + (b + 2\tau)\, \boldsymbol{u}_0^2 \right\}.$$

Show that for $b = 1 - \varepsilon$, where $\varepsilon \ll 1$, one has

$$\sigma_1 \gg \sigma_2 ,$$

a result mentioned in the text of Chapter IX, § 8.

9. Prove from the expression of the preceding problem for σ_1/σ_2, valid for $\tau \ll 1$, and with the numerical values given in problem 7, that in the case $b = 0$ one obtains

$$\sigma_1 \ll \sigma_2 ,$$

if $|\boldsymbol{u}_0| < 8$ cm sec^{-1}. This is again a result mentioned at the end of section 8 in Chapter IX.

10. Show from the expressions of problem 2 and the results of problem 6 that for $\tau \gg 1$

$$\frac{|\mathrm{d}\boldsymbol{v}/\mathrm{d}t|}{|\operatorname{grad} \mu|} = \frac{1}{a + b - 4 + 2\tau} \ll 1 ,$$

so that the inertia term may be neglected with respect to the grad μ term, as mentioned in Chapter IX, § 8.

CHAPTER X

1. Prove that formula (X.150) remains valid in the presence of an external magnetic field.

CHAPTER XI

1. Prove that the angle between the heat flux and the temperature gradient in the Righi–Leduc effect (Ch. XI, § 1) is given by

$$\operatorname{arctg}\left(-\frac{\lambda_{xy}}{\lambda_{xx}}\right).$$

2. Check that the relation $\mathsf{D}^{ax} = T^{-1}\,\mathsf{L}^a\cdot\mathsf{A}^a\cdot\mu^x$ and the analogous formula with index b instead of a can be transformed into each other with the help of the transformation properties of the matrices D. L and A (Ch. XI, § 2).

3. Check that from the forms (XI.23) and (XI.32), (XI.33) for the entropy source strength, using as weights $a_k = \delta_{kn}$ (*i.e.* making the choice of "relative flows", indicated with the index r for a), it follows that $\mathsf{A}^r = \mathsf{U}$, where U is the unit matrix. Then, with the help of this result, find the expression (XI.34) for A^a from formulae (XI.45) and (XI.49) with the choice r for b.

4. Prove that the "molar" $(n-1)$-dimensional matrix $\bar{\mathsf{B}}^{ab}$ is related to the analogous "mass" matrix B^{ab} as

$$\bar{\mathsf{B}}^{ab} = \mathsf{M}^{-1}\cdot\mathsf{B}^{ab}\cdot\mathsf{M}.$$

Note that this transformation formula does not have the same form as the corresponding transformation formula (XI.41) for $\bar{\mathsf{A}}^a$ and A^a.

5. Prove the relation

$$\mathsf{D}^{ay} = \mathsf{D}^{ax}\cdot\frac{\partial x}{\partial y}$$

from (XI.51), applied for x and for y, and (XI.54).

6. From (XI.67) it follows that

$$\mu^c\cdot(\tilde{\mathsf{A}}^c)^{-1} = (\mathsf{A}^c)^{-1}\cdot\tilde{\mu}^c.$$

Derive from this formula with the help of (XI.47) the relation

$$\mu^c_{ik} - \sum_{j=1}^{n-1}\mu^c_{ij}c_j = \mu^c_{ki} - \sum_{j=1}^{n-1}\mu^c_{kj}c_j\,,\quad (i, k = 1, 2, \ldots, n-1)\,,$$

which is equivalent to (XI.70).

7. Prove that

$$G^{0\rho}_{ik} = -\rho \frac{\partial}{\partial \rho_k}\left(\frac{\partial g}{\partial \rho_i}\right) - \rho \sum_{j=1}^{n-1} (\mu_j - \mu_n) \frac{\partial}{\partial \rho_k}\left(\frac{\partial c_j}{\partial \rho_i}\right),$$

$$(i, k = 1, 2, \ldots, n-1),$$

where the specific Gibbs function g is considered as a function of $p, T, \rho_1, \rho_2, \ldots, \rho_{n-1}$. Note that the matrix $G^{0\rho}$ is symmetric.

Hints: From (XI.64) it follows that

$$\frac{\partial g}{\partial \rho_i} = \sum_{j=1}^{n-1} \frac{\partial g}{\partial c_j} \frac{\partial c_j}{\partial \rho_i} = -\sum_{j=1}^{n-1} \frac{\partial c_j}{\partial \rho_i} (\mu_j - \mu_n),$$

and hence

$$\frac{\partial}{\partial \rho_k}\left(\frac{\partial g}{\partial \rho_i}\right) = -\sum_{j, m=1}^{n-1} \frac{\partial c_j}{\partial \rho_i} \frac{\partial(\mu_j - \mu_n)}{\partial c_m} \frac{\partial c_m}{\partial \rho_k} - \sum_{j=1}^{n-1} (\mu_j - \mu_n) \frac{\partial}{\partial \rho_k}\left(\frac{\partial c_j}{\partial \rho_i}\right).$$

Finally one must transform the first term at the right-hand side with the help of the formulae

$$\frac{\partial(\mu_j - \mu_n)}{\partial c_m} = (\mathsf{A}^c \cdot \mu^c)_{jm}, \quad \frac{\partial c_j}{\partial \rho_i} = \rho^{-1} B^{c0}_{ji},$$

$$\mathsf{A}^0 = \widetilde{\mathsf{B}^{c0}} \cdot \mathsf{A}^c, \quad \mu^\rho = \mu^c \cdot \left(\frac{\partial \mathbf{c}}{\partial \rho}\right), \quad \mathsf{G}^{0\rho} = \mathsf{A}^0 \cdot \mu^\rho,$$

which occur in Chapter XI, § 2 and § 3.

8. Write down explicitly formulae (XI.78)–(XI.85) for the ternary system ($n = 3$; $i, k = 1, 2$).

9. Prove that

$$\frac{\partial \boldsymbol{N}}{\partial \boldsymbol{n}} = N \bar{\mathsf{B}}^{0m}$$

and

$$\frac{\partial \boldsymbol{n}}{\partial \boldsymbol{N}} = \frac{1}{N} \bar{\mathsf{B}}^{m0}.$$

[These are relations between $(n-1)$-dimensional matrices.]

10. Construct the table of Chapter XI, § 4 for $x_i = c_i$ instead of $x_i = \rho_i$.

11. Complete the table of Chapter XI, § 5 for all four examples of basic weights b_k, as was done in the table of Chapter XI, § 4. Compile the complete table also for $x_i = c_i$.

12. Prove that if the velocity field

$$\boldsymbol{v}(\boldsymbol{r}\,;t) = \boldsymbol{a} + \boldsymbol{b} \wedge \boldsymbol{r} + c\boldsymbol{r} + 2\boldsymbol{r}(\boldsymbol{d}\cdot\boldsymbol{r}) - \boldsymbol{d}r^2\,,$$

with $\boldsymbol{a}(t)$, $\boldsymbol{b}(t)$, $c(t)$ and $\boldsymbol{d}(t)$, is inserted into the expression for mechanical equilibrium

$$\frac{\mathrm{d}\boldsymbol{v}}{\mathrm{d}t} \equiv \frac{\partial \boldsymbol{v}}{\partial t} + (\boldsymbol{v}\cdot\text{grad})\,\boldsymbol{v} = 0\,,$$

one finds conditions

$$\boldsymbol{b} = 0\,,\quad \boldsymbol{d} = 0\,,\quad \boldsymbol{a} + c\boldsymbol{a} = 0\,,\quad c + c^2 = 0\,,$$

by using the fact that the mechanical equilibrium condition is fulfilled everywhere in the system, *i.e.* for arbitrary vector $\boldsymbol{r}$ with components x, y, z.

13. Prove the identity

$$\frac{\partial \boldsymbol{c}}{\partial \rho} = \frac{1}{\rho}\mathsf{B}^{co}$$

from (XI.53), (XI.129), both with $a_i = c_i$, $b_i = \rho_i v_i$ and $x_i = \rho_i$, and the results (XI.136) and $\mathsf{D}^{ay} = \mathsf{D}^{ax}\cdot(\partial\boldsymbol{x}/\partial\boldsymbol{y})$ of problem 5, choosing $a_i = \rho_i v_i$, $y_i = \rho_i$ and $x_i = c_i$.

14. Prove for the ideal mixture that

$$(\bar{\mu}^n)^{-1}_{ik} = \frac{n_i}{RT}\delta_{ik}\,,\quad (i, k = 1, 2, \ldots, n-1)\,,$$

$$(\bar{\mu}^N)^{-1}_{ik} = \sum_j \left(\frac{\partial N_i}{\partial n_j}\right)(\bar{\mu}^n)^{-1}_{jk} = \frac{N}{RT}n_k\,\{\delta_{ik} + N_i\,(\bar{v}_n - \bar{v}_k)\}\,,$$

$$(\bar{v}^{0N})^{-1}_{ik} \equiv \{(\mu^N)^{-1}\cdot(\bar{A}^0)^{-1}\}_{ik}$$

$$= \frac{1}{RT}\,[N_k\delta_{ik} + N_iN_k\{\bar{v}_n - \bar{v}_i - \bar{v}_k - \sum_j N_j\bar{v}_j(\bar{v}_n - \bar{v}_j)\}]\,,$$

where the index -1 denotes a reciprocal matrix. Note that the last matrix is symmetric.

15. Prove for the ideal gas mixture that

$$(\bar{v}^{0N})^{-1}_{ik} = \frac{N}{RT}(n_k\delta_{ik} - n_i n_k), \quad (i, k = 1, 2, \ldots, n-1).$$

16. There exist useful alternative forms for the entropy source strength (XI.175) in a rotating system:

$$\sigma = \frac{1}{T}\sum_{k=1}^{n} \boldsymbol{J}_k^a \cdot \{\omega^2 \boldsymbol{r} + 2\boldsymbol{v}^a \wedge \boldsymbol{\omega} - \rho v_k(\omega^2 \boldsymbol{r} + 2\boldsymbol{v} \wedge \boldsymbol{\omega}) - (\text{grad}\,\mu_k)_{T,p}\},$$

$$\sigma = \frac{1}{T}\sum_{k=1}^{n} \boldsymbol{J}_k^0 \cdot \{(\omega^2 \boldsymbol{r} + 2\boldsymbol{v}^0 \wedge \boldsymbol{\omega}) - (\text{grad}\,\mu_k)_{T,p}\}$$

$$= \frac{1}{T}\{\rho(\boldsymbol{v} - \boldsymbol{v}^0)\cdot\boldsymbol{\omega r}^2 + 2\rho\boldsymbol{v}\cdot(\boldsymbol{v}^0 \wedge \boldsymbol{\omega}) - \sum_{k=1}^{n} \boldsymbol{J}_k^0 \cdot (\text{grad}\,\mu_k)_{T,p}\},$$

where $\boldsymbol{v}^a$ is an arbitrary mean velocity $\sum_{k=1}^{n} a_k \boldsymbol{v}_k$, and $\boldsymbol{v}^0$ the mean volume velocity $\sum_{k=1}^{n} \rho_k v_k \boldsymbol{v}_k$. Derive these expressions.

17. Prove that for vanishing volume velocity $\boldsymbol{v}^0$ one has the identity

$$\sum_{k=1}^{n} \rho_k \boldsymbol{v}_k (1 - \rho v_k) = \sum_{k=1}^{n-1} \rho_k \boldsymbol{v}_k \left(1 - \frac{v_k}{v_n}\right).$$

Note that the first member occurs in (XI.175) and the second in (XIII.158).

18. Prove that

$$\sum_{k=1}^{n-1} A_{ik}^0 (1 - \rho v_k) = 1 - \frac{v_i}{v_n},$$

where A^0 is the matrix (XI.34) with $a_i = \rho_i v_i$. Note that the left-hand side occurs in (XI.177) and that forms with the right-hand side are used in Chapter XIII, § 8.

19. The tensors of phenomenological coefficients describing diffusion in rotating systems (Ch. XI, § 6) of arbitrary anisotropy obey Onsager relations of the form [*cf.* (IV.60)]

$$\mathsf{L}_{ik}(\boldsymbol{\omega}) = \tilde{\mathsf{L}}_{ki}(-\boldsymbol{\omega}), \quad (i, k = 1, 2, \ldots, n-1),$$

(where $\boldsymbol{\omega}$ is the angular velocity of rotation) or, for the symmetric (index s) and antisymmetric (index a) parts of these Cartesian tensors:

$$L^{s}_{ik}(\omega) = L^{s}_{ki}(-\omega)\,, \quad (i, k = 1, 2, \ldots, n-1)\,,$$

$$L^{a}_{ik}(\omega) = -L^{a}_{ki}(-\omega)\,, \quad (i, k = 1, 2, \ldots, n-1)\,.$$

Check that for an isotropic system in an ω-field the parities (in ω) of L^{s}_{ik} and L^{a}_{ik} are even and odd respectively. Show that the Onsager relations then become

$$L_{ik}(\omega) = L_{ki}(\omega)\,, \quad (i, k = 1, 2, \ldots, n-1)\,.$$

For the isotropic system all tensors have a form like (XI.13). This reduces the number of coefficients from $9(n-1)^2$ to $3(n-1)^2$. Show that the Onsager relations further diminish the number of independent coefficients to $\frac{3}{2}n(n-1)$.

20. Prove that the heat conductivity κ' for the binary system is positive. Prove the same for the multi-component system.

21. Prove that the heat of transfer Q'^{*}_{i} of Chapter XI, § 7 is equal to the quotient of the complementary minors of L_{iq} and L_{qq} in the n by n determinant of all phenomenological coefficients L_{qq}, L_{kq}, L_{qk} and L_{ik} $(i, k = 1, 2, \ldots, n-1)$.

22. Prove that the absolute quantities of transfer introduced in Chapter XI, § 7 are related as follows

$$Q^{*}_{k,\,\mathrm{abs}} = Q'^{*}_{k,\,\mathrm{abs}} + h_k\,,$$

$$S^{*}_{k,\,\mathrm{abs}} = \frac{Q'^{*}_{k,\,\mathrm{abs}}}{T} + s_k = \frac{Q^{*}_{k,\,\mathrm{abs}} - \mu_k}{T}\,.$$

23. Prove that for a multi-component, reacting system

$$\mathrm{Tr}\,\boldsymbol{\Lambda} = \boldsymbol{a}^{-1} : \boldsymbol{b}\,,$$

where Tr indicates the trace of a matrix, and where $\boldsymbol{\Lambda}$, $\boldsymbol{a}$ and $\boldsymbol{b}$ were defined in Chapter XI, § 8.

24. For a multi-component, single reaction $(r = 1)$ system the result of the preceding problem gives, with the notations of Chapter XI, § 8,

$$\Lambda_1 = a^{-1} : b = l_{11} \sum_{i,k=1}^{n} \nu_i \nu_k a^{-1}_{ik}$$

Show that this formula leads to (XI.353) for a binary system ($n = 2$).

25. Show that the temperature gradient in a binary, single reaction system (Ch. XI, § 8) is given by

$$\frac{\partial T(x)}{\partial x} = -\left(\frac{T}{h}\right)^2 J_q \left\{ (c_1^*)^2 \frac{\cosh (2x/d_1)}{\cosh (d/d_1)} + (c_2^*)^2 \right\} .$$

Discuss this form by plotting it as a function of the space coordinate x for various values of d_1 ranging from 0 (chemical equilibrium) to ∞ (no chemical reaction).

26. Prove by means of the method used in Chapter XI, § 8 that the heat conductivity of a multi-component, non-reacting system is given by

$$\frac{1}{{}_0} = \left(\frac{T}{h}\right)^2 \boldsymbol{c}^* \cdot \boldsymbol{c}^* = \left(\frac{T}{h}\right)^2 \boldsymbol{c} \cdot a^{-1} \cdot \boldsymbol{c} .$$

Check that this result leads to (XI.343) and (XI.359) for a binary system.

27. Show that the characteristic depth d_1 introduced in Chapter XI, § 8 is related as

$$d_1 \equiv \frac{2}{\sqrt{\Lambda_1}} = \frac{2}{\nu_1} \sqrt{\frac{D \kappa_0 T \rho_2}{l_{11} \kappa_\infty \mu_{11}^c}}$$

to the quantities l_{11}, D, κ_0 and κ_∞, which characterize irreversible processes.

28. Perform the transition (*cf.* Ch. XI, § 8) from the transport quantities a_{11}, a_{12}, a_{21} and a_{22} to the transport quantities λ, D', D'' and D without the use of the Onsager relations. Show that one obtains

$$\lambda = \{a_{11} h_1^2 + (a_{12} + a_{21}) h_1 h_2 + a_{22} h_2^2 \} / T^2 ,$$

$$D' = (a_{22} c_1 h_2 + a_{21} c_1 h_1 - a_{12} c_2 h_2 - a_{11} c_1 h_1) / c_1 c_2 \rho T^2 ,$$

$$D'' = (a_{22} c_1 h_2 + a_{12} c_1 h_1 - a_{21} c_2 h_2 - a_{11} c_2 h_1) / c_1 c_2 \rho T^2 ,$$

$$D = \{a_{11} c_2^2 - (a_{12} + a_{21}) c_1 c_2 + a_{22} c_1^2 \} \mu_{11}^c / c_2 \rho T .$$

Check that the Onsager relation $a_{12} = a_{21}$ implies the Onsager relation $D' = D''$, as it should be for a linear transformation of fluxes and thermodynamic forces.

29. Show that for a multi-component, reacting system the heat conduction $\kappa(x)$, defined by

$$J'_q = -\kappa'(x) \frac{\partial T(x)}{\partial x},$$

where J'_q is the reduced heat flow, is given by

$$\kappa'(x) = \kappa(x) \left\{ 1 + \sum_{i=1}^{n} \frac{h_i - h}{h} \left(\sum_{k=1}^{r} Q_{ki}^{-1} c_k^* \frac{\cosh(2x/d_k)}{\cosh(d/d_k)} + \sum_{k=r+1}^{n} Q_{ki}^{-1} c_k^* \right) \right\}$$

$$= \kappa(x) \left\{ -\sum_{i=1}^{n} \frac{h_i - h}{h} \sum_{k=1}^{r} \left(1 - \frac{\cosh(2x/d_k)}{\cosh(d/d_k)} \right) Q_{ki}^{-1} c_k^* \right\}.$$

Hint: Use (III.24), (XI.280), (XI.305), (XI.315), (XI.324), (XI.326) and (XI.328).

30. The heat conduction $\kappa'(x)$ for a binary ($n = 2$), single reaction ($r = 1$) system is

$$\kappa'(x) = \kappa(x) \left\{ 1 - \frac{(a_{12} + a_{22})c_1 - (a_{11} + a_{12})c_2}{a_{11} + 2a_{12} + a_{22}} \frac{\Delta h}{h} \left(1 - \frac{\cosh(2x/d_1)}{\cosh(d/d_1)} \right) \right\},$$

where $\Delta h = h_1 - h_2$. Derive this result from the preceding problem and the expressions for $\mathbf{Q}^{-1}$ and c_1^*, given in Chapter XI, § 8.

31. Prove with the help of the result obtained in problem 30, that in the limiting cases of $d_1 = \infty$ (no chemical reaction) and $d_1 = 0$ (chemical equilibrium) the heat conductivity $\kappa'(x)$ of the binary, single reaction system becomes

$$\kappa'_0 = \kappa_0$$

and (in the bulk)

$$\kappa'_\infty = \kappa_\infty \left\{ 1 - \frac{(a_{12} + a_{22})c_1 - (a_{11} + a_{12})c_2}{a_{11} + 2a_{12} + a_{22}} \frac{\Delta h}{h} \right\}$$

$$= \frac{h}{T^2} \{ h_1(a_{11} + a_{12}) + h_2(a_{12} + a_{22}) \}.$$

Note that both κ'_0 and κ'_∞ are independent of the coordinate x.

32. Prove that the heat conductivity $\kappa'(x)$ of a binary, single reaction system, for arbitrary chemical reaction rate, can be expressed in terms of $\kappa_0, \kappa_\infty, \kappa'_\infty, d_1, d$ and x as

$$\kappa'(x) = \kappa(x)\left[1 - \left(1 - \frac{\kappa'_\infty}{\kappa_\infty}\right)\left\{1 - \frac{\cosh(2x/d_1)}{\cosh(d/d_1)}\right\}\right]$$

$$= \frac{1 - \left(1 - \frac{\kappa'_\infty}{\kappa_\infty}\right)\left\{1 - \frac{\cosh(2x/d_1)}{\cosh(d/d_1)}\right\}}{\frac{1}{\kappa_0} - \left(\frac{1}{\kappa_0} - \frac{1}{\kappa_\infty}\right)\left\{1 - \frac{\cosh(2x/d_1)}{\cosh(d/d_1)}\right\}}.$$

Check that the limiting cases of no chemical reaction and chemical equilibrium follow from this result.

33. Prove that the (reduced) heat conductivity of a binary, single reaction system at chemical equilibrium is given by

$$\kappa'_\infty = \lambda + \rho c_1 c_2 D' \, \Delta h \,,$$

by expressing both sides as functions of a_{11}, $a_{12} = a_{21}$ and a_{22}. See problem 31 and formulae (XI.369) and (XI.370). Write also the general coefficient $\kappa'(x)$ of the preceding problem in terms of λ, D', D and l_{11}.

34. Derive the formula

$$\kappa'_\infty = \frac{h}{T^2}\{h_1(a_{11} + a_{12}) + h_2(a_{12} + a_{22})\},$$

which is valid for a binary, single reaction system at chemical equilibrium, from

$$J'_q = -\kappa'_\infty \operatorname{grad} T \,,$$

(XI.282), (XI.283), the Gibbs–Duhem relation and the condition of chemical equilibrium $\psi_1 = \psi_2$ (*i.e.* vanishing affinity A).

35. The condition (XI.313) of chemical equilibrium for a binary, single reaction system reads

$$\psi_1^* = 0$$

Show with the help of (XI.306) and the properties of Q that this formulation is equivalent with the condition of the vanishing of the affinity $A = T\sum_{i=1}^{2} \nu_i \psi_i = T\,\nu_1(\psi_1 - \psi_2)$.

36. Show that the heat flow $\boldsymbol{J}_q$ for a binary, non-reacting system at time $t = 0$ (the heat conductivity χ of Ch. XI, § 7) is equal to the reduced heat flow $\boldsymbol{J}'_q$ for a binary, single reaction system at chemical equilibrium and in the stationary state (the heat conductivity κ'_∞ of problem 33) as a result of the validity of the Onsager relation $D' = D''$.

37. Prove that formula (XI.390) can be written in terms of κ, a_{11}, a_{12} and a_{22} as

$$\frac{\Delta c_1}{\Delta T} = \frac{(T/h)\kappa - (1/T)\{h_1 a_{11} + (h_1 + h_2)a_{12} + h_2 a_{22}\}}{(\mu^c_{11}/c_2)\{c_1 a_{22} + (c_1 - c_2)a_{12} - c_2 a_{11}\}}$$

by expressing the transport quantities as functions of a_{11}, a_{12} and a_{22}.

38. Prove the relation

$$\frac{\operatorname{grad} c_1}{\operatorname{grad} T} = \frac{c_2\, \Delta h}{\mu^c_{11}\, T},$$

which is valid for a binary, single reaction system at chemical equilibrium, from the condition of chemical equilibrium $\psi_1 = \psi_2$ (vanishing affinity).

[I. Prigogine and R. Buess, Bull. Acad. roy. Belg., Cl. Sc. [5] **38** (1952) 711.]

39. Find κ_∞ and κ'_∞ (heat conductivities at chemical equilibrium) in terms of λ, D' and D [see (XI.376) and problem 33] by using the result of the preceding problem in the phenomenological equations for $\boldsymbol{J}_q$ and $\boldsymbol{J}'_q$. Prove also that

$$\boldsymbol{J}_1 = -\left\{\frac{c_2 \rho D\, \Delta h}{\mu^c_{11}} + \rho c_1 c_2 D'\right\} \operatorname{grad} T$$

is true at chemical equilibrium. Note that in practical cases the second term is usually negligible with respect to the first term, because D'/D is of the order of 10^{-3} to 10^{-5} reciprocal degrees.

CHAPTER XII

1. Show that for incompressible fluids (div $\boldsymbol{v} = 0$) the coupled differential equations (XII.32) and (XII.37) can be written in the form

$$\frac{\mathrm{d}\omega}{\mathrm{d}t} = \frac{4\eta_r}{\rho\Theta}(\boldsymbol{\Omega} - \omega),$$

$$\rho\frac{\mathrm{d}\boldsymbol{\Omega}}{\mathrm{d}t} = -(\eta + \eta_r)\,\mathrm{rot\,rot}\,\boldsymbol{\Omega} + \eta_r\,\mathrm{rot\,rot}\,\boldsymbol{\omega},$$

where the abbreviation $\boldsymbol{\Omega} = \frac{1}{2}\,\mathrm{rot}\,\boldsymbol{v}$ and the equality (XII.39) have been used.

2. Find back the phenomenological equations (XII.25) and (XII.26) for the viscous flow of an isotropic fluid from the scheme (XII.53) by letting the magnetic field disappear.

Hint: Note first that for complete isotropy one finds

$$L_{12} = L_{21} = L_{23}, \quad L_{11} = L_{22}, \quad L_{55} = \tfrac{1}{2}(L_{22} - L_{23}),$$

$$L_{24} = 0, \quad L_{56} = 0$$

as relations between the coefficients in the scheme (XII.48).

3. Prove that the phenomenological coefficients L and l_{vv} introduced in (IV.16) and (IV.17) are given by

$$L = L_{11} - L_{12} \quad (= 2T\eta),$$

$$l_{vv} = \tfrac{1}{3}(L_{11} + 2L_{12}) \; (= T\eta_v),$$

where L_{11} and L_{12} are the phenomenological coefficients of the preceding problem, describing the viscous behaviour of an isotropic system.

4. Derive the following formulae for the relaxation times at constant (v, s), (p, s), (v, T) and (p, T) respectively:

$$\tau_{v,s} = \frac{1}{\beta}\left(\frac{\partial\xi}{\partial A}\right)_{v,s} = \frac{1}{\beta}\left(\frac{\partial\xi}{\partial A}\right)_{v,u}, \quad \tau_{p,s} = \frac{1}{\beta}\left(\frac{\partial\xi}{\partial A}\right)_{p,s} = \frac{1}{\beta}\left(\frac{\partial\xi}{\partial A}\right)_{p,u},$$

$$\tau_{v,T} = \frac{1}{\beta}\left(\frac{\partial\xi}{\partial A}\right)_{v,T}, \quad \tau_{p,T} = \frac{1}{\beta}\left(\frac{\partial\xi}{\partial A}\right)_{p,T}.$$

5. Note that all these relaxation times contain the same rate constant β, which occurs in the phenomenological equation (XII.57), but different thermodynamic quantities $\partial\xi/\partial A$.

Note that one can define four compressibilities

$$\chi_{s,A} = -\frac{1}{v_0}\left(\frac{\partial v}{\partial p}\right)_{s,A}, \quad \chi_{T,A}, \quad \chi_{s,\xi}, \quad \chi_{T,\xi},$$

and four specific heats

$$c_{p,A} = T\left(\frac{\partial s}{\partial T}\right)_{p,A}, \quad c_{v,A}, \quad c_{p,\xi}, \quad c_{v,\xi},$$

analogues to the four quantities τ of the preceding problem.

Derive the following six thermodynamic relations which exist amongst quantities involving the compressibilities, the specific heats and the $\partial A/\partial\xi$:

$$\frac{(\partial p/\partial v)_{s,A}}{(\partial p/\partial v)_{T,A}} = \frac{(\partial s/\partial T)_{p,A}}{(\partial s/\partial T)_{v,A}}, \quad \frac{(\partial p/\partial v)_{s,\xi}}{(\partial p/\partial v)_{T,\xi}} = \frac{(\partial s/\partial T)_{p,\xi}}{(\partial s/\partial T)_{v,\xi}},$$

$$\frac{(\partial p/\partial v)_{A,s}}{(\partial p/\partial v)_{\xi,s}} = \frac{(\partial A/\partial\xi)_{p,s}}{(\partial A/\partial\xi)_{v,s}}, \quad \frac{(\partial p/\partial v)_{A,T}}{(\partial p/\partial v)_{\xi,T}} = \frac{(\partial A/\partial\xi)_{p,T}}{(\partial A/\partial\xi)_{v,T}},$$

$$\frac{(\partial A/\partial\xi)_{s,p}}{(\partial A/\partial\xi)_{T,p}} = \frac{(\partial s/\partial T)_{A,p}}{(\partial s/\partial T)_{\xi,p}}, \quad \frac{(\partial A/\partial\xi)_{s,v}}{(\partial A/\partial\xi)_{T,v}} = \frac{(\partial s/\partial T)_{A,v}}{(\partial s/\partial T)_{\xi,v}}.$$

Hint concerning the first relation: Write $-(\partial s/\partial v)_{p,A}/(\partial s/\partial p)_{v,A}$ for the numerator of the first member and then transform $(\partial s/\partial v)_{p,A}$ into $(\partial s/\partial T)_{p,A}\,(\partial T/\partial v)_{p,A}$ and proceed similarly for $(\partial s/\partial p)_{v,A}$.

6. Prove that

$$\frac{\tau_{p,s}}{\tau_{v,s}} = \frac{(\partial p/\partial v)_{\xi,s}}{(\partial p/\partial v)_{A,s}} = \frac{\chi_{A,s}}{\chi_{\xi,s}},$$

and derive similar relations involving also the other relaxation times of problem 4.

7. One can define "internal specific heats" as

$$c_{p,i} \equiv c_{p,A} - c_{p,\xi},$$

$$c_{v,i} \equiv c_{v,A} - c_{v,\xi}.$$

Prove that

$$c_{p,\,i} = T\left(\frac{\partial A}{\partial T}\right)_{\xi,\,p}\left(\frac{\partial \xi}{\partial T}\right)_{A,\,p} = -\,T\left(\frac{\partial A}{\partial \xi}\right)_{T,\,p}\left(\frac{\partial \xi}{\partial T}\right)^2_{A,\,p},$$

$$c_{v,\,i} = T\left(\frac{\partial A}{\partial T}\right)_{\xi,\,v}\left(\frac{\partial \xi}{\partial T}\right)_{A,\,v} = -\,T\left(\frac{\partial A}{\partial \xi}\right)_{T,\,v}\left(\frac{\partial \xi}{\partial T}\right)^2_{A,\,v},$$

in complete analogy to

$$c_{p,\,A} - c_{v,\,A} = T\left(\frac{\partial p}{\partial T}\right)_{v,\,A}\left(\frac{\partial v}{\partial T}\right)_{p,\,A} = -\,T\left(\frac{\partial p}{\partial v}\right)_{T,\,A}\left(\frac{\partial v}{\partial T}\right)^2_{p,\,A},$$

$$c_{p,\,\xi} - c_{v,\,\xi} = T\left(\frac{\partial p}{\partial T}\right)_{v,\,\xi}\left(\frac{\partial v}{\partial T}\right)_{p,\,\xi} = -\,T\left(\frac{\partial p}{\partial v}\right)_{T,\,\xi}\left(\frac{\partial v}{\partial T}\right)^2_{p,\,\xi},$$

which are well-known thermodynamical formulae, taken at A and ξ constant respectively.

8. Derive the relation

$$\frac{c_{p,\,A}}{c_{v,\,A}} - 1 = \frac{\alpha_A^2 v T}{c_{p,\,A}\chi_{A,\,s}},$$

where α_A is the expansion coefficient at constant A.

Hint: Apply the first relation of problem 5 and the penultimate of problem 7.

9. Derive the following formulae, inverting the roles of ξ and A in the equations of state of Chapter XII, § 4:

$$\hat{\kappa}(\omega) = \left(\frac{\partial p}{\partial v}\right)_{A,\,s} + \left(\frac{\partial p}{\partial A}\right)_{v,\,s}\left(\frac{\partial A}{\partial \xi}\right)_{v,\,s}\left(\frac{\partial \xi}{\partial v}\right)_{A,\,s}\frac{\mathrm{i}\omega\tau}{1-\mathrm{i}\omega\tau}$$

$$= \left(\frac{\partial p}{\partial v}\right)_{A,\,s} - \left(\frac{\partial p}{\partial A}\right)_{v,\,s}\left(\frac{\partial A}{\partial v}\right)_{\xi,\,s}\frac{\mathrm{i}\omega\tau}{1-\mathrm{i}\omega\tau}.$$

Prove the equivalence of these formulae with (XII.115) and prove also

$$\hat{\kappa}(0) = \left(\frac{\partial p}{\partial v}\right)_{A,\,s} \leqslant 0,$$

$$\hat{\kappa}(\infty) = \left(\frac{\partial p}{\partial v}\right)_{A,\,s} + \left(\frac{\partial p}{\partial A}\right)_{v,\,s}\left(\frac{\partial A}{\partial v}\right)_{\xi,\,s} = \left(\frac{\partial p}{\partial v}\right)_{\xi,\,s} \leqslant 0.$$

10. Derive the following set of formulae for a perfect gas, in which vibration-translation relaxation is possible and which has the equation of state

$$p = \frac{RT}{vM} = \frac{kT}{vm},$$

where $M = Nm$ is the molar mass (m is the mass of a molecule) and where $R = Nk$ is the gas constant (note that the pressure p is independent of ξ and of A):

$$\left(\frac{\partial p}{\partial v}\right)_{s,A} = \frac{c_{p,A}}{c_{v,A}}\left(\frac{\partial p}{\partial v}\right)_{T,A} = -\frac{c_{p,A}}{c_{v,A}}\frac{RT}{v^2 M} \tag{1}$$

and an analogous formula with ξ instead of A. Furthermore:

$$c_{p,A} = c_{v,A} + \frac{R}{M}, \quad c_{p,\xi} = c_{v,\xi} + \frac{R}{M}. \tag{2}$$

With molar specific heats, such as $\bar{c}_{p,A} = c_{p,A}M$,

$$\begin{aligned} \hat{\kappa}(0) &= \left(\frac{\partial p}{\partial v}\right)_{s,A} = -\left(1 + \frac{R}{\bar{c}_{v,A}}\right)\frac{RT}{v^2 M}, \\ \hat{\kappa}(\infty) &= \left(\frac{\partial p}{\partial v}\right)_{s,\xi} = -\left(1 + \frac{R}{\bar{c}_{v,\xi}}\right)\frac{RT}{v^2 M}, \end{aligned} \tag{3}$$

where the so-called "total" and "external" specific heats $\bar{c}_{v,A}$ and $\bar{c}_{v,\xi}$ appear.

11. Show with the help of the results of problem 10 that for the case of vibration-translation relaxation in a perfect gas, in the approximation where $\hat{\kappa}(0) - \hat{\kappa}(\infty) \ll |\hat{\kappa}(\infty)|$ (*cf.* Ch. XII, § 4), the velocity of sound is given by

$$c^2(\omega) = \left\{1 + R\,\frac{\bar{c}_{v,A}\omega^2\tau^2 + \bar{c}_{v,\xi}}{\bar{c}_{v,A}\bar{c}_{v,\xi}(\;+\omega^2\tau^2)}\right\}\frac{RT}{M},$$

or

$$c^2(\omega) = \tfrac{1}{2}\{c^2(\infty) + c^2(0)\} + \tfrac{1}{2}\{c^2(\infty) - c^2(0)\}\,\mathrm{tgh}\,\ln\omega\tau\,.$$

Show also that the attenuation over one wave length in this system is given by

$$\mu = \frac{\pi}{2} \frac{(\bar{c}_{v,A} - \bar{c}_{v,\xi})R}{\bar{c}_{v,\xi}(\bar{c}_{v,A} + R)} \frac{1}{\cosh \ln \omega\tau} .$$

(Note that here the "internal specific heat" $\bar{c}_{v,i} \equiv \bar{c}_{v,A} - \bar{c}_{v,\xi}$ occurs.) It is convenient to plot c^2 and μ as functions of the logarithm of ω. Prove that in such graphs c^2 increases from $c^2(0)$ to $c^2(\infty)$, passing through an inflexion point at $\omega = \tau^{-1}$, and that μ starts at 0 for $\omega = 0$, passes through a maximum at $\omega = \tau^{-1}$ and decreases to zero again at $\omega = \infty$.

12. Find the thermodynamic connexions between each two of the four susceptibilities $\hat{\kappa}(\omega)$, $\hat{\kappa}_T(\omega)$, $\hat{\kappa}_\xi(\omega)$ and $\hat{\kappa}_A(\omega)$ (introduced in Chapter XII, § 4), by eliminating the relaxation time τ. *Example*:

$$\hat{\kappa}(\omega) = \left(\frac{\partial p}{\partial v}\right)_{\xi, s} + \left(\frac{\partial p}{\partial \xi}\right)_{v, s} \kappa_\xi(\omega) .$$

13. Prove the two following formulae for the relaxation susceptibilities of Chapter XII, § 4:

$$\hat{\kappa}_A(\omega) = \left(\frac{\partial \xi}{\partial v}\right)_{A, s} \frac{1}{\beta} \frac{\mathrm{i}\omega}{1 - \mathrm{i}\omega\tau} ,$$

$$\hat{\kappa}_A(\omega) = \frac{1}{\beta} \mathrm{i}\omega \, \hat{\kappa}_\xi(\omega) .$$

14. Prove that, if in Chapter XII, § 4 one has $\hat{\kappa}''(\omega) \gg | \hat{\kappa}'(\omega) |$, the following results are obtained:

$$k = \gamma = \frac{\omega \rho_0}{\sqrt{2\hat{\kappa}''(\omega)}} .$$

Check that the inequality is realized when $| \hat{\kappa}(0) | \ll | \hat{\kappa}(\infty) |$ and at the same time $\omega\tau \ll 1$.

15. Derive formula (XII.168) for the effective volume viscosity $\bar{\eta}_v$ with the help of (XII.98) and the condition $\omega\tau \ll 1$.

Hint: For small values of $\omega\tau$ the susceptibility $\hat{\kappa}_\xi(\omega)$ is simply equal to $(\partial\xi/\partial v)_{A,s}$ as (XII.136) shows. This leads to

$$\frac{\partial\xi}{\partial t} = \left(\frac{\partial\xi}{\partial v}\right)_{A,s} \frac{\partial v}{\partial t}.$$

Use this result and the form of the entropy production

$$\sigma = \frac{\rho_0}{T}\frac{1}{\beta}\left(\frac{\partial\xi}{\partial t}\right)^2$$

which follows from (XII.68) and (XII.164).

16. Derive for the dissipation caused by acoustic relaxation of a harmonic sound wave per period and per unit mass the following expression

$$\pi\kappa''(\omega_0)\,|v|^2\,\mathrm{e}^{-2\gamma\cdot r} = \frac{2\pi}{\omega\rho_0}\left|\left(\frac{\partial p}{\partial v}\right)_{A,s}\right|^{\frac{3}{2}} |v|^2\,\gamma\,\mathrm{e}^{-2\gamma\cdot r},$$

which is valid in the approximation $\hat{\kappa}''(\omega) \ll |\hat{\kappa}'(\omega)|$ and $\omega\tau \ll 1$ (*cf.* Ch. XII, § 4). Note the dependency of the result on the attenuation factor γ.

17. Derive formula (XII.169), which is valid for small values of the heat conductivity, proceeding in the following way. First eliminate $\hat{u}$ from (XII.88) and (XII.89). This gives $\hat{s}$ in terms of $\hat{T}$. With the help of this relation eliminate $\hat{s}$ from (XII.91)–(XII.93). Furthermore eliminate $\hat{\xi}$ from (XII.90)–(XII.93). The right-hand sides of (XII.91)–(XII.93) have now become linear expressions in $\hat{v}$, $\hat{A}$ and $\hat{T}$. One now inserts first (XII.92) and subsequently (XII.93) into (XII.91), neglecting terms quadratic in λ. Then one obtains an equation of the form (XII.96) with the desired susceptibility (XII.169), if the Maxwell relations (XII.74) are used.

18. Prove the following thermodynamic relation

$$\frac{c_{p,A} - c_{v,A}}{c_{p,A}c_{v,A}} = -\frac{1}{T}\left(\frac{\partial v}{\partial p}\right)_{A,s}\left(\frac{\partial T}{\partial v}\right)^2_{A,s},$$

which is used in the transition from (XII.172) to (XII.173). *Hint*: One must apply the first formula of the set of six relations in problem 5, the penultimate result of problem 7 and furthermore the relation

$$\left(\frac{\partial v}{\partial T}\right)_{s,A} = -\frac{c_{v,A}}{T}\left(\frac{\partial T}{\partial p}\right)_{v,A} = \frac{c_{v,A}}{T}\left(\frac{\partial T}{\partial v}\right)_{p,A}\left(\frac{\partial v}{\partial p}\right)_{T,A},$$

which can be derived if ds is expressed in terms of dv and dT at A constant.

CHAPTER XIII

1. Derive the equation of conservation of electric charge

$$\frac{\partial \rho z}{\partial t} = -\operatorname{div} \boldsymbol{I}$$

from the balance equations for the mass of component k

$$\frac{\partial \rho_k}{\partial t} = -\operatorname{div}(\rho_k \boldsymbol{v} + \boldsymbol{J}_k), \quad (k = 1, 2, \ldots, n),$$

(*cf.* Chapter II).

2. Prove that at mechanical equilibrium formula (XIII.43) (thermodynamical equilibrium) gets the form

$$\operatorname{grad} \tilde{\mu}_k = -\frac{z_k}{c}\left(\frac{\partial \boldsymbol{A}}{\partial t} - \boldsymbol{v} \wedge \operatorname{rot} \boldsymbol{A}\right), \quad (k = 1, 2, \ldots, n).$$

3. Prove that the energy $\rho(\tilde{u} + \frac{1}{2}\boldsymbol{v}^2)$, [where $\tilde{u}$ is defined by (XIII.46)] is conserved under the conditions

$$\rho z \frac{\partial \varphi}{\partial t} = 0, \quad \frac{1}{c}\boldsymbol{I}\cdot\frac{\partial \boldsymbol{A}}{\partial t} = 0.$$

4. On the basis of the expression (XIII.155) for the entropy production, one can establish the phenomenological equations

$$\boldsymbol{I} = L_{11}\boldsymbol{E} + L_{12}\omega^2 \boldsymbol{r},$$

$$\boldsymbol{J} = L_{21}\boldsymbol{E} + L_{22}\omega^2 \boldsymbol{r},$$

valid in isotropic fluids, with the Onsager relation

$$L_{12} = L_{21} .$$

One can define the following three physical quantities:

$\left(\frac{\boldsymbol{I}}{\boldsymbol{E}}\right)_{\boldsymbol{g}=0}$, the specific electric conductivity,

$\left(\frac{\boldsymbol{E}}{\boldsymbol{g}}\right)_{\boldsymbol{I}=0}$, the sedimentation potential,

$\left(\frac{\boldsymbol{J}}{\boldsymbol{E}}\right)_{\boldsymbol{g}=0}$, the barycentric electrophoresis coefficient,

where $\boldsymbol{g} = \omega^2 \boldsymbol{r}$.

Prove that the following connexion exists between them

$$\left(\frac{\boldsymbol{J}}{\boldsymbol{E}}\right)_{\boldsymbol{g}=0} = -\left(\frac{\boldsymbol{I}}{\boldsymbol{E}}\right)_{\boldsymbol{g}=0}\left(\frac{\boldsymbol{E}}{\boldsymbol{g}}\right)_{\boldsymbol{I}=0},$$

as a consequence of the Onsager relation.

CHAPTER XV

1. Prove that if the potential energy is not conserved in chemical reactions, *i.e.* if formula (II.27) is not employed in the derivation of Chapter XV, § 2 and § 3 [see (XV.24)], one obtains for the entropy production an expression as (XV.57), except that instead of (XV.63) one has now

$$x_j^\alpha = -\frac{\tilde{A}_j^\alpha}{T} \quad \text{with} \quad \tilde{A}_j^\alpha = A_j^\alpha + \sum_{k=1}^n \psi_k^\alpha \nu_{kj} = \sum_{k=1}^n \tilde{\mu}_k \nu_{kj} ,$$

(terms of higher order than the bilinear terms written down in σ_{tot} have been neglected).

2. Prove with the help of (XV.15) and (XV.48) that an alternative form of (XV.52) is

$$\frac{\mathrm{d_i}S^\alpha}{\mathrm{d}t} = \frac{1}{T^\alpha}\frac{\mathrm{d_i}U^\alpha}{\mathrm{d}t} - \sum_{k=1}^{n} \frac{\mu_k^\alpha}{T^\alpha}\frac{\mathrm{d}M_k^\alpha}{\mathrm{d}t}, \quad (\alpha = \mathrm{I, II}),$$

where

$$\frac{\mathrm{d}M_k^\alpha}{\mathrm{d}t} = \frac{\mathrm{d_e}M_k^\alpha}{\mathrm{d}t} + \frac{\mathrm{d_i}M_k^\alpha}{\mathrm{d}t},$$

as in (XV.12), (XV.14) and (XV.15).

3. With the help of the progress variable ξ_j of the chemical reaction j, introduced in Chapter X, one can write for (XV.60):

$$j_j^\alpha = J_j^\alpha V^\alpha = \frac{\mathrm{d}\xi_j^\alpha}{\mathrm{d}t} M^\alpha .$$

Check that then the phenomenological equation (XV.98) can be written in the form

$$\frac{\mathrm{d}\xi_j^\alpha}{\mathrm{d}t} = \sum_{j'=1}^{r} \Lambda_{jj'}^\alpha x_{j'}^\alpha = -\sum_{j'=1}^{r} \Lambda_{jj'}^\alpha \frac{A_{j'}^\alpha}{T^\alpha}, \quad (\alpha = \mathrm{I, II}),$$

where the mass M^α does not appear.

4. Prove that alternative forms for (XV.83) and (XV.117) are

$$j_v = \int^{\omega} \sum_{k=1}^{n} v_k \rho_k \boldsymbol{v}_k \cdot \mathrm{d}\boldsymbol{\Omega}$$

and

$$i = \int^{\omega} \sum_{k=1}^{n} z_k \rho_k \boldsymbol{v}_k \cdot \mathrm{d}\boldsymbol{\Omega} ,$$

respectively.

5. Show why the matrix (XV.84) follows from the general expression (X.34), if for the weights the choice $a_k = c_k$ is made.

6. Derive the relations (XV.106)–(XV.108) from the expressions (XV.99)–(XV.104).

7. Derive the relations (XV.109) and (XV.110) from (XV.106)–(XV.108).

8. Establish the formalism for the general case which results from the combination of the first and second example, discussed in the end of § 3 of Chapter XV and in § 4 of that chapter.

9. Prove that if in the formalism of Chapter XV, § 5 the index v is

replaced by 0 and Δp written as $\Delta\mu_0$, one finds in the stationary state with fixed ΔT

$$\frac{\Delta\mu_m}{\Delta T} = -\sum_{i=0}^{n-1} \Lambda_{mi}^{-1} \Lambda_{iq} \frac{1}{T}, \quad (m = 0, 1, \ldots, n-1),$$

where Λ_{mi}^{-1} is the reciprocal of the n by n matrix Λ_{mi} $(m, i = 0, 1, \ldots, n-1)$.

10. The result of the preceding problem contains both the thermo-molecular pressure difference and the thermal effusion effect. Derive formula (XV.138) from this result, using standard rules of determinant calculus.

11. Prove that from (XV.76)–(XV.78) it follows (the terms $\boldsymbol{\Pi}\cdot\boldsymbol{v}$ were proved to be negligible) that

$$j_u = \int^{\omega} (\boldsymbol{J}_q + h\rho\boldsymbol{v})\cdot \mathrm{d}\boldsymbol{\Omega} = \int^{\omega} \boldsymbol{J}_q^{\mathrm{abs}} \cdot \mathrm{d}\boldsymbol{\Omega},$$

$$j_k = \int^{\omega} (\boldsymbol{J}_k + \rho_k\boldsymbol{v})\cdot \mathrm{d}\boldsymbol{\Omega} = \int^{\omega} \rho_k\boldsymbol{v}_k\cdot \mathrm{d}\boldsymbol{\Omega} = \int^{\omega} \boldsymbol{J}_k^{\mathrm{abs}} \cdot \mathrm{d}\boldsymbol{\Omega},$$

$$j'_q = \int^{\omega} \left(\boldsymbol{J}_q - \sum_{k=1}^{n} h_k \boldsymbol{J}_k\right)\cdot \mathrm{d}\boldsymbol{\Omega} = \int^{\omega} \boldsymbol{J}'_q\cdot \mathrm{d}\boldsymbol{\Omega},$$

where in the last members notations of Chapter XI, §7 have been used. Find the relation (XV.70):

$$j'_q = j_u - \sum_{k=1}^{n} h_k j_k$$

also from the formalism of Chapter XI, § 7 and the first three formulae of this problem.

12. Prove that the transfer quantities defined by

$$j_u = \sum_{k=1}^{n} u_k^* j_k, \quad j'_q = \sum_{k=1}^{n} q_k'^* j_k, \quad (\Delta T = 0),$$

are related as

$$q_k'^* = u_k^* - h_k,$$

and show that

$$u_k^* = Q_{k,\,\mathrm{abs}}^*\,, \quad q_k'^* = Q_{k,\,\mathrm{abs}}'^*\,,$$

where the quantities at the right-hand sides were defined in Chapter XI, § 7.

13. In Chapter XV, § 5 we found the relation

$$j_q = j_u - hj\,.$$

Prove that the transfer quantities, defined by

$$j_q = \sum_{k=1}^{n} q_k^* j_k\,, \quad (\Delta T = 0)\,,$$

satisfy the relation

$$q_k^* = u_k^* - h\,.$$

14. Prove that the transfer quantities u^* and q^*, defined by

$$j_u = u^* j\,, \quad j_q = q^* j\,, \quad (\Delta T = 0)\,,$$

are related as

$$q^* = u^* - h\,.$$

15. Show that in a Knudsen gas (a perfect gas enclosed in two reservoirs with a hole between them, the diameter of which is small compared with the mean free path in the gas) the energy of transfer is equal to

$$u^* = \frac{2kT}{m} = \frac{2RT}{M}\,, \tag{1}$$

where k is Boltzmann's constant, m the mass of a molecule, R is the gas constant and M the molecular mass ($R = kN$ and $M = mN$, with N Avogadro's number).

Hints: Since every molecule which arrives at the hole will freely pass through it, the mean energy per molecule passing through the hole is

$$\overline{\tfrac{1}{2}mv^2} = \frac{\int\limits_0^\infty \tfrac{1}{2}mv^2 n_v \,\mathrm{d}v}{\int\limits_0^\infty n_v \,\mathrm{d}v}\,, \tag{2}$$

where $n_v\,dv$ is the number of molecules, passing through the hole per unit surface and per unit time with velocities between v and $v + dv$. Show that this number is

$$n_v\,dv = \int_{\varphi=0}^{2\pi} \int_{\theta=0}^{\frac{1}{2}\pi} v_x n f \boldsymbol{v}^2\,dv\,d\omega\,, \tag{3}$$

when x is the coordinate perpendicular to the plane of the hole, θ the polar angle counted from the x-axis and φ the azimuth angle. Furthermore n is the number density, f is the Maxwell distribution function

$$f = \left(\frac{m}{2\pi\,kT}\right)^{\frac{3}{2}} \exp\left(-\frac{\frac{1}{2}m\boldsymbol{v}^2}{kT}\right), \tag{4}$$

and $nf\,\boldsymbol{v}^2\,dv\,d\omega$ is the number of molecules per unit volume with velocities between v and $v + dv$ in the solid angle $d\omega = \sin\theta\,d\theta\,d\varphi$. The proof consists in showing first that

$$n_v\,dv = \pi n f v^3\,dv \tag{5}$$

follows from (3), (4) and $v_x = v\cos\theta$. (Note that the integration over the directions is to be carried out over a half sphere.) Then (2) leads to

$$\overline{\tfrac{1}{2}m\boldsymbol{v}^2} = 2\,kT\,, \tag{6}$$

which yields immediately (1), the transfer of energy per unit mass.

16. Check that formula (XV.162):

$$q^* = -\frac{1}{2}\frac{kT}{m} = -\frac{1}{2}\frac{RT}{M}$$

follows from the results of the two preceding problems with the value $\frac{5}{2}kT/m$ for the enthalpy per unit mass of a perfect gas.

17. Prove that in the n-component, non-reacting discontinuous system (Ch. XV, § 5) with $\Delta T = 0$ and fixed Δp the following relations exist

$$u^* = \sum_{k=1}^{n} c_k u_k^*\,,\quad q^* = \sum_{k=1}^{n} c_k q_k^* = \sum_{k=1}^{n} c_k q_k'^*\,.$$

Hint: Use the definitions of the quantities of transfer (problems 12, 13 and 14) and formulae (XV.149) and (XV.156).

18. The entropy flow j_s can be defined as

$$j_s = -\frac{\mathrm{d_i}S^{\mathrm{I}}}{\mathrm{d}t}.$$

Prove from (XV.52) or the expression in problem 2 that

$$j_s = \frac{j_u - \sum_{k=1}^{n} \mu_k j_k}{T} = \frac{j_q'}{T} + \sum_{k=1}^{n} s_k j_k\,,$$

noticing that the contributions of the chemical reactions can be neglected in the capillary. Compare the result with the corresponding formula of Chapter XI, § 7.

19. Prove that the entropy flow is given by

$$j_s = \int^{\omega} \left\{ \frac{\boldsymbol{J}_q'}{T} + \sum_{k=1}^{n} s_k (\boldsymbol{J}_k + \rho_k \boldsymbol{v}) \right\} \cdot \mathrm{d}\boldsymbol{\Omega}$$

$$= \int^{\omega} (\boldsymbol{J}_s + s\rho \boldsymbol{v}) \cdot \mathrm{d}\boldsymbol{\Omega} = \int^{\omega} \boldsymbol{J}_{s,\,\mathrm{abs}} \cdot \mathrm{d}\boldsymbol{\Omega}\,,$$

where in the last member a notation of Chapter XI, § 7 is employed.

20. Show that the entropy of transfer s_k^* defined by

$$j_s = \sum_{k=1}^{n} s_k^* j_k\,, \quad (\Delta T = 0)\,,$$

is equal to $S_{k,\mathrm{abs}}^*$ of Chapter XI, § 7.

21. Prove that the solutions of the equations for the stationary state of order p (*i.e.* with fixed $x_1, x_2, \ldots, x_p$):

$$j_k = \sum_{i=1}^{N} A_{ki} x_i = 0\,, \quad (k = p+1, p+2, \ldots, N)\,,$$

(*cf.* Ch. XV, § 5) are given by

$$x_k = - \sum_{i=p+1}^{N} \sum_{m=1}^{p} \Lambda_{ki}^{-1} \Lambda_{im} x_m , \quad (k = p + 1, p + 2, \ldots , N) ,$$

where Λ_{ki}^{-1} are the elements of the reciprocal matrix of Λ_{ki} with $k, i = p + 1, p + 2 , \ldots , N$.

22. Prove that the heat conductivity κ' in a one-component system (Ch. XV, § 4 and § 5), defined by

$$j'_q = - \kappa' \Delta T$$

in the stationary state with fixed ΔT, is given by

$$\kappa' - \frac{\Lambda_{qq}}{T^2}$$

in case *a* (wide capillary), and by

$$\kappa' = \frac{\Lambda_{qq}\Lambda_{vv} - \Lambda_{vq}\Lambda_{qv}}{\Lambda_{vv} T^2}$$

in case *b* (narrow capillary).

23. Prove that the heat conductivity κ' in a binary, non-reacting system (Ch. XV, §§ 4, 5) is in the stationary state with fixed ΔT

$$\kappa' = \frac{\Lambda_{qq}\Lambda_{11} - \Lambda_{q1}\Lambda_{1q}}{\Lambda_{11} T^2} , \quad (\text{case } a) ,$$

$$\kappa' = \frac{\begin{vmatrix} \Lambda_{qq} & \Lambda_{q1} & \Lambda_{qv} \\ \Lambda_{1q} & \Lambda_{11} & \Lambda_{1v} \\ \Lambda_{vq} & \Lambda_{v1} & \Lambda_{vv} \end{vmatrix}}{\begin{vmatrix} \Lambda_{11} & \Lambda_{1v} \\ \Lambda_{v1} & \Lambda_{vv} \end{vmatrix} T^2} , \quad (\text{case } b) .$$

24. Prove (XV.247) for an *n*-component, *r*-reaction system. *Hint:* The proof of (XV.160) for the non-reacting system can be simply generalized to the case of a reacting system by remarking that the affinities play formally the same role as the $\Delta\mu_i$, and similarly the reaction rates play formally the same role as the diffusion flows j'_i.

25. Prove that the heat conductivity κ' of a binary, single reaction system at uniform pressure (and fixed temperature difference ΔT), which is defined by the equation $j'_q = -\kappa' \Delta T$, is given by

$$\kappa' = \frac{\Lambda_{qq}\Lambda_{11} - \Lambda_{q1}\Lambda_{1q} + \{c_2^{-2}\Lambda_{qq} - c_2^{-1}(h_2 - h_1)\Lambda_{q1}\}\Lambda}{(\Lambda_{11} + c_2^{-2}\Lambda)\, T^2},$$

where the stoichiometric coefficient ν_2 has been chosen as 1.

Hint: Check that the equations (XV.234) and (XV.235) are valid for this case. Use furthermore (XV.95), (XV.96), (XV.98), (XV.203) and (XV.219).

26. Prove that the heat conductivity κ of a binary, single reaction system at uniform pressure (and fixed temperature difference ΔT), which is defined by the equation $j_q = -\kappa \Delta T$, is given by

$$\kappa = \frac{\Lambda_{qq}\Lambda_{11} - \Lambda_{q1}\Lambda_{1q} + \{c_2^{-2}\Lambda_{qq} - c_2^{-1}(h_2 - h_1)(\Lambda_{q1} + \Lambda_{1q}) + (h_2 - h_1)^2\Lambda_{11}\}\Lambda}{(\Lambda_{11} + c_2^{-2}\Lambda)\, T^2},$$

where the stoichiometric coefficient ν_2 has been chosen as 1.

Hint: The result follows from the preceding problem, using also (XV.242) and again (XV.96), (XV.234) and (XV.235).

27. Prove that in a binary, single reaction system, in which no reduced heat flow can occur in the isothermal state ($j'_q = 0$ if $\Delta T = 0$, *i.e.* $\Lambda_{q1} = 0$, $\Lambda_{qv} = 0$), the heat conductivities derived in the preceding two problems reduce to

$$\kappa' = \frac{\Lambda_{qq}}{T^2}, \quad \kappa = \frac{\Lambda_{qq}}{T^2} + \frac{(h_2 - h_1)^2\Lambda_{11}\Lambda}{(\Lambda_{11} + c_2^{-2}\Lambda)\, T^2},$$

if an Onsager relation is used.

28. Prove that in the limiting case $\Lambda = 0$ (no chemical reactions) the results of problems 25 and 26 reduce to

$$\kappa' = \kappa = \frac{\Lambda_{qq}\Lambda_{11} - \Lambda_{q1}\Lambda_{1q}}{\Lambda_{11}\, T^2}, \quad (\Lambda = 0),$$

which is also the result of problem 23 for systems at uniform pressure.

29. Prove that in the limiting case $\Lambda = \infty$ (chemical equilibrium)

the results of problems 25 and 26 reduce to

$$\kappa' = \frac{\Lambda_{qq} - c_2 (h_2 - h_1) \Lambda_{q1}}{T^2}, \quad (\Lambda = \infty),$$

and

$$\kappa = \frac{\Lambda_{qq} - c_2 (h_2 - h_1) (\Lambda_{q1} + \Lambda_{1q}) + c_2^2 (h_2 - h_1)^2 \Lambda_{11}}{T^2}, \quad (\Lambda = \infty),$$

respectively.

30. Prove that the results of problems 25 and 26 become in the limit of Λ_{11} tending to infinity ("superfluidity"):

$$\kappa' = \frac{\Lambda_{qq}}{T^2}$$

and

$$\kappa = \frac{\Lambda_{qq} + (h_2 - h_1)^2 \Lambda}{T^2},$$

respectively.

NAME INDEX

SUBJECT INDEX

A CATALOGUE OF SELECTED DOVER BOOKS
IN ALL FIELDS OF INTEREST

A CATALOGUE OF SELECTED DOVER BOOKS IN ALL FIELDS OF INTEREST

THE ANATOMY OF THE HORSE, George Stubbs. Often considered the great masterpiece of animal anatomy. Full reproduction of 1766 edition, plus prospectus; original text and modernized text. 36 plates. Introduction by Eleanor Garvey. 121pp. 11 x 14¾. 23402-9 Pa. $6.00

BRIDGMAN'S LIFE DRAWING, George B. Bridgman. More than 500 illustrative drawings and text teach you to abstract the body into its major masses, use light and shade, proportion; as well as specific areas of anatomy, of which Bridgman is master. 192pp. 6½ x 9¼. (Available in U.S. only) 22710-3 Pa. $3.50

ART NOUVEAU DESIGNS IN COLOR, Alphonse Mucha, Maurice Verneuil, Georges Auriol. Full-color reproduction of *Combinaisons ornementales* (c. 1900) by Art Nouveau masters. Floral, animal, geometric, interlacings, swashes—borders, frames, spots—all incredibly beautiful. 60 plates, hundreds of designs. 9⅜ x 8-1/16. 22885-1 Pa. $4.00

FULL-COLOR FLORAL DESIGNS IN THE ART NOUVEAU STYLE, E. A. Seguy. 166 motifs, on 40 plates, from *Les fleurs et leurs applications decoratives* (1902): borders, circular designs, repeats, allovers, "spots." All in authentic Art Nouveau colors. 48pp. 9⅜ x 12¼. 23439-8 Pa. $5.00

A DIDEROT PICTORIAL ENCYCLOPEDIA OF TRADES AND INDUSTRY, edited by Charles C. Gillispie. 485 most interesting plates from the great French Encyclopedia of the 18th century show hundreds of working figures, artifacts, process, land and cityscapes; glassmaking, papermaking, metal extraction, construction, weaving, making furniture, clothing, wigs, dozens of other activities. Plates fully explained. 920pp. 9 x 12. 22284-5, 22285-3 Clothbd., Two-vol. set $40.00

HANDBOOK OF EARLY ADVERTISING ART, Clarence P. Hornung. Largest collection of copyright-free early and antique advertising art ever compiled. Over 6,000 illustrations, from Franklin's time to the 1890's for special effects, novelty. Valuable source, almost inexhaustible.
Pictorial Volume. Agriculture, the zodiac, animals, autos, birds, Christmas, fire engines, flowers, trees, musical instruments, ships, games and sports, much more. Arranged by subject matter and use. 237 plates. 288pp. 9 x 12. 20122-8 Clothbd. $14.50

Typographical Volume. Roman and Gothic faces ranging from 10 point to 300 point, "Barnum," German and Old English faces, script, logotypes, scrolls and flourishes, 1115 ornamental initials, 67 complete alphabets, more. 310 plates. 320pp. 9 x 12. 20123-6 Clothbd. $15.00

CALLIGRAPHY (CALLIGRAPHIA LATINA), J. G. Schwandner. High point of 18th-century ornamental calligraphy. Very ornate initials, scrolls, borders, cherubs, birds, lettered examples. 172pp. 9 x 13. 20475-8 Pa. $7.00

HOLLYWOOD GLAMOUR PORTRAITS, edited by John Kobal. 145 photos capture the stars from 1926-49, the high point in portrait photography. Gable, Harlow, Bogart, Bacall, Hedy Lamarr, Marlene Dietrich, Robert Montgomery, Marlon Brando, Veronica Lake; 94 stars in all. Full background on photographers, technical aspects, much more. Total of 160pp. 8⅜ x 11¼. 23352-9 Pa. **$6.00**

THE NEW YORK STAGE: FAMOUS PRODUCTIONS IN PHOTOGRAPHS, edited by Stanley Appelbaum. 148 photographs from Museum of City of New York show 142 plays, 1883-1939. *Peter Pan, The Front Page, Dead End, Our Town,* O'Neill, hundreds of actors and actresses, etc. Full indexes. 154pp. 9½ x 10. 23241-7 Pa. **$6.00**

DIALOGUES CONCERNING TWO NEW SCIENCES, Galileo Galilei. Encompassing 30 years of experiment and thought, these dialogues deal with geometric demonstrations of fracture of solid bodies, cohesion, leverage, speed of light and sound, pendulums, falling bodies, accelerated motion, etc. 300pp. 5⅜ x 8½. 60099-8 Pa. **$4.00**

THE GREAT OPERA STARS IN HISTORIC PHOTOGRAPHS, edited by James Camner. 343 portraits from the 1850s to the 1940s: Tamburini, Mario, Caliapin, Jeritza, Melchior, Melba, Patti, Pinza, Schipa, Caruso, Farrar, Steber, Gobbi, and many more—270 performers in all. Index. 199pp. 8⅜ x 11¼. 23575-0 Pa. $7.50

J. S. BACH, Albert Schweitzer. Great full-length study of Bach, life, background to music, music, by foremost modern scholar. Ernest Newman translation. 650 musical examples. Total of 928pp. 5⅜ x 8½. (Available in U.S. only) 21631-4, 21632-2 Pa., Two-vol. set **$11.00**

COMPLETE PIANO SONATAS, Ludwig van Beethoven. All sonatas in the fine Schenker edition, with fingering, analytical material. One of best modern editions. Total of 615pp. 9 x 12. (Available in U.S. only) 23134-8, 23135-6 Pa., Two-vol. set $15.50

KEYBOARD MUSIC, J. S. Bach. Bach-Gesellschaft edition. For harpsichord, piano, other keyboard instruments. English Suites, French Suites, Six Partitas, Goldberg Variations, Two-Part Inventions, Three-Part Sinfonias. 312pp. 8⅛ x 11. (Available in U.S. only) 22360-4 Pa. **$6.95**

FOUR SYMPHONIES IN FULL SCORE, Franz Schubert. Schubert's four most popular symphonies: No. 4 in C Minor ("Tragic"); No. 5 in B-flat Major; No. 8 in B Minor ("Unfinished"); No. 9 in C Major ("Great"). Breitkopf & Hartel edition. Study score. 261pp. 9⅜ x 12¼. 23681-1 Pa. $6.50

THE AUTHENTIC GILBERT & SULLIVAN SONGBOOK, W. S. Gilbert, A. S. Sullivan. Largest selection available; 92 songs, uncut, original keys, in piano rendering approved by Sullivan. Favorites and lesser-known fine numbers. Edited with plot synopses by James Spero. 3 illustrations. 399pp. 9 x 12. 23482-7 Pa. **$9.95**

A MAYA GRAMMAR, Alfred M. Tozzer. Practical, useful English-language grammar by the Harvard anthropologist who was one of the three greatest American scholars in the area of Maya culture. Phonetics, grammatical processes, syntax, more. 301pp. 5⅜ x 8½. 23465-7 Pa. $4.00

THE JOURNAL OF HENRY D. THOREAU, edited by Bradford Torrey, F. H. Allen. Complete reprinting of 14 volumes, 1837-61, over two million words; the sourcebooks for *Walden*, etc. Definitive. All original sketches, plus 75 photographs. Introduction by Walter Harding. Total of 1804pp. 8½ x 12¼. 20312-3, 20313-1 Clothbd., Two-vol. set $70.00

CLASSIC GHOST STORIES, Charles Dickens and others. 18 wonderful stories you've wanted to reread: "The Monkey's Paw," "The House and the Brain," "The Upper Berth," "The Signalman," "Dracula's Guest," "The Tapestried Chamber," etc. Dickens, Scott, Mary Shelley, Stoker, etc. 330pp. 5⅜ x 8½. 20735-8 Pa. $4.50

SEVEN SCIENCE FICTION NOVELS, H. G. Wells. Full novels. *First Men in the Moon, Island of Dr. Moreau, War of the Worlds, Food of the Gods, Invisible Man, Time Machine, In the Days of the Comet.* A basic science-fiction library. 1015pp. 5⅜ x 8½. (Available in U.S. only)
20264-X Clothbd. $8.95

ARMADALE, Wilkie Collins. Third great mystery novel by the author of *The Woman in White* and *The Moonstone*. Ingeniously plotted narrative shows an exceptional command of character, incident and mood. Original magazine version with 40 illustrations. 597pp. 5⅜ x 8½.
23429-0 Pa. $6.00

MASTERS OF MYSTERY, H. Douglas Thomson. The first book in English (1931) devoted to history and aesthetics of detective story. Poe, Doyle, LeFanu, Dickens, many others, up to 1930. New introduction and notes by E. F. Bleiler. 288pp. 5⅜ x 8½. (Available in U.S. only)
23606-4 Pa. $4.00

FLATLAND, E. A. Abbott. Science-fiction classic explores life of 2-D being in 3-D world. Read also as introduction to thought about hyperspace. Introduction by Banesh Hoffmann. 16 illustrations. 103pp. 5⅜ x 8½.
20001-9 Pa. $2.00

THREE SUPERNATURAL NOVELS OF THE VICTORIAN PERIOD, edited, with an introduction, by E. F. Bleiler. Reprinted complete and unabridged, three great classics of the supernatural: *The Haunted Hotel* by Wilkie Collins, *The Haunted House at Latchford* by Mrs. J. H. Riddell, and *The Lost Stradivarious* by J. Meade Falkner. 325pp. 5⅜ x 8½.
22571-2 Pa. $4.00

AYESHA: THE RETURN OF "SHE," H. Rider Haggard. Virtuoso sequel featuring the great mythic creation, Ayesha, in an adventure that is fully as good as the first book, *She*. Original magazine version, with 47 original illustrations by Maurice Greiffenhagen. 189pp. 6½ x 9¼.
23649-8 Pa. $3.50

UNCLE SILAS, J. Sheridan LeFanu. Victorian Gothic mystery novel, considered by many best of period, even better than Collins or Dickens. Wonderful psychological terror. Introduction by Frederick Shroyer. 436pp. 5⅜ x 8½. 21715-9 Pa. $6.00

JURGEN, James Branch Cabell. The great erotic fantasy of the 1920's that delighted thousands, shocked thousands more. Full final text, Lane edition with 13 plates by Frank Pape. 346pp. 5⅜ x 8½. 23507-6 Pa. $4.50

THE CLAVERINGS, Anthony Trollope. Major novel, chronicling aspects of British Victorian society, personalities. Reprint of Cornhill serialization, 16 plates by M. Edwards; first reprint of full text. Introduction by Norman Donaldson. 412pp. 5⅜ x 8½. 23464-9 Pa. $5.00

KEPT IN THE DARK, Anthony Trollope. Unusual short novel about Victorian morality and abnormal psychology by the great English author. Probably the first American publication. Frontispiece by Sir John Millais. 92pp. 6½ x 9¼. 23609-9 Pa. $2.50

RALPH THE HEIR, Anthony Trollope. Forgotten tale of illegitimacy, inheritance. Master novel of Trollope's later years. Victorian country estates, clubs, Parliament, fox hunting, world of fully realized characters. Reprint of 1871 edition. 12 illustrations by F. A. Faser. 434pp. of text. 5⅜ x 8½. 23642-0 Pa. $5.00

YEKL and THE IMPORTED BRIDEGROOM AND OTHER STORIES OF THE NEW YORK GHETTO, Abraham Cahan. Film *Hester Street* based on *Yekl* (1896). Novel, other stories among first about Jewish immigrants of N.Y.'s East Side. Highly praised by W. D. Howells—Cahan "a new star of realism." New introduction by Bernard G. Richards. 240pp. 5⅜ x 8½. 22427-9 Pa. $3.50

THE HIGH PLACE, James Branch Cabell. Great fantasy writer's enchanting comedy of disenchantment set in 18th-century France. Considered by some critics to be even better than his famous *Jurgen*. 10 illustrations and numerous vignettes by noted fantasy artist Frank C. Pape. 320pp. 5⅜ x 8½. 23670-6 Pa. $4.00

ALICE'S ADVENTURES UNDER GROUND, Lewis Carroll. Facsimile of ms. Carroll gave Alice Liddell in 1864. Different in many ways from final Alice. Handlettered, illustrated by Carroll. Introduction by Martin Gardner. 128pp. 5⅜ x 8½. 21482-6 Pa. $2.50

FAVORITE ANDREW LANG FAIRY TALE BOOKS IN MANY COLORS, Andrew Lang. The four Lang favorites in a boxed set—the complete *Red, Green, Yellow* and *Blue* Fairy Books. 164 stories; 439 illustrations by Lancelot Speed, Henry Ford and G. P. Jacomb Hood. Total of about 1500pp. 5⅜ x 8½. 23407-X Boxed set, Pa. $15.95

YUCATAN BEFORE AND AFTER THE CONQUEST, Diego de Landa. First English translation of basic book in Maya studies, the only significant account of Yucatan written in the early post-Conquest era. Translated by distinguished Maya scholar William Gates. Appendices, introduction, 4 maps and over 120 illustrations added by translator. 162pp. 5⅜ x 8½.
23622-6 Pa. $3.00

THE MALAY ARCHIPELAGO, Alfred R. Wallace. Spirited travel account by one of founders of modern biology. Touches on zoology, botany, ethnography, geography, and geology. 62 illustrations, maps. 515pp. 5⅜ x 8½.
20187-2 Pa. $6.95

THE DISCOVERY OF THE TOMB OF TUTANKHAMEN, Howard Carter, A. C. Mace. Accompany Carter in the thrill of discovery, as ruined passage suddenly reveals unique, untouched, fabulously rich tomb. Fascinating account, with 106 illustrations. New introduction by J. M. White. Total of 382pp. 5⅜ x 8½. (Available in U.S. only) 23500-9 Pa. $4.00

THE WORLD'S GREATEST SPEECHES, edited by Lewis Copeland and Lawrence W. Lamm. Vast collection of 278 speeches from Greeks up to present. Powerful and effective models; unique look at history. Revised to 1970. Indices. 842pp. 5⅜ x 8½. 20468-5 Pa. $8.95

THE 100 GREATEST ADVERTISEMENTS, Julian Watkins. The priceless ingredient; His master's voice; 99 44/100% pure; over 100 others. How they were written, their impact, etc. Remarkable record. 130 illustrations. 233pp. 7⅞ x 10 3/5. 20540-1 Pa. $5.95

CRUICKSHANK PRINTS FOR HAND COLORING, George Cruickshank. 18 illustrations, one side of a page, on fine-quality paper suitable for watercolors. Caricatures of people in society (c. 1820) full of trenchant wit. Very large format. 32pp. 11 x 16. 23684-6 Pa. $5.00

THIRTY-TWO COLOR POSTCARDS OF TWENTIETH-CENTURY AMERICAN ART, Whitney Museum of American Art. Reproduced in full color in postcard form are 31 art works and one shot of the museum. Calder, Hopper, Rauschenberg, others. Detachable. 16pp. 8¼ x 11.
23629-3 Pa. $3.00

MUSIC OF THE SPHERES: THE MATERIAL UNIVERSE FROM ATOM TO QUASAR SIMPLY EXPLAINED, Guy Murchie. Planets, stars, geology, atoms, radiation, relativity, quantum theory, light, antimatter, similar topics. 319 figures. 664pp. 5⅜ x 8½.
21809-0, 21810-4 Pa., Two-vol. set $11.00

EINSTEIN'S THEORY OF RELATIVITY, Max Born. Finest semi-technical account; covers Einstein, Lorentz, Minkowski, and others, with much detail, much explanation of ideas and math not readily available elsewhere on this level. For student, non-specialist. 376pp. 5⅜ x 8½.
60769-0 Pa. $4.50

SECOND PIATIGORSKY CUP, edited by Isaac Kashdan. One of the greatest tournament books ever produced in the English language. All 90 games of the 1966 tournament, annotated by players, most annotated by both players. Features Petrosian, Spassky, Fischer, Larsen, six others. 228pp. 5⅜ x 8½. 23572-6 Pa. $3.50

ENCYCLOPEDIA OF CARD TRICKS, revised and edited by Jean Hugard. How to perform over 600 card tricks, devised by the world's greatest magicians: impromptus, spelling tricks, key cards, using special packs, much, much more. Additional chapter on card technique. 66 illustrations. 402pp. 5⅜ x 8½. (Available in U.S. only) 21252-1 Pa. $4.95

MAGIC: STAGE ILLUSIONS, SPECIAL EFFECTS AND TRICK PHOTOGRAPHY, Albert A. Hopkins, Henry R. Evans. One of the great classics; fullest, most authorative explanation of vanishing lady, levitations, scores of other great stage effects. Also small magic, automata, stunts. 446 illustrations. 556pp. 5⅜ x 8½. 23344-8 Pa. $6.95

THE SECRETS OF HOUDINI, J. C. Cannell. Classic study of Houdini's incredible magic, exposing closely-kept professional secrets and revealing, in general terms, the whole art of stage magic. 67 illustrations. 279pp. 5⅜ x 8½. 22913-0 Pa. $4.00

HOFFMANN'S MODERN MAGIC, Professor Hoffmann. One of the best, and best-known, magicians' manuals of the past century. Hundreds of tricks from card tricks and simple sleight of hand to elaborate illusions involving construction of complicated machinery. 332 illustrations. 563pp. 5⅜ x 8½. 23623-4 Pa. $6.00

MADAME PRUNIER'S FISH COOKERY BOOK, Mme. S. B. Prunier. More than 1000 recipes from world famous Prunier's of Paris and London, specially adapted here for American kitchen. Grilled tournedos with anchovy butter, Lobster a la Bordelaise, Prunier's prized desserts, more. Glossary. 340pp. 5⅜ x 8½. (Available in U.S. only) 22679-4 Pa. $3.00

FRENCH COUNTRY COOKING FOR AMERICANS, Louis Diat. 500 easy-to-make, authentic provincial recipes compiled by former head chef at New York's Fitz-Carlton Hotel: onion soup, lamb stew, potato pie, more. 309pp. 5⅜ x 8½. 23665-X Pa. $3.95

SAUCES, FRENCH AND FAMOUS, Louis Diat. Complete book gives over 200 specific recipes: bechamel, Bordelaise, hollandaise, Cumberland, apricot, etc. Author was one of this century's finest chefs, originator of vichyssoise and many other dishes. Index. 156pp. 5⅜ x 8. 23663-3 Pa. $2.75

TOLL HOUSE TRIED AND TRUE RECIPES, Ruth Graves Wakefield. Authentic recipes from the famous Mass. restaurant: popovers, veal and ham loaf, Toll House baked beans, chocolate cake crumb pudding, much more. Many helpful hints. Nearly 700 recipes. Index. 376pp. 5⅜ x 8½. 23560-2 Pa. $4.50

HISTORY OF BACTERIOLOGY, William Bulloch. The only comprehensive history of bacteriology from the beginnings through the 19th century. Special emphasis is given to biography-Leeuwenhoek, etc. Brief accounts of 350 bacteriologists form a separate section. No clearer, fuller study, suitable to scientists and general readers, has yet been written. 52 illustrations. 448pp. 5⅝ x 8¼. 23761-3 Pa. $6.50

THE COMPLETE NONSENSE OF EDWARD LEAR, Edward Lear. All nonsense limericks, zany alphabets, Owl and Pussycat, songs, nonsense botany, etc., illustrated by Lear. Total of 321pp. 5⅜ x 8½. (Available in U.S. only) 20167-8 Pa. $3.95

INGENIOUS MATHEMATICAL PROBLEMS AND METHODS, Louis A. Graham. Sophisticated material from Graham *Dial,* applied and pure; stresses solution methods. Logic, number theory, networks, inversions, etc. 237pp. 5⅜ x 8½. 20545-2 Pa. $4.50

BEST MATHEMATICAL PUZZLES OF SAM LOYD, edited by Martin Gardner. Bizarre, original, whimsical puzzles by America's greatest puzzler. From fabulously rare *Cyclopedia,* including famous 14-15 puzzles, the Horse of a Different Color, 115 more. Elementary math. 150 illustrations. 167pp. 5⅜ x 8½. 20498-7 Pa. $2.75

THE BASIS OF COMBINATION IN CHESS, J. du Mont. Easy-to-follow, instructive book on elements of combination play, with chapters on each piece and every powerful combination team—two knights, bishop and knight, rook and bishop, etc. 250 diagrams. 218pp. 5⅜ x 8½. (Available in U.S. only) 23644-7 Pa. $3.50

MODERN CHESS STRATEGY, Ludek Pachman. The use of the queen, the active king, exchanges, pawn play, the center, weak squares, etc. Section on rook alone worth price of the book. Stress on the moderns. Often considered the most important book on strategy. 314pp. 5⅜ x 8½. 20290-9 Pa. $4.50

LASKER'S MANUAL OF CHESS, Dr. Emanuel Lasker. Great world champion offers very thorough coverage of all aspects of chess. Combinations, position play, openings, end game, aesthetics of chess, philosophy of struggle, much more. Filled with analyzed games. 390pp. 5⅜ x 8½. 20640-8 Pa. $5.00

500 MASTER GAMES OF CHESS, S. Tartakower, J. du Mont. Vast collection of great chess games from 1798-1938, with much material nowhere else readily available. Fully annoted, arranged by opening for easier study. 664pp. 5⅜ x 8½. 23208-5 Pa. $7.50

A GUIDE TO CHESS ENDINGS, Dr. Max Euwe, David Hooper. One of the finest modern works on chess endings. Thorough analysis of the most frequently encountered endings by former world champion. 331 examples, each with diagram. 248pp. 5⅜ x 8½. 23332-4 Pa. $3.75

THE AMERICAN SENATOR, Anthony Trollope. Little known, long unavailable Trollope novel on a grand scale. Here are humorous comment on American vs. English culture, and stunning portrayal of a heroine/villainess. Superb evocation of Victorian village life. 561pp. 5⅜ x 8½.
23801-6 Pa. $6.00

WAS IT MURDER? James Hilton. The author of *Lost Horizon* and *Goodbye, Mr. Chips* wrote one detective novel (under a pen-name) which was quickly forgotten and virtually lost, even at the height of Hilton's fame. This edition brings it back—a finely crafted public school puzzle resplendent with Hilton's stylish atmosphere. A thoroughly English thriller by the creator of Shangri-la. 252pp. 5⅜ x 8. (Available in U.S. only)
23774-5 Pa. $3.00

CENTRAL PARK: A PHOTOGRAPHIC GUIDE, Victor Laredo and Henry Hope Reed. 121 superb photographs show dramatic views of Central Park: Bethesda Fountain, Cleopatra's Needle, Sheep Meadow, the Blockhouse, plus people engaged in many park activities: ice skating, bike riding, etc. Captions by former Curator of Central Park, Henry Hope Reed, provide historical view, changes, etc. Also photos of N.Y. landmarks on park's periphery. 96pp. 8½ x 11. 23750-8 Pa. $4.50

NANTUCKET IN THE NINETEENTH CENTURY, Clay Lancaster. 180 rare photographs, stereographs, maps, drawings and floor plans recreate unique American island society. Authentic scenes of shipwreck, lighthouses, streets, homes are arranged in geographic sequence to provide walking-tour guide to old Nantucket existing today. Introduction, captions. 160pp. 8⅞ x 11¾. 23747-8 Pa. $6.95

STONE AND MAN: A PHOTOGRAPHIC EXPLORATION, Andreas Feininger. 106 photographs by *Life* photographer Feininger portray man's deep passion for stone through the ages. Stonehenge-like megaliths, fortified towns, sculpted marble and crumbling tenements show textures, beauties, fascination. 128pp. 9¼ x 10¾. 23756-7 Pa. $5.95

CIRCLES, A MATHEMATICAL VIEW, D. Pedoe. Fundamental aspects of college geometry, non-Euclidean geometry, and other branches of mathematics: representing circle by point. Poincare model, isoperimetric property, etc. Stimulating recreational reading. 66 figures. 96pp. 5⅝ x 8¼.
63698-4 Pa. $2.75

THE DISCOVERY OF NEPTUNE, Morton Grosser. Dramatic scientific history of the investigations leading up to the actual discovery of the eighth planet of our solar system. Lucid, well-researched book by well-known historian of science. 172pp. 5⅜ x 8½. 23726-5 Pa. $3.50

THE DEVIL'S DICTIONARY. Ambrose Bierce. Barbed, bitter, brilliant witticisms in the form of a dictionary. Best, most ferocious satire America has produced. 145pp. 5⅜ x 8½. 20487-1 Pa. $2.25

TONE POEMS, SERIES II: TILL EULENSPIEGELS LUSTIGE STREICHE, ALSO SPRACH ZARATHUSTRA, AND EIN HELDENLEBEN, Richard Strauss. Three important orchestral works, including very popular *Till Eulenspiegel's Marry Pranks,* reproduced in full score from original editions. Study score. 315pp. 9⅜ x 12¼. (Available in U.S. only)
23755-9 Pa. $8.95

TONE POEMS, SERIES I: DON JUAN, TOD UND VERKLARUNG AND DON QUIXOTE, Richard Strauss. Three of the most often performed and recorded works in entire orchestral repertoire, reproduced in full score from original editions. Study score. 286pp. 9⅜ x 12¼. (Available in U.S. only)
23754-0 Pa. $7.50

11 LATE STRING QUARTETS, Franz Joseph Haydn. The form which Haydn defined and "brought to perfection." *(Grove's).* 11 string quartets in complete score, his last and his best. The first in a projected series of the complete Haydn string quartets. Reliable modern Eulenberg edition, otherwise difficult to obtain. 320pp. 8⅜ x 11¼. (Available in U.S. only)
23753-2 Pa. $7.50

FOURTH, FIFTH AND SIXTH SYMPHONIES IN FULL SCORE, Peter Ilyitch Tchaikovsky. Complete orchestral scores of Symphony No. 4 in F Minor, Op. 36; Symphony No. 5 in E Minor, Op. 64; Symphony No. 6 in B Minor, "Pathetique," Op. 74. Bretikopf & Hartel eds. Study score. 480pp. 9⅜ x 12¼.
23861-X Pa. $10.95

THE MARRIAGE OF FIGARO: COMPLETE SCORE, Wolfgang A. Mozart. Finest comic opera ever written. Full score, not to be confused with piano renderings. Peters edition. Study score. 448pp. 9⅜ x 12¼. (Available in U.S. only)
23751-6 Pa. $11.95

"IMAGE" ON THE ART AND EVOLUTION OF THE FILM, edited by Marshall Deutelbaum. Pioneering book brings together for first time 38 groundbreaking articles on early silent films from *Image* and 263 illustrations newly shot from rare prints in the collection of the International Museum of Photography. A landmark work. Index. 256pp. 8¼ x 11.
23777-X Pa. $8.95

AROUND-THE-WORLD COOKY BOOK, Lois Lintner Sumption and Marguerite Lintner Ashbrook. 373 cooky and frosting recipes from 28 countries (America, Austria, China, Russia, Italy, etc.) include Viennese kisses, rice wafers, London strips, lady fingers, hony, sugar spice, maple cookies, etc. Clear instructions. All tested. 38 drawings. 182pp. 5⅜ x 8.
23802-4 Pa. $2.50

THE ART NOUVEAU STYLE, edited by Roberta Waddell. 579 rare photographs, not available elsewhere, of works in jewelry, metalwork, glass, ceramics, textiles, architecture and furniture by 175 artists—Mucha, Seguy, Lalique, Tiffany, Gaudin, Hohlwein, Saarinen, and many others. 288pp. 8⅜ x 11¼.
23515-7 Pa. $6.95

THE SENSE OF BEAUTY, George Santayana. Masterfully written discussion of nature of beauty, materials of beauty, form, expression; art, literature, social sciences all involved. 168pp. 5⅜ x 8½. 20238-0 Pa. $3.00

ON THE IMPROVEMENT OF THE UNDERSTANDING, Benedict Spinoza. Also contains *Ethics, Correspondence,* all in excellent R. Elwes translation. Basic works on entry to philosophy, pantheism, exchange of ideas with great contemporaries. 402pp. 5⅜ x 8½. 20250-X Pa. $4.50

THE TRAGIC SENSE OF LIFE, Miguel de Unamuno. Acknowledged masterpiece of existential literature, one of most important books of 20th century. Introduction by Madariaga. 367pp. 5⅜ x 8½.
20257-7 Pa. $4.50

THE GUIDE FOR THE PERPLEXED, Moses Maimonides. Great classic of medieval Judaism attempts to reconcile revealed religion (Pentateuch, commentaries) with Aristotelian philosophy. Important historically, still relevant in problems. Unabridged Friedlander translation. Total of 473pp. 5⅜ x 8½. 20351-4 Pa. $6.00

THE I CHING (THE BOOK OF CHANGES), translated by James Legge. Complete translation of basic text plus appendices by Confucius, and Chinese commentary of most penetrating divination manual ever prepared. Indispensable to study of early Oriental civilizations, to modern inquiring reader. 448pp. 5⅜ x 8½. 21062-6 Pa. $5.00

THE EGYPTIAN BOOK OF THE DEAD, E. A. Wallis Budge. Complete reproduction of Ani's papyrus, finest ever found. Full hieroglyphic text, interlinear transliteration, word for word translation, smooth translation. Basic work, for Egyptology, for modern study of psychic matters. Total of 533pp. 6½ x 9¼. (Available in U.S. only) 21866-X Pa. $5.95

THE GODS OF THE EGYPTIANS, E. A. Wallis Budge. Never excelled for richness, fullness: all gods, goddesses, demons, mythical figures of Ancient Egypt; their legends, rites, incarnations, variations, powers, etc. Many hieroglyphic texts cited. Over 225 illustrations, plus 6 color plates. Total of 988pp. 6⅛ x 9¼. (Available in U.S. only)
22055-9, 22056-7 Pa., Two-vol. set $16.00

THE STANDARD BOOK OF QUILT MAKING AND COLLECTING, Marguerite Ickis. Full information, full-sized patterns for making 46 traditional quilts, also 150 other patterns. Quilted cloths, lame, satin quilts, etc. 483 illustrations. 273pp. 6⅞ x 9⅝. 20582-7 Pa. $4.95

CORAL GARDENS AND THEIR MAGIC, Bronsilaw Malinowski. Classic study of the methods of tilling the soil and of agricultural rites in the Trobriand Islands of Melanesia. Author is one of the most important figures in the field of modern social anthropology. 143 illustrations. Indexes. Total of 911pp. of text. 5⅝ x 8¼. (Available in U.S. only)
23597-1 Pa. $12.95